Bioelectricity
A Quantitative Approach

BIOELECTRICITY
A QUANTITATIVE APPROACH

ROBERT PLONSEY
and
ROGER C. BARR
Duke University
Durham, North Carolina
USA

THIRD EDITION

Roger C. Barr
Duke University
Durham, North Carolina 27708
USA
rbarr@potassium.egr.duke.edu

Robert Plonsey
Duke University
Durham, North Carolina 27708
USA
rplonsey@acpub.duke.edu

Additional material to this book can be downloaded from http://extra.springer.com.

ISBN 978-1-4899-8408-1

ISBN 978-0-387-48865-3 (eBook)

Softcover re-print of the Hardcover 3rd edition 2007

9 8 7 6 5 4 3 2 1

springer.com

To our unseen co-authors, our wives:
VIVIAN PLONSEY
JEAN BARR
and our unnamed co-authors:
The students in BME 101

About The Authors

Robert Plonsey is Pfizer-Pratt Professor Emeritus of Biomedical Engineering at Duke University. He received the PhD in Electrical Engineering from the University of California in 1955. He received the Dr. of Technical Science from the Slovak Academy of Science in 1995 and was Chair, Department of Biomedical Engineering, Case Western Reserve, University, 1976-1980, Professor 1968-1983. Awards: Fellow of AAAS, William Morlock Award 1979, Centennial Medal 1984, Millennium Medal 2000, from IEEE Engineering in Medicine and Biology Society, Ragnar Granit Prize 2004, (First) Merit Award, 1997, International Union for Physiological & Engineering Science in Medicine, the Theo Pilkington Outstanding Educator Award, 2005, Distinguished Service award, Biomedical Engineering Science, 2004, ALZA distinguished lecturer, 1988. He was elected Member, National Academy of Engineering, 1986 ("For the application of electromagnetic field theory to biology, and for distinguished leadership in the emerging profession of biomedical engineering").

Roger C. Barr is Professor of Biomedical Engineering and Associate Professor of Pediatrics at Duke University. In past years he served as the Chair of the Department of Biomedical Engineering at Duke, and then as Vice President and President of the IEEE Engineering in Medicine and Biology Society. He received the Duke University Scholar-Teacher Award in 1991. He is the author of more than 100 research papers about topics in bioelectricity and is a Fellow of the IEEE and American College of Cardiology. This text is a product of interactions with students, and in this regard he has taught the bioelectricity course sequence numerous times.

PREFACE

The study of electrophysiology has progressed rapidly because of the precise, delicate, and ingenious experimental studies of many investigators. The field has also made great strides by unifying these experimental observations through mathematical descriptions based on electromagnetic field theory, electrochemistry, etc., which underlie these experiments. In turn, these quantitative materials provide an understanding of many electrophysiological applications through a relatively small number of fundamental ideas.

This text is an introduction to electrophysiology, following a quantitative approach. The first chapter summarizes much of the mathematics required in the following chapters. The second chapter presents a very concise overview of the principles of electrical fields and the concomitant current flow in conducting media. It utilizes basic principles from the physical sciences and engineering but takes into account the biological applications. The following six chapters are the core material of this text. Chapter 3 includes a description of how voltages/currents exist across membranes and how these are evaluated using the Nernst–Planck equation. The membrane channels, which are the basis for cell excitability, are described in Chapter 4. An examination of the time course of changes in membrane voltages that produce action potentials are considered in Chapter 5. Propagation of action potentials down fibers is the subject of Chapter 6, and the response of fibers to artificial stimuli, such as those used in cardiac pacemakers, is treated in Chapter 7. The voltages and currents produced by these active processes in the surrounding extracellular space is described in Chapter 8. The subsequent chapters present more detailed material about the application of these principles to the study of the electrophysiology of cardiac and skeletal muscle with a modest inclusion of neural electrophysiology.

The material of this text was designed as an introduction to bioelectricity (electrophysiology), and one might think that fundamentals change very slowly. In fact the rapid growth of the field has reflected back changes in the underlying material. Since a quantitative approach to electrophysiology is a precursor to the various new applications; it is, in fact, a real challenge keeping things up-to-date. The second edition is the authors' effort to bring the text more into line with the current new applications found in recent texts.

In particular, we have introduced a few underlying factors in molecular biology as it interacts with electrophysiology. While the result is a very modest introduction it is hoped that the treatment will outline the importance of this topic in bioelectricity. In other applications we have also endeavored to bring matters up-to-date. This is done in both the chapters on applications as well as those devoted to fundamentals. We hope this conveys to the reader our excitement with this field.

In this third edition, we respond to the many requests from students and faculty colleagues that the book include more exercises with solutions. Thus the exercises have been reorganized, and many more exercises and solutions added. Additionally, Chapter 8 on extracellular potentials has been revised and extended, with many new figures, as we recognize that this chapter is key to understanding many clinical measurements. In addition a number of other chapters have been revised, with more information now included for the reader about the reasons why different topics are considered important and how they are related, information that allows one to better focus on those topics most important to particular instructors and students.

Each time we consider the material in the text we become aware, once again, of how many talented and energetic investigators and students of the field have made substantial contributions to its progress. It is the nature of a textbook to reflect the integrated ideas of many individuals over more than century, so only a few of the many contributors are recognized by citation. Even so, a wealth of additional material is available to the reader, and that material provides a much more complete picture. We have included a few citations in the text on particular points and at the end of each chapter as additional material, so that the student has a entryway to the extensive library of published work that now is available.

The revisions also include many corrections and focused responses to suggestions received from colleagues, readers elsewhere, and especially from our students. We hope they will find the revisions to their liking. For the future we continue to invite comments and criticisms from students and faculty colleagues.

ROBERT PLONSEY
ROGER C. BARR

Contents

1

VECTOR ANALYSIS

1.1. INTRODUCTION

This text is directed to presenting the fundamentals of electrophysiology from a quantitative standpoint. The treatment of a number of topics in this book is greatly facilitated using vectors and vector calculus. This chapter reviews the concepts of vectors and scalars and the algebraic operations of addition and multiplication as applied to vectors. The concepts of gradient and divergence also are reviewed, since they will be encountered more frequently.[1]

1.2. VECTORS AND SCALARS

In any experiment or study of biophysical phenomena one identifies one or more variables that arise in a consideration of the observed behavior. For physical observables, variables are classified as either *scalars* or *vectors*, that is, the variable is defined by a simple value (e.g., temperature, conductivity, voltage) or both a magnitude plus direction (e.g., current density, force, electric field).

In a given preparation a scalar property might vary as a function of position (e.g., the conductivity as a function of position in a body). The collection of such values at all positions is referred to as a *scalar field.* A vector function of position (e.g., blood flow at different points in a major artery) is similarly a *vector field.* We designate scalars by unmodified letters, while vectors are designated with a bar over the letter. Thus T is for temperature, but $\overline{J}$ is for current density.

As mentioned, a vector that has a value at every position in a region is referred to as a vector field. $\overline{J}(x, y, z)$ is a vector field where at each (x, y, z) a particular vector $\overline{J}$ exists. Physiological vector fields are usually considered to be well behaved, continuous, with continuous derivatives.

1.3. VECTOR ALGEBRA

1.3.1. Sum

The sum of two vectors is also a vector. Thus,

$$\overline{C} = \overline{A} + \overline{B} \tag{1.1}$$

where $\overline{C}$ is the resultant or sum of $\overline{A}$ plus $\overline{B}$. Vectors are added by application of the *parallelogram law* (let $\overline{A}$ and $\overline{B}$ be drawn from a common origin; if the parallelogram is completed then $\overline{C}$ is the diagonal drawn from the common origin).

1.3.2. Vector Times Scalar

The result of multiplying a vector $\overline{A}$ by a scalar m is a new vector with the same orientation but a magnitude m times as great. If we designate this by $\overline{B}$ then

$$\overline{B} = m\overline{A} \tag{1.2}$$

and

$$|\overline{B}| = m|\overline{A}| \text{ or } B = mA \tag{1.3}$$

1.3.3. Unit Vector

A unit vector is one whose magnitude is unity. It is sometimes convenient to describe a vector ($\overline{A}$) by its magnitude (A) times a unit vector ($\overline{a}$) that supplies the direction. Thus $\overline{A} = A\overline{a}$.

1.3.4. Dot Product

The scalar product (or dot product) of two vectors is defined as the product of their magnitudes times the cosine of the angle between the vectors (assumed drawn from a common origin). From Figure 1.1 we note that the scalar product of $\overline{A}$ and $\overline{B}$ is the product of the magnitude of one of them (say, $|\overline{B}|$) times the projection of the other on the first ($|\overline{A}|\cos\theta$). We designate the dot product as $\overline{A}\cdot\overline{B}$, so that

$$\overline{A}\cdot\overline{B} = AB\cos\theta \tag{1.4}$$

Clearly from the definition,

$$\overline{A}\cdot\overline{B} = \overline{B}\cdot\overline{A}$$

so that the *commutative* law of multiplication is satisfied. Note that if $\overline{A}$ and $\overline{B}$ are orthogonal ($\theta = 90°$) then their dot product is zero. Considering $\overline{A}\cdot\overline{A}$, since in this case $\theta = 0°$, then $\overline{A}\cdot\overline{A} = A^2$.

In bioelectricity the dot product often is used to find the component of one vector in the direction of another, e.g., the component of the electric field along the axial direction of a fiber.

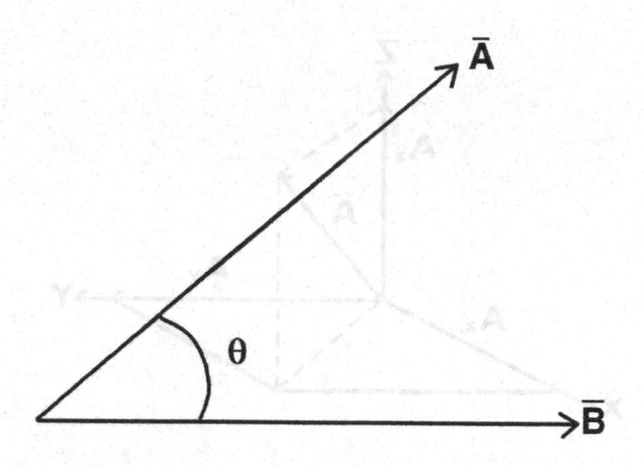

Figure 1.1. Dot Product. The dot product $\overline{A} \cdot \overline{B}$ is given by $AB \cos\theta$.

1.3.5. Cross Product

The cross product (also called the vector product) of two vectors $\overline{A} \times \overline{B}$ differs from the dot product in its geometrical meaning and in its form. Geometrically the cross product corresponds to the area of the parallelogram whose sides are defined by $\overline{A}$ and $\overline{B}$.

If we designate the resultant vector as $\overline{C}$ then

$$\overline{C} = \overline{A} \times \overline{B}$$

where

$$|C| = |A||B| \sin\theta \tag{1.5}$$

and angle θ is between $\overline{A}$ and $\overline{B}$.

The direction of $\overline{C}$ is orthogonal to the plane defined by $\overline{A}$ and $\overline{B}$ and is the direction that a normal, right-handed screw advances if turned from $\overline{A}$ to $\overline{B}$, i.e., the direction follows the "right-hand rule." For example, in Figure 1.2

$$\overline{A_x} = \overline{A_y} \times \overline{A_z}$$

Returning to Figure 1.1, if $\overline{C} = \overline{A} \times \overline{B}$, then the direction of vector $\overline{C}$ will be into the page. In terms of components,

$$\overline{A} \times \overline{B} = (A_y B_z - A_z B_y)\overline{a}_x + (A_z B_x - A_x B_z)\overline{a}_y + (A_x B_y - A_y B_x)\overline{a}_z \tag{1.6}$$

This result can be verified by replacing each vector $\overline{A}$ and $\overline{B}$ by its rectangular components, and expanding the result. A convenient way of describing the operations in the cross product, and an aid in remembering it, is to use the notation of determinants.

Cross products arise less frequently in bioelectricity than dot products. Most frequently cross products appear when dealing with geometrical surfaces, as in taking into account the shape of the body surface, in electrocardiography.

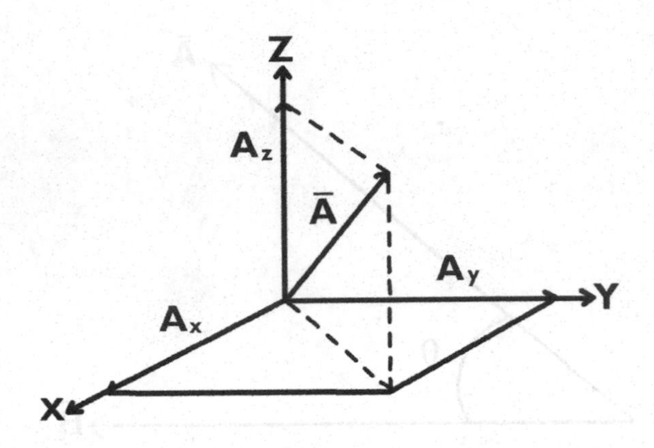

Figure 1.2. Vector $\overline{A}$ and its rectangular components.

1.3.6. Resolution of Vectors into Components

Vector $\overline{A}$ is the sum of its rectangular components A_x, A_y, A_z, as described in Figure 1.2. That is,

$$\overline{A} = A_x\overline{a}_x + A_y\overline{a}_y + A_z\overline{a}_z$$

Similarly, we may describe

$$\overline{B} = B_x\overline{a_x} + B_y\overline{a}_y + B_z\overline{a}z$$

Using the distributive law of algebra, the dot product of $\overline{A}$ and $\overline{B}$ can be formulated as

$$\begin{aligned}\overline{A}\cdot\overline{B} = \quad & A_xB_x\overline{a}_x\cdot\overline{a}_x + A_yB_y\overline{a}_y\cdot\overline{a}_y + A_zB_z\overline{a}_z\cdot\overline{a}_z \\ & +A_xB_y\overline{a}_x\cdot\overline{a}_y + A_xB_z\overline{a}_x\cdot\overline{a}_z + A_yB_x\overline{a}_y\cdot\overline{a}_x \\ & +A_yB_z\overline{a}_y\cdot\overline{a}_z + A_zB_x\overline{a}_z\cdot\overline{a}_x + A_zB_y\overline{a}_z\cdot\overline{a}_y\end{aligned} \tag{1.7}$$

Now terms such as $\overline{a}_x\cdot\overline{a}_x = 1$, since the angle between the vectors is zero and the cosine of zero is unity. On the other hand, terms such as $\overline{a}_x\cdot\overline{a}_y = \overline{a}_y\cdot\overline{a}_z = \overline{a}_z\cdot\overline{a}_x = 0$, since the angle between the unit vectors is 90°. Consequently, (1.7) becomes

$$\overline{A}\cdot\overline{B} = A_xB_x + A_yB_y + A_zB_z \tag{1.8}$$

The result expressed by (1.8) is, of course, a scalar.

1.4. GRADIENT

Let $\Phi(x, y, z)$ be a scalar field (scalar function of position) and assume that it is single-valued, continuous, and a differentiable function of position. (Physiological fields normally satisfy these requirements.) We define a surface on which this field has a constant value by

$$\Phi(x, y, z) = C \tag{1.9}$$

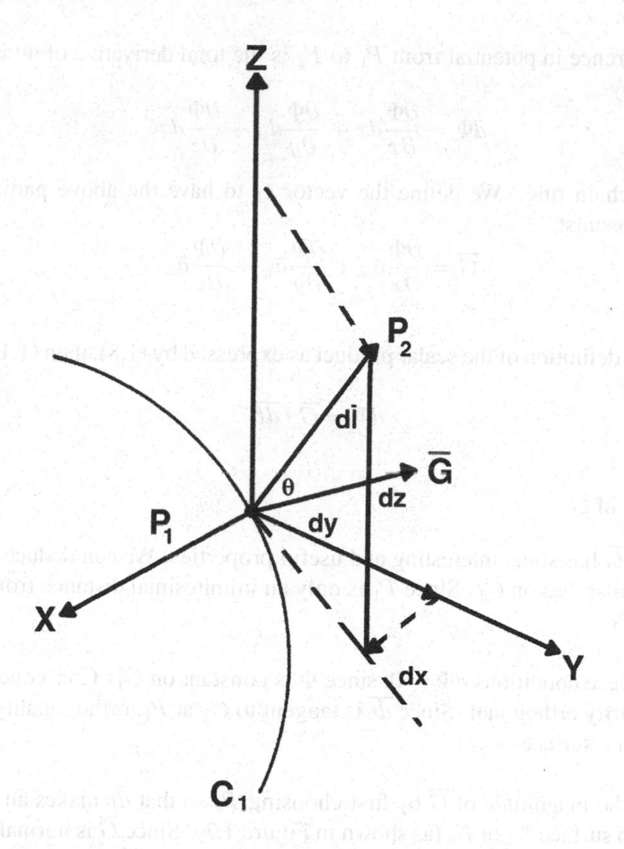

Figure 1.3. Equipotential surface C_1 along with points P_1 (on C_1) and P_2 (located arbitrarily). The components of $\overline{d\ell}$ given by (1.10) are shown; these also enter (1.11).

where C is a constant. Frequently, in this book, the symbol Φ is a potential (electrical, chemical) in which case the surface of constant value is referred to as an *equipotential* surface (biologists prefer the designation *isopotential*).

If we let C take on a succession of increasing values, a family of nonintersecting isopotential surfaces results. The geometrical shape of this set of isopotential surfaces is a reflection of the character of the potential field and is useful for at least this reason.

1.4.1. Gradient to Potential Difference

Consider two closely spaced points P_1 and P_2. Point P_1 lies on the surface $\Phi(x, y, z) = C_1$, and P_2, which is close by, may not lie on this surface (see Figure 1.3). Let the coordinates of P_1 be (x, y, z). Then the coordinates of P_2 could be described as $(x + dx, y + dy, z + dz)$.

If $\overline{a}_x, \overline{a}_y, \overline{a}_z$ are unit vectors along the *x,y,z* axes, then the displacement (a vector) from P_1 to P_2 may be expressed as the vector sum of its rectangular components, namely,

$$\overline{d\ell} = \overline{a}_x dx + \overline{a}_y dy + \overline{a}_z dz \tag{1.10}$$

Now the difference in potential from P_1 to P_2 is the total derivative of $\Phi(x, y, z)$ evaluated at P_1. It is given by

$$d\Phi = \frac{\partial \Phi}{\partial x}dx + \frac{\partial \Phi}{\partial y}dy + \frac{\partial \Phi}{\partial z}dz \tag{1.11}$$

according to the chain rule. We define the vector $\overline{G}$ to have the above partial derivatives as rectangular components:

$$\overline{G} = \frac{\partial \Phi}{\partial x}\overline{a}_x + \frac{\partial \Phi}{\partial y}\overline{a}_y + \frac{\partial \Phi}{\partial z}\overline{a}_z \tag{1.12}$$

In view of the definition of the scalar product as expressed by (1.8), then (1.11) can be written as

$$d\Phi = \overline{G} \cdot \overline{d\ell} \tag{1.13}$$

1.4.2. Properties of $\overline{G}$

The function $\overline{G}$ has some interesting and useful properties. We can deduce these as follows. First, suppose P_2 also lies on C_1. Since P_1 is only an infinitesimal distance from P_2, $\overline{d\ell}$ must be tangent to C_1 at P_1.

Now under these conditions $d\Phi = 0$, since Φ is constant on C_1. Consequently, in (1.13) $\overline{d\ell}$ and $\overline{G}$ are necessarily orthogonal. Since $\overline{d\ell}$ is tangent to C_1 at P_1, orthogonality means that $\overline{G}$ is perpendicular to the surface C_1.

We can find the magnitude of $\overline{G}$ by first choosing P_2 so that $\overline{d\ell}$ makes an arbitrary angle θ with the normal to surface C_1 at P_1 (as shown in Figure 1.3). Since $\overline{G}$ is normal to C_1, then from (1.13)

$$d\Phi = \overline{d\ell} \cdot \overline{G} = G\cos\theta d\ell \tag{1.14}$$

Consequently,

$$\frac{d\Phi}{d\ell} = G\cos\theta \tag{1.15}$$

and therefore the derivative of Φ in the direction ℓ (the *directional derivative*) depends on the direction of $\overline{d\ell}$ and is maximum when $\theta = 0$. The condition $\theta = 0$ means that $\overline{d\ell}$ in the direction of the surface normal, $\overline{n}$, so the maximum derivative of Φ is along the normal to the equipotential surface. (Those familiar with contour maps are not surprised at this result.) Accordingly, Eq. (1.15), with $\theta = 0$, yields

$$G = \frac{d\Phi}{dn} \tag{1.16}$$

Thus, from the above, $\overline{G}$ is in the direction of the maximum rate of increase in Φ and has a magnitude equal to that maximum rate; and this maximum is achieved along the direction which is normal to the equipotential surface.

1.4.3. The Del Operator ∇ and the Gradient

The vector $\overline{G}$, defined in (1.12), is known as the *gradient*. Rather than being given the symbol $\overline{G}$, the gradient of Φ usually is written $\nabla\Phi$, where ∇ is an operator.

With this change in notation,

$$\nabla \equiv \bar{a}_x \frac{\partial}{\partial x} + \bar{a}_y \frac{\partial}{\partial y} + \bar{a}_z \frac{\partial}{\partial z} \tag{1.17}$$

The gradient operation $\nabla\Phi$ is executed by considering each term in (1.17) to be acting on Φ. Thus the gradient is found by taking each partial derivative and appending the corresponding unit vector. One can verify that this process leads, correctly, to the right-hand side of (1.12).

Consequently $\nabla\Phi$ not only symbolizes the gradient of Φ but describes the operation leading to its correct evaluation (though only in *rectangular coordinates*). Thus from (1.17) we get, corresponding to (1.12):

$$\nabla\Phi = \frac{\partial\Phi}{\partial x}\bar{a}_x + \frac{\partial\Phi}{\partial y}\bar{a}_y + \frac{\partial\Phi}{\partial z}\bar{a}_z \tag{1.18}$$

The magnitude of $\nabla\Phi$ is evaluated by taking the square root of $\nabla\Phi \cdot \nabla\Phi$. From (1.18) and (1.8) the magnitude is found as

$$|\nabla\Phi| = \sqrt{\left(\frac{\partial\Phi}{\partial x}\right)^2 + \left(\frac{\partial\Phi}{\partial y}\right)^2 + \left(\frac{\partial\Phi}{\partial x}\right)^2} \tag{1.19}$$

1.4.4. Comments about the Gradient

One way to gain an intuitive concept[2] of the gradient is as follows: If $\Phi(x, y)$ describes the elevation of points on the surface of a hill [corresponding to each coordinate (x, y)], then the height (Φ) will vary from place to place in the same way as in a conventional contour map. The gradient of Φ evaluates the slope of the hill at each point. The slope is represented by a magnitude and direction. The magnitude signifies how steep the slope is at a particular point. The direction of the gradient points in the most *uphill* (steepest) direction. On most hills, both the magnitude and direction of the slope will vary considerably from place to place.

As will be seen in the sections below, one of the reasons that the gradient is an important mathematical construct in electrophysiology is that the negative of the gradient of the electrical potential is normally proportional to the strength of the associated electrical current. In a similar way, the flow of water on the surface of a hill is closely related to the hill's slope. Since water flows downhill, on the smooth surface we are assuming, it will thereby flow in the direction of the *negative* gradient.

1.5. DIVERGENCE

Connect a low-voltage battery through two terminals to the body: this establishes a current flow field (a vector field). Current enters the body through the plus terminal and exists through the minus terminal.

Within the body, the configuration of the current field will depend on body shape, electrode positions, and the body's electrical inhomogeneities; the result could be described by the current density function, $\overline{J}(x, y, z)$, that results. Suppose we turn the problem around and are given only the function $\overline{J}$.

In this case, at the least, the electrode locations (i.e., source and sink for $\overline{J}$) should be evident from the features of the field. In a formal procedure we could discretize the body into small cubical elements and evaluate the net flow across the bounding surface of each volume element. If a result is zero then no source is enclosed, but if the value is positive (i.e., a net outflow) then a net positive source lies within the volume; similarly, a net inflow identifies a net sink.

A more familiar problem is where currents arise, as they do, from excitable tissue (say, the heart as a source of the electrocardiogram). We could measure this current flow field and then examine it to determine the sources of that field as in the above example. However, if the field $\overline{J}$ can be described analytically, then an analytic expression for the source density can be evaluated; the procedure will be developed in the following material.

A typical vector field arising in electrophysiology (and one that will be of great interest to us) is the current density, $\overline{J}$, in a volume conductor. The structure of the $\overline{J}$ field depends on the presence of sites at which current is either introduced (sources) or withdrawn (sinks). In this respect the behavior of $\overline{J}(x, y, z)$ is analogous to the vector field describing fluid flow that arises from a distribution of sources and sinks, or of heat flow, etc.

This class of vector fields has in common certain general properties, which we will discuss now in terms of a current flow field. In the following we use the term "sources" to include "sinks" (which are, simply, negative sources).

For an arbitrary, physically realizable source distribution giving rise to a flow field, $\overline{J}$, the latter will be a possibly complicated but well-behaved vector function of position. In particular, for a region that contains no sources the net flow of $\overline{J}$ across the bounding surface of any arbitrary volume within the source-free region (e.g., choosing inflow to be negative and outflow to be positive) must be zero. This result is a consequence of the conservation of charge. (This requirement is also stated to result from the *continuity of current*.)

The evaluation of the net flow across a closed surface may be taken as a measure of the net source (or sink) within the region enclosed by that surface. If the net flow across any (every) surface is zero, then the region is source free, as discussed above. If there is a net outflow, then within the surface there must lie sources whose net magnitude equals the (net) outflow. For a differential rectangular parallelepiped we can derive an expression that evaluates this net outflow. This expression will prove useful when the current density can be described analytically.

1.5.1. Outflow through Surfaces 1 and 2 in Figure 1.4

Referring to Figure 1.4 and assuming a flow field $\overline{J}(x, y, z)$ to be present, one sees that the outflow through surface 2 must be

$$\text{outflow}_2 = dydz\left[J_x + \left(\frac{1}{2}\right)\left(\frac{\partial J_x}{\partial x}\right)dx\right] \tag{1.20}$$

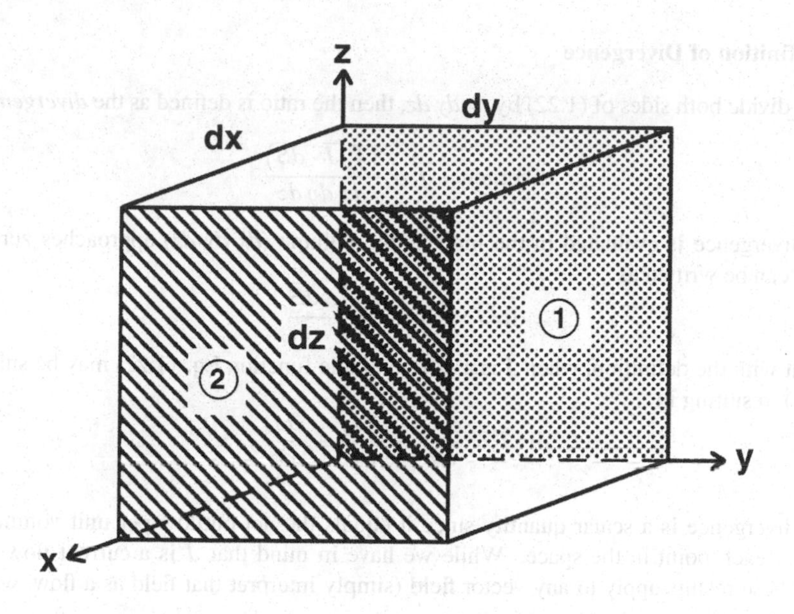

Figure 1.4. Divergence evaluated with a rectangular parallelepiped of differential size.

plus a higher-order term, where J_x is the value of $J_x(x, y, z)$ at the center of the parallelepiped (which accounts for the factor of 1/2 in the expression).

For surface 1 the outflow is

$$\text{outflow}_1 = -dy\,dz\left[J_x - \left(\frac{1}{2}\right)\left(\frac{\partial J_x}{\partial x}\right)dx\right] \tag{1.21}$$

Note the sign: in (1.21) the minus sign (in front of the bracket) arises because in the evaluation of the outflow the area, 1, is oriented in the negative *x* direction. The sum of the above two terms is then $dx\,dy\,dz\,\partial J_x/\partial x$.

1.5.2. Outflow through All Six Surfaces

In the same way the remaining two pairs of faces contribute $dx\,dy\,dz\,\partial J_y/\partial y$ and $dx\,dy\,dz\,\partial J_z/\partial z$. Consequently, the net outflow, which is the sum of the previous three terms, is

$$\oint_S \overline{J}\cdot\overline{dS} = \left(\frac{\partial J_x}{\partial x} + \frac{\partial J_y}{\partial y} + \frac{\partial J_z}{\partial z}\right) dx\,dy\,dz \tag{1.22}$$

Note that $\overline{J}\cdot d\overline{S}$ is the outflow of $\overline{J}$ across an arbitrary surface element. The special integral symbol with a circle in the middle ($\oint$) indicates an integral over a closed surface. In this example the rectangular parallelepiped described in Figure 1.4 is the designated closed surface, and this is evaluated in the right-hand side of (1.22). Note that $\overline{J}\cdot d\overline{S}$ correctly evaluates the flow of $\overline{J}$ across $d\overline{S}$ because the dot product selects the component of $\overline{J}$ in the direction of the surface normal.

1.5.3. Definition of Divergence

If we divide both sides of (1.22) by *dx dy dz*, then the ratio is defined as the *divergence* of $\overline{J}$:

$$\text{divergence}\,\overline{J} = \frac{\left(\oint_s \overline{J} \cdot d\overline{S}\right)}{dx\,dy\,dz} \tag{1.23}$$

Strictly, divergence is evaluated in the limit as the volume (*dx dy dz*) approaches zero. This definition can be written as

$$\text{div}\overline{J} = \lim_{V \to 0} \frac{\oint_S \overline{J} \cdot d\overline{S}}{V} \tag{1.24}$$

Consistent with the definition, when **J** is a differentiable function Eq. (1.22) may be substituted into (1.24), resulting in

$$\text{div}\overline{J} = \frac{\partial J_x}{\partial x} + \frac{\partial J_y}{\partial y} + \frac{\partial J_z}{\partial z} \tag{1.25}$$

The divergence is a scalar quantity since it equals the net outflow per unit volume of the vector $\overline{J}$ at each point in the space. While we have in mind that $\overline{J}$ is a current flow (current density), these results apply to any vector field (simply interpret that field as a flow, whether it actually is or not).

If we treat the ∇ operator, defined in (1.17), as having vector-like properties, then in view of (1.25) and the properties of the dot product given in (1.8) we have (formally)

$$\nabla \cdot \overline{J} \equiv \text{div}\overline{J} = \frac{\partial J_x}{\partial x} + \frac{\partial J_y}{\partial y} + \frac{\partial J_z}{\partial z} \tag{1.26}$$

establishing $\nabla \cdot \overline{J}$ as both a symbol for the divergence operation and a description of its evaluation (in rectangular coordinates).

1.5.4. Comments about the Divergence

It is important to keep in mind that the divergence is a quantity that varies from point to point. As an analogy, note that water flowing on the surface of a hill has a flow pattern that varies from point to point. At most points on the hillside, the water arrives from the uphill side and departs to the downhill side. At these points the water's "divergence" is zero. At a few sites water emerges onto the surface from an underground spring (or maybe rain falls on that spot). At such a point, from the viewpoint of the two-dimensional surface flow function, water is emerging and flowing out from the point. At these points, the "divergence" is positive, and the point is called a "source." At a few other points, water disappears from the surface (maybe it goes down a drain pipe). As one might expect, the "divergence" at that position becomes negative, and the point is called a "sink."

1.5.5. Laplacian

We have seen that the gradient operation on a scalar field results in a vector field. The vector field may be subjected in turn to a divergence operation—which returns a new scalar field. This successive application of the ∇ operator is called the *Laplacian* and is symbolized by ∇^2. That is,

$$\nabla^2\Psi = \nabla \cdot \nabla\Psi \tag{1.27}$$

Since, from (1.18), we have

$$\nabla\Psi = \overline{a}_x \frac{\partial \Psi}{\partial x} + \overline{a}_y \frac{\partial \Psi}{\partial y} + \overline{a}_z \frac{\partial \Psi}{\partial z} \tag{1.28}$$

then, by virtue of (1.26), we obtain

$$\nabla \cdot \nabla\Psi = \frac{\partial}{\partial x}\left(\frac{\partial \Psi}{\partial x}\right) + \frac{\partial}{\partial y}\left(\frac{\partial \Psi}{\partial y}\right) + \frac{\partial}{\partial z}\left(\frac{\partial \Psi}{\partial z}\right)$$

or, more simply

$$\nabla^2\Psi = \frac{\partial^2 \Psi}{\partial x^2} + \frac{\partial^2 \Psi}{\partial y^2} + \frac{\partial^2 \Psi}{\partial z^2} \tag{1.29}$$

Equation (1.29) evaluates the Laplacian of any scalar function in rectangular coordinates.

1.5.6. Laplace's Equation

If there are no sources or sinks of Ψ within a region, then throughout that region the divergence is zero, so at every point one has

$$\nabla^2\Psi = 0 \tag{1.30}$$

With the right-hand side zero, the equation is called *Laplace's equation*. Often the goal of a problem in bioelectricity is to find an analytical or numerical function that obeys Laplace's equation within some specified region, e.g., around an electrically active fiber.

1.5.7. Comments about the Laplacian

The Laplacian deals with the divergence of the gradient. If the gradient is proportional to the flow, as of water on the surface of the hill, the Laplacian will find the divergence of that flow. There will be a nonzero divergence of the flow at those points where water is emerging from a spring or falling into drainpipes, i.e., at sources and sinks, but the divergence of the surface flow will be zero elsewhere. That is, evaluating the Laplacian of a flow function is a means of identifying the presence and magnitude of sources and sinks of that function. A very important special case occurs when Laplace's equation is satisfied everywhere in a region, since that means there are no sources or sinks within that region.

1.6. VECTOR IDENTITIES

Vector identities describe relationships that are true for all well-behaved scalar and vector functions. That is, while the identity expression looks like an equality, it does not simply hold for certain values of the variables but rather for all values of the variable. In subsequent chapters we shall refer to the vector identities listed here. The proof of the first expression will be given. The reader is invited to use this as a model for confirming the others.

In the following expressions Φ and Ψ are well-behaved scalar functions:

$$\nabla \cdot (\Phi\overline{A}) = \overline{A} \cdot \nabla\Phi + \Phi\nabla \cdot \overline{A} \tag{1.31}$$

$$\nabla(\Phi\Psi) = \Phi\nabla\Psi + \Psi\nabla\Phi \tag{1.32}$$

$$\nabla^2(r) = 0, \quad r = \sqrt{x^2 + y^2 + z^2} \tag{1.33}$$

1.6.1. Verification of Eq. (1.31)

To verify (1.31) we replace $\overline{A}$ by its rectangular components $(A_x\overline{a}_x + A_y\overline{a}_y + A_z\overline{a}_z)$ leading to

$$\nabla \cdot (\Phi\overline{A}) = \nabla \cdot (\Phi A_x\overline{a}_x + \Phi A_y\overline{a}_y + \Phi A_z\overline{a}_z) \tag{1.34}$$

Now using the definition of divergence given in (1.26) results in the expression

$$\nabla \cdot (\Phi\overline{A}) = \frac{\partial}{\partial x}(\Phi A_x) + \frac{\partial}{\partial y}(\Phi A_y) + \frac{\partial}{\partial z}(\Phi A_z) \tag{1.35}$$

By chain rule we have

$$\nabla \cdot (\Phi\overline{A}) = \Phi\frac{\partial A_x}{\partial x} + A_x\frac{\partial \Phi}{\partial x} + \Phi\frac{\partial A_y}{\partial y} + A_y\frac{\partial \Phi}{\partial y} + \Phi\frac{\partial A_z}{\partial z} + A_z\frac{\partial \Phi}{\partial z} \tag{1.36}$$

Collecting terms gives

$$\nabla \cdot (\Phi\overline{A}) = \Phi\left(\frac{\partial A_x}{\partial x} + \frac{\partial A_y}{\partial y} + \frac{\partial A_z}{\partial z}\right) + A_x\frac{\partial \Phi}{\partial x} + A_y\frac{\partial \Phi}{\partial y} + A_z\frac{\partial \Phi}{\partial z} \tag{1.37}$$

We now identify

$$\begin{aligned}\overline{A} \cdot \nabla\Phi &= \left(\frac{\partial \Phi}{\partial x}\overline{a}_x + \frac{\partial \Phi}{\partial y}\overline{a}_y + \frac{\partial \Phi}{\partial z}\overline{a}_z\right) \cdot \overline{A} \\ &= A_x\frac{\partial \Phi}{\partial x} + A_y\frac{\partial \Phi}{\partial y} + A_z\frac{\partial \Phi}{\partial z}\end{aligned} \tag{1.38}$$

so that substituting (1.38) and (1.26) into (1.37) leads to

$$\nabla \cdot (\Phi\overline{A}) = \Phi\nabla \cdot \overline{A} + \overline{A} \cdot \nabla\Phi \tag{1.39}$$

which confirms (1.31).

1.7. SOURCE AND FIELD POINTS

Many problems in electrophysiological modeling require a vector $\overline{r}$ that extends from a "source point" (x, y, z) to a "field point" (x', y', z') (Figure 1.5). These names arise when considering active tissues lying in passive volume conductors where we shall be interested in the electric potential field at (x', y', z') established by current sources at (x, y, z).

The use of primed and unprimed variables will be seen later to be a useful way of distinguishing source and field points. It is important to distinguish, because sometimes one needs to perform a mathematical operations on the one or on the other, without confusion and while utilizing a common coordinate system.

The radius, r, from source to field is a scalar function whose magnitude is

$$r = [(x - x')^2 + (y - y')^2 + (z - z')^2]^{1/2} \tag{1.40}$$

Since $r(x, y, z, x', y', z')$ is a scalar field we can examine its gradient. In this case, since it depends on both the source and field, we can evaluate the gradient with respect to either the field coordinates (primed) or the source coordinates (unprimed) while holding the other coordinate fixed.

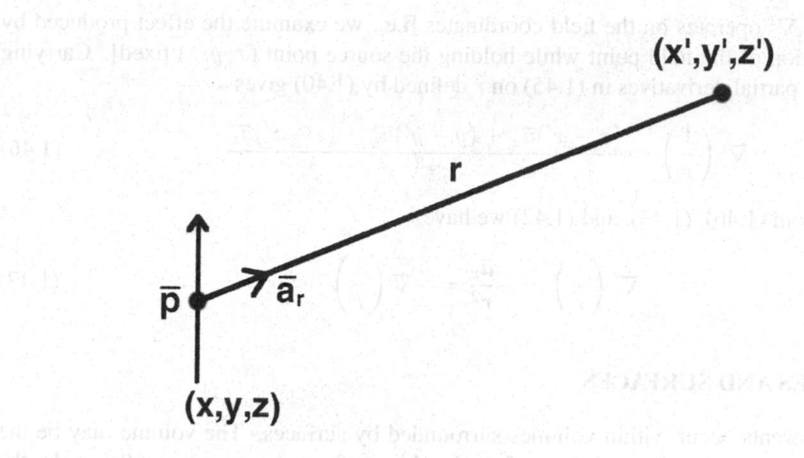

Figure 1.5. Dipole Field. Source at (x, y, z) and field at (x', y', z').

1.7.1. Gradient of $(1/r)$ with Respect to Source Coordinates

In source field problems it is frequently necessary to examine the gradient of the scalar function $(1/r)$. With care and some patience, finding $\nabla(1/r)$ can be accomplished by carrying out the gradient (derivative) operations in x, y, z coordinates. That is, from (1.17) we recall that

$$\nabla \equiv \overline{a}_x \frac{\partial}{\partial x} + \overline{a}_y \frac{\partial}{\partial y} + \overline{a}_z \frac{\partial}{\partial z} \tag{1.41}$$

Assuming that the gradient is desired at the source point, then we apply the ∇ operation only on the unprimed coordinate variables in r [see Eq. (1.40)], and the result is

$$\nabla\left(\frac{1}{r}\right) = -\frac{(x - x')\overline{a}_x + (y - y')\overline{a}_y + (z - z')\overline{a}_z}{r^3} \tag{1.42}$$

Because the unit radial vector from the source point to the field point, $\overline{a}_r$, is (see Figure 1.5)

$$\overline{a}_r = -((x - x')\overline{a}_x + (y - y')\overline{a}_y + (z - z')\overline{a}_z)/|r| \tag{1.43}$$

then (1.42) can be written as

$$\nabla\left(\frac{1}{r}\right) = \frac{\overline{a}_r}{r^2} \tag{1.44}$$

Note that our choice of unprimed variables to describe source geometry and primed variables to describe the field geometry is arbitrary; the reverse definition could equally well be made.

1.7.2. Gradient of $(1/r)$ with Respect to Field Coordinates

In some cases we will be interested in applying the operator

$$\nabla' \equiv \overline{a}_x \frac{\partial}{\partial x'} + \overline{a}_y \frac{\partial}{\partial y'} + \overline{a}_z \frac{\partial}{\partial z'} \tag{1.45}$$

In this case, ∇' operates on the field coordinates [i.e., we examine the effect produced by varying the position of the field point while holding the source point (x, y, z) fixed]. Carrying out the indicated partial derivatives in (1.45) on r defined by (1.40) gives

$$\nabla'\left(\frac{1}{r}\right) = \frac{(x - x')\overline{a}_x + (y - y')\overline{a}_y + (z - z')\overline{a}_z}{r^3} \tag{1.46}$$

Consequently, from (1.46), (1.43), and (1.42) we have

$$\nabla'\left(\frac{1}{r}\right) = -\frac{\overline{a}_r}{r^2} = -\nabla\left(\frac{1}{r}\right) \tag{1.47}$$

1.8. VOLUMES AND SURFACES

Bioelectric events occur within volumes surrounded by surfaces. The volume may be the whole torso volume of a human, and the surface the skin surface. At a much smaller scale, the volume may be a cell volume and the surface a cell membrane. The extensions of vector analysis to some of the mathematics of volumes and surfaces thus prove useful in bioelectricity.

1.8.1. Gauss's Theorem

In a previous section we saw that the net outflow of current from a given volume is a measure of the net source contained in the volume. For a volume *V* bounded by a surface *S* the outflow is given by

$$\text{outflow} = \oint_s \overline{J} \cdot d\overline{S} \tag{1.48}$$

where $d\overline{S}$ is a surface element whose direction is the outward normal. The divergence $\nabla \cdot \overline{J}$ also evaluates the net outflow in each unit of the volume.

Thereby, the outflow evaluated in (1.48) can also be found by integrating $\nabla \cdot \overline{J}$ through the volume bounded by *S*. In fact,

$$\int_V \nabla \cdot \overline{J}\, dV = \oint_s \overline{J} \cdot d\overline{S} \tag{1.49}$$

This relationship is true for any well-behaved vector field. It is known as *Gauss's theorem* or *the divergence theorem.*

1.8.2. Green's First Identity

Suppose that

$$\overline{J} = \Phi \nabla \Psi \tag{1.50}$$

where Φ and Ψ are two scalar fields. Substituting (1.50) in (1.49) gives

$$\int_V \nabla \cdot \Phi \nabla \Psi dV = \oint_s \Phi \nabla \Psi \cdot d\overline{S} \tag{1.51}$$

Expanding (1.51) with the help of Eq. (1.31) produces *Green's first identity*, which is:

$$\int_V \Phi \nabla^2 \Psi dV + \int_V \nabla \Phi \cdot \nabla \Psi dV = \oint_s \Phi \nabla \Psi \cdot d\overline{S} \tag{1.52}$$

1.8.3. Green's Second Identity

From Green's first identity one gets Green's second identity, also called *Green's Theorem.* To do so, one observes that there is no special relationship required in (1.50) between scalars Φ and Ψ and hence (1.52) describes a vector identity.

It is consequently also valid if Φ and Ψ are interchanged; specifically

$$\int_V \Psi\nabla^2\Phi\, dV + \int_V \nabla\Psi\cdot\nabla\Phi\, dV = \oint_s \Psi\nabla\Phi\cdot d\overline{S} \tag{1.53}$$

Equations (1.52) and (1.53) are now two equations, different from each other, but both involving the same scalar fields Φ and Ψ. If Eq. (1.53) is subtracted from (1.52), the result is *Green's Theorem:*

$$\int_V (\Phi\nabla^2\Psi - \Psi\nabla^2\Phi)\, dV = \oint_s (\Phi\nabla\Psi - \Psi\nabla\Phi)\cdot d\overline{S} \tag{1.54}$$

1.8.4. Comment on Green's Theorem

Green's Theorem may be seen as an abstract theorem (with, perhaps, an austere beauty) since it shows relationships between scalar fields Φ and Ψ, and their gradients and divergences, without assigning any specific physical or biological meaning to either one.

To view Green's Theorem as having significance limited to the abstract is a mistake, however, since Green's Theorem can be used as a powerful tool in analyzing real problems. (Later on in this book, for example, Green's Theorem is used as a way of examining how currents in the heart affect voltages on the body surface.)

Exploitation of Green's Theorem often proceeds by choosing specific forms of the scalar fields Φ and Ψ. For example, Φ may be interpreted as an electric potential while Ψ may be the reciprocal distance from source to field, $1/r$. Once such assignments are made and used, the seemingly abstract equation (1.54) quickly becomes a specific equation relating the physically real variables of the chosen problem itself.[3]

1.9. THE GRADIENT AND DIVERGENCE OF (1/r)

This section uses the mathematics of Chapter 1 to examine the extraordinary nature of $(1/r)$, its gradient, and its divergence, whether or not $r = 0$. The results are used to relate currents to potentials, as presented in Chapter 2, and then used routinely in the chapters thereafter.

Specifically, the electric potential field of a current point source of strength I_0, located at the coordinate origin, and lying in a uniform volume conductor of infinite extent and conductivity σ, is

$$\Phi_e(r) = \frac{I_o}{4\pi\sigma}\frac{1}{r} \tag{1.55}$$

The above equation is frequently used to find the potential for a point source, based on the $1/r$ function. It is implicit that the potentials arise from a single current source located at that one single point. Thus, it must be the case that $\nabla^2(1/r) = 0$ at all points where $r \neq 0$.

On the other hand, we have established that whenever $\nabla^2\Phi$ is greater than zero, there are sources. Thus, for (1.55) to be true, it must be that $\nabla^2(1/r) \neq 0$ at $r = 0$.

It is not obvious that the $1/r$ function has these properties in its divergence.

Restating the issue in a different way: because $\nabla^2\Phi$ evaluates the volume source distribution of Φ, then $\nabla^2(1/r)$ should be zero everywhere except at the origin. Conversely, at the origin, where a point source corresponds to an infinite source density (i.e., a finite source within an infinitesimal volume), the divergence must be nonzero and in fact infinite in a special way.

Demonstrating this special property of $\nabla^2(1/r)$ is of further interest when one recognizes that any arbitrary distribution of electrical sources can be considered to consist of a collection of point sources, whose collective effect is the linear sum of the effect of each one individually. Thus the properties of the scalar field $(1/r)$ are of special interest, and a discussion of this problem has far-reaching ramifications.

We consider first the value of $\nabla^2(1/r)$ for $r \neq 0$. Writing r in terms of x, y, z coordinates, and using direct differentiation, we have:

$$r = \sqrt{(x-x')^2 + (y-y')^2 + (z-z')^2} \tag{1.56}$$

Using (1.30) and (1.55) we have

$$\begin{aligned}
\nabla^2(1/r) &= \frac{\partial^2}{\partial x^2}\left(\frac{1}{r}\right) + \frac{\partial^2}{\partial y^2}\left(\frac{1}{r}\right) + \frac{\partial^2}{\partial z^2}\left(\frac{1}{r}\right) \\
&= \left(\frac{1}{r^3} - \frac{3(x-x')^2}{r^5}\right) + \left(\frac{1}{r^3} - \frac{3(y-y')^2}{r^5}\right) + \left(\frac{1}{r^3} - \frac{3(z-z')^2}{r^5}\right) \\
&= 0, r \neq 0
\end{aligned} \tag{1.57}$$

This result confirms the expected behavior of a point-source field at the origin at any finite radial distance.

$\nabla^2\Phi$ describes the *negative* of the source density of Φ. Consequently, the total source contained in a small concentric sphere of radius a around the point-source field described by Φ_e in (1.55) should equal the (negative of the) point-source strength (*SS*). We may evaluate SS by finding the volume integral of the source density, namely,

$$\begin{aligned}
SS &= -\int_V \nabla^2\Phi_e dV \\
&= -\frac{I_0/\sigma}{4\pi}\left(\int_0^a \nabla^2(\frac{1}{r})\,(4\pi r^2)dr\right) \\
&= -\frac{I_0}{\sigma}\int_V \nabla\cdot\nabla(\frac{1}{r})\,dV
\end{aligned} \tag{1.58}$$

The last integral of (1.58) can be carried out by applying the divergence theorem. Because of symmetry and uniformity on $r = a$ we get, using (1.64) below,

$$SS = -\frac{I_0/\sigma}{4\pi}\int_S \nabla(\frac{1}{r})\Big|_a \cdot d\vec{S} = -\left(\frac{I_0/\sigma}{4\pi}\right)4\pi a^2\nabla(\frac{1}{r})\Big|_a \cdot \vec{a}_r \tag{1.59}$$

The gradient may be evaluated by recognizing that it depends only on the variable r and hence, from its fundamental definition, requires only a derivative with respect to r, giving

$$\nabla(1/r)\Big|_a = -(1/r^2)\Big|_a \overline{a}_r = -(1/a^2)\overline{a}_r \tag{1.60}$$

The outcome is that

$$SS = I_0/\sigma \tag{1.61}$$

We note that, as might be expected, the outcome did not depend on a no matter how small it might be chosen (consistent with the radius of a point source being zero). This result confirms that the source is, indeed, a point source and also that its magnitude equals I_0/σ, where σ converts to electric potential.

In the evaluation of $\nabla(1/r)$ above, we could have applied (1.18), which finds the gradient in rectangular coordinates. The result would have been the same as in (1.60), but the derivation would be more lengthy. One can, in fact, derive general expressions for gradient, divergence, and the Laplacian in coordinate systems other than rectangular; these may be particularly advantageous in applications where the geometry is cylindrical, spherical, etc.

For example, in spherical coordinates these expressions are

$$\nabla\Phi = \overline{a}_r \frac{\partial\Phi}{\partial r} + a_\theta \frac{1}{r}\frac{\partial\Phi}{\partial\theta} + \frac{\overline{a}_\phi}{r\sin\theta}\frac{\partial\Phi}{\partial\phi} \tag{1.62}$$

$$\nabla\cdot\overline{A} = \frac{1}{r^2}\frac{\partial}{\partial r}(r^2 A_r) + \frac{1}{r\sin\theta}\frac{\partial}{\partial\theta}(\sin\theta A_\theta) + \frac{1}{r\sin\theta}\frac{\partial A_\phi}{\partial\phi} \tag{1.63}$$

$$\nabla^2\Phi = \frac{1}{r^2}\frac{\partial}{\partial r}(r^2\frac{\partial\Phi}{\partial r}) + \frac{1}{r^2\sin\theta}\frac{\partial}{\partial\theta}(\sin\theta\frac{\partial\Phi}{\partial\theta}) + \frac{1}{r^2\sin^2\theta}\frac{\partial^2\Phi}{\partial\phi^2} \tag{1.64}$$

Note that the gradient operation deduced for (1.60) from basic principles could also be obtained from (1.62) (noting that the function has only a radial component).

The results found in this example can be summarized in the following useful equation:

$$\nabla^2(1/r) = -4\pi\delta(r) \tag{1.65}$$

where $\delta(r)$ is a delta function. This function is defined to be zero everywhere except where $r = 0$, in which case $\delta(0) = \infty$. However, the singularity is integrable, so that the volume integral over a volume containing the singular point is finite; in fact (by definition)

$$\int \delta(r)\, dv = 1 \tag{1.66}$$

Equation (1.65) could have been used in the analysis of the divergence $(1/r)$, which was examined in a preceding section. Use of (1.65) there would have resulted in the same conclusions, achieved by means of a shorter sequence of steps.

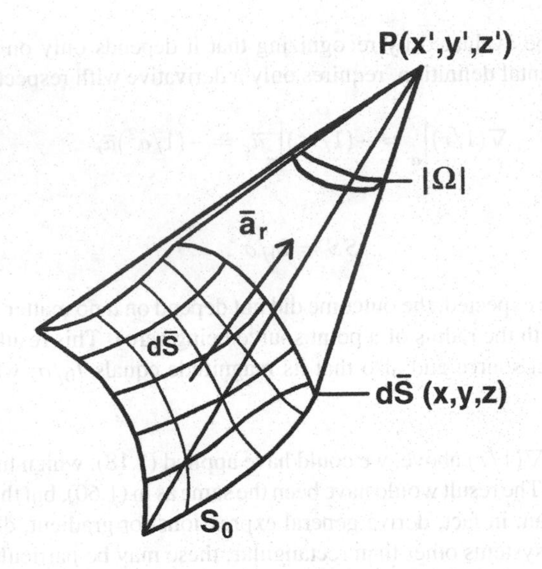

Figure 1.6. The solid angle of the surface S_0 is evaluated at P. The magnitude of Ω can be interpreted as the area intercepted on a unit sphere.

1.10. SOLID ANGLES

Just as analysis of arcs and lengths in a 2D plane requires the use of angles, the analysis of surfaces in 3D requires the use of *solid angles*. In bioelectricity, such angles are essential in finding potentials generated from three dimensional objects (such as cells) or organs (such as the heart).

Angles in two dimensions can be considered fractions of a unit circle (with maximum angle in radian measure of 2π). In an analogous way, angles in three dimensions can be considered as fractions of a sphere of unit radius, with a maximum solid angle of 4π steradians.

Vector analysis provides the necessary operations for the definition and understanding of solid angles. In Figure 1.6, the element of the solid angle, $d\Omega$, subtended at the point P is

$$d\Omega = -\nabla\left(\frac{1}{r}\right)\cdot d\overline{S} \tag{1.67}$$

where r is the distance from an element of surface $d\overline{S}$ to P. That is, if P is at the coordinate (x', y', z') and $d\overline{S}$ is at (x, y, z) then

$$r = \sqrt{(x'-x)^2 + (y'-y)^2 + (z'-z)^2} \tag{1.68}$$

and

$$\nabla\left(\frac{1}{r}\right) = \frac{\overline{a}_r}{r^2} \tag{1.69}$$

as can be verified by expanding the gradient [or by reference to Eq. (1.44)]. In Eq. (1.69), $\overline{a}_r$ is from $d\overline{S}$ to P, as illustrated in Figure 1.6.

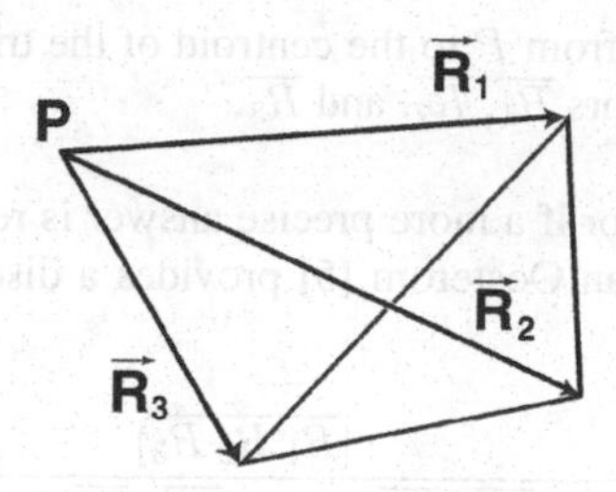

Figure 1.7. Vectors to Corners of a Triangle. Vectors $\overline{R_1}$, $\overline{R_2}$, and $\overline{R_3}$ extend from point P to the three corners of a triangle. The vectors touch the corners in clockwise order when seen from the "inside" of the surface defined by the triangle's surface vector.

If (1.69) is substituted into (1.67), then one obtains

$$d\Omega = -\frac{(\overline{a}_r \cdot d\overline{S})}{r^2} \tag{1.70}$$

One can interpret the magnitude of $d\Omega$ evaluated in (1.70) as the area intercepted on a unit sphere by the rays drawn to the periphery of the area element dS from P. And, consequently, the magnitude of the total solid angle Ω given by

$$\Omega = \int_{S_O} d\Omega = -\int_{S_O} \frac{(\overline{a}_r \cdot d\overline{S})}{r^2} \tag{1.71}$$

is the area intercepted on a unit sphere by the rays drawn to the periphery of S_O.

The interpretation of the solid angle as the subtended area on a unit sphere follows because $\overline{a}_r \cdot d\overline{S}$ is the component of the area $d\overline{S}$ that lies on an included sphere. At the same time, the area magnitude of $\overline{a}_r \cdot d\overline{S}$ is scaled by the factor $1/r^2$, a scaling that brings the area to that of the *unit* sphere.

The solid angle Ω is negative when surface vector $\overline{S}$ points toward P, as in Figure 1.6. On a closed surface (e.g., a cell), the surface vector is most often chosen to point *outward*.

Numerical evaluation of solid angles often is done by dividing the surface into a set of triangles, and then computing the solid angle for the surface as a whole as the sum of the solid angles of each triangle. An example of such a triangle is shown in Figure 1.7.

The integral of (1.71) then has to be evaluated numerically. (The integral can be done analytically for special cases, but not in general.) Then numerical evaluation usually can be divided into two categories, triangles that are far enough away from P to use an approximate method, and those that are close to it, which require a vector method.

If the triangle is far enough away from P (far enough that the distance r to any point on the triangle is approximately constant), then the r can be factored out of the integral of the solid angle, and (1.71) becomes

$$\Omega = \int_{S_O} d\Omega = -\int_{S_O} \frac{(\overline{a}_r \cdot d\overline{S})}{r^2} = -\frac{1}{r^2}\int_{S_O} (\overline{a}_r \cdot d\overline{S}) \approx \frac{(\overline{a}_r \overline{S})}{r^2} \tag{1.72}$$

Here r is nominally the distance from P to the centroid of the triangle. Note that both $\overline{S}$ and r can be found from the set of vectors $\overline{R_1}$, $\overline{R_2}$, and $\overline{R_3}$.

If the triangle is close to P, or if a more precise answer is required, a more detailed vector analysis is required, e.g., [1]. Van Oosterom [5] provides a discussion of alternatives and the superior vector formula

$$\tan\left(\frac{1}{2}\Omega\right) = \frac{[\overline{R_1}\,\overline{R_2}\,\overline{R_3}]}{R_1\,R_2\,R_3 + (\overline{R_1}\cdot\overline{R_2})R_3 + (\overline{R_1}\cdot\overline{R_3})R_2 + (\overline{R_2}\cdot\overline{R_3})R_1} \tag{1.73}$$

where R_1 (no overbar) is the magnitude, and the operation [] is defined as

$$[\overline{R_1}\,\overline{R_2}\,\overline{R_3}] \equiv \overline{R_1}\cdot(\overline{R_2}\times\overline{R_3}) \tag{1.74}$$

Equation (1.73) requires taking the inverse tangent to find Ω. The equation proves satisfactory if care is taken when finding the inverse tangent regarding the signs of the numerator and denominator.

1.11. OPERATIONS SUMMARY

Below is a summary of the mathematical operations reviewed in this chapter:

Symbol	Definition
$\cdot$	Dot (scalar) product
∇	Del operator (partial derivatives)
$\nabla\Phi$	Gradient of the scalar field Φ
$\nabla\cdot\overline{J}$	Divergence of the vector field $\overline{J}$
$\nabla^2\Psi$	Laplacian of scalar field Ψ
Ω	Solid Angle

1.12. NOTES

1. Because this book is about bioelectricity, an extensive discussion of vector analysis would be inappropriate. Many readers will have studied vector analysis before coming to this text. For those who wish additional material, several texts are suggested at the end of the chapter.
2. Here and there in this text there are digressions about the subject at hand that describe ways of thinking about the mathematical points, often using analogies. These are offered as an aid in developing an intuitive feel for the subject. While this can be a valuable asset, it is important to be cautious and not substitute such intuition for an actual analysis of the subject since the analogy may be loosely, but not precisely, true.
3. Green's Theorem, with its upside down triangles, also can be used to impress your friends and family, when they asked you what you learned today.

1.13. REFERENCES

1. Barnard ACL, Duck IM, Lynn MS, Timberlake WP. 1967. The application of electromagnetic theory to electrocardiology, II. *Biophys J* **7**:463–491.

2. Davis HF. 1995. *Introduction to vector analysis*, 7th ed. Dubuque, IA: William C. Brown.
3. Lewis PE, Ward JP. 1987. *Vector analysis for engineers and scientists*. Reading, MA: Addison-Wesley.
4. Stein FM. 1963. *An introduction to vector analysis*. London: Harper and Row.
5. van Oosterom A, Strackee J. 1983. The solid angle of a plane triangle. *IEEE Trans Biomed Eng* **30:**125–126.
6. Young EC. 1993. *Vector and tensor analysis*. New York: Marcel Dekker.

2

Sources and Fields

Understanding electricity in living tissue—where it comes from, what it does, and how it does it—has been a goal actively pursued since the 1700s, at the inception of the scientific study of electricity. Such famous investigators as Luigi Galvani, who startled the scientific world in 1791 with his descriptions of the effects of currents in frogs, and his critic Alessandro Volta (for whom the volt is named) were extensively involved with "animal electricity."

Some theories of that time were fantastic in light of present-day knowledge (such as the speculation by Galvani that there was an electrical fluid prepared by the brain, flowing through nerve tubes into the muscles). Nonetheless, the careful experimental work of Galvani, Volta, and other investigators of that time laid the foundations of the field of bioelectricity, as well as electricity more broadly.

In the 1700s some investigators thought that animal electricity was different in fundamental ways from the electricity observed in nonliving objects. That was wrong. One thing that now is certain is that animal electricity is not a different kind of electricity. Rather, bioelectricity is based on the same fundamental laws that describe electricity in the atmosphere, in solid-state materials such as silicon, in television sets, or lighting systems.

There are at the same time many substantial differences between the elements of electrical systems that exist in living tissue as compared to man-made electrical systems, and in the ways they work. One of the major differences is that the living systems derive their electrical energy from the ionic concentration differences that exist across cell membranes.

Consequently the energy sources are distributed in space along the membrane. Use of this energy involves a flow of current across the membrane. As a corollary, current in living systems necessarily and desirably flows both inside and outside electrically active cells, and in a controlled fashion crosses over the membrane separating the one from the other.

In contrast, systems designed by humans usually have a localized energy source, such as a battery, that drives currents through a restricted conductor, such as a wire. In such engineered systems, currents outside the wire are usually due to leakage or other imperfections, rather than being an important part of the system itself.

The goal of this chapter is to describe, concisely, the fundamental mathematical relationships linking sources and the electric potentials to the current fields they produce. These relationships are presented mainly in the form used when considering current sources in a conducting medium, the form most often of value in bioelectricity.

The most basic relationships of sources, currents, and potentials are given below in only a few paragraphs, but their ramifications are extensive. Much of the rest of this chapter (and indeed much of the rest of this book) may be seen as concerned with their detailed applications.

2.1. FIELDS

The perspective of electrical sources and fields as used in bioelectricity visualizes space as filled with potentials and currents. Both have values that are functions of position, but both exist more or less everywhere throughout the region. This view corresponds to the recognition that animals and people are large volumes, filled with conducting solutions, with ionic currents moving extensively throughout.

Some readers will be familiar with the quantitative properties of electrical circuits. Such circuits are characterized through the behavior of discrete (lumped parameter) elements connected together by lossless wires. The perspective of fields as used in bioelectricity differs from the perspective of circuits in fundamental ways, and is more akin to subjects such as antennas. In this text the language and symbols of circuits are used from time to time, but one has to keep in mind the limits of such a description, because the distinctly different nature of the bioelectric environment changes everything.

2.2. TISSUE RESISTANCE AND CONDUCTANCE

One of the goals of this book is the elucidation of electrical sources, potentials, and currents in biological tissues. The existence of currents throughout a *volume conductor* implies the existence of an electric field, $\vec{E}$. The electric field is important because it describes the force that is exerted upon a unit charge. Thus it quantifies the force that moves the ions, the constituent elements of the current. Furthermore, for inhomogeneous materials we will expect a resistivity ρ (or inverse conductivity σ) to be a function of position. We will discuss this subsequently.

The resistive property of materials is included in electrical circuits by means of lumped elements with pure resistance. Physically we understand that the *resistance* measures the magnitude of voltage across the element when passing the circuit current, as expressed in Ohm's law $V = IR$. The *resistor* is the physical element. For a uniform cylindrically shaped rod current can be assumed uniform across the resistor's cross-section; hence, the resistor may be treated as one dimensional, or simply as lumped. Its resistive value can be evaluated by dividing the total voltage across the element by the current, using $R = V/I$.

Biological materials, a major focus of this book, have resistive properties. In general these are not lumped. Biological materials are cells or organs that have significant spatial dimensions, and often their properties change from one place to another. Instead of a lumped total resistance R, there is a property of the biological material, the *resistivity*, often denoted by ρ, in units of *Ohm-cm*. The resistance of a particular element of the material then is determined as $R = \rho A/L$, where A is the cross-sectional area through which current is flowing, and L is the length through which current flows.

The concept of resistivity applies to a uniform medium, but that is not required, as resistivity also allows an inhomogeneous medium. In the latter case ρ is a function of position.

In the analysis of many biological situations, it is more convenient (and established practice) to use *conductivity*, denoted σ, instead of resistivity. For example, that is so for the fundamental equations presented in the next section of this chapter. Conductivity is simply the reciprocal of resistivity, i.e., $\sigma = 1/\rho$. The units of σ are *Siemens/cm*. The use of conductivity is more convenient when there are multiple current pathways in parallel. In this case the conductivities simply can be added, an intuitively and computationally simple step not possible with resistivities.

A further discussion of resistivity and conductivity and their related units appears near the end of this chapter. Tissue also has substantial capacitance, which is discussed there also.

2.3. FIELDS AND CURRENTS

As noted, the existence of currents throughout a *volume conductor* implies the presence of an electric field $\overline{E}$. In electrophysiological problems, even under normal time-varying conditions, $\overline{E}$ behaves like a static field at each instant of time (we call it *quasi-static*).[1]

Consequently, $\overline{E}$ can be described, as for electrostatic fields, as the negative gradient of a scalar potential, Φ, that is,

$$\overline{E} = -\nabla\Phi \tag{2.1}$$

in a conducting medium. (A conducting medium has charged ions or other particles than can move.) The force exerted by the electric field results in the flow of charge (i.e., a current).

The current density $\overline{J}$ (current per unit of cross-sectional area) is related to the electric field, E, by Ohm's law, namely,

$$\overline{J} = \sigma\overline{E} = -\sigma\nabla\Phi \tag{2.2}$$

In (2.2), σ is the *conductivity* of the conducting medium through which the current is flowing. Inspection of (2.2) shows that the current density $\overline{J}$ is in the same direction as the electric field $\overline{E}$, if σ is a scalar as assumed here. Conversely, $\overline{J}$ may be large or small, for a fixed value of $\overline{E}$, depending on the value of the conductivity. (For physiological volume conductors the charge carriers are ions, in contrast with electrons in the case of electric wires.)

The conducting region, in general, may be considered to contain current sources described by a source density $I_v(x, y, z)$. Sources may occur naturally, as in a membrane, or artificially, as from a stimulus electrode.

From the divergence properties of the current density, $\overline{J}$, we require

$$\nabla \cdot \overline{J} = I_v \tag{2.3}$$

Equation (2.3) is true because divergence, being a measure of outflow per unit volume, is equivalent to the source density.

When we consider point sources, the volume distribution function I_v will be singular at those sources because, as already noted, such source densities are infinite (consisting, as they do, of a finite source strength at an infinitesimal volume); the source density function, while singular, is necessarily integrable (since the integral evaluates the source magnitude).

When a volume conducting region is evaluated, we might find that the volume integral of I_v is zero. Finding a result of zero, we may conclude that the volume is either source free or contains no net source (the total current across the bounding surface is zero) because its sources equal its sinks. If the volume integral is nonzero, then the region is net positive or net negative, and compensating sources lie outside the region. In bioelectricity, compensating sources are necessary to satisfy the requirement that the sum of all sources be equal to zero, a condition that preserves overall current conservation.

2.3.1. Poisson's Equation

Potentials link directly to the current sources and sinks that produce them. Taking the divergence of (2.2) and applying (2.3) gives

$$\nabla \cdot \overline{J} = I_v = -\sigma \nabla^2 \Phi \tag{2.4}$$

Thus, for a region where the conductivity is homogeneous but which contains a source density I_v, *Poisson's equation* for Φ results, namely [from (2.4)],

$$\nabla^2 \Phi = -\frac{I_v}{\sigma} \tag{2.5}$$

2.3.2. Laplace's Equation

An important special case of Poisson's equation occurs when the source density I_v is zero everywhere in a region of interest (i.e., sources lie outside or at the boundary of this identified region). For this case, that of a homogeneous conducting region that is free of sources (i.e., sources lie outside the identified region), conservation of current requires that $\nabla \cdot \overline{J} = 0$. Equation (2.4), along with the condition that I_v be zero, results in

$$\nabla \cdot \overline{J} = -\sigma \nabla \Phi = 0 \tag{2.6}$$

Under these conditions (2.5) requires that Φ satisfy the partial differential equation called *Laplace's equation*, namely,

$$\nabla^2 \Phi = 0 \tag{2.7}$$

A solution for the electric potential Φ in Poisson's equation (2.5) can be written in integral form. The solution is

$$\begin{aligned} \Phi(x', y', z') &= \frac{1}{4\pi\sigma} \int \frac{I_v dV}{r} \\ &= \frac{1}{4\pi\sigma} \sum \frac{I_o^j}{r_j} \end{aligned} \tag{2.8}$$

The solution presented in 2.8 is given in two forms: (a) the integral form in terms of I_v applies when the sources are distributed; (b) the summation form in terms of I_o^j (a point source at distance r_j applies when there is a collection of point sources). That Eq. (2.8) is a solution to (2.5) can be verified by returning to the section in Chapter 1 on the special nature of the $(1/r)$ function, where this question is examined.

2.4. FIELDS FROM SOURCES, AND VICE VERSA

Note that Eq. (2.8) provides an expression for the electrical potential from a known source configuration I_v, whereas Eq. (2.4) permits an evaluation of the sources, I_v, assuming it is the electric potential Φ that is known. In other words, when one knows sources, then one can get potentials, and vice versa.

2.5. DUALITY

The equations of the previous section are similar to those found in the study of electrostatics. The electrostatic equations may already be familiar to some readers since they appear in introductory physics courses. Electrostatics is concerned with electric charges and fields in a dielectric (i.e., insulating) medium while our interest lies in currents in conducting media.

Electrostatics and bioelectricity are different physical environments. In spite of the differences we will show below the similarity of governing equations of electrostatics and the equations that arise when there is steady current, and we will describe how mathematical solutions found in one context can be transformed into the other.

For electrostatic fields the basic equations are

$$\overline{E} = -\nabla\Phi \tag{2.9}$$

$$\overline{D} = \varepsilon\overline{E} \tag{2.10}$$

$$\nabla^2\Phi = -\rho/\varepsilon \tag{2.11}$$

$$\Phi = \frac{1}{4\pi\varepsilon}\int\frac{\rho dV}{r} \tag{2.12}$$

where ρ is the charge (source) density, and ε is the dielectric permittivity. In fact, with $\overline{D}$ as the dielectric displacement,

$$\nabla\cdot\overline{D} = \rho \tag{2.13}$$

Equation (2.12) is seen as an extension of Coulomb's law, but this expression also is the solution of Poisson's equation (2.11) in integral form. (The adventurous reader can check that this is so.)

Now Eq. (2.9) is identical with our Eq. (2.1), while Eqs. (2.10), (2.11), and (2.13) correspond precisely to Eqs. (2.2), (2.5), and (2.3), provided we replace

$$\varepsilon \to \sigma \tag{2.14}$$

$$\overline{D} \to \overline{J} \tag{2.15}$$

$$\rho \to I_v \tag{2.16}$$

These correspondences are an application of the "principle of duality."

The fact that the mathematics of currents in a volume conductor aligns so closely with that of charges in a dielectric is widely recognized and frequently advantageous. The advantage is that results (theoretical solutions, computer programs, etc.) learned in one context (e.g., physics) can be readily transferred to another (e.g., electrophysiology).

It is important to keep in mind, however, that the duality *does not imply equivalence.* For example, conductivity σ has an altogether different physical meaning than permittivity ε. In fact, while $\infty < \sigma < 0, k\varepsilon_0 < \varepsilon < \varepsilon_0$.

2.6. MONOPOLE FIELD

A "monopole" is a single pole. In the context of current fields a monopole is a single (point) source or sink of current within a conducting medium. It is quite rare that natural sources in bioelectricity involve monopoles, since sources arising from excitable tissues consist of differentially spaced source and sink combinations. Nonetheless, an understanding of the field generated by a monopole is important, because monopole fields are building blocks for more complicated and realistic configurations, and sometimes they are directly useful for stimulating electrodes.

Suppose a point source of electric current, a monopole, is embedded in a uniform conducting medium of conductivity σ. We assume the medium is infinite in extent to make the analysis simpler. Let the position of the monopole be (x, y, z), as illustrated in Figure 2.1. Because the medium is uniform, currents are radial.

Furthermore, the current density will be uniform on spheres centered on the source. In view of the continuity of current, the total current crossing a spherical surface of arbitrary radius r must equal the current (source) strength I_0; consequently, the current density $\overline{J}$ on r equals I_0 divided by the area of the sphere, namely, $4\pi r^2$.

An expression for $\overline{J}$ as a vector field requires only the additional notation that $\overline{J}$ is directed radially outward. Thus we may write $\overline{J}$ as

$$\overline{J} = \frac{I_0}{4\pi r^2}\, \overline{a}_r \tag{2.17}$$

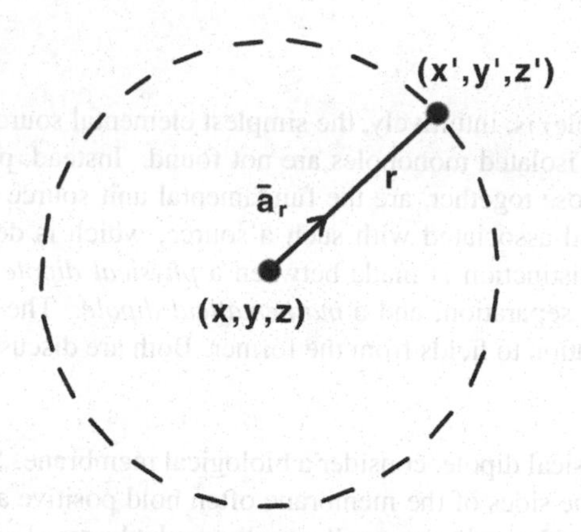

Figure 2.1. Current from a Point Source (monopole). The source at (x, y, z) creates currents along radial lines r toward surfaces such as the mathematical surface identified with the dashed line.

where $\overline{a}_r$ is a unit vector in the outward radial direction, and

$$r^2 = (x - x')^2 + (y - y')^2 + (z - z')^2 \tag{2.18}$$

In (2.18), (x', y', z') is the location of the field point at which the current density is evaluated (described in Figure 2.1).

The potential field may now be evaluated if we apply Eq. (2.2) to (2.17). The result is

$$\nabla\Phi = -\frac{I_0}{4\pi\sigma r^2}\overline{a}_r \tag{2.19}$$

so

$$\frac{d\Phi}{dr} = -\frac{I_0}{4\pi\sigma r^2} \tag{2.20}$$

Integration with respect to r gives an expression for the electric scalar potential, Φ_m, arising from a *monopole source* (a point source), namely,

$$\Phi_m = \frac{I_0}{4\pi\sigma r} \tag{2.21}$$

Sometimes the stimulus electrode embedded within a biological preparation is considered to act as a monopole source, to a reasonable approximation. The source location is the electrode's tip, and to a good approximation such a source may produce the potential field given by (2.21).

2.7. DIPOLE FIELD

A point source (monopole) is, intuitively, the simplest elemental source. It turns out that in electrically excitable tissue, isolated monopoles are not found. Instead, pairs of monopoles of equal and opposite signs, close together, are the fundamental unit source in electrophysiology. Thus in this section the field associated with such a source, which is designated a *dipole*, is considered. Sometimes a distinction is made between a *physical dipole*, which consists of a source–sink pair with small separation, and a *mathematical dipole*. The fields from the latter often give a good approximation to fields from the former. Both are discussed more precisely in a section below.

As an example of a physical dipole, consider a biological membrane. Such a membrane has a thickness under 100 Å. The sides of the membrane often hold positive and negative elements of equal magnitude, separated by only the small membrane thickness. For a membrane patch[2], such a configuration identifies a dipole source element. An active membrane source, therefore, consists of a distribution of such dipoles. Thus the electrical properties of dipoles are studied here both as a technical example of how the monopole building block can be combined into a more complicated source, and as an introduction to a specific source that is directly applicable to biomedical problems.

2.7.1. Dipole Analysis

Suppose we place at the origin of coordinates a point source of strength I_0 and a point source of strength $-I_0$. These sources cancel, and the result is that the potential field is zero.

If, now, the source I_0 is displaced a small distance d, as illustrated in Figure 2.2, incomplete cancellation results. The total field under these conditions is precisely the change in the field resulting from the displacement of I_0 by d. (The change in the field at the field point P resulting from the displacement of I_0 by $\overline{d}$ is precisely the amount by which cancellation of $-I_0$ fails; the change therefore is the required field.) Thus, the dipole field, Φ_d, is given by

$$\Phi_d = \frac{\partial}{\partial d}\left(\frac{I_0}{4\pi\sigma r}\right)\bigg|_0 \times d \tag{2.22}$$

where $r = [(x-x')^2 + (y-y')^2 + (z-z')^2]^{1/2}$ and the zero by the vertical bar indicates that evaluation of the derivative takes place at $x = y = z = 0$.

2.7.2. Comparison with Potential from Two Monopoles

We can obtain the same result from a more formal approach by noting that, as described in Figure 2.2, the distance from the sink to the field point is r_0 while that from the source to the field point is r_1, and consequently,

$$\Phi_d = -\frac{I_0}{4\pi\sigma}\frac{1}{r_0} + \frac{I_0}{4\pi\sigma}\frac{1}{r_1} \tag{2.23}$$

The disadvantage of (2.23) in comparison to (2.22) is the need for two terms, rather than one, and especially needing to find the small difference between two relatively large values. Thus one wishes to combine both terms into one, by expressing r_1 in terms of r_0.

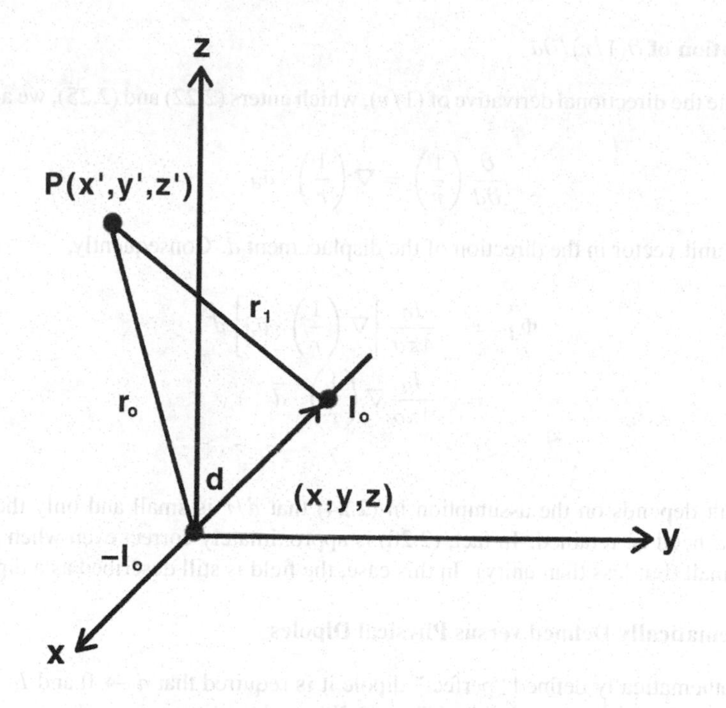

Figure 2.2. Dipole Configuration. A current source (I_o) and current sink ($-I_o$) have equal magnitude but opposite sign. The sources are separated by distance d. The question is what potential P they produce at point (x', y', z').

2.7.3. Expressing r_1 in Terms of r_0

Because distances r_0 and r_1 are large in comparison to the displacement d, then r_1 can be expressed in terms of r_0 as

$$\frac{1}{r_1} = \frac{1}{r_0} + \frac{\partial}{\partial d}\left(\frac{1}{r}\right)\bigg|_0 \times d \tag{2.24}$$

Equation(2.24) is the leading term of a Taylor series expansion of $(1/r_1)$, where the vertical bar and subscript zero indicate an evaluation at the origin of the value of the derivative. Substituting (2.24) into (2.23) yields the total dipole field Φ_d, and this again is seen as a residual (non-canceling) component, namely,

$$\Phi_d = \frac{I_0}{4\pi\sigma}\frac{\partial(1/r)}{\partial d}\, d \tag{2.25}$$

We note that (2.25) is the same as (2.22). The partial derivatives in (2.25), as in (2.22), constitute directional derivatives, as in Eq. (1.15).

2.7.4. Evaluation of $\partial(1/r)/\partial d$

To evaluate the directional derivative of $(1/r)$, which enters (2.22) and (2.25), we apply (1.15) to yield

$$\frac{\partial}{\partial d}\left(\frac{1}{r}\right) = \nabla\left(\frac{1}{r}\right)\cdot \overline{a}_d \tag{2.26}$$

where $\overline{a}_d$ is a unit vector in the direction of the displacement $\overline{d}$. Consequently,

$$\Phi_d = \frac{I_0}{4\pi\sigma}\left[\nabla\left(\frac{1}{r}\right)\cdot \overline{a}_d\right] d \tag{2.27}$$

$$= \frac{I_0}{4\pi\sigma}\nabla\left(\frac{1}{r}\right)\cdot \overline{d} \tag{2.28}$$

This result depends on the assumption in (2.24) that d/r is small and only the linear dependence on d need be retained. In fact, (2.28) is approximately correct even when d/r is only moderately small (but less than unity). In this case, the field is still described as a dipole.

2.7.5. Mathematically Defined versus Physical Dipoles

For a mathematically defined "perfect" dipole it is required that $d \to 0$ and $I_0 \to \infty$ such that $I_0 d = p$ remains constant and finite. Then (2.28) can be written as

$$\Phi = \frac{1}{4\pi\sigma}\nabla\left(\frac{1}{r}\right)\cdot \overline{p} \tag{2.29}$$

While the expression (2.29) is based on $d/r \to 0$, that is never the case in electrophysiology. Nevertheless, the expression often is used for physical dipoles (sometimes called "real" dipoles) as an approximation. The approximation is a good one when d/r is small, say < 0.1.

The first neglected term in (2.24), being quadratic, suggests that the linear approximation will be satisfactory if $d/r \leq 0.1$. In a later section a comparison is presented of potentials computed first as two monopoles and then as one dipole.

2.8. EVALUATING ∇(1/r) WITH RESPECT TO SOURCE VARIABLES

In Figure 2.3 there is a dipole $\overline{p}$ at position (x, y, z) and a field point at (x', y', z') at which the dipole field is to be evaluated. The distance between these points is r. The gradient operator in (2.29) takes partial derivatives with respect to the source (unprimed) variable. That the unprimed variables are involved is evident from the way the gradient was introduced to replace the directional derivative at the *source* point.

Here we show how to evaluate this derivative. Using the definition of r,

$$r = \sqrt{(x-x')^2 + (y-y')^2 + (z-z')^2} \tag{2.30}$$

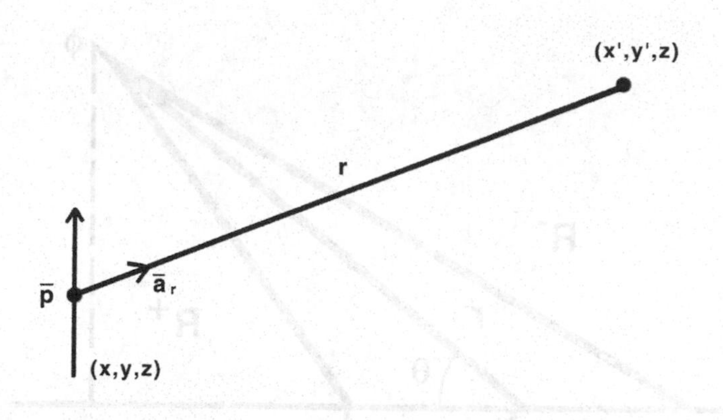

Figure 2.3. Dipole Field. The source is at (x, y, z) and the field is to be determined at point (x', y', z').

and carrying out the gradient operation (1.17) we have

$$\begin{aligned} \nabla\left(\frac{1}{r}\right) &= \frac{\partial}{\partial x}\left(\frac{1}{r}\right)\bar{a}_x + \frac{\partial}{\partial y}\left(\frac{1}{r}\right)\bar{a}_y + \frac{\partial}{\partial z}\left(\frac{1}{r}\right)\bar{a}_z \\ &= -\frac{1}{r^2}\left[\frac{(x-x')}{r}\bar{a}_x + \frac{(y-y')}{r}\bar{a}_y + \frac{(z-z')}{r}\bar{a}_z\right] \\ &= \frac{1}{r^2}\left[\frac{(x'-x)\bar{a}_x + (y'-y)\bar{a}_y + (z'-z)\bar{a}_z}{r}\right] \end{aligned} \tag{2.31}$$

or

$$\nabla\left(\frac{1}{r}\right) = \frac{\bar{a}_r}{r^2} \tag{2.32}$$

where $\bar{a}_r$ is a unit vector from *source to field.* Consequently, from (2.29), where $p = I_0 d$ and $\bar{p} = p\,\bar{a}_d$,

$$\Phi_d = \frac{\bar{a}_r \cdot \bar{p}}{4\pi\sigma r^2} \tag{2.33}$$

It is frequently convenient to orient the z axis along $\bar{p}$, in which case

$$\Phi_d = \frac{1}{4\pi\sigma}\frac{p\cos\theta}{r^2} \tag{2.34}$$

since $\bar{a}_r \cdot \bar{a}_z = \cos\theta$, and θ is the polar angle.

2.9. MONOPOLE PAIRS TO DIPOLES?

Electrical sources in biological preparations often arise in pairs. An earlier section developed equations for the potential from the pair when the pair was taken as two monopoles, or as two dipoles. Biological dipoles never fulfill the exact conditions required of mathematical dipoles, so the question arises as to the degree of error present when pairs of physical sources are represented as a dipole. When is so doing a good approximation? In this section we follow up on that question with a specific example that forms the basis for more general conclusions.

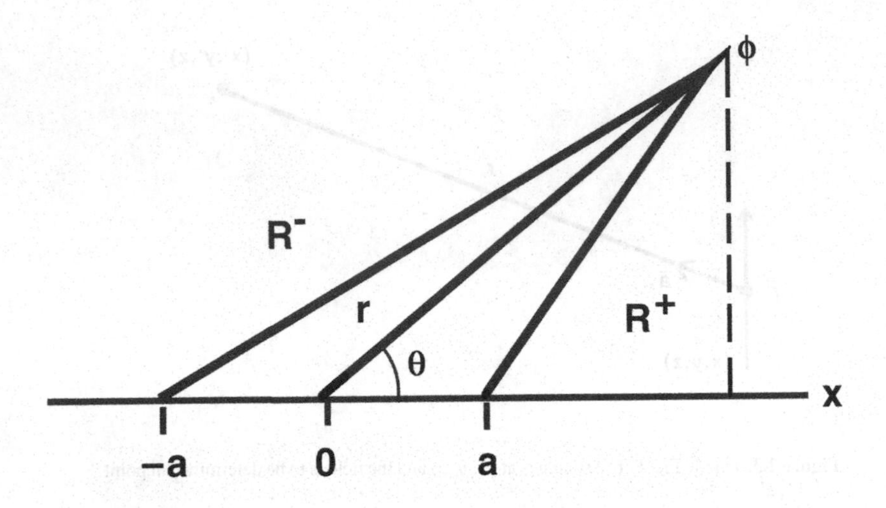

Figure 2.4. Two Monopole Sources. The +1 mA monopole is at $x = a$, and the –1 mA monopole is at $x = -a$. The distance from the field point (r, θ) to the positive pole is R^+, from the negative pole R^-, and the distance from the center of the pair is r. Potential ϕ is at the field point. The dashed line is $r \sin\theta$, and is the y coordinate for r, R^+, and R^-.

2.9.1. Sample Problem

We have point current sources of 1 mA located at $x = 1$ mm and -1 mA at $x = -1$ mm (Figure 2.4). These two sources form a "physical dipole," i.e., a dipole with a finite monopole displacement between the two poles. Calculate and plot the normalized potential as a function of polar angle for a radius (from the origin) of 2, 4, and 8 mm. Compare with the field from an idealized dipole. Explain any differences. Use $\sigma = 0.001$ S/mm.

2.9.2. Solution for Two Monopoles

Because of symmetry about the polar axis, all results are independent of azimuth angle, and we may choose this to be 0°. The distance from the positive source to an arbitrary field point, (r, θ), is designated R^+ (for somewhat greater initial generality, we let each monopole be spaced a mm from the origin).

In Cartesian coordinates the field point is $(r\cos\theta, r\sin\theta)$, while the source is at $(a, 0)$, so

$$R^+ = \sqrt{(r\cos\theta - a)^2 + (r\sin\theta)^2} = \sqrt{r^2 + a^2 - 2ra\cos\theta} \tag{2.35}$$

(a result that can also be obtained from the law of cosines). For the negative monopole the source field distance is

$$R^- = \sqrt{r^2 + a^2 + 2ra\cos\theta} \tag{2.36}$$

The total field, by superposition using (2.21), is given by

$$\Phi(r, \theta) = \frac{1000}{4\pi}\left[(r^2 + a^2 - 2ra\cos\theta)^{-1/2} - (r^2 + a^2 + 2ra\cos\theta)^{-1/2}\right] \tag{2.37}$$

To compare more easily the potential found with two monopoles to the potential found with one dipole, it is helpful to factor (2.37) and get

$$\Phi = \frac{1000}{4\pi}\frac{2a}{r^2 - a^2}\left[\frac{r^2 - a^2}{2a\sqrt{r^2 + a^2 - 2ra\cos\theta}} - \frac{r^2 - a^2}{2a\sqrt{r^2 + a^2 + 2ra\cos\theta}}\right] \tag{2.38}$$

The solution as expressed in (2.38) for the potential for two monopoles will be compared to the solution given below for the sources expressed as a mathematical dipole.

The reason the form (2.38) is advantageous is that in this form the bracketed term is equal to one when $\cos\theta = 0$. Also, note the behavior of the magnitude (coefficient) term (the part outside the brackets). In the denominator the component $r^2 - a^2$ is well approximated by r^2 alone, when $a << r$, a point to keep in mind for the comparison below.

2.9.3. Solution for One Dipole

At "great enough" distances the monopole source pair is expected to behave as a dipole. In this example, the dipole strength is $p = 1$ mA $\times 2a$ mm. Hence, using (2.34), we have

$$\Phi(r,\theta)_d = \frac{2000a}{4\pi r^2}[\cos\theta] \tag{2.39}$$

The solution (2.39) for the sources expressed as a dipole is in a good form for comparison to the solution computed from two monopoles, as it also is divided into a magnitude part, outside the brackets, and an angular part, inside.

2.9.4. Numerical Comparison

The solution for potentials from two monopoles (2.38) can now be compared to the solution for the potentials when the two sources are approximated as a single dipole (2.39).

Magnitude dependence: the magnitude (coefficient) ratio obtained by dividing the coefficient of (2.39) with that of (2.38) is $(r^2 - a^2)/r^2 = 1 - a^2/r^2$. The difference from unity decreases inversely with the square of the ratio of dipole separation to source–field distance. If $a/r = 1/10$, then the dipole magnitude error is under 1%.

Angular dependence: as noted, the bracketed term in (2.38) is the normalized field at a constant radius. The normalized field is plotted in Figure 2.5 with $a = 1$ and for $r = 2, 4, 8$ mm (as specified in the problem statement).

Also plotted is the normalized variation of the idealized dipole (which is simply $\cos\theta$).

2.9.5. Numerical comparison as a function of r

In Figure 2.5 the $r = 8$ mm curve is virtually the same for the solution expressed as one dipole with that of the solution for two monopoles. Visually it is difficult to separate the solutions.

Consequently, one concludes from this example that when the source–field distance is five to ten times the monopole separation, the dipole is a good approximation to the magnitude and to

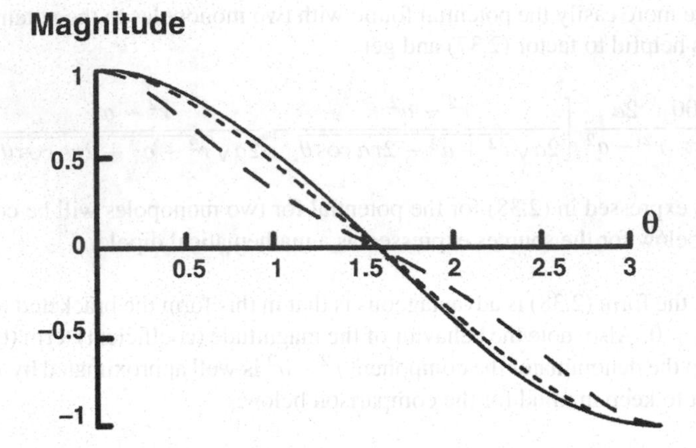

Figure 2.5. Normalized Potential Field at constant r from monopoles lying on the polar axis and equally spaced from the origin by 1 mm. Dashed curve for $r = 2$, dotted for $r = 4$ mm, The solid curves describe $r = 8$ mm.

the pattern of potentials computed as two monopoles. Although this conclusion has been shown here only for a particular example, it is a conclusion that is true more generally.

2.9.6. Analytical comparison

One can also investigate the goodness of the dipole approximation analytically, by carrying the Taylor series representation of (2.24) into higher terms. Considering the positive monopole at $x = a$ (the negative monopole is at $x = -a$) and letting

$$R_d = [r^2 + a^2 - 2ra\cos\theta]^{-1/2}$$

we obtain

$$1/R^+ = 1/r + \frac{\partial R_d}{\partial a}|_0 a + \frac{\partial^2 R_d}{\partial a^2}|_0 \frac{a^2}{2} + \frac{\partial^3 R_d}{\partial a^3}|_0 \frac{a^3}{6} + \cdots \tag{2.40}$$

where $|_0$ means that evaluation is at $a = 0$.

By inspection of (2.40) it can be seen that an expression for $1/R^-$ is given by replacing a by $-a$, resulting in a similar expression except that odd terms are negative. In forming the total potential (superposition of positive and negative monopole contributions), the even terms in the Taylor series expansion drop out and only the odd terms are left. In this remaining expansion the leading term is the one identified as the dipole.

The next term serves as a correction to the dipole term and is examined here. From (2.40) we have

$$\Phi(r,\theta) = \frac{p\cos\theta}{4\pi r^2}\left[1 + \left(\frac{5\cos^2\theta - 3}{2}\right)\frac{a^2}{r^2} + \cdots\right] \tag{2.41}$$

Actually, the second term in (2.41) is useful as a dipole correction term provided that a/r is sufficiently small compared to unity that the first neglected term, which depends on $(a/r)^4$, can

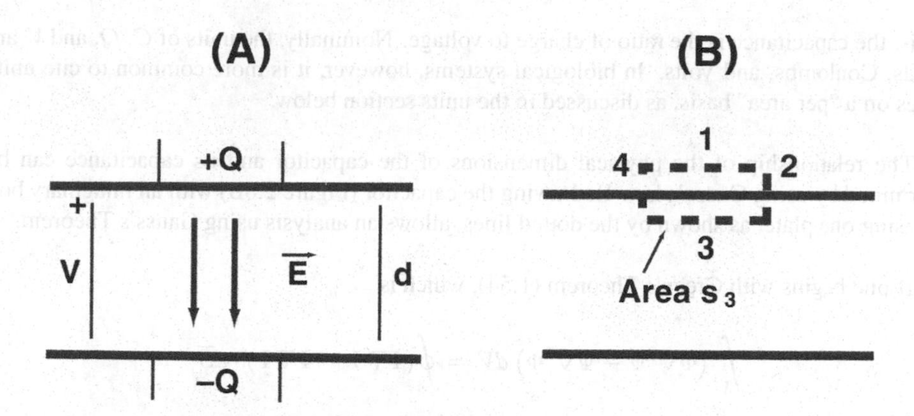

Figure 2.6. Parallel Plate Capacitor. Panel A shows schematically a parallel plate capacitor with voltage V, charge Q, and separation d identified. These quantities are linked through the electric field $\overline{E}$ between the plates. In panel B the capacitor is redrawn with a closed surface added, with dotted lines. The box is an imaginary surface used for analysis. The product $\overline{E} \cdot \overline{dS}$ is zero on all six sides of the box, except for side 3.

be ignored. Under these conditions we note that, as above, the first-order magnitude correction varies as $(a/r)^2$.

An angular correction is seen to contain a dependence on 2θ. This double angle factor can be seen in Figure 2.4 for $r = 2$ mm.

The reader might wish to plot (2.41) for $r = 2$ mm and compare it with the exact result plotted in Figure 2.4. The result demonstrates that the two terms in (2.41) would be just about adequate for $r/a = 2$ and should improve for increasing values of this ratio. An expression for the error in using only the two terms in (2.41) could be found by evaluating the third term, i.e., the first neglected term, in (2.41).]

2.10. CAPACITANCE

Somewhat surprisingly, biological structures exhibit high values of capacitance, mostly arising from cell membranes. The presence of this capacitance markedly affects the tissue's natural electrical behavior, as well as response to external stimulation. Thus a brief review of capacitance and its relation to structure is presented here, as it will be pertinent to material presented in later chapters.[3]

Suppose a voltage V is present between two parallel conductors that are separated by a distance d, as shown diagrammatically in Figure 2.6A. The voltage will be accompanied by a charge Q (per unit area) on each of the conducting plates.

The insulating material separating the conductors, the dielectric, may be air, cell membrane, or some other nonconductor. The capacitance C between the plates is defined as

$$C \equiv \frac{Q}{V} \tag{2.42}$$

that is, the capacitance is the ratio of charge to voltage. Nominally, the units of C, Q, and V are Farads, Coulombs, and Volts. In biological systems, however, it is more common to cite units values on a "per area" basis, as discussed in the units section below.

The relationship of the physical dimensions of the capacitor and its capacitance can be determined by using Gauss's law. Redrawing the capacitor (Figure 2.6B) with an imaginary box enclosing one plate, as shown by the dotted lines, allows an analysis using Gauss's Theorem.

If one begins with Green's Theorem (1.54), which is

$$\int_V (\Phi\nabla^2\Psi - \Psi\nabla^2\Phi)\, dV = \oint_s (\Phi\nabla\Psi - \Psi\nabla\Phi) \cdot d\overline{S}$$

and makes the choices $\Psi = 1$ and $\Phi = \phi$, the electric potential between the plates of the capacitor, then Green's Theorem simplifies to

$$\int_v \nabla^2\phi\, dv = \oint_S \nabla\phi \cdot d\overline{S} \tag{2.43}$$

On the left one substitutes Poisson's equation for electrostatics (2.11), so that

$$\int_v \nabla^2\phi\, dv = -\frac{Q}{\epsilon} \tag{2.44}$$

On the right of (2.43) one uses the relation (2.9), which identifies the gradient of potential as the electric field. Because the electric field exists only between the plates, and the direction of the electric field is perpendicular to the plates, five of the six sides of the dashed box (of Figure 2.6B) make no contribution to the surface integral.

On the remaining side, side 3, the magnitude of the electric field $\overline{E}$ within a parallel plate capacitor (away from the edges) is $-V/d$ (ignoring any edge effects). Thus

$$\oint_S \nabla\phi \cdot d\overline{S} = \sum_6 \overline{E} \cdot \overline{S}_j = -\frac{V}{d} s_3 \tag{2.45}$$

Substituting into (2.43) the relationships given in (2.44) and (2.45),

$$\frac{Q}{\epsilon} = \frac{V s_3}{d} \tag{2.46}$$

Rearranging and making use of the definition of capacitance(2.42), one finds that

$$C \equiv \frac{Q}{V} = \frac{\epsilon A}{d} \tag{2.47}$$

thus

$$C_m = \frac{C}{A} = \frac{\epsilon}{d}$$

Following convention, we have written the area of the capacitor plate now as A instead of s_3. The permittivity ϵ of a dielectric often is given as $k\epsilon_0$, where ϵ_0 is the permittivity of free space, and k, the dielectric constant, is a value specific to the material.

The lumped parameter treatment of capacitance in electric circuits is associated physically with two parallel conducting plates between which there is a dielectric. In biology the membrane of a nerve or muscle fiber constitutes a distributed capacitance. The conductive extracellular and intracellular media are associated with the two plates of a capacitor. The membrane itself furnishes the intervening dielectric. The cell membrane often is approximated with $k = 3$.

The capacitance often is described per unit area, as such a description reflects a membrane characteristic that may hold true across membranes of different size, e.g., for cells of different size.

2.11. UNITS FOR RESISTANCE AND CAPACITANCE

Earlier in this chapter we noted that the resistance of biological materials is *distributed.* With lumped parameters as used in circuit theory, individual units have resistances R or capacitances C that do not involve geometric dimensions. In contrast, with distributed resistance, conductance, or capacitance the determination of resistance or capacitance values often is slightly more complicated. The additional complications occur because the dimensions of the structure as well as the characteristics of the material have to be taken into account.

A corollary is that resistance, conductance, and capacitance are specified using several different sets of units. All reflect, of course, basic properties of the materials involved. However, usually they are also chosen to be convenient and helpful for the particular geometric shapes and structures involved.[4]

2.11.1. Resistance and Resistivity, Conductance, and Conductivity

A double-box geometric structure that provides some interesting examples is depicted in Figure 2.7. Consider first the small box, inside of which the resistivity is $\rho = R_i$ Ohm-cm. The resistance of the box to a current passing through it will depend on the direction of the current. First suppose there is a current in the direction $\overline{a}_x$. What is the resistance of the small box to a current in this direction? For such a current, the end-to-end resistance of the small box, R (in Ohms), is

$$R = R_i \cdot L/A_x = R_i \cdot L/a^2 \tag{2.48}$$

that is, the resistance R (in Ohms) is found by dividing the resistivity R_i (in Ohm-cm) by the cross-sectional area to the direction of the current, A_x (in cm^2), and multiplying by the length L (in cm) through which current must pass.

Numerical example: If $R_i = 200$ Ohm-cm, $a = 10\ \mu$m, and $L = 100\ \mu$m, what is the small box's resistance, R? Answer: Using (2.48),

$$R = 200\Omega\text{–cm} \cdot (100 \times 10^{-4})\text{cm}/(10 \times 10^{-4})^2\text{cm}^2 = 2 \times 10^6\,\Omega \tag{2.49}$$

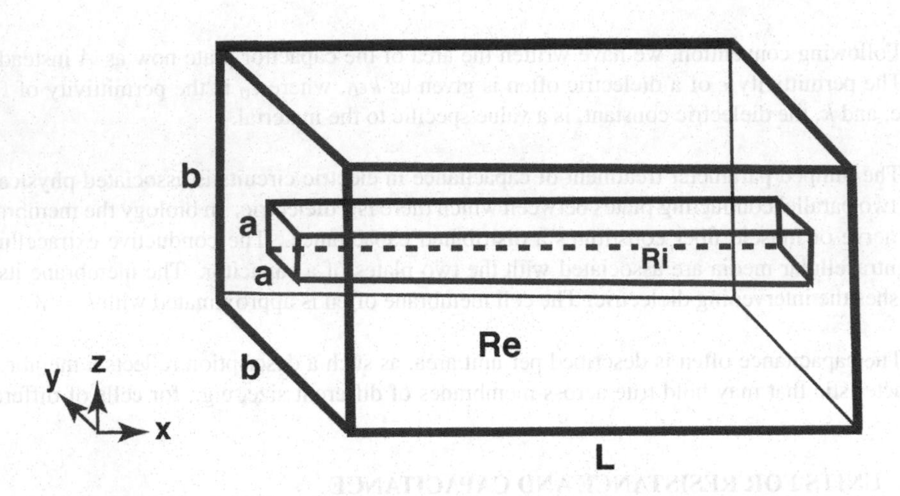

Figure 2.7. Box in Box. A smaller box, with sides of length a, is inside a bigger box that has sides of length b. Both boxes have length L. The boxes have different resistivities: R_i inside the small box, and R_e outside the small box.

In biological structures, it is commonly the case that resistances of this magnitude are involved and not unusual that units conversions (such as μm to cm) are required in resistance calculation.

Taking the reciprocal of (2.48), one gains the corresponding equation for the tissue's conductance, G, as related to conductivity σ and the tissue's dimensions. Specifically,

$$G = \sigma_i \cdot A_x / L = \sigma_i \cdot a^2 / L \tag{2.50}$$

Numerical example: If $\sigma_i = 1/R_i = 0.005$ S/cm, with $a = 10$ μm, and $L = 100$ μm, what is the small box's conductance, G, along $\overline{a}_x$?

$$G = .005\text{S/cm} \cdot (10 \times 10^{-4})^2 \text{cm}^2 / 100 \times 10^{-4} \text{cm} = 0.5 \times 10^{-6} \text{Siemens} \tag{2.51}$$

Note that the conductivity is the reciprocal of the resistivity, and the conductance is the reciprocal of the resistance.

2.11.2. RGC and a Surface Boundary

The paragraphs that follow present a multiplicity of ways of defining resistance, conductance, and capacitance (R, G, C). Each of these definitions is used because it is the most convenient for a certain class of situations. The key to resolving confusion about the different ways of specifying resistance and resistivity is to return to (2.48) or (2.50) and see how terms are factored to join different parts of the geometry with the resistivity (or conductivity) of the material. A similar strategy can be used consistently for conductance (as reciprocal resistance), or for capacitance.

For commonality with the main applications in later chapters, we will call the surface separating the small box from the big box the "membrane." Here we are thinking of it simply as a thin resistive material with a uniform but tiny thickness t and a resistivity ρ_m. Then finding the resistance between the inside and outside of the small box can be done in the same fashion as (2.48), specifically

$$R = \rho_m \cdot t/A_m = R_m/A_m \tag{2.52}$$

where the membrane resistance R_m is defined implicitly. Nominally $R_m = \rho_m t$, but in practice it is more likely to be determined from independent experimental measurements. Note that R_m has the units of Ohm-cm^2.

Conductance per unit area can be defined using the reciprocal of (2.52) as

$$G = A_m/(\rho_m t) = G_m A_m \tag{2.53}$$

As for resistivity, the thickness can be merged into the conductance per unit area, G_m. The units of G_m are Siemens/cm^2.

For capacitance, we can find the total capacitance C, keeping (2) in mind, as

$$C = \epsilon \cdot A_m/t = C_m \cdot A_m \tag{2.54}$$

Here A_m is the surface area of the inner box, i.e., $4aL$. The units of C_m, the capacitance per unit area, most often are microfarads per cm^2.

2.11.3. Axial Properties per Unit Length

Specialized structures that have axial uniformity often may be more simply described with a resistance or conductance per unit length. Considering again the small box along the x axis, one notes that the cross-sectional area does not change as a function of x. Thus one can find the resistance for a length L along x as

$$R = R_i \cdot L/A_x = r_i \cdot L \qquad r_i = R_i/A_x \tag{2.55}$$

Defining r_i in this way, as a resistance per unit length, avoids the repetitive specification of the cross-sectional area. It is also the case that r_i can be specified without reference to (or need to know) the length L. The units of r_i are Ohms/cm.

2.11.4. Membrane Resistance and Capacitance per unit length

For resistance, we can define the cable membrane resistance per unit length r_m implicitly by considering the resistance of a length of membrane.

One step is to describe the surface area of a length of membrane as $s \cdot L$, its circumference s times length L.[5]

From (2.52)

$$R = \frac{R_m}{A_m} = \frac{R_m}{sL} = r_m/L \quad \text{where} \quad \mathrm{r_m} \equiv \mathrm{R_m/s} \tag{2.56}$$

In the case of the small box used here, $s = 4a$. For fibers with a circular cross-section, it would be $s = 2\pi a$, where a is the fiber radius.

Membrane capacitance per unit length c_m is defined implicitly by

$$C = C_m A_m = C_m sL = c_m L \qquad \text{where} \quad \mathrm{c_m} \equiv \mathrm{C_m s} \tag{2.57}$$

where again s is the circumference. The units of c_m are microFarads per cm. That is, knowing c_m for a segment, one need only multiply by the length of the segment to get the capacitance, making it easy to adjust for segments of varying length.

2.12. UNITS FOR SOME ELECTRICAL QUANTITIES

Here are units for some of the variables presented in this chapter.

Symbol	Units	Definition
r	centimeters (cm)	distance
R	Ohms	Resistance
G	Siemens	Conductance
ρ	Ohm-cm	Resistivity
σ	Siemens/cm	Conductivity
$\overline{J}$	Ampere/cm^2	Current density
I	Ampere = Coulomb/sec	Current
I_v	Ampere/cm^3	Current source density
Φ	Volt = Joule/ Coulomb	Electric potential
C	Farad = Coulomb / Volt	Capacitance

2.13. NOTES

1. In thinking about the quasi-static nature of most bioelectric situations, it is helpful to focus on the *rate* of change. Certainly there is a constant change in bioelectric fields at a millisecond and microsecond level, e.g., throughout every heartbeat for a cardiac field. On the other hand, the rate of change of voltages is not fast enough to generate to initiate detectable radio waves, or to prevent currents on the body surface from being observed at virtually the same time they are created within the heart.
2. By "patch" we mean a membrane area so small that variables of interest have constant values over its surface.
3. The reader can find much more comprehensive presentations of capacitance and dielectrics in the references, such as the text by Smythe [5].
4. Experience shows that many errors result from misunderstanding the way resistance or capacitance was given, or the significance of the units used. Thus this topic may seem tedious or boring, but it is not difficult to get the main ideas, and it is worth a few minutes to get them straight in one's mind.
5. Here using s instead of c for circumference so as to avoid overlap with C for capacitance.

2.14. REFERENCES

1. Jeans JH. 1927. *The mathematical theory of electricity and magnetism*, 5th ed. Cambridge: Cambridge UP.
2. Plonsey R, Collin RE. 1961. *Principles and applications of electromagnetic fields*. New York: McGraw-Hill.
3. Pugh EM, Pugh EW. 1960. *Principles of electricity and magnetism*. Reading, MA: Addison-Wesley.
4. Skilling HH. 1948. *Fundamentals of electric waves*. New York: John Wiley and Sons.
5. Smythe WR. 1968. *Static and dynamic electricity*, 3rd ed. New York: McGraw-Hill.
6. Stratton JA. 1941. *Electromagnetic theory*. New York: McGraw-Hill.

3

Bioelectric Potentials

In this book we will be examining the behavior of excitable cells, notably nerve and muscle, both descriptively and quantitatively. The behavior is described mostly in terms of the potentials and currents that excitable cells produce. These potentials and currents are observed in the cells' interior volume, across their membranes, and in their surrounding conducting volume from the cell surface to the body surface. Such electrical signals are vital to the transmission of information in nerves, the initiation of contraction in muscles, and hence essential to vision, hearing, the heartbeat, digestion, and other biological processes. Despite the tremendously different functions of these organ systems, it is remarkable how extensively their underlying electrical systems share many basic principles of organization, and how fundamentally similar they remain in almost all living creatures.

This chapter has two major divisions. The first deals with potentials and currents in an extensive solution, the goal of which is establishing the terminology and relationships of ionic flows from diffusion or electric fields, and the commonly used symbols, nomenclature, and mathematical operations used for describing them. As will be seen, flow occurs because of electric fields and due to the effects of diffusion, sometimes acting in tandem, and other times in opposition. This half of the chapter applies directly to fields such as *electrochemotherapy*, where electric fields cause the movement of charged particles such as DNA [8, 11].

The second half of the chapter deals with flow, ionic concentrations, and electric potentials across a membrane. Information on membrane structure is included, though for most electrical analysis the membrane is treated in terms of its electrical properties of capacitance and resistance. Membrane resistance is described as coming from the membrane's channels, which are specialized structures that allow the movement of particular ions. Certain critical relationships are established, such as the Nernst equation, which gives the potential at which flow across a membrane due to electric forces in one direction offsets the flow in the other direction due to concentration differences. This latter portion of the chapter is used most often in the subsequent chapters of the book, as it is fundamental to the mechanisms of operation of excitable nerves and muscle.

It was almost a century after the work of Volta and Galvani when Svante August Arrhenius showed (1887) that solutions which readily conduct electric current, such as the solutions within the body of a human or animal, contain electrically charged particles. These particles now are called ions (from the Greek word for "going"), and among those of particular significance are ions of sodium and potassium.

We begin by first describing the ionic composition inside and outside well-known excitable cells. We proceed to develop the equations showing how the flow of currents is related to potential gradients and to ionic concentrations in physiological media. In their fundamental respects these equations are true in any solution containing ions, but in our description we anticipate the material about biological membranes that appears in the latter part of the chapter. There, the focus is on the important special case of potential differences and currents that accompany the concentration differences across cell membranes, first at equilibrium, and then with a membrane model that can handle a wider range of circumstances.

3.1. CURRENTS IN SOLUTIONS

The kind of charge carriers that are present in living tissues are ions within the electrolytes (solutions of acids, bases, and salts, which conduct electricity). Such charge-carrying ions are present both inside and outside of cells, especially ions of sodium and potassium, allowing current to flow extensively throughout both intracellular and extracellular volumes. Such charge movement within living tissue is similar to that in sea water, with its high dissolved salt content, but not very similar to the flow of current in wires, where the charge carriers are electrons that move within the metallic structure in the wire, but not through the insulation into the surrounding space.

As noted by Harned and Bereton [2], the science of electrical currents in solutions is complex. In this chapter our ambitions are necessarily limited—to provide some basic information about the underlying physical and electrochemical basis of bioelectric potentials and currents. Our goal is not a comprehensive study of electrolytic solutions, a topic well beyond and different from the scope of this text. Rather, it is to provide enough information that the subsequent chapters can be connected to the movements of ions, and the forces that act upon them, in a way that is based on sound physical principles, even if these are described in a simplified manner.

3.2. MOLES AND AMPERES

One aspect of the study of bioelectricity that gives it a special fascination (and sometimes a special frustration) is that it brings together subjects that have developed independently, each with its own conventions, terminology, units, and even cultures. In this chapter such a juxtaposition arises immediately in the study of the movement of ions, since such movement can be described in either of two ways.

When ionic movements are described from the perspective of chemistry, the quantities of ions are naturally expressed in terms of moles. One mole is simply an amount of pure substance in grams that is numerically equal to its atomic weight. Consequently, a mole includes a fixed number of molecules, namely, Avogadro's number, which is 6.0225×10^{23} molecules. Flows are then described in terms of moles per second, and fluxes by moles per second per unit area. Here, such fluxes, counted in terms of the number of moles, are denoted by a lowercase letter j.

When ionic movements are described from the perspective of the study of electricity, the quantities of ions are expressed in terms of Coulombs. This unit derives from the forces between charges, rather than their number. From this origin, the magnitude of the charge on one electron (and thus on one ion with valance equal to one) is determined to be 1.6×10^{-19} Coulombs. Flows are then described in terms of Coulombs per second, or Amperes. Thus flux (flow per unit area) is in the units of Amperes per unit area, e.g., Amperes per cm^2. Here, such fluxes, when expressed in terms of the electrical charge movement, are denoted by an uppercase letter J.

It is important to realize that an example of ionic movement, a single physical phenomenon, can be described in either of two ways—as a particle flow or as an electrical current. The numerical values of the flow will be different (with different units) depending on which way is chosen, e.g., moles per second versus Amperes.

A conversion from units of particle movement is required to get a value of electrical current. The conversion factor is called Faraday's constant, F. The conversion is

$$F = (6.02 \times 10^{23})\,\frac{\text{particles}}{\text{mole}} \times (1.6 \times 10^{-19})\,\frac{\text{Coulombs}}{\text{particle}} = 96,487\,\frac{\text{Coulombs}}{\text{mole}}$$

Joos [6] points out that the Faraday can be understood independently of arguments about atomic structure as the amount of charge required for the electrolytic liberation of one chemical equivalent (for a univalent ion, one mole) of a substance, irrespective of which element is ionized. The constant value that was measured demonstrated that each ion, on the average, carried the same amount of charge and anticipated the understanding that each ion carries a single charge (if univalent).

Faraday's constant appears in many different equations in this text, sometimes for reasons that initially seem obscure. In studying these expressions, it is often helpful to keep in mind that the introduction of Faraday's constant normally comes from the need to convert current flows from particles per second to electrical current, or vice versa.

3.3. IONIC COMPOSITION

From the viewpoint of electrical currents the most important ions are sodium and potassium. Those of calcium and chloride also play a significant role in some circumstances, as do other ions.

As examples of ionic concentrations, literature values for frog muscle and squid nerve axon are given in Table 3.1. There are wide variations in concentration that exist from ion to ion and intracellular versus extracellular. Note that these large differences exist despite the tendency of concentration to average out, due to diffusion. The concentrations of ions in either the intracellular or extracellular volumes allow significant currents to flow in either place.

The concentrations between the intracellular and extracellular volumes are especially important to excitable cells. For all excitable cells the concentration of intracellular potassium greatly exceeds extracellular potassium. Such is the case in Table 3.1 for both frog muscle and nerve axon, even though the concentration values are quite different. (The relative ratios of intracellular to extracellular K^+, Na^+, and Cl^- in Table 3.1 are similar to those generally found in other excitable muscle and nerve.)

Table 3.1. Ionic Concentrations[a]

	Muscle (frog)		Nerve (squid axon)	
	Intracellular mM	Extracellular mM	Intracellular mM	Extracellular mM
K^+	124.0	2.2	397	20
Na^+	4.0	109.0	50	437
Cl^-	1.5	77.0	40	556
A^-	126.5			

[a]The A^- ion is large and impermeable.

In contrast to potassium, the extracellular sodium and chloride concentrations greatly exceed intracellular sodium and chloride. The different ratios of intracellular to extracellular concentrations for sodium and potassium ions are of great importance to transmembrane voltage, and to how it changes.

3.4. NOTATION FOR ION SPECIES

Much of the evaluation of the movement of ions that is described in the sections that follow has the same physical basis and mathematical form no matter which ion is being considered. This situation could produce a tremendous amount of largely redundant text, which we have tried to avoid in the following way: Where we subsequently refer to "the *p*th ion," this notation means that the same argument can be made for each of the ion species individually (sodium or potassium or whatever). We have used this slightly abstract way of identifying ionic species so as to avoid having to repeat the same argument over and over, once for sodium, once for potassium, etc.

On the other hand, it is important to keep in mind that while the mathematical argument (and the form of the equations) may be the same for each ion individually, the numerical values will be different, and thus the electrophysiological effects will be quite different, maybe even the opposite.

3.5. NERNST–PLANCK EQUATION

The Nernst–Planck equation relates the flow of ions to spatial differences in concentration or in the electric potential. It is helpful to think first about these effects separately.

3.5.1. Diffusion and the Diffusion Coefficient

If a drop of blue ink is placed in a beaker of water, then ink molecules will, on average, move away from the highly concentrated region into the surrounding water. The process will continue until the ink is uniformly distributed in the water (which will become a uniform light blue color). The process is known as *diffusion* and arises because of the thermal energy of the molecules. At first, the dense dye concentration in the drop is surrounded by a concentration of dye that is much lower. Consequently, there is a net flow from the center outward. We say that diffusion is in a direction of decreasing concentration, i.e., that flow takes place "down the concentration gradient."

Table 3.2. Numerical Values for Several Diffusion Coefficients

Ion	D	Units	Conditions	Ref
Na^+	1.33×10^{-5}	cm^2/sec	at 25 °C	[7]
K^+	1.96×10^{-5}	cm^2/sec	at 25 °C	[7]
Cl^-	2.03×10^{-5}	cm^2/sec	at 25 °C	[7]
KCl	2.03×10^{-5}	cm^2/sec	0.002 mole/l, 25 °C	[10]
NaCl	1.58×10^{-5}	cm^2/sec	0.002 mole/l, 25 °C	[10]

No electric field is required for diffusion to occur as diffusion is not an electrical phenomenon. Rather, diffusion arises as a consequence of the pronounced random motion of molecules that occurs at ordinary temperatures. This random motion scatters the blue ink. At first, more move away from the center of the ink drop than toward the center, simply because more ink molecules are near the center. Such random movement results in a net movement from the original site, until the ink becomes diffused throughout the beaker.

A quantitative description of diffusion is Fick's law. That law is

$$\bar{j}_d = -D\nabla C \tag{3.1}$$

In Fick's law, C is the concentration of some substance, such as the blue ink, as a function of position. D is a proportionality constant (called Fick's constant or the diffusion constant). Fick's constant is sometimes called "Fick's coefficient" since its value is not quite independent of concentration but increases slightly with increases in C.

Usually, D is determined from experiment rather than from basic principles. Some physical insight into D can be gained by noting that for a gas $D = \ell v/3$, where ℓ is the mean free path and v is the average molecular velocity. (In a liquid, this relationship is complicated by ionic interactions.) A sample of experimental values of diffusion coefficients is given in Table 3.2.

The flux, $\bar{j}_d$, is the number of particles (ions) moving per unit time through a cross-section of unit area. A lowercase j is used to describe ion flow. As noted earlier, an upper case J refers to the associated electric current density, that is, j embodies the movement of particles, while J embodies the movement of charges.

3.6. MOBILITY

3.6.1. Electric Field and Mobility

Because of their charge, ions are subject to electric field forces. Because of collisions, the force exerted by a given field will move ions with a finite velocity. This velocity is denoted by u_p, termed the *mobility*, which is the velocity achieved under a unit field for the *pth* type ion.

If the valence of this ion is Z_p, then the ionic flux is given by the product of ion concentration and its velocity, namely,

$$\bar{j}_e = -u_p \frac{Z_p}{|Z_p|} C_p \nabla\Phi \tag{3.2}$$

Table 3.3. Faraday's Constant F and the Gas Constant R

Constant	Value
F	96,487 Coulombs/mole
R	8.314 Joules/degree K-mole
RT/F	8.314×.300/96487 = 25.8 mV at 27 °C

where $\nabla\Phi$ is the electric field, $Z_p/|Z_p|$ the sign of the force on the pth ion [positive for positively charged ions (cations) and negative for negatively charged ions (anions)], and consequently $-u_p(Z_p/|Z_p|)\nabla\Phi$ is the mean ion velocity. The ionic mobility depends on the viscosity of the solvent, the size of the pth ion, and its charge.

Equation (3.2) gives the ion flux per unit area. The units of flux depend on the units of ion concentration. Often flux is expressed as moles per unit area per second.

3.6.2. The Diffusion Coefficient and Mobility

The mobility relates the force due to the electric field $(-\nabla\Phi)$ to the ionic flux it produces. With a similar mathematical form, Fick's constant relates spatial changes in concentration (∇C), sometimes thought of as the "force" due to diffusion, to the movement of ions down the concentration gradient.

Since both flows are impeded by the same molecular processes (collisions with solvent molecules), a physical connection exists between parameters u_p and D. The mathematical description of the connection was worked out by Einstein, and the resulting equation thus bears his name. Einstein's equation is

$$D_p = \frac{u_p RT}{|Z_p|F} \tag{3.3}$$

As before, p signifies the pth ion species with valence $|Z_p|$, and u_p is its mobility, T is the absolute temperature, F is Faraday's constant, and R is the gas constant. Numerical values are given in Table 3.3.

3.7. TEMPERATURE VARIATIONS

Temperature is commonly held uniform in experimental studies, and many animal systems (including humans) are evaluated at their normal body temperature, about 37°C. Temperature may, however, vary in human or animal systems, with disease or under special circumstances, such as surgery. Furthermore, temperature in cold-blooded living systems, such as the squid, varies with that of the environment, and some famous experimental results are reported at 6.3°C.

Thus it is important to take note of the presence of the temperature term in Eq. (3.3) and to keep in mind that the temperature dependence arises from the term RT/F. Temperature sensitivity of some of the active membrane processes described in later chapters is even greater. Temperature differences between nominal room temperature, normal human body temperature, and temperatures reported for various experimental or natural circumstances are sufficient to produce noticeable (and sometimes pronounced) differences in numerical results.

Such temperature differences may produce noticeable differences in experimental values and in the overall performance of the living tissue itself.

3.8. FLUX DUE TO DIFFUSION PLUS ELECTRIC FIELD

The total flux when both diffusional and electric field forces are present is

$$\overline{j}_p = \overline{j}_d + \overline{j}_e \tag{3.4}$$

or, using (3.1), (3.2), and (3.3),

$$\overline{j}_p = -D_p\left(\nabla C_p + \frac{Z_p C_p F}{RT}\nabla\Phi\right) \tag{3.5}$$

Equation (3.5) is known as the *Nernst–Planck* equation.

Equation (3.5) describes the flux of the *p*th ion under the influence of diffusion and an electric field. Its dimensions, which depend on those used to express the concentration and velocity, normally are moles per cross-sectional area per unit time.

This flux can be converted into an electric current density when multiplied by FZ_p, the number of charges (Coulombs) carried by each mole; $\overline{J}_p$ is the resulting electric current density (in Coulombs per second per cm^2 or Amperes per cm^2).

Applied to (3.5), conversion results in

$$\overline{J}_p = FZ_p\overline{j}_p = -D_p\, FZ_p\left(\nabla C_p + \frac{Z_p\, C_p\, F\,\nabla\Phi}{RT}\right) \tag{3.6}$$

As noted earlier, a capital *J* designates the electric current density while a lowercase *j* describes the flux. Alternatively, using Einstein's equation (3.3) to substitute for D_p, one has

$$\overline{J}_p = -\left(u_p\, RT\,\frac{Zp}{|Z_p|}\,\nabla C_p + u_p|Z_p|\, C_p\, F\nabla\Phi\right) \tag{3.7}$$

In Eq. (3.7), it seems at first paradoxical that the sign of the valence (positive for cations) should be required in the diffusion term (the first term), whereas only the magnitude of the valence is required in the electric field term (second term).

The resolution of the paradox comes from realizing that in the diffusion term the concentration gradient controls the direction of the flow, but the current will be in the same or opposite direction, depending on whether Z_p is negative or positive. Conversely, in the electric field term the electric field itself determines the direction of positive current flow, so knowledge of Z_p is necessary only to determine current magnitude.

3.8.1. Conductivity

Materials that conduct electricity are commonly characterized in terms of their resistance, or its reciprocal, conductance. How do factors such as mobility and concentration relate to the overall electrical conductance of the intracellular or extracellular space? One imagines that all these parameters must be closely connected, because conductance is a parameter that characterizes the ease of movement of charge through a conducting medium, an idea fundamentally a part of the preceding sections.

Such a connection does indeed exist. The thrust of this section is to identify it explicitly. Thereby one is able to tie the aggregate characteristic of conductance to the parameters that describe the movement of ions in the medium.

The electric current in an electrolyte arising from the movement of an ion under the influence of an electric field, $\overline{E} = -\nabla\Phi$, according to (3.7), is

$$\overline{J}_p^e = -u_p|Z_p|\, C_p\, F\, \nabla\, \Phi \tag{3.8}$$

The ion type is designated by the subscript p, while the superscript e signifies that the current is due solely to an electric field. For a KCl electrolyte, for example, the total current (density) is given by

$$\overline{J}_{\mathrm{KCl}}^e = FC_{\mathrm{KCl}}[u_{\mathrm{K}} + u_{\mathrm{Cl}}]\overline{E} \tag{3.9}$$

where the contribution to current flow from both K^+ and Cl^- is accounted for; the concentration of K^+ and Cl^- is given by that of C_{KCl}, assuming complete dissociation.

One can compare the preceding equation to a standard form of Ohm's law, $\overline{J} = \sigma\overline{E}$, where σ is the effective electrical conductivity. These two equations are the same, if the electrolyte conductivity is defined as

$$\sigma = FC_{\mathrm{KCl}}[u_{\mathrm{K}} + u_{\mathrm{Cl}}] \tag{3.10}$$

This equation is the connection between the conductivity value and its constituent parameters, which was the goal of this section. Note that for a particular current path, the *conductivity* (in units such as Siemens per cm) can be converted into a *conductance*. One does so by taking into account the dimensions of the path, e.g., using an expression such as conductance equaling conductivity times the cross-sectional area, divided by the path length.

Equation (3.10) identifies conductivity σ as proportional to ionic concentration. A limitation of this equation is that it arises from the assumption that the salt is completely dissociated, an assumption that must at some point deviate from reality, as the concentration increases and dissociation is no longer virtually complete. In our context, the issue that arises is how well the assumption holds for the ionic concentrations in excitable tissue.

This question can be more precisely addressed mathematically by slightly modifying the definition of σ to let the degree of dissociation appear explicitly. In particular, if only α percent dissociates, then

$$\sigma = \alpha\, FC_{\mathrm{KCl}}[u_{\mathrm{K}} + u_{\mathrm{Cl}}] \tag{3.11}$$

To examine more carefully the effects of the degree of dissociation, it is helpful to define a quantity that describes its effect on conductivity apart from the effect of concentration increases. This quantity is called "equivalent conductance," Λ. With α included, equivalent conductance is defined as

$$\Lambda = \alpha\, F\, [u_{\mathrm{K}} + u_{\mathrm{Cl}}] \times 1000 \tag{3.12}$$

Table 3.4. Equivalent Conductance vs Concentration

Concentration of KCL (mM)	Measured Equivalent conductance
0.0001	148.9
0.0010	146.9
0.0100	141.3
0.1000	128.9

Table 3.5. Equivalent Conductance

Ion	Λ	Units	T	Ref
Na^+	50.08×10^{-4}	m^2 s/mol	25 °C	[7]
K^+	73.48×10^{-4}	m^2 s/mol	25 °C	[7]
Cl^-	76.31×10^{-4}	m^2 s/mol	25 °C	[7]

Note that the expression for Λ is the same as the expression for conductivity, omitting the concentration term. (The factor of 1000 may be included to place numbers in more convenient units.) With this definition, Λ is affected by concentration only to the extent that α diminishes from unity for increasing concentration.

As an example, Table 3.4 tabulates the effect of concentration on equivalent conductance, as measured experimentally. From the table one sees that Λ does not change very much over several orders of magnitude of concentration change, but it does begin to decline significantly as the concentration of KCL grows to values of 0.0100 mM and above.

Equivalent conductance is also important since it is a quantity that can be measured experimentally and tabulated for reference. Table 3.5 gives some numerical examples. The conductance values are valuable in part because they can be used to obtain other parameters, e.g., by use of the equation obtained by combining (3.3) with (3.12) for $\alpha = 1$:

$$D = \frac{RT}{F^2} \frac{\Lambda}{|Z|} \tag{3.13}$$

3.8.2. Transference Numbers

In general, the contributions to the net conductivity come from all mobile ions, but the contributions of each will be in different proportions. The proportionality factors are known as *transference numbers* (or transport numbers). Using the above example, we define

$$t_K = \frac{u_K}{u_K + u_{Cl}}, \qquad t_{Cl} = \frac{u_{Cl}}{u_K + u_{Cl}}, \qquad t_K + t_{Cl} = 1 \tag{3.14}$$

where t_K and t_{Cl} are the transference numbers for potassium and chloride, respectively.

Table 3.6. Transference Number[a]

Ion	Range (mole/L)	t
KCl	0.02–3.0	$t_K = 0.49$
HCl	0.01–0.2	$t_H = 0.83$
LiCl	0.01–0.2	$t_{Li} = 0.32$
NaCl	0.01–0.2	$t_{Na} = 0.39$

[a]In each case the transference number for chloride is found by subtracting the cation value, a given above, from unity.

Some sense of the relative magnitude of transference numbers is illustrated by the chloride electrolytes in Table 3.6. For an electrolyte with a more complex ionic composition, the transference numbers depend on relative concentrations as well as mobilities. Later in this book, when we consider action currents, a detailed account of the charge carriers in the intracellular and extracellular spaces will depend on their respective transference numbers.

3.9. MEMBRANE STRUCTURE

Excitable cells are surrounded by a *plasma membrane*, whose main function is to control the passage of ions and molecules into and out of the cell. This membrane behavior will be found to underlie the tissue's electrical properties.

The plasma membrane is a structure that bounds the cell. The membrane is mainly made of *lipid*, which often represents as much as 70% of the membrane volume, depending on cell type. The membrane lipid itself prevents the passage of ions through the membrane. The membrane is heterogeneous, with numerous large, complex proteins (on the order of 2,500 amino acids) embedded within it. Some of these proteins are the constituents of *pumps* and *channels* that exchange ions between intracellular and extracellular space.

The ions themselves have radii on the order of 1 Å. Knowledge of the size and three-dimensional structure of channels remains incomplete. As diagrammed hypothetically by Hille ([3], p. 71), the channel structure is on the order of 100 Å, with an internal pore that is much wider than an ion over most of its length, narrowing to atomic dimensions only in the ion selective areas.

The membrane is about 75 Å thick. In this and the following chapter we focus on membrane structure as it affects the membrane's capacitative and ionic properties, in which thickness plays a significant role.

The membrane's thickness is, however, usually much less than other dimensions of interest. Thus in later chapters the membrane often will be considered as an interface without thickness but having net resistive and capacitive values.

3.9.1. Transmembrane Potential

If the electrical potential at the inside surface of the membrane of an excitable cell is compared to the potential at the outside surface, then, at rest, a potential difference, called the *transmembrane potential*, on the order of 0.1 volt will be found. Mathematically, the definition of V_m is

$$V_m \equiv \Phi_i - \Phi_e$$

Because the membrane has a resistance (i.e., is not a perfect insulator), there will be a transmembrane current, I_m. By definition, this current is considered to have a positive sign when it flows across the membrane in the direction from the inside to the outside.

Excitable membranes have periodic resting and active phases, during which the transmembrane potential fluctuates in the range ± 0.1 Volts. The 0.1-volt difference is not very much in comparison to ordinary household voltages, but it becomes enormous because the distance across which the potential changes is so small, just the membrane thickness. That is, this transmembrane voltage produces an enormous electric field across and within the membrane.

In the subsequent sections of this chapter we will consider explanations for the origin of the resting potential, and explanations of its magnitude and sign. We will see that the potential at rest depends on the *selective permeability* of the membrane to the several major ions that are present and to the different ionic composition of intracellular and extracellular space.

3.9.2. Pumps and Channels in the Membrane

The electrical behavior of nerve and muscle depends on the movement of sodium, potassium, calcium, and other ions across their membranes through the pumps and channels which lie therein. Pumps are active processes (consuming energy) that move ions against the concentration gradient. The sodium–potassium pump tends to operate at a slow but steady rate and maintains the concentration differences of Na^+ and K^+ between the intracellular and extracellular regions.

Channels make use of the energy stored, in effect, in the concentration differences, to allow the flow of each ion type down its concentration gradient. They do so in a way that is highly controlled as to when and to what degree the flow is allowed. The flow of ions through channels results in changes in transmembrane potential, sometimes quite rapidly.

The *selective permeability* of the membrane to individual ions, and the ability to rapidly increase or decrease the permeability selectively, is a truly astonishing property. Such control seems to be accomplished by means of channel gates which respond to the presence of electric fields or to certain ligands. Channels are the means by which rapid changes in transmembrane potentials occur. These rapid changes are associated with information transmission in nerves and mechanical contractions in muscle.

The basic structure of the membrane, including its lipid and protein (channel) content, is depicted in Figure 3.1. Under steady-state conditions (e.g., at rest) a fixed fraction of each channel type will be open and the membrane can be considered (macroscopically) to provide a particular ionic conductance to each ion species present. Ion movement across the membrane is subject to both diffusion and electric field forces, and we will describe the application of physical principles to evaluate the net transmembrane flow.

3.9.3. Lipid Content

A lipid such as olive oil that is placed on a water surface will spread out. If given enough room, it will reach a thickness of a single molecule (monolayer). At the same time, on a calm

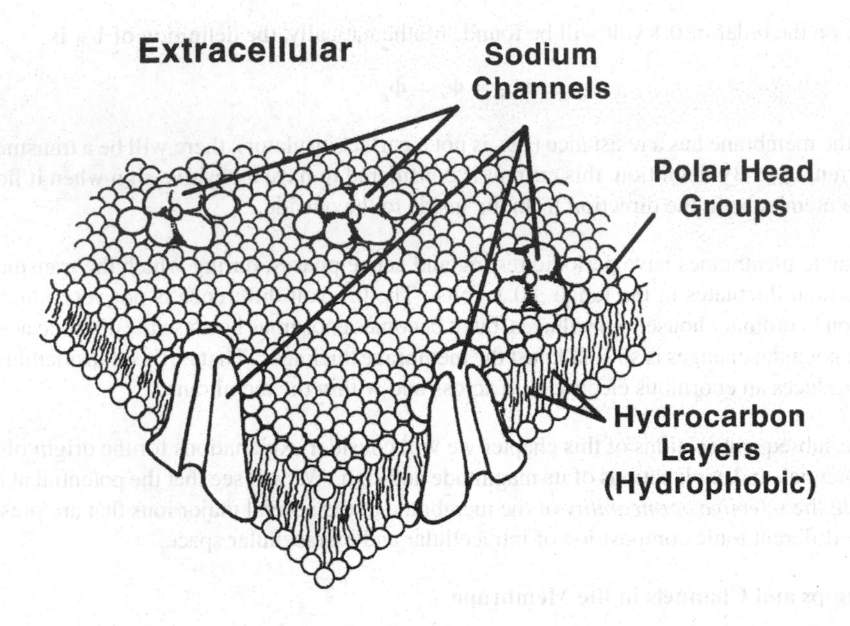

Figure 3.1. Schematic Representation of the Model of Membrane Structure, showing sodium channel proteins embedded in the lipid bilayer matrix of the membrane. The channel density is un physiologically high, for illustrative purposes. Drawing based on Catterall WA, et al. 1994. Structure and modulation of voltage-gated sodium channels. In *Ion channels in the cardiovascular system*, Ed PM Spooner, AM Brown. Armonk, NY: Futura.

surface the monolayer will remain contiguous. In this condition the polar heads of the lipid (which are hydrophilic) face the water while the non-polar tails (hydrophobic) lift away.

The biological membrane consists, basically, of two such layers of lipid. (Naturally, these two layers are called "the lipid bilayer.") The structure is shown in Figure 3.1. These layers organize themselves so that the polar group of each layer faces the intracellular or extracellular aqueous medium. Conversely, their non-polar tails are in contact, and form the interior of the membrane.

Figure 3.1 illustrates the hydrophobic lipid tails in the membrane interior. This inner portion of the membrane behaves like a dielectric (insulator) of perhaps 30 Åthickness.

One aspect of the membrane that is critical to its electrical function is its electrical capacitance. If we determine the capacitance of such a parallel plane structure with a high dielectric constant (estimated as $k = 3$, using the value for oil), then, using the parallel plane capacitance formula of chapter 2, we get

$$C_m = \frac{k\varepsilon_0}{d} = \frac{(3 \times 10^{-9})}{36\pi(3 \times 10^{-9})} = 0.009\,\text{F/m}^2 = 0.9\,\mu\text{F/cm}^2$$

where $\varepsilon_0 = 10^{-9}/36\pi$ Farads per meter is the permittivity of free space and d is the membrane thickness. The computed value of capacitance is an estimate, but the resulting value is similar to the measured value of most excitable membranes. The resulting value of roughly 1 μF/cm^2 is an extremely high value as compared to most materials used in ordinary factory-made capacitors. Inspection of the result shows that the high value comes about because of the remarkable combination of high membrane resistance and dielectric constant within a membrane that is very thin, a combination hard to construct in a non-biological capacitance.

After staining and fixing, the membrane is seen under electron microscopy to be characterized by two dense lines separated by a clear space and aggregating 75 Å. Presumably the polar groups take up the stain (along with the associated protein) but not the non-polar region. This membrane appearance is readily recognized in electron micrographs.

3.10. NERNST POTENTIAL

The unequal concentration of ions in the intracellular versus extracellular spaces causes diffusion of ions from high to low concentrations; the rate of diffusion depends on the difference in concentration and the membrane permeability (which depends on the open channel density and channel resistance).

Charged ions accumulate on the membrane because of its capacitance. These charged ions set up an electric field across and internal to the membrane. The electric field will, in turn, exert forces on all charged particles lying within the membrane. Consequently any quantitative description of ion flow within or across the membrane must take into account both forces of diffusion and electric field.

At a particular moment, the movement of ions of type p across a membrane will depend on the relative density of p channels, the relative probability of such a channel being open, the conductance of the channel, and the net driving force (diffusional plus electric field) for this ion species. Living membranes have the property of *selective permeability* to various ion species, as determined based on the relative ease of movement of the respective ions across the membrane, considered macroscopically. The selective permeability can be described by an effective conductance (per unit area) for the transmembrane flux of p.

In the following we treat the permeability of the membrane as, in its effect, uniformly distributed, that is, we determine a macroscopic description arising from the averaged microscopic channel behavior. Furthermore, we assume the movement of a particular ion species to be independent of that of others. To the extent that channels are relatively uniformly distributed, that our interest lies in their macroscopic effects, and that the pth ion's flow is confined to the pth ion's channels, this assumption is satisfactory.

As noted, for the excitable cell the membrane has the property of selective permeability. In other words, certain ions pass readily across the membrane while other ion species flow with more difficulty or not at all. Because the ionic compositions of the intracellular and extracellular regions are quite different, an initial diffusional ion movement of permeable ions takes place. The result is a net transfer of charge and the establishment of a membrane electric field. Consequently, both diffusional and electric field forces are always expected with biological membrane systems. Accordingly, the Nernst–Planck equation is the appropriate expression for the examination of ion flow across biological membranes. The equilibrium conditions that are determined correspond to those that could also be found from thermodynamic considerations.

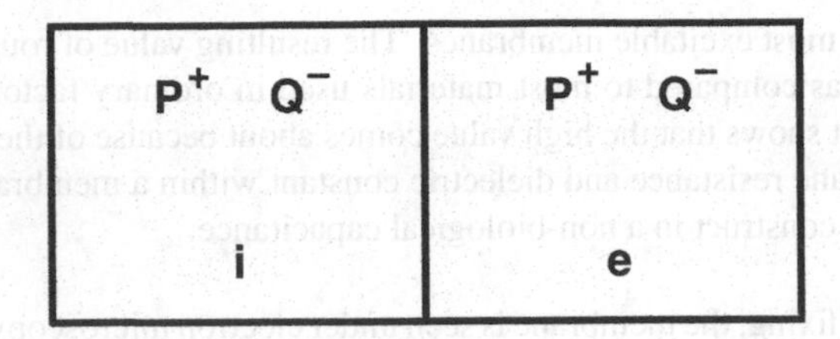

Figure 3.2. Concentration Cell. The concentration cell has a calibrated membrane area separating two compartments, designated i and e. Known ionic concentrations are present in each compartment. Here two hypothetical ions, P^+ and Q^-, are present in each compartment. The compartments are well mixed, so that concentration differences within a compartment are insignificant. The concentration cell provides well-defined geometry within which concentration changes over time can be examined.

3.10.1. Single Ion Permeability

We analyze first the movement of ions in a *concentration cell*. The results will be seen to be useful in the examination of the biological membrane. A concentration cell is illustrated in Figure 3.2. Such a cell is a two-compartment system separated by a selectively permeable membrane. We assume the concentration of P^+ in compartment i to exceed that in compartment e. We also assume the membrane impermeable to Q^-. Consequently, P^+ will diffuse from i to e, (but Q^- cannot diffuse from i to e).

Ions in solution experience significant diffusion wherever there is a significant change in concentration between nearby locations. Since the ionic compositions outside and inside excitable cells are quite different, and since these regions are quite close together (separated only by a thin membrane), a high concentration gradient is often present. Correspondingly, diffusion must necessarily play a strong role in transmembrane current behavior, for all permeable ions.

The diffusion results in accumulation of positive charge in e (electrostatic forces cause these charges to reside on the membrane). This leaves behind in i an excess of negative charge of similar magnitude (these reside on the i side of the membrane due to electrostatic forces). The result is a difference of potential, V_m, related to the charge ΔP^+ by the membrane capacitance C_m (where $V_m = \Delta P^+/C_m$).

For a membrane thickness of d there is also an electric field $E = V_m/d = \Delta P^+/(Cd)$. The electric field is directed from e to i, and it clearly increases in magnitude as ΔP^+ diffuses from i to e. The growing electric field increasingly hinders further diffusion until it brings about its termination and equilibrium is reached.

3.10.2. Nernst Equilibrium

At equilibrium the electric field force (from e to i) just balances the diffusion force (from i to e). Under these conditions we obtain from (3.6)

$$\bar{J}_p = 0 = -D_p F Z_p \left[\nabla C_p + \frac{Z_p C_p F \nabla \Phi}{RT} \right] \tag{3.15}$$

hence

$$\nabla C_p = -\frac{Z_p C_p F}{RT} \nabla \Phi \tag{3.16}$$

We assume that quantities vary in the direction perpendicular to the membrane only. Calling this coordinate x then simplifies (3.16) to

$$\frac{dC_p}{dx} = -\frac{Z_p C_p F}{RT} \frac{d\Phi}{dx} \tag{3.17}$$

and rearranging gives

$$\frac{dC_p}{C_p} = -\left(\frac{Z_p F}{RT}\right) d\Phi \tag{3.18}$$

We can integrate (3.18) across the membrane from compartment e to compartment i:

$$\int_e^i \frac{dC_p}{C_p} = -\frac{Z_p F}{RT} \int_e^i d\Phi \tag{3.19}$$

The result is

$$\ln\left(\frac{[C_p]i}{[C_p]e}\right) = -\frac{Z_p F}{RT} \{\Phi_i - \Phi_e\} \tag{3.20}$$

Thus the potential difference at equilibrium across the membrane, V_m^{eq}, equals

$$V_m^{\text{eq}} = \Phi_i - \Phi_e = \frac{-RT}{Z_p F} \ln\left(\frac{[C_p]_i}{[C_p]_e}\right) \tag{3.21}$$

where the transmembrane potential has been defined as the intracellular (i) minus extracellular (e) potentials following present-day convention.

The following numerical coefficient replaces *RT/F* in (3.21) for the case that T is chosen to be at a cool temperature (17 °C):[1]

$$V_m^{\text{eq}} = E_p = \frac{-25}{Z_p} \ln\left(\frac{[C_p]_i}{[C_p]_e}\right) mV = \frac{25}{Z_p} \ln\left(\frac{[C_p]_e}{[C_p]_i}\right) \tag{3.22}$$

or, using base 10 instead of natural logarithms,

$$V_m^{\text{eq}} = E_p = \frac{58}{Z_p} \log_{10}\left(\frac{[C_p]_e}{[C_p]_i}\right) mV \tag{3.23}$$

The *Nernst potential* for an ion is the V_m given by these equations. When ion p is in equilibrium, the Nernst potential is the transmembrane voltage. At equilibrium the ion's rate of movement in one direction, due to diffusion, is equal to its rate of movement in the opposite direction, due to the electric field associated with the transmembrane voltage.

Another way of thinking about the Nernst potential is that it is an electrical measure of the strength of diffusion, as it is the potential required to provide an exact counterbalance to a particular concentration ratio.

3.10.3. Symbolic Names for Nernst Potentials

The Nernst potentials for the major ion species are important quantities and are usually assigned special symbols. We have made such an assignment in (3.22) and (3.33), where E_p is defined to be V_m^{eq}. Even though Nernst potentials are given special symbols, it is important to keep in mind that they are simply the transmembrane potential that exists for particular circumstances, the equilibrium of a particular ion species.

E_p may be a constant, or it may change slowly. More specifically, for potassium E_p becomes E_{K}. In most mathematical expressions involving E_{K}, we treat it as a constant, recognizing that so doing implies that the intracellular and extracellular concentrations of potassium (but not the transmembrane potential) remain the same. (Or, at least, E_{K} changes slowly compared to other variables of interest.) As a notational device, use of the symbol E_{K} allows one to readily differentiate between V_m, the transmembrane voltage at a certain moment (but perhaps changing rapidly), as compared to E_{K}, a value of V_m at potassium equilibrium.

3.10.4. Examples for Potassium and Sodium Ions

Suppose frog muscle had the concentrations of Table 3.1, and suppose its membrane was permeable to potassium ions but to no others. At what transmembrane potential would there be equilibrium?

Answer: Using (3.22),

$$V_m = 25 \ln \frac{2.2}{124} = -100.8\,\text{mV}$$

If the same membrane were permeable only to sodium ions, at what transmembrane potential would there be equilibrium?

Answer:

$$V_m = 25 \ln \frac{109}{4} = +82.6\,\text{mV}$$

Note that the polarity of V_m is different for sodium versus potassium ions because of the ratio of concentrations, even though both ions are cations.

3.11. ELECTROLYTES

3.11.1. Relative Charge Depletion

The equilibrium state at the Nernst potential is achieved by the net movement of charge from one side of the membrane to the other. For biological membranes, the resting potential depends on the charge magnitude and on the membrane capacitance. A reasonable question is, then, how much the intracellular or extracellular concentrations are modified by this charge movement. To answer this question, one needs (a) to assess the amount of charge required to charge the membrane to equilibrium voltage, and to compare that amount to (b) the amount of charge in the electrolyte.

(a) To evaluate the amount of charge $\mathcal{Q}$ moved, consider as an example a muscle fiber of radius 10 μm. The total charge movement associated with the resting condition is given by

$$\mathcal{Q} = CV$$

If we take the nominal values $C = 1\ \mu\text{F cm}^2$ and $V = 100$ mV, then for an axon 1 cm in length the amount of charge on the membrane at equilibrium is

$$\mathcal{Q} = 1 \times 10^{-6} \times 2\pi \times 0.001 \times 1 \times 0.1 = 0.628 \times 10^{-8}\,\text{Coulombs/cm} \qquad (3.24)$$

(b) For an intracellular potassium concentration of 124 mM/liter we can find the amount of charge in the electrolyte as

$$\mathcal{Q}_K = 0.124 \times \left(\frac{\pi \times (0.001)^2 \times 1}{1000} \right) \times 96487 = 0.375 \times 10^{-4}\,\text{Coulombs/cm} \quad (3.25)$$

Taking the ratio of $\mathcal{Q}$ to $\mathcal{Q}_K$ in this example, we see that the relative charge depletion is only ≈ 0.00017, that is, only this tiny fraction of the charge in the solution is necessary to charge the membrane. The conclusion is that the change in concentration caused by the movement of these ions is inconsequential insofar as a single charging cycle. (An even smaller fraction would be seen for larger-diameter fibers.)

On the other hand, over a longer time period with many charging cycles, there could be a significant cumulative effect.

3.11.2. Electroneutrality

In any region within an electrolyte, it is expected that the concentration of anions equals that of cations, the condition being known as *electroneutrality*. Electroneutrality is expected because any net charge brings into play strong electrostatic forces that tend to restore the zero net charge condition.

The movement of charge to achieve the Nernst potential in the above example does not violate the electroneutrality of intracellular and extracellular electrolytes. The reasons there is no violation are because the charge involved is so small, relatively, and because the excess positive or negative charges reside on the membrane, not within the solution.

While the above results for the Nernst potential show only a tiny relative charge depletion, one should not apply this result to every situation without careful evaluation. First, it is worth noting that the analysis for the Nernst potential applies directly only to a specific case involving a membrane permeable to only a single ion. Many biological membranes evaluated in other parts of this text have multi-ion flows. In addition, intracellular and extracellular volumes are highly variable. In some tissues volumes may be sufficiently small such that significant concentration changes do result. In such cases electroneutrality may not be maintained, and a more sophisticated analysis, such as Debye–Hückel [9] theory, is necessary.

3.12. SUMMARY SO FAR

The developments in this chapter have so far provided two major results along with a number of secondary ones. The two major results are:

1. The Nernst–Planck equation allows one to compute the resulting current density wherever the concentration and potential gradients are known, if the characteristics of the medium (such as temperature) are also known. Current flow in the interior and exterior of cells can be evaluated beginning with the Nernst–Planck equation, and (as shown above) it provides the foundation step for finding the Nernst potential, across membranes.

2. For membranes, the equation for the Nernst potential is the second major result. Knowledge of the Nernst potential allows one to know, for given concentrations, the potential at equilibrium that would exist across real membranes if the membrane were permeable to only one ion. It is particularly useful for knowing the limiting transmembrane potential that can be reached if the membrane approaches a state where significant permeability remains for only one ion. An advantage of the equations derived is that they do not require that the permeability itself be known.

We now need to proceed to evaluate current flow across membranes in more detail. We thus need to take into account the more normal situation where the membrane is permeable in various degrees to several ions simultaneously. Also, we need to evaluate how much the membrane current is, when it is not zero. In principle, the Nernst–Planck equation, as an expression applicable for any concentration gradient and any potential gradient, might be expected to provide us with a starting point for such an evaluation.

The Nernst–Planck equation fails, however, to do so. The reasons for its failure lies not in a deficiency in the equation itself but in our ability to apply it. The problem is that we do not know, for a channel, the needed quantities required to use the Nernst–Planck equation effectively.

In other words, we know neither the concentration nor the potential as a function of distance as one moves along a pathway from the inner edge of the membrane to its outer edge. We do know that all these quantities have quite different values on one side of the membrane as compared to the other, so that large changes must occur somewhere. Although one makes simplifying assumptions about the transmembrane potential, as was done by Goldman [1] for the GHK equations (below), we are uncertain as to how far such results can be relied upon in view of the actual complexities.

On the other hand, some quantities about excitable membranes are fairly readily measured, such as their conductivities to various ions. Thus, to take advantage of what can be done and to avoid what is unknown, it is at this point advantageous to "erase the board," so to speak, and to begin again with a new model of the membrane, the parallel conductance model. The parallel-conductance model is consistent in principle with all the developments considered so far, but quite different in its appearance and mathematics. It begins again in a way that takes advantage of what has been done so far, while looking at things from quite a different perspective.

3.13. PARALLEL-CONDUCTANCE MODEL

The parallel-conductance model embodies the bold mental step of asserting that different ions pass through the membrane independently one of another, as shown diagrammatically in

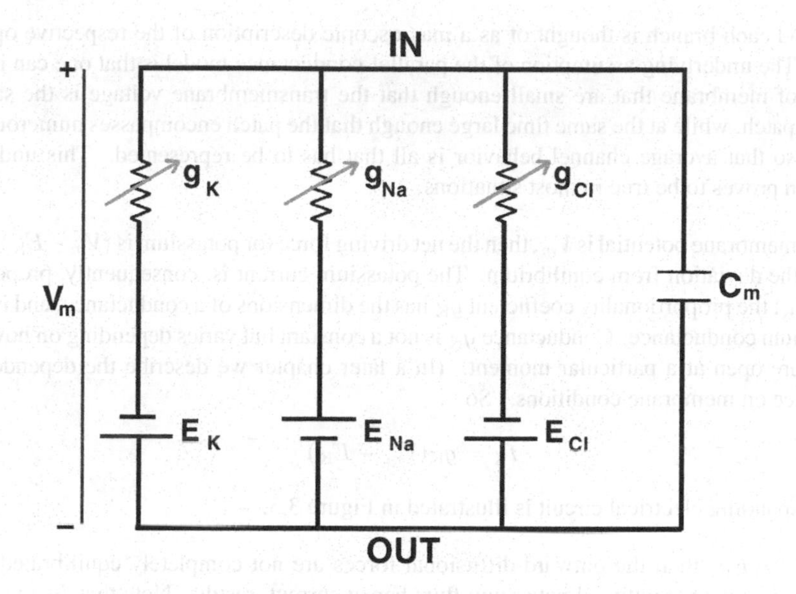

Figure 3.3. The Parallel-Conductance Model of an Excitable Membrane (IN = intracellular, OUT = extracellular). Independent conductance channels are present for K^+, Na^+, and Cl^-. Transmembrane potential V_m is positive when the inside has higher potential than the outside. The battery polarity is chosen to show that usually the Nernst potentials of E_K and E_{Cl} are negative (inside more negative than outside) and that of E_{Na} is positive (inside more positive than outside).

Figure 3.3, that is, different pathways across the membrane operate in parallel and simultaneously. Under that premise, the issues are now subdivided into how individual pathways operate, and how they combine together to create composite effects such as the membrane resting potential.

Resting Potential

The excitable membrane resting potential is one that brings the membrane into a *steady state*. It is that value at which the total membrane current is zero (otherwise, we have a changing potential and the membrane is not at rest). Before intracellular electrodes, it was supposed that all permeable ions were individually in equilibrium (i.e., in which case there would be a total absence of transmembrane ion flow). This is not the case, as is clear because the Nernst potential of each permeable ion is different (there is no single transmembrane potential that brings all ions to equilibrium).

To be able to address such dynamic mechanisms, we introduce the membrane representation of Figure 3.3. This representation is called the "parallel-conductance model." The model is intended to represent the flow of ions through their respective ionic channels in a small area of membrane, often called a membrane element, or a *membrane patch*,

Ionic Currents

As shown, the model assumes that the significant ions are potassium, sodium, and chloride. Each branch determines the contribution to the total transmembrane current from a specific ion species, and each branch is thought of as a macroscopic description of the respective open ion channels. The underlying assumption of the parallel-conductance model is that one can identify segments of membrane that are small enough that the transmembrane voltage is the same all across the patch, while at the same time large enough that the patch encompasses numerous ionic channels, so that average channel behavior is all that has to be represented. This underlying assumption proves to be true in most situations.

If the membrane potential is V_m, then the net driving force for potassium is $(V_m - E_K)$, which evaluates the deviation from equilibrium. The potassium current is, consequently, proportional to $V_m - E_K$; the proportionality coefficient g_K has the dimensions of a conductance, and is called the potassium conductance. Conductance g_K is not a constant but varies depending on how many channels are open at a particular moment. (In a later chapter we describe the dependence of conductance on membrane conditions.) So

$$I_K = g_K(V_m - E_K) \tag{3.26}$$

The corresponding electrical circuit is illustrated in Figure 3.3.

If $V_m > E_K$, then the outward diffusional forces are not completely equilibrated by the electric field, and a net outward potassium flux, hence current, results. Note that I_K in (3.26) is appropriately positive.

For the sodium ion

$$I_{Na} = g_{Na}(V_m - E_{Na}) \tag{3.27}$$

Here if $V_m > E_{Na}$, then because E_{Na}, is positive V_m must be positive and even larger than E_{Na}. The result is an outward sodium flux driven by an electric field which exceeds the inward diffusional forces. Equation (3.27) provides the correct sign for I_{Na}.

For the chloride ion, analogous to (3.26), one obtains

$$I_{Cl} = g_{Cl}(V_m - E_{Cl}) \tag{3.28}$$

If $V_m > E_{Cl}$, the inward chloride diffusion is not completely equilibrated and a net influx occurs. Since this influx is of ions with a negative charge, it constitutes an outward electric current; I_{Cl} should be positive, and from (3.28) it is.

3.13.1. Parallel Conductance in Alternative Forms

It is worth keeping in mind that the parallel-conductance model is included in this chapter to present the main concept of such models in a simple form.

Parallel-conductance models as used in particular contexts (for example, those of later chapters in this book) are normally modified in significant ways from the form above. For example, Hodgkin and Huxley used a parallel-conductance model that had sodium and potassium pathways, and a third pathway identified as leakage. They inferred the properties of the leakage channel from measurements of the transmembrane potential, rather than by modeling a specific ionic flow.

Other investigators have used parallel-conductance models with pathways to represent additional ions, such as calcium ions, and additional mechanisms, such as ionic pumps.

3.13.2. Capacitive Current

To complete the list of contributions to the transmembrane current we add the capacitive (or displacement) current, which is simply

$$I_C = C_m \frac{dV_m}{dt} \tag{3.29}$$

At rest (that is, at steady state), $I_C = 0$ since $dV_m/dt = 0$ (because if dV_m/dt was not zero, V_m would be changing and thus not at steady state).

Capacitive current is much more important than one might at first suspect, because the membrane is very thin and thus highly capacitive.

3.13.3. Resting V_m from Steady-State Constraints

As noted above, a resting membrane requires steady-state conditions, namely, zero net transmembrane current. The membrane current I_m is

$$I_m = I_C + I_{\mathrm{K}} + I_{\mathrm{Cl}} + I_{\mathrm{Na}}$$

At steady state, $I_C = 0$ since $dV_m/dt = 0$. Thus, at steady state,

$$I_m = 0 = 0 + I_{\mathrm{K}} + I_{\mathrm{Cl}} + I_{Na}$$

Substituting the expressions for each ionic current from (3.26), (3.27), and (3.28) gives

$$g_{\mathrm{Na}}(V_m - E_{\mathrm{Na}}) + g_{\mathrm{K}}(V_m - E_k) + g_{\mathrm{Cl}}(V_m - E_{\mathrm{Cl}}) = 0 \tag{3.30}$$

One can solve for V_m from (3.30) and obtain the resting value, V_{rest}, given by

$$V_{rest} = \frac{g_K E_K + g_{\mathrm{Cl}} E_{\mathrm{Cl}} + g_{Na} E_{Na}}{g_K + g_{Na} + g_{Cl}} \tag{3.31}$$

Equation (3.31) is known as the *parallel-conductance equation* for the resting transmembrane potential. It describes how V_m arises as a weighted average of E_k, E_{Cl}, and E_{Na} depending on their relative conductivities.

3.13.4. Example for Squid Axon

The above ideas can be illustrated using values of E_{k}, E_{Cl}, and E_{Na} for the squid axon from Table 3.1 and (3.23). We choose resting values of the conductances to be

$$g_{\mathrm{K}} = 0.415\,\mathrm{mS/cm}^2, \quad g_{\mathrm{Cl}} = 0.582\,\mathrm{mS/cm}^2, \quad g_{\mathrm{Na}} = 0.010\,\mathrm{mS/cm}^2$$

These values permit an examination of the relative influence of the Nernst potential and conductivity on resting conditions from (3.31). One obtains from Table 3.1 and (3.23) the following Nernst potentials:

$$E_K = -74.7\,\text{mV}, \quad E_{Na} = 54.2\,\text{mV}, \quad E_{Cl} = -65.8\,\text{mV}.$$

Substituting the above values into (3.31) yields $V_m = -68.0$ mV. This resting potential gives rise to a steady efflux of potassium.

The efflux is driven by the difference between V_m and E_K of 6.7 mV. An influx of sodium will also occur driven by the difference in V_m from its equilibrium Nernst potential, which in this example equals 122.2 mV. This large driving force acts on a relatively low conductivity, so the efflux of potassium and influx of sodium are roughly in balance, preserving the steady state. (For simplicity, we have here assumed chloride to be, essentially, in equilibrium.)

The parallel-conductance model is based on membrane conductances, and these must be found either from experiment or from yet another model. The Hodgkin–Huxley model, described in a subsequent chapter, utilizes (3.26)–(3.28) and gives expressions that evaluate the ionic conductances.

3.14. CONTRIBUTIONS FROM CHLORIDE

3.14.1. Chloride–Potassium Equilibrium

The role of the chloride ion in determining the resting potential appears to be secondary to that of potassium. This comes about because the intracellular chloride concentration is very small and undergoes a large percentage change with small amounts of chloride influx or efflux (which is not true for potassium). Consequently, chloride ion movements can be expected to occur that accommodate it to the potassium ion ratio to bring both ion ratios into consonance (i.e., the same Nernst potential), namely, when

$$[K^+]_i[Cl^-]_i = [K^+]_e[Cl^-]_e \tag{3.32}$$

Since the chloride ion ratio tracks the potassium, one need only follow the latter to evaluate the resting membrane potential (as a rough approximation). Changes in resting potential may similarly be thought due solely to the potassium ion ratio, namely,

$$V_m \approx 25 \ln \left(\frac{[K^+]_e}{[K^+]_i} \right) \tag{3.33}$$

3.14.2. Resting Conditions and Behavior of Chloride

An experimental investigation of the behavior of chloride under resting conditions was performed by Hodgkin and Horowicz [4, 5]. Their frog muscle preparation was placed in a normal extracellular medium.

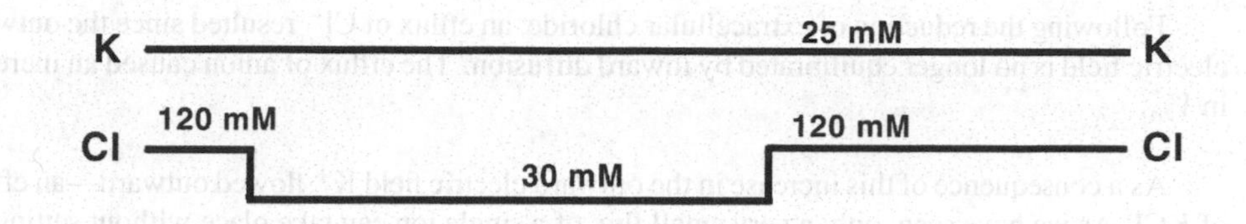

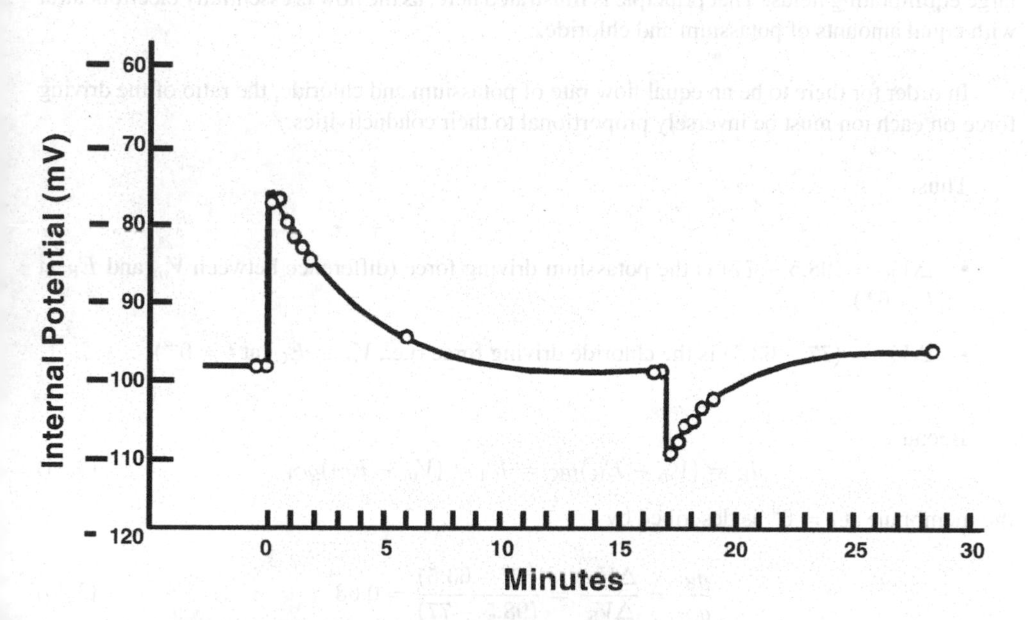

Figure 3.4. Effect of a Sudden Reduction in the External Chloride Concentration on the membrane potential of an isolated frog muscle fiber. Reprinted with permission from Hodgkin AL, Horowicz P. 1959. The influence of potassium and chloride ions on the membrane potential of single muscle fibers. *J Physiol* **148**:127–160.

These conditions applied:

- Extracellular $[\mathrm{Cl}^-]_e = 120$ mM.
- Extracellular $[\mathrm{K}^+]_e = 2.5$ mM.
- Intracellular potassium concentration $[\mathrm{K}^+]_i = 140$ mM.
- Intracellular chloride concentration $[\mathrm{Cl}^-]_i = 2.4$ mM.
- The resting potential was nominally the chloride equilibrium potential of −98.5 mv.

At $t = 0$ they rapidly reduced the extracellular chloride concentration from 120 mM to 30 mM (see Figure 3.4).

The effect was to increase E_{Cl} by 58 $\log_{10}$ (120/30) = 34.9 mV, so E_{Cl} increased from −98.5 to −63.6 mV. The membrane potential is found, experimentally, to rise to −77 mV (see Figure 3.4). (Note that this value lies between $E_K \approx -98.5$ mV and $E_{\mathrm{Cl}} = -63.6$ mV).[2]

Following the reduction of extracellular chloride, an efflux of Cl^- resulted since the outward electric field is no longer equilibrated by inward diffusion. The efflux of anion caused an increase in V_m.

As a consequence of this increase in the outward electric field K^+ flowed outward—an efflux of KCl. As we have seen, only a very small flux of a single ion can take place without setting up large equilibrating fields. That principle is illustrated here, as the flow is essentially electroneutral with equal amounts of potassium and chloride.

In order for there to be an equal flow rate of potassium and chloride, the ratio of the driving force on each ion must be inversely proportional to their conductivities.

Thus,

- $\Delta V_K = (98.5 - 77)$ is the potassium driving force (difference between V_m and E_K at $t = 0^+$).
- $\Delta V_{Cl} = (77 - 63.5)$ is the chloride driving force (i,e., $V_m - E_{Cl}$ at $t = 0^+$).

Because

$$J_K = (V_m - E_K)g_K = J_{Cl} = (V_m - E_{Cl})g_{Cl} \tag{3.34}$$

the membrane at $t = 0^+$ is described by

$$\frac{g_K}{g_{Cl}} = \frac{\Delta V_{Cl}}{\Delta V_K} = \frac{(77 - 63.5)}{(98.5 - 77)} = 0.63 \tag{3.35}$$

Equation (3.34) may also be thought of as an illustration of the parallel-conductance membrane model. In this case the experiment provides a direct measure of the relative potassium/chloride conductivity. Note the following experimental observations:

- The effect of an efflux of KCl is to diminish $[K^+]_i$, but we shall confirm that this amount is a small change from 140 to 138 mM. However, the drop of 2 mM $[Cl^-]_i$ results in a large percentage change of chloride.
- Ultimately (in 15 min or so) E_{Cl} diminishes to the value held by E_K (as $[Cl^-]_i$ diminishes due to chloride efflux). For the new steady state $E_{Cl} \approx -98.5$ mV, assumed unchanged, so $[Cl^-]_i$ must drop by a factor of 1/4 to offset the reduction in $[Cl^-]_e$ by a factor of 1/4 (i.e., from 120 to 30 mM).
- This new concentration occurs when $[Cl^-]_i$ reaches 0.6 mM from its initial value of 2.4 mM, a loss of 1.8 mM.
- To balance, the KCl efflux thus consists of 1.8 mM. This efflux results in a decrease in intracellular potassium concentration from 140 to 138.2 mM. As noted, this amount of movement has a negligible effect on E_K.

Consequently, it appears that, indeed, chloride accommodates to changing conditions so that equilibrium is restored. In effect, chloride leaves the fixing of the resting potential to potassium.

Why is it that the efflux of 1.8 mM was not considered to affect the values of $[K^+]e$? It was assumed there was no effect because extracellular space is assumed to represent a very large volume of electrolyte (a condition that is often true, but not always). With a large extracellular volume the ions that are flowing out produce an insignificant change in the overall concentration in the extracellular space. In this regard, the assumed large extracellular volume contrasts sharply with the confined intracellular space.

3.15. REFERENCE VALUES

Symbol	Name	value
N_0	Avogadro	6.02×10^{23} ions per mole
0°C	Celsius temp	corresponds to 273.16 °K (absolute)
F	Faraday	96,487 absolute Coulombs/gram equivalent
R	Gas Constant	8.314 J/K mole at 27 °C
RT/F	coefficient	$8.314 \times 300/96487 = 25.8$ mV ($\approx 27\,°C$)
ϵ_0	permittivity	$(10^{-9})/(36\pi)$ (Coulombs^2 sec^2) / (kg m^2)

3.16. NOTES

1. 17°C≈ 63°F. See also the reference values above.
2. For simplicity we let the initial rest values of $E_K = E_{Cl}$.

3.17. REFERENCES

1. Goldman DE. 1943. Potential, impedance, and rectification in membranes. *J Gen Physiol* **27**:37–60.
2. Harned HS, Bereton BO. 1950. *The physical chemistry of electrolytic solutions*, 2nd ed. New York: Reinhold.
3. Hille B. 2001. *Ionic channels of excitable membranes*, 3rd ed. Sunderland, MA: Sinauer Associates.
4. Hodgkin AL, Horowicz P. 1959. The influence of potassium and chloride ions on the membrane potential of single muscle fibers. *J Physiol* **148**:127–160.
5. Hodgkin AL, Katz B. 1949. The effect of sodium ions on the electrical activity of the giant axon of the squid. *J Physiol* **108**:37–77.
6. Joos G, Freeman IM. 1958. *Theoretical physics*, 3rd ed. London: Blackie.
7. Lide DR, ed. 1993. *Handbook of chemistry and physics*, 74th ed. Boca Raton, FL: CRC Press.
8. Mossop BJ, Barr RC, Zaharoff DA, Yuan F. 2004. Electric fields within cells as a function of membrane resistivity. *IEEE Trans NanoBiosci* **3**:225–231.
9. Plonsey R. 1969. *Bioelectric phenomena*. New York: McGraw-Hill.
10. Robinson RA, Stokes RH. 1970. *Electrolyte solutions*, 2nd ed. London: Butterworths.
11. Zaharoff DA, Barr RC, Li C-Y, Yuan F. 2002. Electromobility of plasmid DNA in tumor tissues during electric field mediated gene delivery. *Gene Ther* **9**:1286–1290.

Additional References

Aidley DJ. 1978. *The physiology of excitable cells*. Cambridge: Cambridge UP.

Junge D. 1981. *Nerve and muscle excitation*. Sunderland, MA: Sinauer Associates.

Keynes RD, Aidley DJ. 1991. *Nerve and muscle*, 2nd ed. Cambridge: Cambridge UP.

Jackson MB. 2005. *Molecular and cellular biophysics*. Cambridge: Cambridge UP.

Weiss TF. 1996. *Cellular biophysics*. Cambridge: MIT Press.

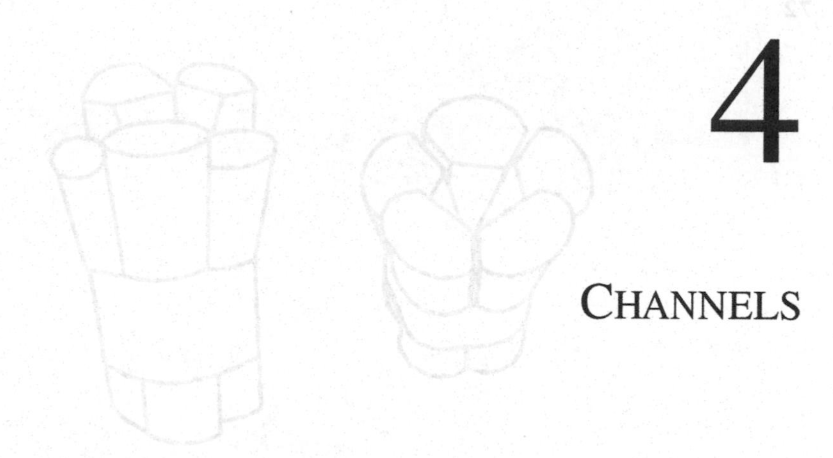

4

CHANNELS

4.1. INTRODUCTION

In the previous chapter it was pointed out that biological cells are enclosed by a plasma membrane. This membrane consists of a lipid bilayer and, as seen in Figure 3.1, the hydrophilic polar heads are oriented facing the intracellular and extracellular water-containing media, while the hydrophobic tails, on the other hand, are internal to the membrane. The biophysical consequence of the lipid is a membrane with the high dielectric constant of oil and the high resistivity of that material. The thin membrane has a high membrane capacitance of $C_m = 1\mu F/\text{cm}^2$; the lipid membrane also has a high specific resistance of $10^9\ \Omega\ \text{cm}^2$.

The very high membrane resistance is essentially an insulator to the movement of ions. Ion flux takes place because of the presence of membrane proteins called channels. These membrane proteins lie transverse to the membrane and contribute an aqueous path for ion movement. Specific resistances of biological membrane of 1,000 to 10,000 Ωcm^2 are observed. Such specific resistances are much lower than that of lipid membranc alone and occur as a consequence of the presence of open membrane channels.

But channels do not simply furnish a passive opening for ions to flow. Rather, channels are generally selective for a particular ion. In addition, a striking property of channels is that they have gates that open and close, and ion flow is controlled through that mechanism.

This chapter is devoted to an examination of channels, their structure and their bioelectrical properties.

4.2. CHANNEL STRUCTURE BY ELECTRON MICROSCOPY

We first review what has been found from electron microscopy (EM), electron diffraction, molecular biological, and biophysical approaches. The use of electron microscopy and x-ray diffraction requires a regular lattice, but general methods for crystallizing membrane proteins are not available as yet. There are several purified channel proteins that do form fairly regular two-dimensional lattices. These lattices have been investigated with x-ray diffraction and EM.

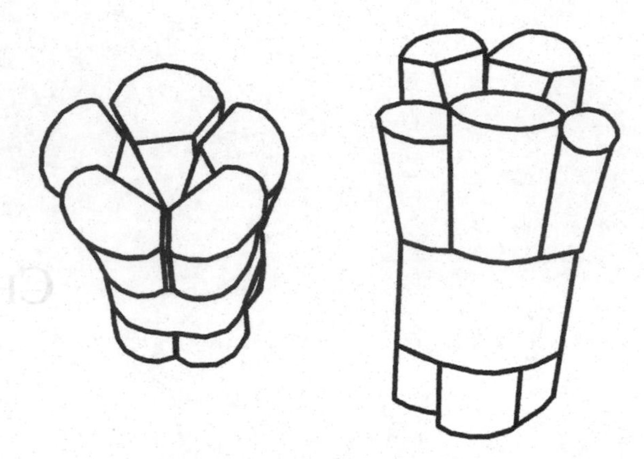

Figure 4.1. Model of the Acetylcholine Receptor that shows the five component subunits and the aqueous pore. The band locates the membrane bilayers through which the molecule passes; the lower part is cytoplasmic. From Stroud RM, Finer-Moore J. 1985. Acetylcholine receptor structure, function, and evolution. Reproduced with permission from *Annu Rev Cell Biol* 1:317–351. Copyright ©1985, Annual Reviews Inc.

The achieved resolution of around 17 Å describes a general structure but is not adequate for many details of interest (e.g., pore cross-section and gates).

A conception of the acetylcholine (ACh) receptor that results is shown in Figure 4.1. We note that it contains five component subunits enclosing an aqueous pore. Also, the total length substantially exceeds the plasma membrane. This molecule has been estimated to be about 120 Å in length, 80 Å in diameter, with a 2.0–2.5 nm central well. The dimensions of other ionic channels are not too different.

4.3. CHANNEL STRUCTURE: MOLECULAR GENETICS

An increasingly important technique for investigating channel structure is based on gene cloning methods that determine the primary amino acid sequence of channel proteins. The results can be tested by determining whether a cell that does not normally make the supposed protein will do so when provided the cloned message or gene.

Oocytes of the African toad *Xenopus laevis* are frequently used to examine *expression* of putative channel mRNA. The resulting channel properties can be evaluated to determine whether the protein synthesized is indeed the desired protein.

Although the *primary structure* of many channels has now been determined, the rules for deducing secondary and tertiary structure are not known. Certain educated guesses on folding of the amino acid chain can be made, however. One involves a search for a run of twenty or so hydrophobic amino acids, since this would just extend across the membrane and have the appropriate intramembrane (intra-lipid) behavior.

In this way the linear amino acid sequence can be converted into a sequence of loops based on the location of the portions lying within the membrane, within the cytoplasm, and within the

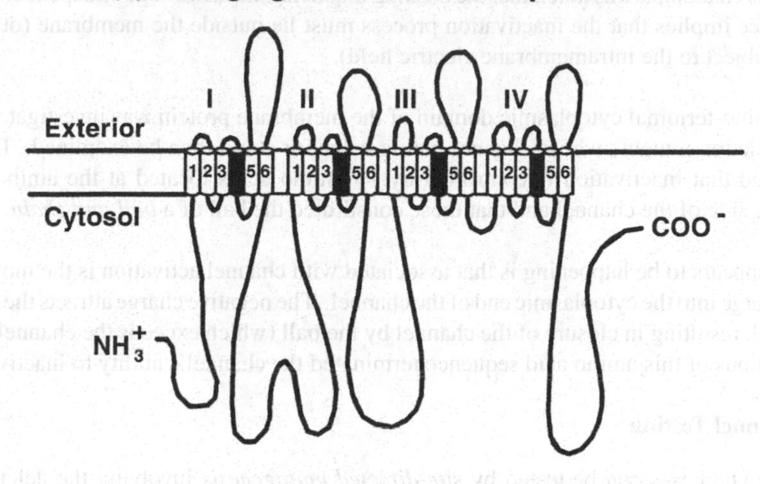

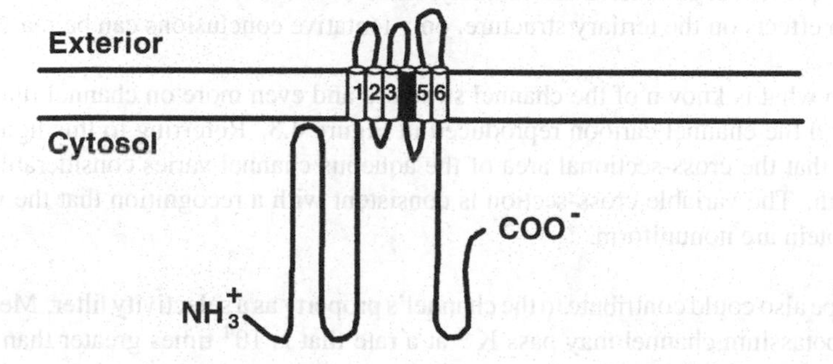

Figure 4.2. Proposed Transmembrane Structure of (a) voltage-gated Na^+ channel protein and (b) voltage-gated K^+ channel protein. The sodium channel arises from a single gene; it contains 1800–2000 amino acids, depending on the source. About 29 percent of the residues are identical to those in the voltage-gated Ca^{++} channel protein. There are four homologous domains indicated by the Roman numerals. Each of these is thought to contain six transmembrane α helices (Arabic numerals). Helix number 4 in each domain is thought to function as a voltage sensor. The shaker K^+ channel protein (b) isolated from *Drosophila* has only 616 amino acids; it is similar in sequence and transmembrane structure to each of the four domains in the Na^+ channel protein. From Darnell J, Lodish H, Baltimore D. 1990. *Molecular cell biology*, 2nd ed. New York: Scientific American Books. Adapted from Catterall WA. 1988. Structure and function of voltage-sensitive ion channels. *Science* **242**:50–61. Copyright ©1988, American Association for the Advancement of Science.

extracellular space (Figure 4.2). From the membrane portion of the sequence the particular run of amino acids gives some clues as to the structure of and the boundaries of the ion-conducting (pore-forming) region, as well as the location of charge groups that could enter into voltage-sensing gating charge movement.

This approach was successfully used in the study of *shaker*[1] K^+ inactivation. Following activation of this channel, it was noted that the ensuing inactivation was voltage-independent. Voltage independence implies that the inactivation process must lie outside the membrane (otherwise it would be subject to the intramembrane electric field).

The amino-terminal cytoplasmic domain of the membrane protein was investigated by constructing deletion mutants whose channel-gating behavior could then be examined. The results demonstrated that inactivation is controlled by 19 amino acids located at the amino-terminal cytoplasmic side of the channel and that these constituted the ball of a *ball and chain*.

What appears to be happening is that associated with channel activation is the movement of negative charge into the cytoplasmic end of the channel. The negative charge attracts the positively charged ball, resulting in closure of the channel by the ball (which exceeds the channel mouth in size). Deletions of this amino acid sequence terminated the channel's ability to inactivate.

4.3.1. Channel Testing

Some hypotheses can be tested by *site-directed mutagenesis* involving the deletion or insertion of specific protein segments (as just noted). By examining the altered properties of the channel expressed in *Xenopus* oocytes, one can make educated guesses concerning the function of the respective protein segments. Unfortunately, since the introduced changes can have complex and unknown effects on the tertiary structure, only tentative conclusions can be reached.

Based on what is known of the channel structure and even more on channel function, Hille [1] constructed the channel cartoon reproduced in Figure 4.3. Referring to this figure we note, for example, that the cross-sectional area of the aqueous channel varies considerably along the channel length. The variable cross-section is consistent with a recognition that the walls of the enclosing protein are nonuniform.

This shape also could contribute to the channel's property as a selectivity filter. Measurements show that a potassium channel may pass K^+ at a rate that is 10^4 times greater than Na^+, even though the latter is 0.4 Å smaller in crystal radius—so that selectivity is not a simple steric property. The observed high channel selectivity could be related to the particular distribution of charges along the walls of the pore, and this possibility can be investigated based on the amino acids lining the pore.

In Figure 4.1 the barrel stave structure has been thought to facilitate rapid gating. Such gating could be accomplished by only a small rotation of the contributing components. Thus, small conformational changes could give rise to large changes in the cross-section of the aqueous channel. The very rapid gating that is observed biophysically requires such structures.

Voltage gating implies that the molecular structure of the channel protein contains effectively embedded charges or dipoles—and these are sought within the amino acid sequence. An applied electric field causes intramolecular forces that can result in a conformational change.

This movement of charges constitutes a *gating current*; the charge displacement through an electrical potential adds or subtracts from the potential energy of the protein, and this can be related to the density of open or closed channels in a large population (through Boltzmann's

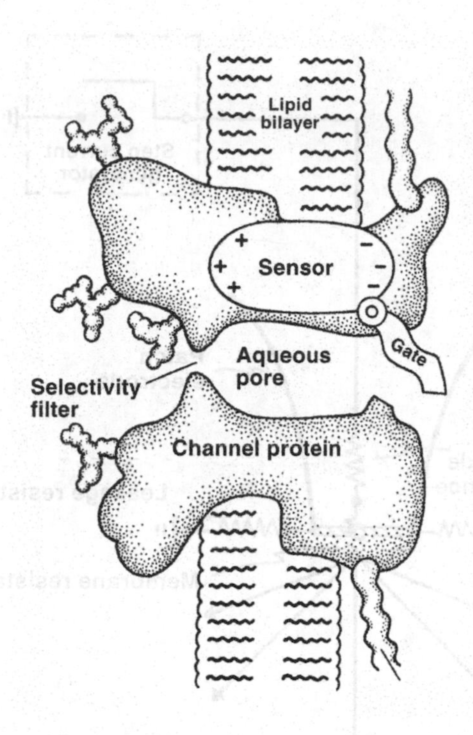

Figure 4.3. Functional Description of Membrane Channel. "The channel is drawn as a transmembrane macromolecule with a hole through the center. The functional regions—selectivity filter, gate, and sensor—are deduced from voltage-clamp experiments and are only beginning to be charted by structural studies. We have yet to learn how they actually look." [From Hille B. 1992. *Ionic channels of excitable membranes*, 2nd ed. Sunderland, MA: Sinauer Associates.]

equation); this will be discussed later. Since the gating current saturates when all channels are open, this constraint can be used to separate its contribution from the total capacitive current.

4.4. ION CHANNELS: BIOPHYSICAL METHODS

4.4.1. Single-Channel Currents and Noise

If an open single channel should behave as an aqueous path for ions, then the open channel contributes a path for electrical current with a fixed conductance of γ. Based on the parallel conductance model, a potassium channel should provide a current i_K given by

$$i_K = \gamma_K(V_m - E_K) \tag{4.1}$$

Assuming $V_m - E_K = 50$ mV and a conservative value of $\gamma_K = 20$ pS, then from (4.1) a current of 2 pA results, a value consistent with experiment. Measurement of this single-channel current with a micropipette having a tip diameter of 1 μm is feasible if other currents (noise in particular) can be minimized.

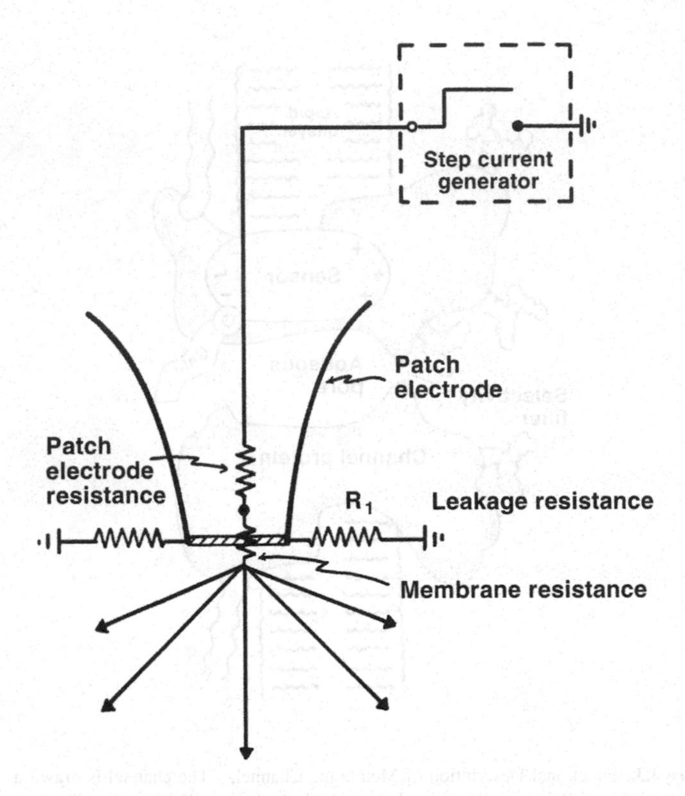

Figure 4.4. Inside–Out Patch Clamp Configuration. The desired current path through the cell is challenged by the alternate (leakage) pathway available in the region of electrode–membrane contact. A single open channel is assumed to give a membrane conductance equal to or greater than 20 pS (a resistance of $\leq$ 50 GΩ). To keep leakage current low (hence minimal loss of signal strength as well as reduced Johnson noise), this resistance should be in the tens of gigaohms; fortunately, patch electrodes with 100 GΩ leakage resistance are currently available.

We note that the Johnson noise current is

$$\sigma_n = \sqrt{4kT\Delta f/R} = 0.0180\,\text{pA} \tag{4.2}$$

based on $\Delta f = 1$ kHz, $R = 1/20$ pS = 5×10^{10} Ω, $T = 293$ K, and k (Boltzmann's constant) $= 1.38 \times 10^{-23}$. Even with a signal current of 1 pA, an entirely acceptable S/N = 56 results.

The problem in single-channel measurements that results from bringing a micropipette in contact with the cell membrane is illustrated in Figure 4.4. Using conventional techniques we would have a leakage resistance of 10 MΩ in parallel with that of the channel (5×10^{10} Ω). With the reduced *R* value of this combination, (4.2) evaluates a leakage noise current of 1.3 pA. This noise current results in a poor signal-to-noise ratio.

To reduce the noise current by tenfold requires a 100-fold increase in leakage resistance! Such a reduction was achieved by Neher and Sakmann by careful preparation of the electrode tip, preparation of the biological material, and application of a small amount of suction. The resultant instrument is called a *patch clamp*. The measurement of these low picoampere-level currents could not have been achieved without the advent of field-effect transistors with low-voltage noise and sub-picoampere input currents.

The choice of tip diameter of around 1 μm is about ten times larger than that used for intracellular micropipettes. The larger tip results in a lower tip resistance, a desirable factor to achieve a lower noise, as noted above. On the other hand, it may mean that more than one channel will find itself under the patch electrode. An examination of the recorded signal will reveal whether, in fact, only a single channel is accessed.

4.4.2. Voltage-Clamp Methods

Investigations of electrically active membrane often make strategic use of a patch clamp, space clamp or some other form of voltage clamp. To understand why, some background is needed, so that one understands the historic difficulties such a strategy is designed to overcome. In essence, what was discovered over time was that in such a membrane the conductivity changes, so that Ohm's law does not hold. Such conductivity changes are now known to arise from the opening or closing of channels. Varying numbers of open channels then produce changes in transmembrane voltage.

Closing the loop, the membrane voltage changes alter the number of channels that are open or closed. There is thus a feedback system, in which transmembrane voltages change channel openings, and changes in channel openings alter voltages. This feedback mechanism is a beautiful engine for cell activity and response, but one that, in its natural form, is hard to analyze in terms of its components, and next to impossible to evaluate experimentally.

To cut through this complexity, the voltage-clamp experiment was developed over a period of years in the mid-1900s. Though its origin had an experimental focus, the voltage clamp was also a powerful analytical concept. A core goal of the voltage clamp is to break the feedback loop where voltages affect channels, and channels affect voltages. Breaking the loop allows one to separate changes in voltage from the changes in the numbers of open channels that result.

The separation is achieved by making transmembrane voltage something the investigator sets as an independent variable, whether mathematically or experimentally, rather than something determined intrinsically by the cell, as happens normally. Thereafter the consequences of setting the voltage are determined, in terms of membrane currents, channels open, or other effects.

In the voltage-clamp protocol (Figure 4.5), the transmembrane voltage is set to two values in succession, here designated V_m^1 and V_m^2. The time periods during which these voltages are applied are called phase 1 and phase 2. In phase 1, the clamp mechanism and its control system supplies enough current of the right polarity to hold the transmembrane voltage at V_m^1 until the cell reaches steady state at time $t = t_1$. Then, at time $t = 0$, there is an abrupt transition from the first to the second transmembrane voltage, i.e., a transition from V_m^1 to V_m^2. In phase 2, transmembrane voltage V_m^2 is maintained until a new steady state is reached at time $t = t_2$.

Often the primary focus is on evaluating changes in the state of the membrane during phase 2 at times $t > 0$ such as $t = t_a$ or $t = t_b$. At such times an evolution in the number of channels

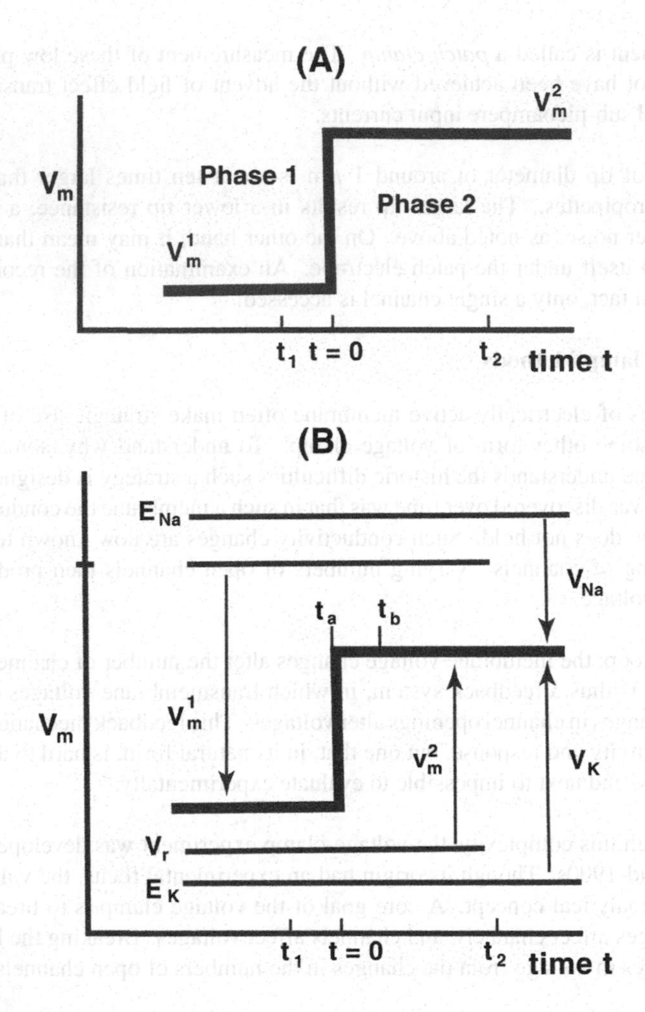

Figure 4.5. Voltage Clamp, V_m versus Time. Panel A, Concept: The transmembrane voltage is held constant at transmembrane voltage V_m^1 until time $t = 0$, when it is abruptly shifted to V_m^2. Times t_1 and t_2 identify the times when the membrane reaches a steady state in phases 1 and 2. Panel B, Detail: The equilibrium voltage for potassium is E_K and that for sodium E_{Na}. Voltage V_m is an absolute value i.e., relative to $V_m = 0$, as shown for V_m^1 on the left, while v_m (note lower case v) is relative to the resting potential (i.e., relative to V_r, as shown by v_m^2 on the right). Time $t = t_a$ marks the end of the voltage transition, while time $t = t_b$ occurs later, but before phase 2's steady state is attained.

that are open and the membrane currents that are flowing through them is taking place, as the membrane evolves into its steady state for phase 2.

These changing numbers of open channels allow changes in ionic currents that are directly observable. In this regard, the driving force for sodium ions, $V_{\mathrm{Na}} = V_m - E_{\mathrm{Na}}$ (shown during phase 2 by a downward arrow) has a different magnitude and sign than the driving force for potassium ions $V_K = V_m - E_K$ (shown during phase 2 by an upward arrow).

In voltage-clamp experiments, transmembrane voltage is set as an independent variable, so the transmembrane voltage comes about in a way that does not occur naturally. However, the consequences of setting that membrane voltage, in terms of numbers of channels open or closed, or in terms of membrane current, can be evaluated as a function of time and as a function of transmembrane voltage. This knowledge then can be integrated into a more complex system where transmembrane voltages as well as channel characteristics evolve in a natural way.

4.4.3. Patch Clamp

If transmembrane currents are measured from a macroscopic cell membrane, the contributing current density will vary with position over the membrane (unless some special effort is made). Such variation may result from propagation of activity, a topic treated in a subsequent chapter.

Because of the nonuniform contributions it may be difficult to interpret the measured current. It will be difficult because the current will originate from multiplicity of channels, and each one may be behaving differently because of a different transmembrane voltage, temporal phase, etc. (Later, we will describe a "space clamp," which ensures identical transmembrane voltages for all channels.)

The patch clamp addresses such difficulty. In the patch clamp, the micropipette tip is small, only around $1\mu m$ diameter. As a consequence its measured current is from only the very small contacted membrane element. A beneficial corollary is that the confounding effects of spatial variation of a large membrane area are avoided. The measurement is unaffected by spatial variations and is hence "space clamped"; the "clamp" in the name "patch clamp" arises from this feature.

A downside of restricting measurement to a patch is that the currents through the patch are smaller than for a larger membrane segment, and thus they are harder to measure. We have noted that for a patch clamp to work satisfactorily careful preparation is required to ensure an adequate signal-to-noise ratio. The following is a brief list of pertinent considerations [1].

1. A high-resistance seal is essential to ensure that the leakage currents (and their noise components) are small compared to the desired transmembrane currents.

2. The pipettes should be fire polished and clean. In general, pipettes may be used only once.

3. Cells should be clean and free of connective tissue, adherent cells, and basement membrane. Good seals are most readily obtained on cultured cells.

4. The application of gentle suction will increase the resistance of the seal to greater than $10^{10}\ \Omega$. This is the desired gigaohm seal (*gigaseal*), permitting the measurement of currents from membrane areas on the order of $1\ \mu m^2$. This resolution makes possible recording currents from single channels.

In Figure 4.6, four configurations of recording from a single-cell membrane using a patch micropipette are depicted. At the upper left, the gigaseal is established and an *on-cell* condition results. If a microelectrode is introduced into the cell, currents between that electrode and the patch electrode must pass through the membrane patch (only). In view of the small size of

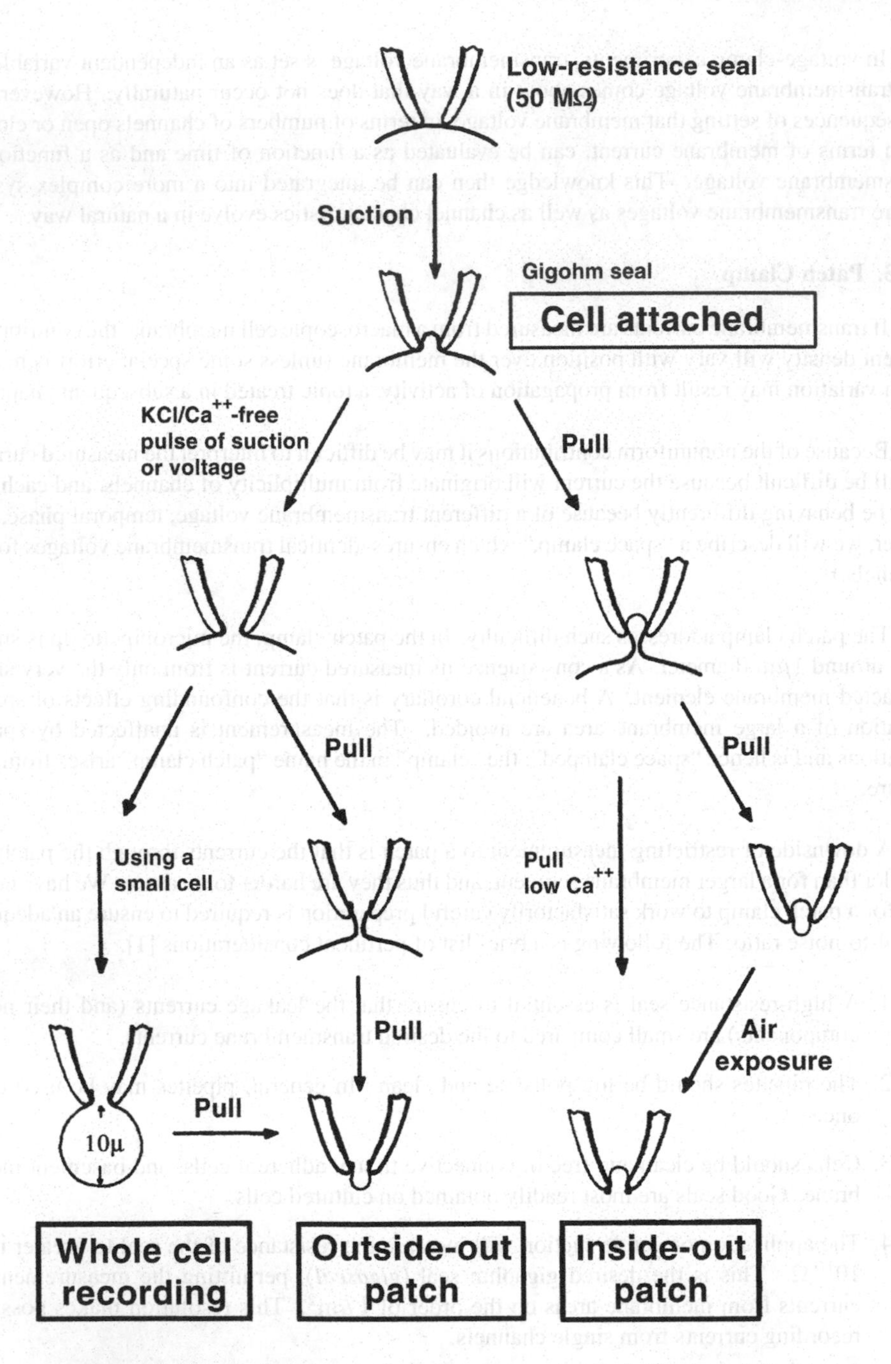

Figure 4.6. Four Configurations for Patch Clamping are described. The clean pipette is pressed against a cell to form a tight seal using light suction, and produces the *cell attached* or *on-cell* configuration. Pulling the pipette away from the cell establishes an *inside–out* patch. Application of a suction pulse disrupts the membrane patch, allowing electrical and diffusional access to the cell interior for *whole-cell* recording. Pulling away from the whole-cell arrangement causes the membrane to re-form into an *outside–out* configuration. From Hamill OP, et al. 1981. Improved patch clamp techniques for high resolution current recording from cells and cell-free membrane patches. *Pflugers Arch* **391**:85–100.

the patch, this size could contain one or only a few channels (or none at all). Application of a constant transmembrane potential permits the study of a single-channel response.

A momentarily elevated suction will rupture the membrane across the pipette (without destroying the seal), yielding the *whole-cell* configuration. In this situation the entire intracellular space is accessible via a low resistance to the patch electrode. It can be shown that the intracellular space is essentially isopotential. Consequently, currents introduced through the pipette flow uniformly across the entire cell while the pipette potential is the same as that at all points on the intracellular cell surface. The macroscopic behavior of the whole cell is examined in this arrangement; small cells in the diameter range of 5–20 μm can only be measured this way.

If, after establishing a gigaseal, the pipette is quickly withdrawn, then a patch of membrane will be found still in contact with the mouth of the pipette. The pipette may then be readily placed in solutions of arbitrary composition and the resulting transmembrane potentials and currents measured. In this arrangement the inside (cytoplasm side) of the membrane is in contact with the bathing solution (i.e., the extracellular or outside); the arrangement is called *inside–out*.

If the pipette is pulled away while first in the whole-cell configuration, the membrane will reform in an *outside–out* configuration, that is, in this case, the outside (extracellular surface) of the cell membrane now faces the outside of the micropipette (i.e., the bathing solution). These comments are also illustrated in Figure 4.6.

4.4.4. Single-Channel Currents

Examination of patch-clamp current reveals discontinuities that directly reflect the opening and closing of channel gates. Thus the concept of gated channels is supported by these experiments and supplements the evidence of gated pores found in EM, x-ray diffraction, and molecular biological studies.

Typical patch-current recordings are shown in Figure 4.7. The current waveform is interpreted as reflecting the opening and closing of a single channel.

The single-channel record is seen to switch to and from an open or closed state. The time in each state varies randomly, but if the ratio of open to closed time is evaluated over a sufficiently long period, then this ratio (with some statistical variation) will be the same over any successive such interval. Such a determination gives the expected value or probability that the channel will be open (closed), a value that is independent of time under these steady-state conditions. An electric circuit representation of the single-channel current is given in Figure 4.8.

If the transmembrane potential is suddenly switched to a new value, the probability of the channel being open also will change, as will be discussed in a later section.

The single-channel behavior is the basis for the macroscopic membrane properties. While the former is statistical in nature, the summation of very large numbers of such contributions results in a continuous functional behavior. We will examine this relationship in this chapter and give further details of macroscopic membrane behavior in subsequent chapters. We will also show that, while the macroscopic properties can be found from the microscopic, to some extent

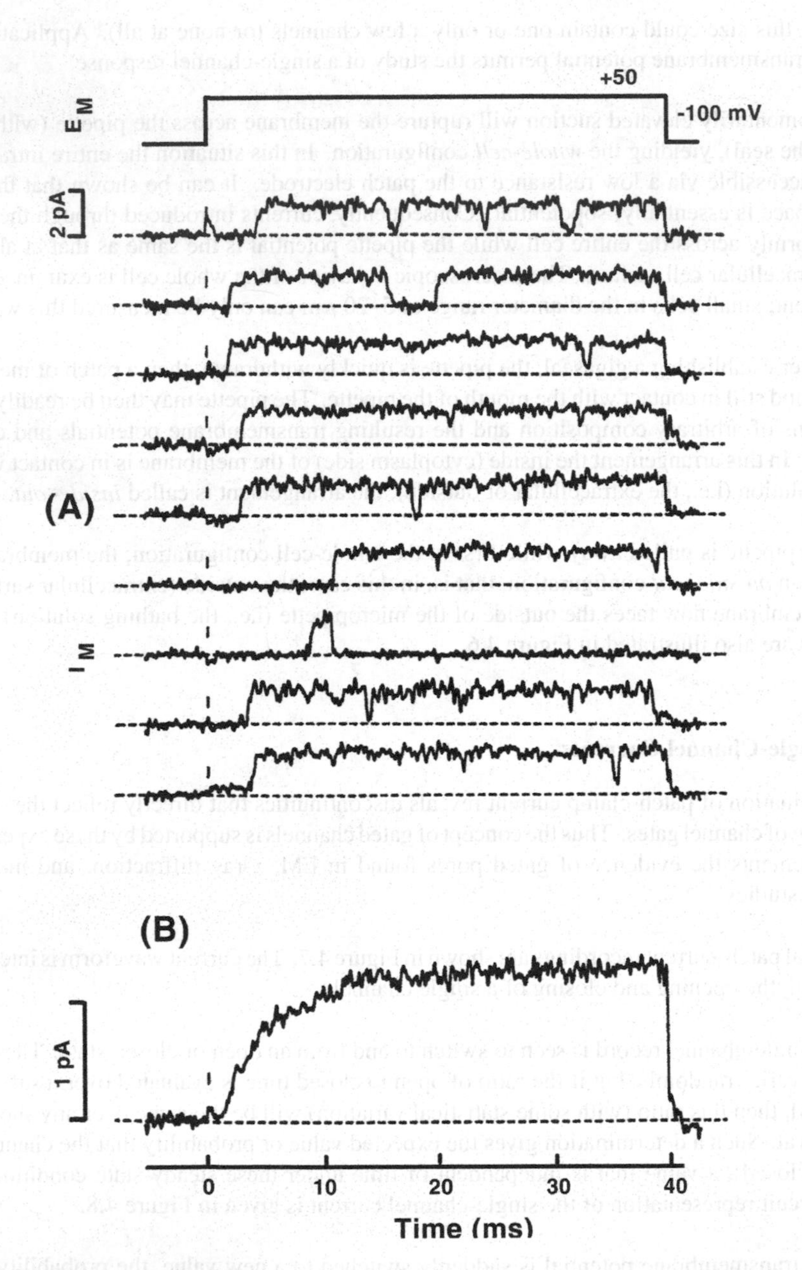

Figure 4.7. Patch-Clamp Recording of unitary K currents in a squid giant axon during a voltage step from –100 to 50 mV. To avoid the overlying Schwann cells, the axon was cut open and the patch electrode sealed against the *cytoplasmic* face of the membrane. (A) Nine consecutive trials showing channels of 20 pS conductance filtered at 2 kHz bandwidth. (B) Ensemble mean of 40 repeats; these reveal the expected macroscopic behavior. $T = 20\,^{\circ}\mathrm{C}$. From Bezanilla F, Augustine GR. 1992. In *Ionic channels of excitable membranes*, 2nd ed. Ed B Hille. Sunderland, MA: Sinauer Associates.

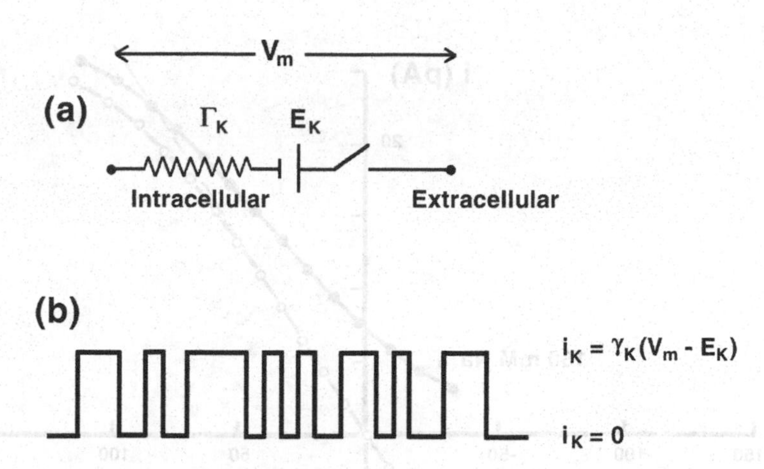

Figure 4.8. (a) Electrical circuit representation for a single (potassium) channel showing fixed resistance r_K, potassium Nernst potential E_K, and the transmembrane potential V_m. The closing and opening of the switch simulates the stochastic opening and closing of the channel gate. (b) Single-channel current corresponding to (a), where $\gamma_K = 1/r_K$. This is an idealization of the recording shown in Figure 4.7.

the inverse is also true, and we will describe how certain microscopic, single-channel statistical parameters can be found from macroscopic measurements.

4.4.5. Single-Channel Conductance

If a channel behaves ohmically, then in (4.1) its conductance γ is expected to be a constant. In an experiment by Yellen [4], shown in Figure 4.9, a single-channel potassium current is evaluated as a function of transmembrane voltage. For the voltage range considered we note that a fairly linear result is obtained supporting the conclusion that γ_K is constant; its value of 265 pS can be found from the slope of the dotted curve in Figure 4.9.

An estimate of the channel conductance can be obtained based on macroscopic ohmic ideas. For the channel shown in Figure 4.1, the pore diameter is on the order of 20 Å . If it is assumed that this is actually a uniform cylinder of length 150 Å (two membrane thicknesses), then the conductance evaluates to

$$\gamma = \pi(10 \times 10^{-8})^2/(250 \times 150 \times 10^{-8}) = 84 \text{ pS} \tag{4.3}$$

based on a bulk resistivity of 250 Ωcm. The resulting value compares with measured values of potassium channel conductance of 20 pS and greater, and might be considered surprisingly close. In this calculation the macroscopic ohmic behavior of a uniform column of electrolyte has been applied to a channel of atomic dimensions that is also likely to be nonuniform. It ignores electrostatic forces between ions and wall charges, possible channel narrowness, etc.

In addition, one should include an access resistance from the mouth of the cylindrical channel into the open regions of intracellular and extracellular space. This resistance (using an idealized model) is given by $\rho/(2a)$. Including this resistance reduces the channel conductance from 84

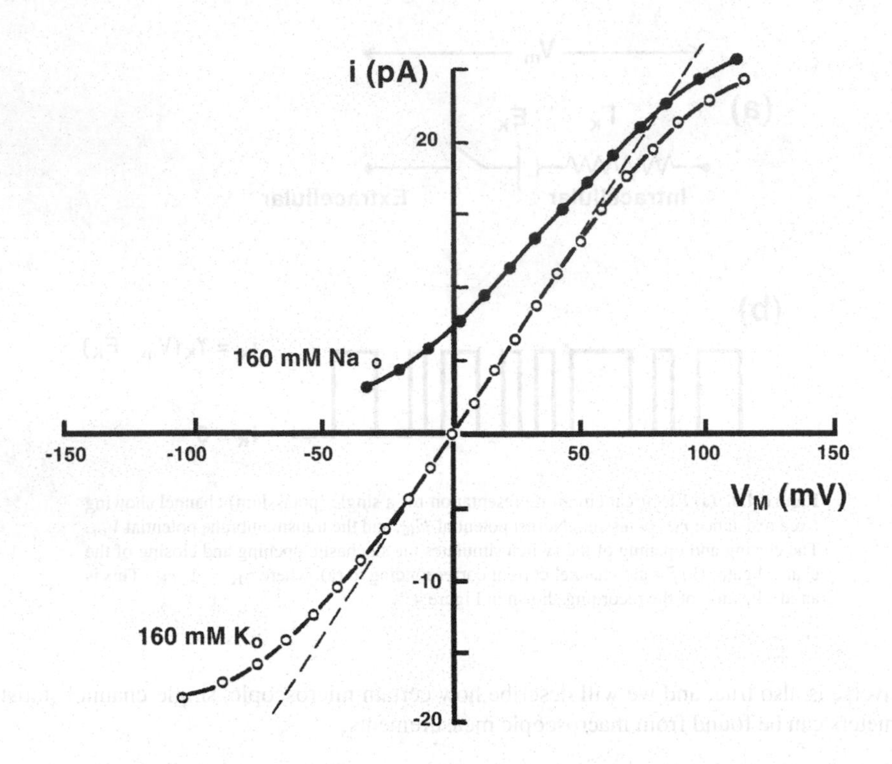

Figure 4.9. Current–Voltage Relations for a single BK K(Ca) channel of bovine chromaffin cell. The excised outside–out patch was bathed in 160 mM KCl or NaCl and the patch pipette contained 160 mM KCl. In symmetrical K solutions the slope of the dashed line is $\gamma = 265$ pS; $T = 23\,^{\circ}$C. From Hille B. 1992. *Ionic channels of excitable membranes*, 2nd ed. Sunderland, MA: Sinauer Associates. Based on measurements of Yellen G. 1984. Ionic permeation and blockade in Ca^{2+}-activated K^{+} channels of bovine chromaffin cells. *J Gen Physiol* **84**:157–186.

[as found in (4.3)] to 76 pS. In comparison, some measured channel conductances are given in Table 4.1 for sodium and potassium.[2]

Table 4.1 also shows the measured channel density, obtained using one of two methods. In one, the macroscopic conductance of a whole cell was measured and then divided by the surface area of the cell and by the single-channel conductance. The result is the number of channels per unit area.

A second approach is based on gating-current measurements, and this approach will be described subsequently. The result as seen in Table 4.1 might be thought to give a sparse density of channels. This conclusion would be reached on Hille's [1] estimate that perhaps 40,000 channels could be physically accommodated in a 1 μm^2 membrane area.

But another reference value for channel density comes from an evaluation of the quantity of charge needed to change the transmembrane potential by 100 mV. Assuming $C = 1\,\mu F/cm^2$ and a voltage change of 100 mV, then from $Q = CV$, one obtains 10^{-7} Coulombs/cm^2 or 10^{-15}

Table 4.1. Conductance of Sodium and Potassium Channels

Preparation	γ	Channels
–	(pS)	(number/μm^2)
Sodium		
Squid giant axon	4	330
Frog node	6–8	400–2000
Rat node	14.5	700
Bovine chromaffin	17	1.5–10
Potassium		
Squid giant axon	12	30
Frog node	2.7–4.6	570–960
Frog skeletal	15	30
Mammalian BK	130–240	—

Coulombs per μm^2. Multiplication by Avogadro's number and division by the Faraday results in 6200 monovalent ions required per μm^2. One channel carrying 1 pA of current moves that many ions in 1 msec. Thus the "sparse" channel density in Table 4.1 is also several orders greater than an absolute minimum.

4.4.6. Channel Gating

We have mentioned that inactivation of the *shaker* K^+ channel is accomplished with a ball and chain configuration (Figure 4.10).

In the study of the ion channel colicin, a radical reconformation accompanies channel opening and closing, described by a "swinging gate" model [3]. In the absence of detailed information on channel protein structure, gating mechanisms require a degree of guesswork.

In any case it is clear that, for voltage gated channels, the influence of a transmembrane potential on a gate is through the force exerted on charged particles by the electric field within the membrane (associated with the gate) in the protein channel. While the total distribution of charges in the macromolecule must be zero, we can have local net charge, though charges are probably organized as dipoles. An adequate force exerted by an electric field will result in a conformational change in which the channel state is switched from closed to open (or vice versa).

At the same time, the charge (dipole) movement contributes to the capacitive current (in much the same way as a dielectric displacement current arises from molecular charge movement or dipole orientation). Such currents are called *gating currents*. Since they saturate at large enough fields, they can be separated from the remaining (non-saturating) capacitive current, which is linearly dependent on $\partial V_m/\partial t$.

Suppose we assume that the energy required to open a closed channel is supplied through the movement of a charge $Q_g = z_q q_e$ through the transmembrane potential V_m, where z_q is the

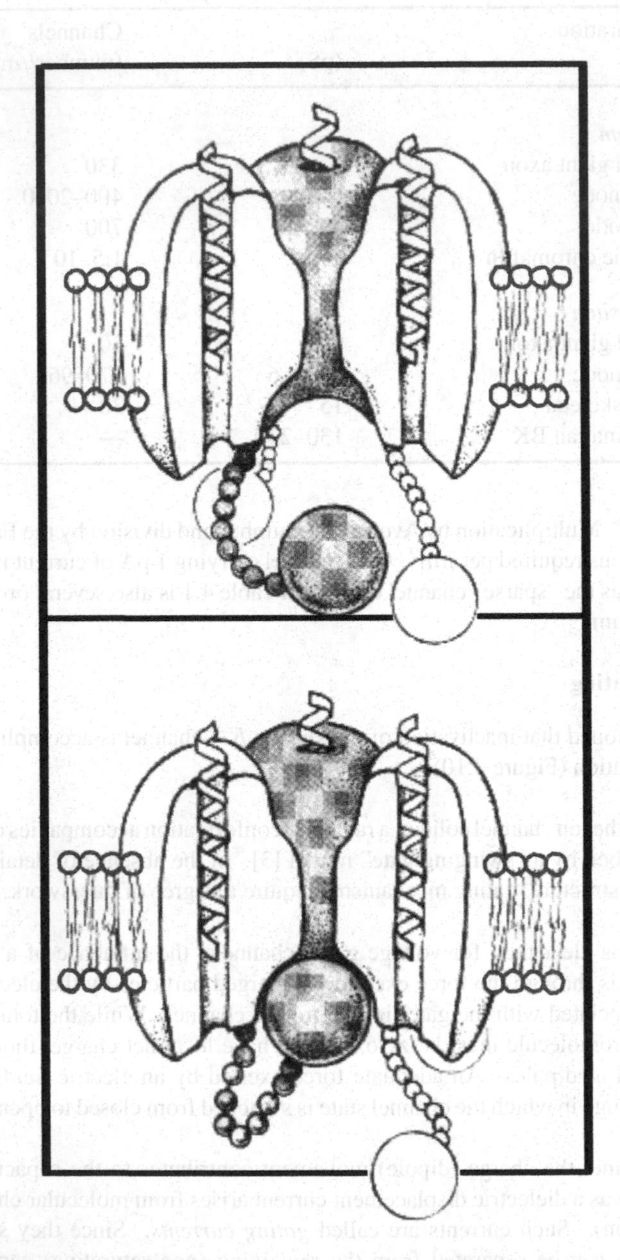

Figure 4.10. A protein ball pops into a pore formed by the bases of four membrane-spanning proteins (one not shown), thereby stopping the flow of potassium ions out of a nerve cell. Based on Hoshi T, Zagotta WW, Aldrich RW. 1990. Biophysical and molecular mechanisms of *Shaker* potassium channel inactivation. *Science* **250**:506–507, 533–538, 568–571.

valence and q_e is the charge. Then Boltzmann's equation expresses the ratio of open to closed channels as

$$\frac{[\text{open}]}{[\text{closed}]} = \exp\left(-\frac{w - z_q q_e V_m}{kT}\right) \tag{4.4}$$

The fraction of open channels is therefore

$$\frac{[\text{open}]}{[\text{open} + \text{closed}]} = \frac{1}{1 + \exp[(w - z_q q_e V_m)/kT]} \tag{4.5}$$

In (4.4) and (4.5) w is the energy required to open the channel when the membrane potential is zero, i.e., with $V_m = 0$, and k is Boltzman's constant. (Recall that the gas constant $R = k/q_e$.)

A plot of (4.5) as a function of V_m for different values of Q_g can be compared with the macroscopic dependence of ionic conductance, as a function of V_m, and in this way Q_g can be estimated [1]. Good fits are achieved, but the model is very simple and the interpretation uncertain.

One complication is that the charged particles may not move across the entire membrane (i.e., through the entire voltage V_m). Another complication is that the transitions may not be smooth but take place in steps. (Step behavior is what is believed to occur.) Moreover, if the force mechanism involves dipole rotation and translation, the energy calculation will necessarily be different from that assumed in (4.4). Nonetheless, the equations are a starting point.

4.5. MACROSCOPIC CHANNEL KINETICS

The membrane functions by changing the number of open channels in response to a changing transmembrane voltage, as well as time. It must do so fast enough to allow eye blinks and escape from predators, but slow enough that the process does not become uncoordinated or out of control. What equations describe the average number of open channels? What equations describe the rates of change of the average number if the transmembrane voltage shifts from one value to another? The equations of macroscopic channel kinetics address these questions.

We consider a large membrane area containing N channels of a given ion type. We assume that each channel's behavior is independent, though governed by similar statistics. We further assume that each channel is either in an open or closed state and that the transition between these states is stochastic. Let the number of closed and open channels *at any instant* be $N_c(t)$ and $N_o(t)$, respectively, where N_c and N_o are random variables; then

$$N = N_c(t) + N_o(t) \tag{4.6}$$

We assume state transitions to follow first-order rate processes. If the rate constant for switching from a closed to an open state is α while that for switching from an open to a closed state is β, then the average behavior is described by

$$N_c \underset{\beta}{\overset{\alpha}{\rightleftharpoons}} N_o \tag{4.7}$$

Based on experience with the measurements of Hodgkin and Huxley (to be described in Chapter 5), we expect α and β to depend on the transmembrane potential (only) and therefore to be constant when the potential is fixed (as assumed at this point).

Based on the relation given in (4.7), we have

$$\frac{dN_c}{dt} = \beta N_o - \alpha N_c \tag{4.8}$$

and similarly

$$\frac{dN_o}{dt} = \alpha N_c - \beta N_o \tag{4.9}$$

If (4.6) is substituted into (4.9); then, after rearranging terms, one has

$$\frac{dN_o}{dt} + (\alpha+\beta)N_o = \alpha N \tag{4.10}$$

The solution of (4.10) is

$$N_o(t) = Ae^{-(\alpha+\beta)t} + \frac{\alpha}{\alpha+\beta}N \tag{4.11}$$

Equation (4.11) is important because it shows how the number of open channels can be determined after, for example, the voltage transition in a voltage clamp. The equation gives the solutions for a time immediately after the voltage change, a long time after the change, or at any time in between.

In this regard, in (4.11) constant A has to be determined by the boundary conditions. Here the boundary condition is the number of open channels at $t = 0$. For a voltage clamp, that would be the number of open channels existing just before the clamp voltage was set to a new value.

The implications of (4.11) can be seen by considering what happens if a voltage step is introduced at $t = 0$. The immediate result of the voltage change will be that α and β switch to new values. Equation (4.11) describes what happens thereafter. If, for example, at $t = 0$ all channels were closed, then from (4.11) we have (for $t > 0$)

$$N_o(t) = \frac{\alpha}{\alpha+\beta}N(1 - e^{-(\alpha+\beta)t}) \tag{4.12}$$

Let $N_o(\infty)$ be the probable number of open channels after a sufficiently long time. ("Sufficiently long time" means long enough for the negative exponential term in (4.12) to go to zero, compared to other terms.) Then consider again the situation following the change to the new rate constants α and β.

From (4.12), we see that N_o moves from a value of zero to a steady-state value of

$$N_o(\infty) = \frac{\alpha}{\alpha+\beta} N \tag{4.13}$$

Thus, at steady state it is only the *average* number of open channels that is constant, as the actual number of open channels will fluctuate around this average value. The average number open depends on the α and β values that are present at the new transmembrane voltage, so after a sufficiently long time any previous history is lost.

In the new steady state, the expected (average) value is of interest, but also of interest is the fluctuation around it. One way of thinking about fluctuations is to note the following somewhat surprising fact: while obtaining (4.11) we assumed a voltage-clamp transition to have occurred.

In fact, expression (4.11) also describes the response to spontaneous fluctuations in the number of open channels. Consequently, the kinetic analysis of fluctuations reflects the same time constants as arise in classical macroscopic analysis. This correspondence is formalized in the fluctuation-dissipation theorem [2], which exhibits the broader and fundamental nature of this correspondence.

4.6. CHANNEL STATISTICS

We assume that each channel in a population of similar channels switches between open and closed states governed by the same rate constants α and β as govern the ensemble (as we discussed in the previous section).

We let C identify a closed channel and O the open channel. Then

$$C \underset{\beta}{\overset{\alpha}{\rightleftharpoons}} O \tag{4.14}$$

For example, if we have $N = 100$ channels and the probability of a channel being open is 50%, then at any instant we have an expected (i.e., steady-state) value of $N_c(\infty) = N_o(\infty) = 50$.

The number 50 is, of course, not the exact value of $N_c(t)$ or $N_o(t)$ at any t, no matter how large. The reason that 50 is not the exact value is because the actual values will fluctuate around 50, over time, because the underlying channel behavior is random and 50 is only the average.

From (4.13) we found that $N_o(\infty)$, the expected number of open channels under steady-state conditions, equals $[\alpha/(\alpha+\beta)]N$. In view of (4.6), we deduce that $N_c(\infty) = [\beta/(\alpha+\beta)]N$. Consequently, for this example, setting $\alpha = \beta = 1$ describes both the ensemble as well as the single channel.

Suppose the total number of channels is N and under steady-state conditions an *average number* $\langle N_o \rangle$ are open and an *average number* $\langle N_c \rangle$ are closed. Under these circumstances the probability, p, of a single channel being open is

$$p = \langle N_o \rangle / N \tag{4.15}$$

Conversely, the probability, q, of a channel being closed is given by the ratio

$$q = \langle N_c \rangle / N \tag{4.16}$$

Since $N = N_o(t) + N_c(t) = \langle N_o \rangle + \langle N_c \rangle$, then

$$p + q = 1 \tag{4.17}$$

The probability of *exactly* N_o channels being open can be found by evaluating the probability of a specific qualifying distribution (i.e., $p^{N_o} q^{N-N_o}$) multiplied by the number of different ways in which that distribution can occur (i.e., which of the exactly N_o channels are open and the remainder closed).

The latter number is given by $N!/[N_o!(N - N_o)!]$, arrived at by recognizing that $N!$ is the total number of rearrangements of N completely different channels. However, interchanging open channels among themselves, $N_o!$ or closed channels among themselves, $(N - N_o)!$ are indistinguishable rearrangements. Such indistinguishable rearrangements are divided out.

Thus, the probability of exactly N_o open channels out of N total channels [which we denote by $B_N(N_o)$] is

$$B_N(N_o) = \frac{N!}{N_o!(N - N_o)!} p^{N_o} q^{N-N_o} \tag{4.18}$$

The distribution (4.18) is given the name *Bernoulli*.

With p and q defined as above and for an arbitrary well-behaved variable y, the following relationship follows from the binomial theorem:

$$(yp + q)^N = \sum_{N_o=0}^{N} B_N(N_o) y^{N_o} \tag{4.19}$$

Equation (4.19) can be confirmed by writing out the series expansion for the left-hand side. The first terms are

$$(yp)^N + N(yp)^{N-1} q + \frac{N(N-1)}{2!} (yp)^{N-2} q^2 + \cdots \tag{4.20}$$

which, using (4.18), can be seen to correspond correctly.

By taking the derivatives of both sides of (4.19) with respect to y one obtains

$$Np(yp + q)^{N-1} = \sum_{N_o=0}^{N} N_o B_N(N_o) y^{N_o - 1}$$

and for $y = 1$ (4.20) gives

$$Np = \sum_{N_o=0}^{N} N_o B_N(N_o) \tag{4.21}$$

Since $B_N(N_o)$ is the probability of N_o, its product with N_o summed over all values of N_o corresponds to the definition of the average value of N_o (i.e., $\langle N_o \rangle$). Consequently

$$pN = \langle N_o \rangle \tag{4.22}$$

Here pN, the probability of a channel being open times the number of channels, is also recognized as the expected (average) number of open channels, hence confirming (4.22).

If the second derivative of (4.19) is taken with respect to y, and y set equal to unity, then one gets

$$\begin{aligned} N(N-1)p^2 &= \sum_{N_o=0}^{N} N_o(N_o-1)B_N(N_o) \\ &= \sum_{N_o=0}^{N} N_o^2 B_N(N_o) - \sum_{N_o=0}^{N} N_o B_N(N_o) \end{aligned} \tag{4.23}$$

The first term on the right-hand side is the second moment of the distribution of N_o, designated $\langle N_o^2 \rangle$.

Using (4.22) permits (4.23) to be written as

$$\langle N_o \rangle^2 - Np^2 = \langle N_o \rangle^2 - \langle N_o \rangle \tag{4.24}$$

Rearranging terms yields

$$Np(1-p) = \langle N_o^2 \rangle - \langle N_o \rangle^2 \tag{4.25}$$

The right-hand side of (4.25) is the variance of N_o, or σ^2. So we have

$$\sigma^2 = Np(1-p) \tag{4.26}$$

The importance of the variance is that it is a measure of the deviation around the average. In the case of channels, it is important to know not only the average number of channels open (or closed) but also much deviation from the average can occur, and how often. This information is provided by the variance and by its square root, the standard deviation. Both are widely used, as the variance tends to be most convenient in mathematical expressions, but the standard deviation is more convenient when comparing numerical values, especially if by hand.

We note that the variance in N_o equals N times the probability of a channel being open times the probability of the channel being closed. This important expression relates the macroscopic quantity σ^2 to the single-channel, microscopic parameter p.

As an illustration, if the aforementioned channels were all potassium channels then the individual open-channel current is

$$i_k = \gamma_K(V_m - E_K) \tag{4.27}$$

as explained in (4.1).

For N channels with p being the probability of a channel being open, the macroscopic current I_K is given by

$$I_K = Np\gamma_K(V_m - E_K) \tag{4.28}$$

Looking at the coefficient, one sees that the macroscopic membrane conductance G_K is given by

$$G_K = Np\gamma_K \tag{4.29}$$

The connection between macroscopic conductance and that of individual channel conductances is made explicit in (4.29).

4.7. THE HODGKIN–HUXLEY MEMBRANE MODEL

Hodgkin and Huxley showed that the total membrane current could be found as the sum of the currents of individual ions. Their mathematical model is presented in Chapter 5. In that chapter we shall review the extensive measurements made on the squid giant axon and the mathematical model developed to simulate that behavior. That work was published in the early 1950s and much has been learned about the underlying single-channel properties since then.

In this section we seek an application of single-channel behavior that leads to the macroscopic behavior that will later be included in the overall Hodgkin–Huxley model. However, the transition from microscopic to macroscopic is still not fully completed, so that at this time one must be guided by the expected macroscopic result.

For the potassium channel, Hodgkin and Huxley assumed that it would be open only if four independent subunits of the channels (which they called "particles") had moved from a closed to an open position. Letting n be the probability that such a particle is in the "open" position, then

$$p_K = n^4 \tag{4.30}$$

is the probability p_K of the potassium channel being open. [As a matter of notation, observe that n in (4.30) is the probability of a potassium particle being in the open state and thus is **not** the number of channels N, as used earlier in this chapter.]

The movement of the particle from closed to open was assumed to be described by a first-order process with rate constant α_n, while the rate constant for going from open to closed is β_n. Consequently,

$$\frac{dn}{dt} = \alpha_n(1 - n) - \beta_n n \tag{4.31}$$

where, of course, $(1 - n)$ is the probability that the particle is in a closed position.[3]

Let us first follow the temporal behavior of a single-channel subunit (particle). We assume it to be closed and investigate the possibility that it will open. To do this we now consider an ensemble of a large number of such closed subunits, all of which are assumed independent.

Then at the moment at which a constant voltage is applied, when α assumes a constant value arising from that voltage, we have for the ensemble

$$\frac{dO}{dt} = \alpha C_N \tag{4.32}$$

where O is the number of particles that switch to the open position, and C_N is the total number of subunits, all initially closed, in the ensemble.

Dividing through (4.32) by C_N gives

$$\frac{\Delta O}{C_N} = \alpha \Delta t \tag{4.33}$$

But $\Delta O/C_N$ is the probability that any closed subunit will open in the Δt interval. We can generalize this so that if a subunit is closed at $t = t_1$, then the probability that it will open by $t = t_1 + \Delta t$ equals $\alpha \Delta t$. Conversely, when the subunit is in the open position, the probability that it will close in the interval Δt is $\beta \Delta t$.

The potassium channel as a whole has several subunits. Specifically, four "*n*" subunits must be in the open position for the channel to be open. The probability of an open potassium channel is thus given by $p = n^4$.

The maximum conductance of *N* potassium channels occurs when they are all open and is

$$\bar{g}_K = N\gamma_K \tag{4.34}$$

where *N* is the number of potassium channels per unit area of membrane, so that $\bar{g}_K$ is a specific conductance.

For large *N*, where expected values can be assumed,

$$g_K = \bar{g}_K n^4 = N\gamma_K n^4 \tag{4.35}$$

If at $t = 0$ a steady voltage (i.e., "voltage clamp") is applied for which the related α_K and β_K are constant, then we can solve (4.31) to give

$$n(t) = n_\infty - (n_\infty - n_o)e^{-t/\tau_n} \tag{4.36}$$

where

$$\tau_n = 1/(\alpha_{K+\beta K}) \quad \text{and} \quad n_\infty = \alpha_K/(\alpha_{K+\beta K}) \tag{4.37}$$

Equations (4.36) and (4.37) describe the temporal behavior of the probability function describing the probability of a subunit being in the open position. It also gives the fraction of all subunits that are expected to be in the open position. But *n* is a random variable and while $\langle n \rangle$ gives its expected (average) value, its actual value will be different. Hodgkin and Huxley will be seen to treat a very large ensemble associated with their macroscopic measurements, in which case

n can be appropriately considered to be a real variable. However, whatever is learned from the macroscopic model can, by a reinterpretation of the meaning of n, be applied to a single channel.

Thus the potassium current through the open channels becomes

$$I_K = g_K(V_m - E_K) \tag{4.38}$$

These comments may be readily extended to sodium, calcium, and other channels. For sodium current, the fundamental change is that the sodium channel is seen as controlled by three particles of type m and one of type h. Thus the probability that a sodium channel is open becomes

$$p_{\text{Na}} = m^3 h \tag{4.39}$$

With p_{Na} so defined, the conductivity for sodium ions has an analogous form to (4.35), namely,

$$g_{\text{Na}} = \bar{g}_{\text{Na}} m^3 h = N \gamma_{\text{Na}} m^3 h \tag{4.40}$$

and the equation for the current from sodium ions is likewise analogous

$$I_{\text{Na}} = g_{\text{Na}}(V_m - E_{\text{Na}}) \tag{4.41}$$

The above equations arise naturally from the understanding of channels as structures within the membrane. Historically, however, these equations originated from observations of current flow across larger segments of tissue, as presented in Chapter 5. It is to the credit of both the earlier and the more recent investigators that there is such a remarkable compatibility of understanding as seen now from both smaller and larger size scales.

4.8. NOTES

1. A mutant of *Drosophila* characterized by shaking; a consequence of the mutation is abnormal inactivation.
2. Data in Table 4.1 come from Hille B. 1992. *Ionic channels of excitable membranes*, 2nd ed. Sunderland, MA: Sinauer Associates, and were based on data from published measurements.
3. Note that n is a continuous variable and hence "threshold" is not seen in a single channel. Threshold is a feature of the macroscopic membrane with, say, potassium, sodium, and other channels, and describes the condition where the collective behavior allows a regenerative process to be initiated that constitutes the upstroke of an action potential. This topic will be developed in Chapter 5.

4.9. REFERENCES

1. Hille B. 2001. *Ion Channels of excitable membranes*, 3rd ed. Sunderland, MA: Sinauer Associates.
2. Kubo R. 1966. Fluctuation dissipation theorem. *Rep Prog Phys London* **29**:255.
3. Simon S. 1994. Enter the "swinging gate." *Nature* **371**:103–104. See also Slatin SL, Qiu XQ, Jakes KS, Finkelstein A. 1994. Identification of a translocated protein segment in a voltage-dependent channel. *Nature* **371**:158–161.
4. Yellen G. 1984. Ionic permeation and blockade in Ca^{2+}-activated K^+ channels in bovine chromaffin cells. *J Gen Physiol* **84**:157–186.

Additional References

De Felice LJ. 1981. *Introduction to membrane noise*. New York: Plenum.

De Felice LJ. 1997. *Electrical properties of cells*. New York: Plenum.

Lodish H, Darnell JE, Edwin J. 1995. *Molecular cell biology*, 3rd ed. New York: Scientific American Books.

Sakmann B, Neher E, eds. 1995. *Single channel recording*, 2nd ed. New York: Plenum.

5

ACTION POTENTIALS

Biological membranes contain a large number of several types of ion channels; these interact through their common transmembrane potential and capacitance. A remarkable result is that such electrically active tissue, by means of a regenerative process, can generate a transient pulse of electrical changes, an action potential, across the cell membrane. The action potential cycle consists of a rapid membrane depolarization (i.e., an increase in transmembrane potential) followed by a slower recovery to resting conditions. Once an action potential is initiated at one site on an extensive membrane, it initiates action potentials at adjacent sites, thus leading to a sequence of action potentials throughout the remaining membrane.

A simple cellular electrophysiological model is that shown in Figure 5.1. Here the cell membrane separates the extracellular and intracellular spaces. Both regions may be idealized as passive and uniformly conducting (though with different conductivities). If an adequate stimulating current is passed between a pair of electrodes across the membrane of the cell, a remarkable series of events ensues, which may be observed by recording the transmembrane voltage across the membrane as a function of time.

Specifically, concurrent with the stimulus current, V_m shows a small direct response. After a short latency a much larger and more energetic second deflection occurs, an *action potential.* The action potential is a consequence of the stimulus, but it is generated by the charged energy stored in the concentration differences that exist across the excitable membrane. The action potential is generated by the membrane's utilizing this stored energy to allow first the flow of sodium ions (to move the voltage up) and then the flow of potassium ions (to move the voltage down). When one examines this phenomenon in detail, it is seen to consisted of a series of remarkably complex events. This chapter is devoted to a quantitative examination of these observations.

Action potentials are nonlinear. If the stimulus current of Figure 5.1 is reduced by half, then no action potential occurs. Conversely, if the stimulus current is doubled, the action potential deflection remains largely unchanged, but the latency is markedly reduced.

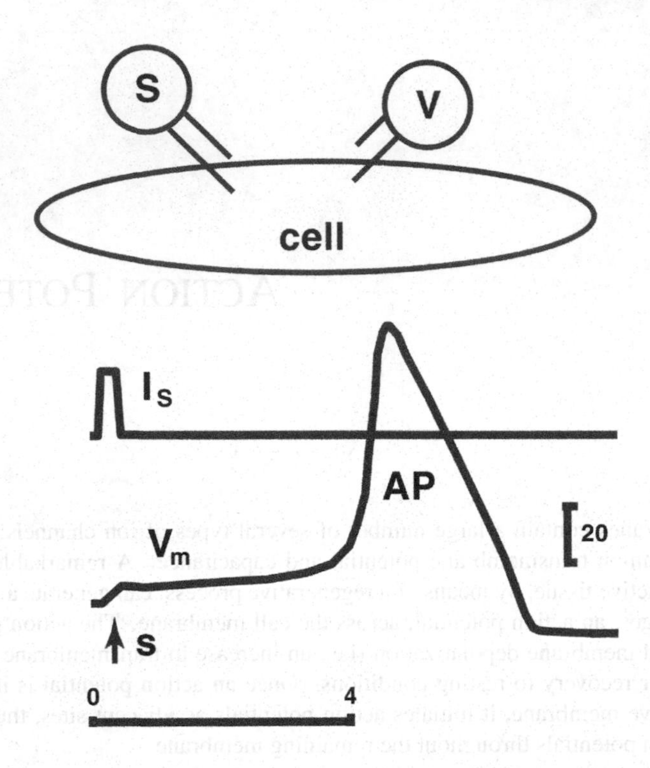

Figure 5.1. Electrical Stimulation of an Excitable Cell. The stimulus elicits an action potential. The top drawing shows a cartoon of a cell with a stimulator S and a voltmeter V attached. The stimulator injects current into the intracellular volume (positive electrode) and removes current from the surrounding extracellular volume. The voltmeter measures transmembrane voltage as a function of time. It is assumed that regions within and around the cell are equipotential, so that a uniform voltage difference exists across all points on the cell membrane. The upper trace shows a stimulus current, I_S, which delivers a short current pulse of 0.3-msec duration, beginning at $t = 0$. The lower trace shows the voltage record, $V_m(t)$. The deflections on the voltage trace are first the direct response to the stimulus, identified with s, and, 4 msec later, a much larger deflection, the action potential (marked AP). The vertical calibration corresponds to 20 μA/cm^2 on the current plot and 20 mV on the voltage plot.

This chapter considers action potentials from several perspectives, while remaining centered on the Hodgkin–Huxley model for membrane current action potentials. One might think of this chapter as having three major divisions. In the first, which includes Sections 5.1 and 5.2, we summarize experimental findings and show how these findings were placed in a quantitative, equation-based framework. In the second division of the chapter (Sections 5.3 and 5.4), we describe the elements of the Hodgkin–Huxley mathematical model, both in terms of equations and in terms of the sequence of steps required to perform HH calculations numerically. The last major division of this chapter (Section 5.5 and thereafter) gives some of the major extensions to the Hodgkin–Huxley framework. Each of these major divisions of this chapter can be understood largely independently of the other, but of course they are tightly linked, in terms of history and concept. The chapter's sections are:[1]

- The first section describes critical experimental observations of action potentials, including those establishing the different ionic compositions present inside and outside an active membrane.

- The second describes the voltage-clamp experimental setup. (The voltage-clamp procedure now has evolved into a conceptual framework as well as experimental platform.) This framework is used to tease out the flow of potassium and sodium ions, selectively as well as overlapping in time. Such flow creates action potential; voltage-clamp experimental data give these events quantitative form.

- The third section develops the mathematical description of membrane ionic conductances. This mathematical description quantifies the selective flow of ions across excitable membrane. The resulting equations are critical links between experimental observations and the general Hodgkin–Huxley mathematical model of transmembrane potentials and currents. One subsection points out some issues of notation for transmembrane potentials and units for transmembrane conductance and current. This notation is used in this section and all later ones.

- The fourth section integrates the ionic current equations into an overall mathematical model for membrane currents and potentials. It also addresses, from a more modern perspective, the sequence of steps that are the basis for simulations of membrane action potentials, and that serve as a foundation for many other computer simulations of electrically active tissue.

- In the fifth section, several topics are presented that are important extensions of the Hodgkin–Huxley theoretical base, including changes with temperature, calcium ion current, and another framework for including ionic pumps needed to recharge the membranes ionic balance.

- This chapter's appendix provides a full derivation of the GHK equation, and shows some consequences in terms of ionic flow and resting potentials.

5.1. EXPERIMENTAL ACTION POTENTIALS

The behavior of a propagating action potential on a single fiber is well illustrated utilizing the giant fibers of the nerve cord of the earthworm. As illustrated in Figure 5.2, individual fibers can he teased out and a single fiber identified at its proximal and distal end.

The fiber can be studied by placing a pair of stimulating electrodes at one end and a pair of recording electrodes at the other. The stimulating electrodes are designated as such since they are connected to a source capable of supplying a current pulse. The recording electrodes, on the other hand, connect to an amplifier and display device (oscilloscope or computer).

The fibers in this preparation consist of one median and two (smaller and equal) lateral fibers, each of which can be thought of as uniform and continuous. (Actually, each fiber consists of multicellular coupled units, but functionally each behaves as if it were cylindrical and bounded by a continuous excitable membrane.)

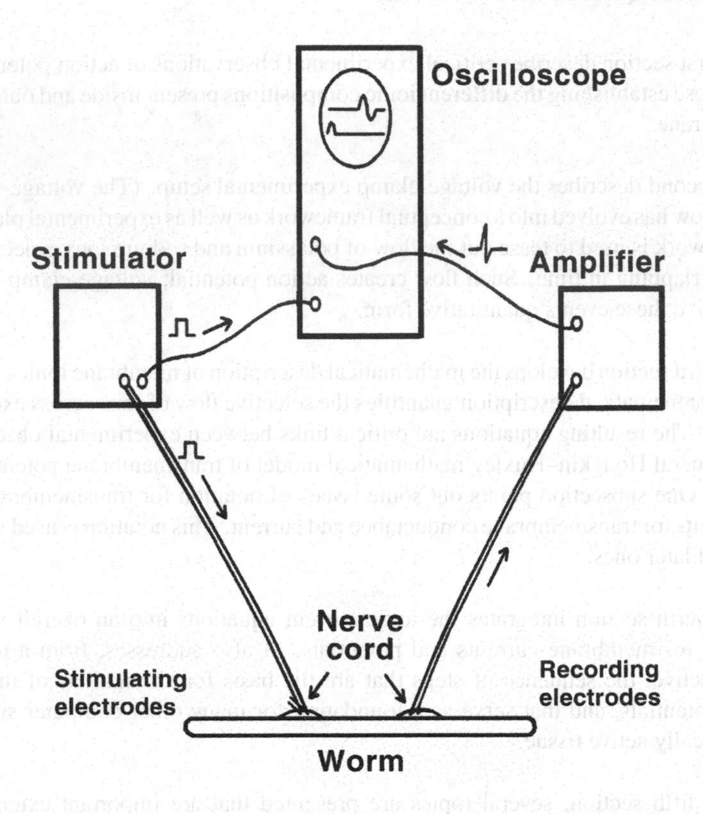

Figure 5.2. Arrangement for Recording Action Potentials from the giant fibers in the nerve cord of the earthworm. From Aidley DJ. 1978. *The physiology of excitable cells.* Cambridge: Cambridge UP. Reprinted with the permission of Cambridge University Press.

5.1.1. Noteworthy AP Attributes

Action potentials have a number of noteworthy characteristics. These unusual characteristics include thresholding, differential response by diameter, and latency, as described in the following sections.

Threshold response

If the amplitude of the stimulus pulse is relatively small, then, as seen in Figure 5.3a, no response is detected in the recording circuit. (The response that coincides with the stimulus, known as the *stimulus artifact*, arises due to direct coupling between signal generator and recorder.) As the stimulus strength is increased, a point is reached (Figure 5.3b) at which a response (a nerve *action potential*) is suddenly seen. This sudden onset illustrates the phenomenon of *threshold* and reflects a discontinuity in response at a specific stimulus amplitude.[2]

As the stimulus is further increased in strength, the response seen in Figure 5.3b remains unchanged. For this reason the action potential is described as *all or none*, i.e., the action potential waveform remains unchanged at the higher level of transthreshold stimulus amplitude.

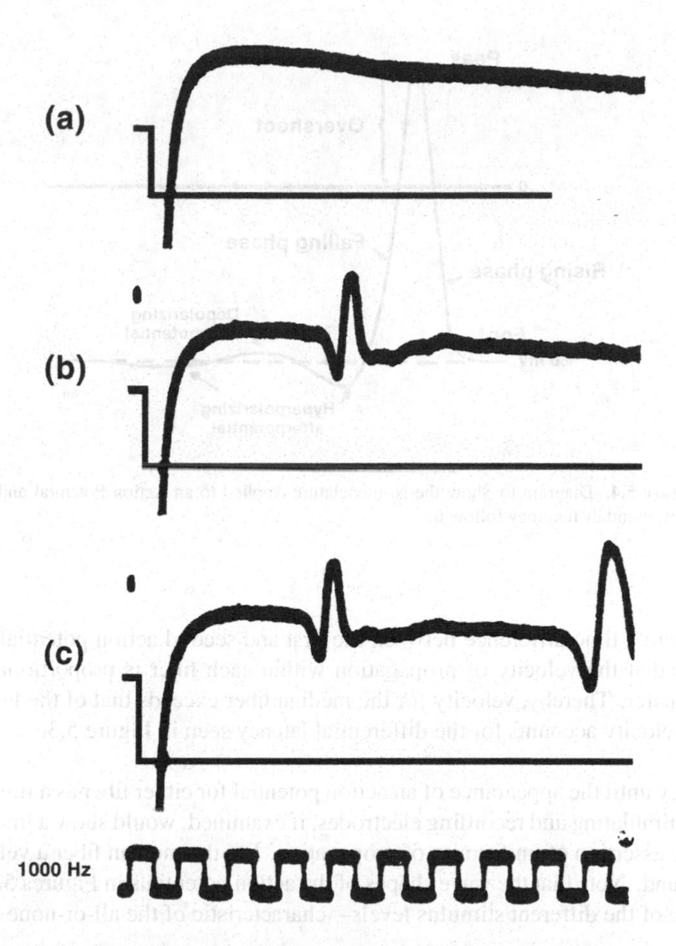

Figure 5.3. Oscilloscope Records from the Experiment Shown in Figure 5.2. In each case the upper trace is a record of the potential changes at the recording electrode and the lower trace (at a much lower amplification) monitors the stimulus pulse. From Aidley DJ. 1978. *The physiology of excitable cells*. Cambridge: Cambridge UP. Reprinted with the permission of Cambridge University Press.

Fiber diameter

When the stimulus reaches a sufficiently higher level, the threshold for a second action potential is achieved (shown in Figure 5.3c).

This second action potential can be explained as follows. The median fiber diameter is larger than the two lateral fibers. Threshold due to an external stimulating source is lower in a large fiber, roughly inversely proportional to the square root of the fiber diameter. Thereby, excitation of the median fiber occurs first, with a lower stimulus magnitude. When the stimulus magnitude increases, it will eventually be large enough to activate both median *and* lateral fibers. Thus there will be two action potentials.

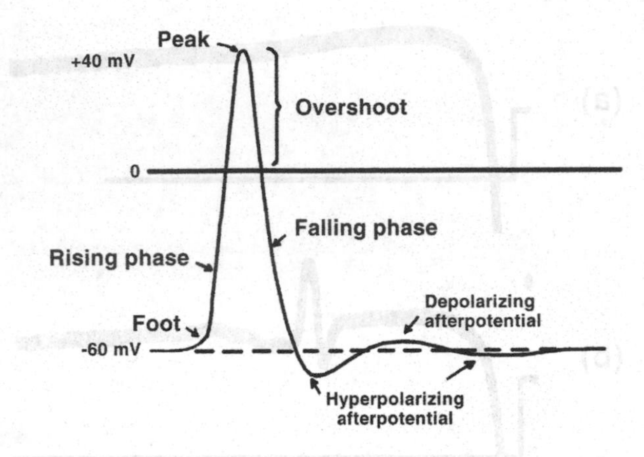

Figure 5.4. Diagram to Show the Nomenclature Applied to an Action Potential and the afterpotentials that may follow it.

Latency

Why is there a time difference between the first and second action potentials, as observed? The answer is that the velocity of propagation within each fiber is proportional to the square root of its diameter. Thereby, velocity for the median fiber exceeds that of the lateral fiber. This difference in velocity accounts for the differential latency seen in Figure 5.3c.

The latency until the appearance of an action potential for either fiber as a function of separation between stimulating and recording electrodes, if examined, would show a linear dependency, confirming the assertion of uniformity of propagation. For the median fiber a velocity of around 12 m/sec is found. Note that the wave shapes of the action potentials in Figures 5.3b and 5.3c are similar in spite of the different stimulus levels—characteristic of the all-or-none behavior.

5.1.2. Extracellular Potential Nomenclature

The action potentials recorded in Figure 5.3 typify those obtained with extracellular electrodes. It is possible, however, to place a microelectrode inside an axon and measure the intracellular versus extracellular (i.e., transmembrane) action potential. Such a measurement more nearly reflects the intrinsic membrane properties and is less dependent on the geometry of the recording electrodes and axon (as will be discussed in detail in a later chapter).

A transmembrane action potential is shown in Figure 5.4 that is typical of those observed on nerve and muscle (though with differences in some details). In all cases the membrane at rest is negative by 60 to 100 mV. The activation process causes a sudden and rapid upstroke, ending in a reversal in this potential to peak values up to +40 mV.

Following activation, a recovery phase restores the resting condition. The potential may, however, return to a more hyperpolarized or depolarized level than the resting value for a period of time. These *afterpotentials*, as illustrated in Figure 5.4, may or may not be observed; if present, usually only one or the other is seen. (If a depolarized or hyperpolarized condition arises, then reference is made to a *depolarizing or hyperpolarizing afterpotential*).

Figure 5.4 depicts the transmembrane potential as a function of time. An action potential propagating on a uniform fiber will also have an action potential as a function of distance along the fiber. Such a spatial action potential has a similar shape to the temporal one, except that it may be a mirror image and have a different horizontal scale.[3]

5.1.3. Nonlinear Membrane Behavior

As shown in Figure 5.3, for excitable membranes no response is elicited unless the stimulus reaches a specific level, called the *threshold*. For all transthreshold stimuli to a single cell, the resulting action potential is identical (all or nothing) as already noted. The threshold level may vary from membrane to membrane or as a function of the stimulus location or duration.

For a stimulating current to activate the membrane, it must be of large enough intensity and have the correct polarity. A given stimulating pulse must also have an adequate duration. The dependence on these parameters will be described later in this chapter.

Membrane response to an increasing stimulus

The transmembrane potential responses from a stimulating current pulse on a crab axon is shown in Figure 5.5, where the zero or reference potential is that at rest. The stimulating pulse duration is shown and is held fixed while the stimulus amplitude and sign is varied.

A stimulus that causes the transmembrane potential (intracellular minus extracellular potential) to be more negative than its value at rest is said to *hyperpolarize* the membrane. With hyperpolarization, there is no excitation no matter what size stimulus, though an increasing passive *(RC)* response arises from an increasing stimulus strength, as described in Figure 5.5. On the other hand, for *depolarizing* stimuli, and for increasing amplitudes, response C in Figure 5.5 is (suddenly) reached. This response shows the lower portion of an elicited action potential (this is the threshold condition).

If one examines the responses to the subthreshold pulses, it is seen that they are essentially those expected from a (passive) *RC* network. In fact, the responses in Figure 5.5 can be simulated from fixed, lumped, *RC* elements. For the hyperpolarizing condition this correspondence is exactly correct, but for depolarizing stimulation a deviation from strictly passive behavior begins to arise beyond 50–80% of threshold.

Note the presence (or absence) of a mirror image in Figure 5.5 (as in parts a and b). For *depolarization* to potentials lying between 50 and 100% of threshold, the response is not a mirror image of the hyperpolarizing response for stimuli of the same magnitude and duration. Under these conditions the depolarization response is not fully passive.

A lack of symmetry is present when a nonlinear active component of the membrane is contributing. Such a response is known as a *local response* and reflects a regenerative phenomena that will arise, though much more strongly, at threshold. The subthreshold passive behavior is called *electrotonic*.

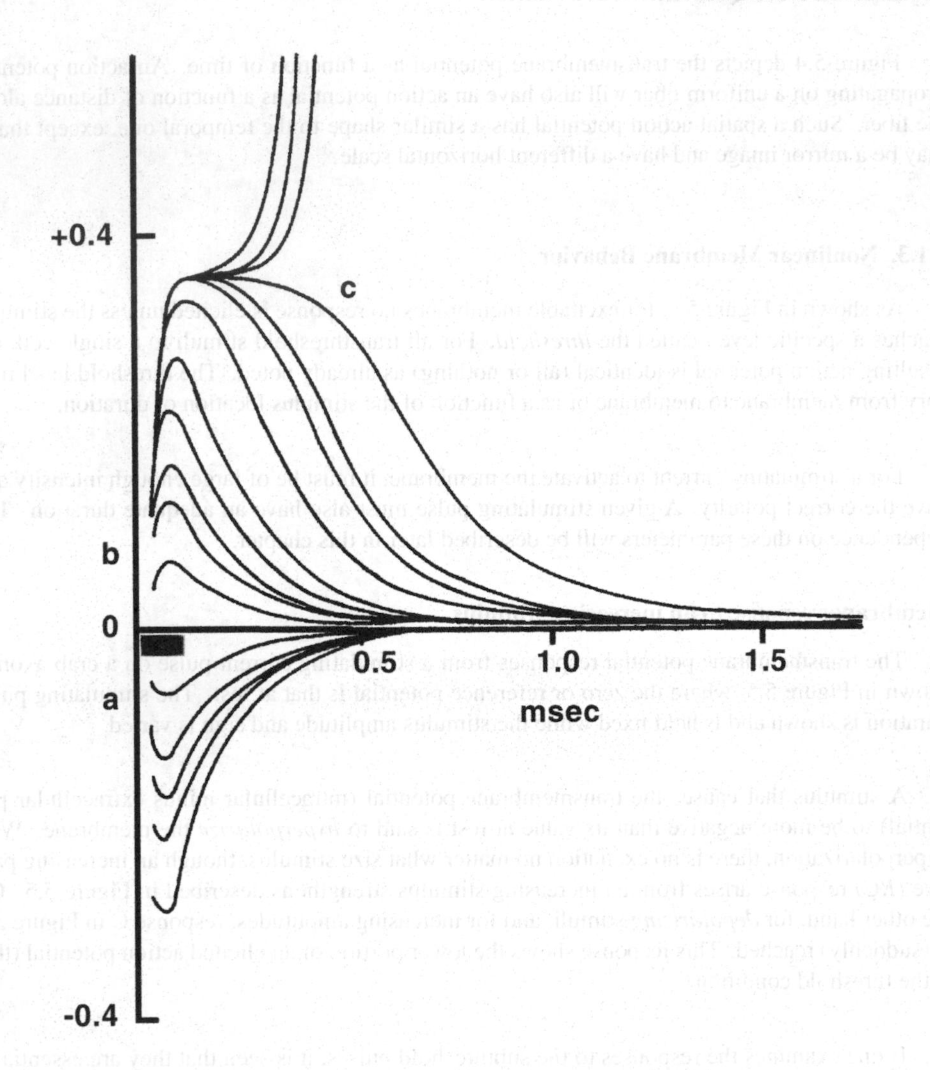

Figure 5.5. Subthreshold Responses Recorded Extracellularly from a crab axon in the vicinity of the stimulating electrodes. The axon was placed in paraffin oil, and, consequently the measured extracellular potential is directly related to the transmembrane potential (according to the linear core-conductor model described in Chapter 6). The heavy bar indicates the stimulus period, which was approximately 50 μsec in duration. The ordinate is a voltage scale on which the height of the action potential is taken as one unit. From Hodgkin AL. 1938. The subthreshold potentials in a crustacean nerve fiber. *Proc R Soc London, Ser B* 126:87–121.

Linear and nonlinear responses to stimulus

Figure 5.6 is derived from Figure 5.5; here the voltage measured at 0.29 msec following the stimulus is plotted. The voltage is expressed as a fraction of the peak action potential amplitude and is shown as a function of stimulus amplitude.

One notes that the relationship between the stimulus and the resulting transmembrane voltage is linear for all hyperpolarizing stimuli. This observation supports a linear passive model.

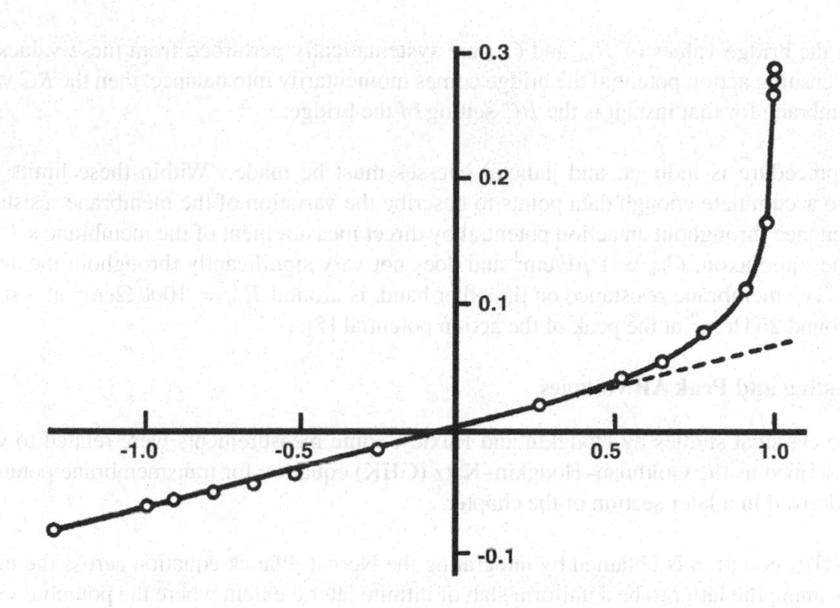

Figure 5.6. Relation between Stimulus and Response in a Crab Axon. This Figure was derived from Figure 5.5. The abscissa shows the stimulus intensity, measured as a fraction of the threshold stimulus. The ordinate shows the recorded potential 0.29 msec after the stimulus, measured as a fraction of the action-potential peak. Reprinted with permission from Hodgkin AL. 1938. The subthreshold potentials in a crustacean nerve fiber. *Proc R Soc London, Ser B* **126**:87–121.

Linearity is also seen for small depolarizing signals, suggesting that in this region the membrane can also be characterized by a passive network.

For depolarizing stimuli of greater magnitude the behavior becomes nonlinear, and an active system description is required.

Nonlinear membrane measurements

Another way of confirming the nonlinear behavior of the membrane is the following. The behavior of an excitable membrane can be explored by placing a small membrane element at one arm of a Wheatstone bridge. The Wheatstone bridge is a well-known device for making precision measurements of resistance and capacitance of a sample. It works by comparing the unknown sample to a reference sample, where the characteristics of the reference are known precisely. By choosing a high frequency ($\approx$ 1000 kHz) and a low-amplitude signal (compared to threshold), the measurement process will not cause active responses of the membrane.

The procedure is to first balance the bridge with the membrane at rest. Balancing the bridge means finding reference elements that have the same characteristics as the membrane. This step identifies the resting values of resistance, R_m, and capacitance, C_m, for the membrane. These resting values can be used when expecting linear behavior.

Then the bridge values of R_m and C_m are systematically perturbed from these values. If during an ensuing action potential the bridge comes momentarily into balance, then the RC value of the membrane for that instant is the RC setting of the bridge.

The procedure is indirect, and judicial guesses must be made. Within these limits it is possible to accumulate enough data points to describe the variation of the membrane resistance and capacitance throughout an action potential by direct measurement of the membrane's R and C. For the squid axon, $C_m \simeq 1\ \mu\text{F/cm}^2$ and does not vary significantly throughout the action potential. The membrane resistance on the other hand, is around $R_m = 1000\,\Omega\text{cm}^2$ at rest and falls to around $25\,\Omega\text{cm}^2$ at the peak of the action potential [5].

5.1.4. Resting and Peak AP Voltages

In the classical studies by Hodgkin and Huxley, some measurements were related to what is now described as the Goldman–Hodgkin–Katz (GHK) equation for transmembrane potential, which is derived in a later section of the chapter.

The GHK equation is obtained by integrating the Nernst–Planck equation across the membrane, assuming the latter to be a uniform slab of infinite lateral extent where the potential varies linearly across it (i.e., a constant electric field). It assumes, in addition, that the electric potential discontinuity across the interface between the membrane and both the intracellular and extracellular space is described by (the same) partition coefficient.

Description of the GHK equation

The result when the total ionic current is zero (i.e., at rest) is

$$V_{\text{rest}} = V_m = \frac{RT}{F} \ln \left[\frac{P_{\text{K}}[\text{K}]_e + P_{\text{Na}}[\text{Na}]_e + P_{\text{Cl}}[\text{Cl}]_i}{P_{\text{K}}[\text{K}]_i + P_{\text{Na}}[\text{Na}]_i + P_{\text{Cl}}[\text{Cl}]_e} \right] \tag{5.1}$$

where

$$P_q = D_q \beta_q / d \tag{5.2}$$

In (5.2), the following symbols are used:

- q is an index for any of the three ions; β is the partition coefficient for the qth ion ($\beta_q = [Q]^+/[Q]^-$).
- The minus superscript designates the ion concentration in the bulk (extracellular) medium just adjacent to the membrane, and the plus superscript the intracellular concentration at the membrane surface.
- The membrane thickness is d.
- D is Fick's coefficient, assumed constant.
- The P_q's are *permeability* coefficients and are seen to play a similar role in determining the resting voltage that is played by the conductances in the parallel-conductance model described in (3.31).

While (5.1) is derived from biophysical principles, in contrast to the parallel-conductance derivation, the permeabilities have in fact been found only from experimental measurements.

The GHK equations for currents are

$$J = J_{\mathrm{K}} + J_{\mathrm{Na}} + J_{\mathrm{Cl}} \tag{5.3}$$

Substituting the expressions for each specific ion enables the summation (5.3) to become

$$J = \frac{V_m F^2 P_{\mathrm{K}}}{RT} \frac{w - y e^{V_m F/RT}}{1 - e^{V_m F/RT}} \tag{5.4}$$

where

$$w = [\mathrm{K}]_e + \frac{P_{\mathrm{Na}}}{P_{\mathrm{K}}}[\mathrm{Na}]_e + \frac{P_{\mathrm{Cl}}}{P_{\mathrm{K}}}[\mathrm{Cl}]_i \tag{5.5}$$

and

$$y = [\mathrm{K}]_i + \frac{P_{\mathrm{Na}}}{P_{\mathrm{K}}}[\mathrm{Na}]_i + \frac{P_{\mathrm{Cl}}}{P_{\mathrm{K}}}[\mathrm{Cl}]_e \tag{5.6}$$

The derivations of these GHK equations for the resting potential and membrane current are given in the appendix to this chapter (Section 5.6).

GHK versus parallel conductance

In the derivation of the parallel-conductance (3.31) and constant-field (GHK) (5.1) expressions for the resting potential, the key constraining condition is that the total *ionic* transmembrane current is zero. The GHK and the parallel-conductance equations both evaluate the resting transmembrane potential as a weighted average of the sodium, potassium, and chloride Nernst potentials.

Recall that a *space-clamped* preparation is one that has been instrumented so that all membrane elements are subject to the same transmembrane potential, all the time. If we study the action potential in a space-clamped preparation, consider the total transmembrane current after the transthreshold stimulus current is over. Thereafter total membrane current (ionic plus capacitive) must necessarily equal zero, because there is no place for current to go, other than across the membrane, or to charge the membrane capacitance.

At rest and at the peak of the space-clamped action potential the total membrane current is zero (because of the space clamp), and additionally $\partial V_m/\partial t = 0$ (i.e., the capacitive current is zero), because V_m is unchanging at rest and at the peak, by definition. Consequently, the total *ionic* transmembrane current also must be zero, because the total current is the sum of its ionic and capacitative components.

Thereby, the parallel-conductance and GHK expressions both apply at rest and at peak. (At the action-potential peak, the conductivities and permeabilities are, however, different from what they are at rest.)

Rest and peak

For a membrane at rest, and assuming that only the potassium channels are carrying current, one notes that both the parallel-conductance and GHK expressions reduce to the (same) potassium

Table 5.1. Mobile Ion Concentrations for *Aplysia*

Ion	Intracellular (mM)	Extracellular (mM)	Nernst Potential (mV)
K	280	10	−83.9
Na	61	485	52.2
Cl	51	485	−56.7

Nernst potential. Similarly, if one assumes that at the action-potential peak only sodium need be considered, then both aforementioned expressions reduce to the sodium Nernst potential.

The peak of the action potential may be seen to approach the sodium Nernst potential but never exceed it. This result is consistent with an elevated sodium permeability. In Hodgkin and Katz [9] a good agreement between theory and experiment for the squid axon was demonstrated by choosing

$$P_{\mathrm{K}} : P_{\mathrm{Na}} : P_{\mathrm{Cl}} = 1.0 : 0.04 : 0.45 \quad \text{for membrane at rest}$$
$$P_{\mathrm{K}} : P_{\mathrm{Na}} : P_{\mathrm{Cl}} = 1.0 : 20.0 : 0.45 \quad \text{at an action potential peak}$$

Note that there is an enormous change—almost three orders of magnitude—in sodium permeability between these two sets of data.

We have noted that the contribution of chloride to membrane behavior is minimal and can be essentially ignored. Doing so is reasonable because chloride is close to equilibrium at rest (see the section on "Contributions from Chloride" in Chapter 3), while during the action-potential peak the chloride permeability is relatively too small to contribute significantly. As a result, to a first approximation,

$$\text{At rest}: \quad V_m \simeq E_{\mathrm{K}} = \frac{RT}{F} \ln\left(\frac{[\mathrm{K}]_o}{[\mathrm{K}]_i}\right) \tag{5.7}$$

$$\text{At the peak}: \quad V_m \simeq E_{\mathrm{Na}} = \frac{RT}{F} \ln\left(\frac{[\mathrm{Na}]_o}{[\mathrm{Na}]_i}\right) \tag{5.8}$$

Nastuk and Hodgkin [15] measured a linear variation of the peak value of V_m against the logarithm of extracellular sodium concentration, when 20 mM $< [\mathrm{Na}^+]_o <$ 200 mM. Their findings support the validity of (5.8) for this range of extracellular sodium.

The intracellular and extracellular ion composition of the *Aplysia* (sea hare mollusk) giant nerve cell is given in Table 5.1, and the corresponding Nernst potentials are shown for each. Given the relative resting permeabilities to be $P_{\mathrm{K}} : P_{\mathrm{Na}} : P_{\mathrm{Cl}} = 1.0 : 0.12 : 1.44$, the application of the GHK equation, (5.1), leads to a resting transmembrane potential of $V_m = -48.8$ mV. Comparison of this value with the Nernst potentials in Table 5.1 shows that no ion is equilibrated, though chloride is somewhat close to this condition.

For sodium, the resting potential is 107 mV from equilibrium, so that a large driving force exists. It results in a sodium influx that is small only because the sodium permeability is small. The resting potential is not negative enough to equilibrate the outward potassium diffusion. Therefore a potassium efflux results, the driving force being equal to (83.9 – 48.8 = 35.1 mV). In the steady state the potassium efflux and sodium influx are essentially equal and opposite.

The conclusion reached here regarding the behavior of the *Aplysia* neuron at rest would apply equally well to other nerve and muscle cells. A question arises as to how the normal intracellular and extracellular compositions are maintained in view of the flux movements both at rest and during action potentials. This question will be addressed subsequently when active transport is considered.

5.1.5. Movements of Ionic Tracers

Membrane ion movement at rest and following an action potential was investigated by Keynes [13]. He measured potassium and sodium ion flux through the use of radioactive tracers. Use of tracers permits a measurement of ion movement directly, i.e., a direct measurement of the particles carrying the current, rather than an indirect inference from changes in transmembrane voltage. Additionally, use of tracers allows a separate determination of influx and efflux, even if these are occurring simultaneously.

Using the cuttlefish *Sepia* giant axon, it was found that, at rest, there was a steady influx of sodium and efflux of potassium, entirely consistent with $E_K < V_m < E_{Na}$, which is seen when evaluating V_m from the GHK or parallel-conductance equation.

During an action potential the transmembrane potential's initial reversal in polarity requires an influx of positive charge (since $Q = CV_m$, a change in sign of V_m requires a similar sign change in Q). This charge is seen in the influx of 3.7 pmoles/cm^2 of sodium per action potential. A subsequent efflux of 4.3 pmoles/cm^2 of potassium per action potential accounts for the restoration of charge and transmembrane potential.

These values can be compared to the charge movement necessary to raise the transmembrane potential from rest to a peak value, something on the order of 125 mV. With this value and a membrane capacitance of $1.0\,\mu\text{F/cm}^2$, one obtains $Q = C_m V = 1.0 \times 10^{-6} \times 0.125 = 1.25 \times 10^{-7}$ Coulombs/cm^2.

Because the ions are monovalent, the number of moles that corresponds to the aforementioned charge (in Coulombs) is found by dividing by the Faraday to give $1.25 \times 10^{-7}/96500 = 1.3$ pmoles/cm^2. This corresponds in order of magnitude with the tracer values.

The use of tracers is not sensitive enough to follow ion movement as a function of time during an action potential. In fact, the aforementioned tracer data on ion movement during an action potential is based on averaged data taken over multiple action potentials. On the other hand, direct measurement of the macroscopic transmembrane current, while providing an instantaneous picture, has the complication that this current includes the capacitive component along with several ionic currents.

These facts, and the limitations following from them, were addressed by the development of voltage-clamp experiments.

5.2. VOLTAGE CLAMP

The voltage and space clamp were techniques introduced by Hodgkin and Huxley and were crucial to the task of separating the capacitive, sodium, and potassium components of membrane current in their measurements on the squid axon. Because of the importance of such clamps, the present section addresses their methodology in some detail.

The *voltage clamp* is a feedback arrangement where the transmembrane potential is held constant electronically during an action potential or subthreshold response. It was conceived as a way to eliminate the complication of the displacement current since, if $dV_m/dt = 0$, then, obviously, $C_m dV_m/dt = 0$ (i.e., the capacitive current is zero). Arranging matters so that the entire membrane under study is activated synchronously further simplifies an analysis of these measurements. This simplification comes about since the confounding effect of spatial (axial) variations of currents and potentials is eliminated.

However, even with a space and voltage clamp the separation of the ionic flux into its sodium and potassium components required an imaginative application of the Nernst–Planck equation (by Hodgkin and Huxley) in a way that will be described presently.

5.2.1. Single-Channel Studies

To the aforementioned classical measurements of the macroscopic membrane undergoing an action potential can be added the results from patch-clamp measurements. It has been pointed out that these are intrinsically "space-clamped" since the spatial extent is negligible. Voltage-clamp measurements eliminate the capacitive component in the same way as described above.

Additionally, assuming that only a single channel is accessed, separation of the several ion components is automatically realized. By repetitive protocol applications and summation of results, the macroscopic behavior is determined, a result that assumes the membrane process to be ergodic (i.e., that successive responses mimic what would have been found from spatial summation).

5.2.2. Voltage Clamp Design

The voltage clamp was carefully designed with particular goals in mind, made possible by a carefully constructed experimental apparatus. Though the apparatus limited the types of cells that could be evaluated, it was highly successful in providing the data needed. These data allowed the investigators to separate the action potential, a composite event, into component currents, which provided a mechanistic basis for understanding what was actually happening during the action potential's time course.

Goal: separation into individual ion components

An action potential is a composite effect of many kinds of currents, including currents involved with charging the membrane capacitance, nonlinear currents associated with sodium and potassium ions, and other currents. Each of the different kinds of current change the transmembrane voltage, so measurement of changes in the voltage are insufficient to allow one to know which components produced any changes. On the other hand, one wishes to know the effects of the components separately, so as to understand the mechanisms by which the action potential was generated. Figuring out how to determine the individual components was the challenge faced by Hodgkin, Huxley, and other investigators of their time.

The components of the transmembrane current during an action potential (or subthreshold transient) includes the ionic flux plus a capacitive (charging) current. Since the capacitance is fixed during an action potential, its current, I_C, is given by

$$I_C = d(C_m V_m)/dt = C_m dV_m/dt \tag{5.9}$$

Consequently, in a circuit arranged to apply a voltage step (i.e., a constant transmembrane potential, hence $dV_m/dt = 0$) across the entire membrane (in a space-clamped configuration), the capacitive current component will be absent after the voltage step. The removal of capacitive current simplifies analysis of the remaining current components, because they must then consist entirely of ionic components.

Hodgkin and Huxley (HH) reasoned successfully, based on experiments described earlier, that the chloride contribution to the total current did not need to be included explicitly. (It was taken into account as a component of a small additional "leakage" current, to be described in a following section.)

The major task that remained was to separate the ionic flux into its sodium and potassium components. This separation turns out also to be facilitated by the measurements of current under constant transmembrane potential conditions. To this end, the voltage-clamp and space-clamp device illustrated in Figure 5.7 was developed. This experimental capability was accomplished independently by Cole and Marmont [6], but particular credit is given to Hodgkin, Huxley, and Katz [10]. In the voltage clamp as designed by Hodgkin and Huxley, a simple proportional controller is used to keep the membrane potential at a preset value.

Clamp electrodes

Controlling the membrane potential is accomplished by controlling the current flow between axial electrode A (inserted into the nerve axoplasm as described in Figure 5.7) and electrode E. Electrode E is a concentric cylindrical electrode in the extracellular fluid. (The axial uniformity of the device results in the elimination of any axial potential changes, hence achieving a space-clamped condition.)

This control system allows the transmembrane potential, as developed between electrodes B and C, to be locked to a preset value. In Figure 5.7 it can be seen how the error signal $V - V_0$ is developed and applied to the current generator. The resultant change in applied current reduces $V - V_0$ toward zero. The radial (transmembrane) current is determined from electrodes C and D, where the known conductivity of the medium is used to convert the measured voltage differences into membrane current.

Electrodes A and B are actually interleaved insulated wire helices wound on a 70-μm glass capillary that are exposed over an axial extent, as shown. In view of its overall size (diameter of 120 μm), the electrode was limited to nerve fibers, such as the giant axon of the squid, with diameters in excess of 300 μm. (The squid giant axon is a nerve fiber whose large size has made it useful in many electrophysiological studies.) In this situation large fibers are required

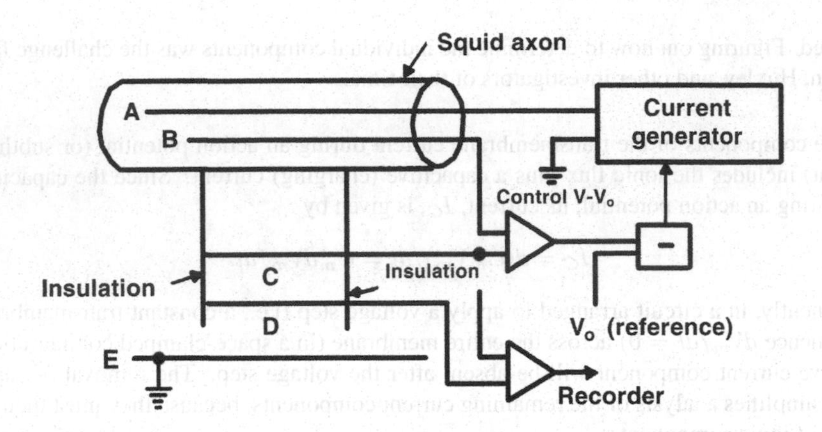

Figure 5.7. Schematic Diagram showing the voltage and space-clamp apparatus as developed by Hodgkin, Huxley, and Katz [10]. Current electrodes are (A) and (E); potential sensing electrodes are (B) and(C). Transmembrane current is determined from the potential between (C) and (D) and the total resistance between these electrodes. (Since the membrane current is uniform and in the radial direction only, the resistance can be calculated if the electrode end-effects are neglected.) Transmembrane voltage V is compared with the desired clamp V_0, and the difference causes a proportional transmembrane current of proper sign so that $(V - V_0)$ becomes 0.

to accommodate this axial electrode. Electrodes C and D are silver wires, while electrode E is a silver cylinder. Exposed portions of the electrodes were coated electrolytically with chloride.

Electrodes B, C, and D are placed within insulating baffles, isolating compartments, and hence confining flow to within an axial region; this eliminates end effects and achieves the desired axial uniformity. Axial uniformity is important since, otherwise, an impulse at one point on an axon is propagated to the remaining resting fiber, resulting in axial variability. In this experiment, one objective is to eliminate the complication of such spatial dependence. Spatial variability is eliminated in the above arrangement by causing the axon to behave synchronously over the spatial extent of the recording electrodes.

Eliminating axial spatial dependence is referred to as *space clamping*. Space clamping results in all potentials and current densities being functions of the radial variable alone (i.e., one dimensional). In effect, all membrane patches (all ion channels) are subject to identical transmembrane potentials.

Suitable cell preparations

The device for achieving a space clamp, described in Figure 5.7, is limited to long cylindrical cells of large diameter. Furthermore, the internal electrodes being of small cross-section and hence high resistance, are not completely isopotential and axial uniformity cannot be fully achieved. For small cells, the patch-clamp electrode in the whole-cell configuration also provides a space-clamped condition, avoiding the aforementioned difficulties.

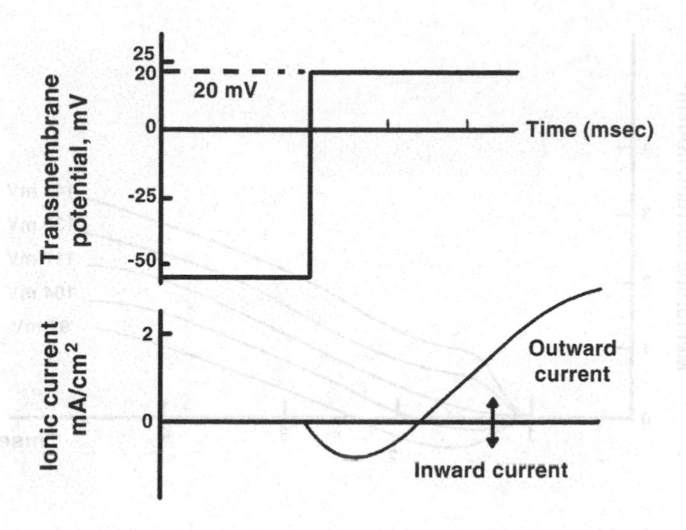

Figure 5.8. Illustrative Example of the Ionic Current for a Squid Axon assuming the application of a voltage clamp of $V_m = 20$ mV at $t = 0$ sec. The assumed parameters are: resting potential of $V_m = V_{\text{rest}} = -60$ mV; sodium and potassium Nernst potentials $E_{\text{K}} = -70$ mV and $E_{\text{Na}} = 57$ mV.

5.2.3. Voltage Clamp Currents

A typical record resulting from the application of a step change in membrane voltage is shown in Figure 5.8. In the lower panel of the figure, one notes an early inward current followed by a rise to an asymptotic outward current. An initial capacitive surge is completed in 20 μsec, corresponding to the presence of a capacitor with $C = 1.0\,\mu\text{F/cm}^2$ Because of the very short time constant, this current drops to zero before the ionic current becomes significant, and hence is normally ignored in studies of the latter.

Illustrative example

The initial flow of ionic current arising from a transthreshold voltage step is due to the sodium ion influx. This behavior is illustrated in Figure 5.8 based on typical parameter values. Here we have a voltage change of 80 mV (so that $V_m = 20$ mV).

As we have seen, the activation process is characterized by a rapid increase in sodium permeability. The net driving force for sodium is the difference between the transmembrane potential of 20 mV and the sodium Nernst potential of 57 mV, resulting in a driving force of 20–57 or –37 mV. Since this is negative, it is inward; consequently, a resultant inward (sodium) current is expected. This inward flow constitutes a bulge, because the elevated sodium permeability is transitory.

In fact, as the sodium permeability falls the potassium permeability rises and remains elevated. This elevated permeability accounts for the "steady-state" or "late" outward current. (Assuming a potassium Nernst potential $E_{\text{K}} = -70$ mV, the potassium driving force is 20 – [–70] or 90 mV and, since the driving force is positive, the current is outward.)

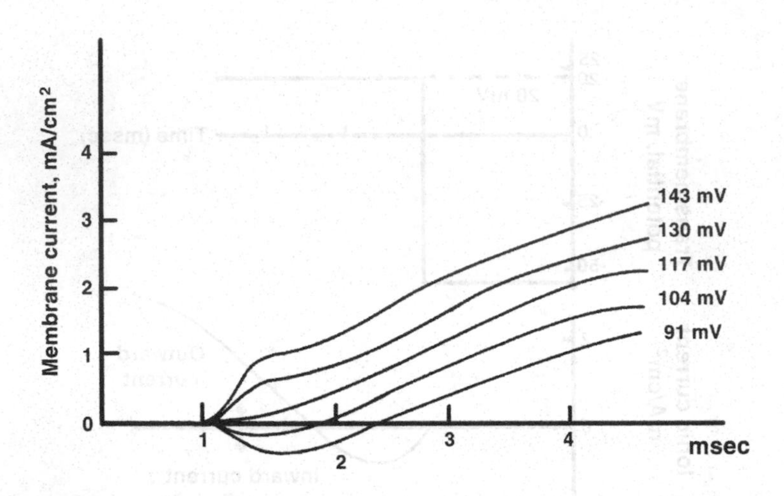

Figure 5.9. Measured Ion Currents for the Squid Axon following the application of a voltage clamp of the value indicated. The sodium Nernst potential is reached with a step change of 117 mV (since the resting potential is −60 mV and E_{Na} = 57 mV). From Hodgkin AL. 1958. Ionic movements and electrical activity in giant nerve fiber. *Proc R Soc* **148**:1–37. After Hodgkin AL, Huxley AF, Katz B. 1952. Measurement of current voltage relations in the membrane of the giant axon of *Loligo*. *J Physiol* **116**:424–448.

Voltage-clamp measurements

Valuable insight on the early membrane current can be achieved by clamping to a value of $V_m = E_{\text{Na}}$. For a short interval there is no current at all, even though in this early phase of the action potential sodium permeability is tremendously elevated. The reason is that there is no net force to cause a sodium current to flow.

In Figure 5.9 the measured transmembrane currents arising from a series of voltage clamps of different magnitude relative to a resting potential of −60 mV is shown. The figure includes a clamp at v_m = 117 mV (or V_m = 57 mV), which corresponds to the sodium equilibrium condition, and we note the early measured current to be zero. The abolition of an early current when V_m is at the sodium Nernst potential confirms that it is the sodium ions that are responsible for this phase of total current. When $v_m >$ 117 mV, the net driving force on sodium $(V_m - E_{\text{Na}})$ is outward. Note that for this condition the early current bulge is outward.

A series of successively larger, early (depolarizing) voltage clamps shows that beyond threshold the magnitude of the peak inward current gets progressively smaller. Ultimately it goes through zero. The *reversal potential* is the value of voltage clamp for which the early inward current equals zero. From the above argument, the reversal potential equals the sodium Nernst potential. This equality proves to be only a good approximation, as will now be explained.

The GHK equation[4] applies in these circumstances, because the total ionic and capacitive current is equal to zero. It provides a more accurate estimate of E_{rev} (the *reversal potential*). Thus, because $\partial/\partial t = 0$, then $\Sigma I_i = 0$. Assuming the chloride contribution to be negligible, we

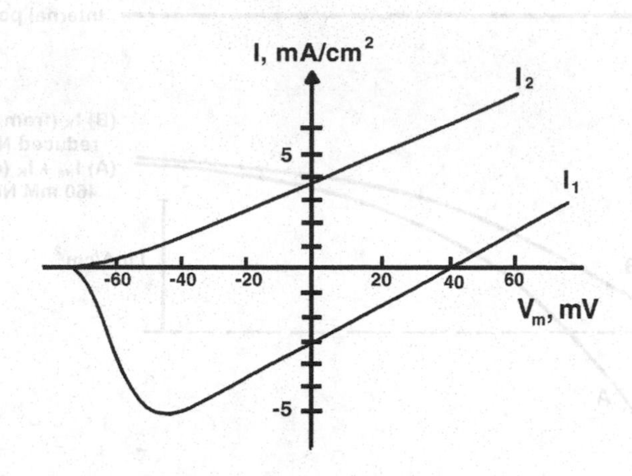

Figure 5.10. A Typical Current–Voltage Relation for Squid Axon. Curve I_1 shows the peak inward current versus clamped transmembrane voltage V_m after holding at rest. Curve I_2 plots the steady-state outward current versus the clamped voltage V_m. The voltage clamp value of V_m is plotted on the abscissa. Note that $I_1 = 0$ at $V_m = V_{\text{rev}} \approx E_{\text{Na}}$.

have

$$E_{\text{rev}} = \frac{RT}{F}\ln\left(\frac{[\text{K}]_o + \frac{P_{\text{Na}}}{P_{\text{K}}}[\text{Na}]_o}{[\text{K}]_i + \frac{P_{\text{Na}}}{P_{\text{K}}}[\text{Na}]_i}\right) \tag{5.10}$$

where P_{Na} and P_{K} are, respectively, the sodium and potassium permeabilities at the time of peak inward current.

This expression permits the introduction of the small but possibly not negligible contribution of the potassium ion component.

Current–voltage curves

Current–voltage curves are another way of describing membrane operating conditions. Examples of such curves are given in Figure 5.10.

The basic data are derived from voltage-clamp experiments carried out over a range of transmembrane potentials V_m where the peak inward current (I_1 in Figure 5.10) and peak outward current (I_2 in Figure 5.10) are the independent variables. From the intersection of I_1 versus V_m with the horizontal axis ($I_1 = 0$), the reversal potential is found.

From the definitions of g_{K} and g_{Na} in (3.26) and (3.27) we can verify that these conductances are always positive. They may be identified graphically in Figure 5.10 as the slope of a line from V_m on the curve to either E_{K} or E_{Na} on the V_m axis. The ordinate is then I_p and the abscissa $V_m - E_p$, where p represents either Na or K.

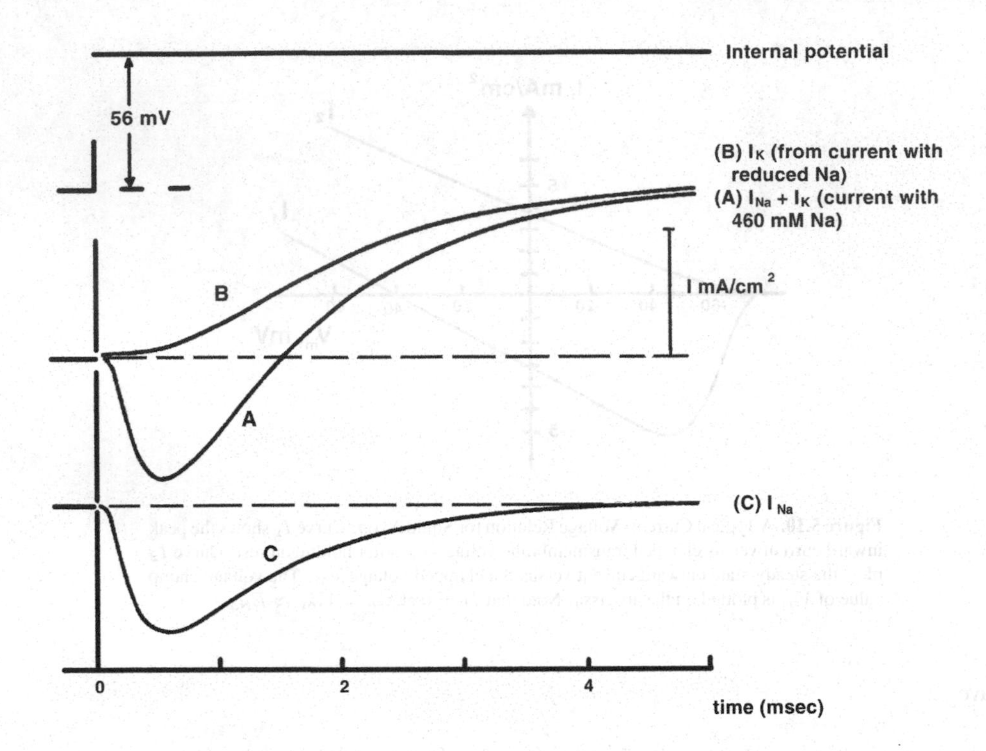

Figure 5.11. Analysis of the Ionic Current in a *Loligo* axon during a voltage clamp. Trace A shows the response to a depolarization of 56 mV with the axon in seawater. Trace B is the response with the axon in a solution comprising 10% seawater and 90% isotonic chloride solution. Trace C is the difference between traces A and B. Normal $E_{\mathrm{Na}} = 57$ mV, and in the reduced seawater $E_{\mathrm{Na}} = -1$ mV. From Hodgkin AL. 1958. Ionic movements and electrical activity in giant nerve fibers. *Proc R Soc* **148**:1–37. After Hodgkin AL, Huxley AF, Katz B. 1952. Measurement of current voltage relations in the membrane of the giant axon of *Loligo*. *J Physiol* **116**:424–448.

This slope is described as a *chord conductance*. It is interesting that for $-75\text{ mV} < V_m - 45$ mV roughly, the sodium *slope conductance* (i.e., dI_{Na}/dV_m) is negative in Figure 5.10.

5.2.4. Strategies for Na/K Ion Separation

The voltage-clamp experiment described in Figure 5.11 is for $v_m = 56$ mV (or $V_m = -60 + 56 = -4$ mV). The transmembrane current that results is shown in Figure 5.11A and, as discussed, contains an early inward current due mainly to sodium.

The transmembrane current also contains a late, steady-state, outward current due mainly to potassium. A separation of the sodium and potassium currents is necessary in order to model the behavior of each ionic component alone. This separation was accomplished through the following procedure.

Second extracellular medium

A second extracellular medium was constructed by replacing 90% of the extracellular sodium by the inert element choline (the introduction of choline was simply to maintain isotonicity). Thereby, extracellular sodium concentration was reduced by a factor of 10. Initially, $E_{\mathrm{Na}} = 57$ mV.

The reduction of extracellular sodium by a factor of ten should lower the Nernst potential by $58 \log_{10} 10 = 58$ mV, so that the new value of $E_{\mathrm{Na}} = -1$ mV. Accordingly, if the voltage-clamp experiment is repeated (with the same $v_m = 56$ mV) in the 10% sodium seawater, the sodium should be essentially in equilibrium. In this case the transmembrane current contains potassium only. This condition is shown in Figure 5.11B.

Independence principle

Hodgkin and Huxley made a key assumption that the sodium and potassium ion fluxes are independent of each other (asserting the *independence principle*). In other words, they assumed that potassium current and sodium current crossed the membrane independently, each in its own pathways. This was a bold assumption, as it would not be true if most sodium channels also leaked potassium or vice versa. If one assumes independence, the potassium component in Figure 5.11A should be precisely that in Figure 5.11B, and subtraction of Figure 5.11B from Figure 5.11A then results in the sodium current alone. This subtracted curve is shown as Figure 5.11C, and its behavior corresponds very closely to what is expected of sodium.

Additional evidence for the independence principle comes from the behavior of the squid axon following the addition of certain toxins to the extracellular medium. For example, it was noted that TTX (tetrodotoxin) blocks the sodium current almost completely while leaving the potassium almost unaffected. Conversely, TEA (tetraethylammonium) blocks potassium but not sodium channels.

These experiments support the existence of separate (independent) sodium and potassium channels. On the other hand, deviation from independence is seen, particularly at higher ion concentrations.

Procedure for Na/K ion separation

Hodgkin and Huxley performed a series of voltage-clamp experiments for increasingly depolarizing values. For each value of clamp voltage two experiments were performed.

The first was for normal composition seawater and the second with a low-sodium seawater (replace 90% sodium chloride by choline chloride while potassium and remaining chloride ions are unchanged).

In the second experiment, depicted in Figure 5.11B, a potassium (only) current arises, so that, given the independence principle, separation into I_K and I_{Na} is relatively easy. But, in general, the voltage clamp voltage in the second experiment does not correspond to the sodium Nernst potential and a sodium current is present along with the potassium.

We now analyze the currents measured as to their significance. In the following, primes designate both currents and Nernst potentials for the low-sodium voltage-clamp experiment.

Key assumptions

The Hodgkin and Huxley ion separation procedure is based on three assumptions:

1. Early current is sodium current alone, that is, $I_K = 0$ for $0 \leq t \leq T/3$, where T is the time of peak inward current. The assumption here is that for small t, due to the rapid response by the sodium system relative to potassium, the earliest current, if there is any, is sodium alone.

2. Outside Na affects I_{Na}. If $I'_{Na}(t)/I_{Na}(t) = A$, that is, for two experiments at the same voltage clamp, but different $[Na^+]_o$, only the driving force changes, going from $(V_m - E_{Na})$ to $(V_m - E'_{Na})$. The driving force is constant with respect to time in each case, although different from case to case. The time course of sodium conductance $g_{Na}(t)$ depends on rate constants, which in turn depend on V_m. But the latter are the same in the two experiments, since both have identical voltage clamps. Since

$$I_{Na}(t) = g_{Na}(t)(V_m - E_{Na})$$

and

$$I'_{Na}(t) = g_{Na}(t)\,(V_m - E'_{Na})$$

then

$$I'_{Na}(t)/I_{Na}(t) = (V_m - E'_{Na})/(V_m - E_{Na}) = A$$

where A is a constant.

3. Ionic Independence. $I_K(t) = I'_K(t)$. Here, Hodgkin and Huxley assumed that since $[K^+]_i$ and $[K^+]_o$ are unchanged, the potassium current (for the same voltage clamp) in normal seawater is the same as in 10% sodium seawater—i.e., they assumed the independence principle.

Deductions

The assumptions above are utilized in the following procedure. One first examines the *early portion* of the total current for a normal voltage clamp $I_m(t)$ and the low-sodium current $I'_m(t)$ curves.

Assuming no potassium contribution, the ratio $I'_m(t)/I_m(t)$ gives the value of A in item 2 above. (One can plot this ratio for small t and confirm its constancy.)

For successive values of time, say at any $t = t_1$,

$$I_m(t_1) = I_{Na}(t_1) + I_K(t_1) \tag{5.11}$$

and

$$I'_m(t_1) = I'_{Na}(t_1) + I'_K(t_1) \tag{5.12}$$

Now, using items 2 and 3 above gives us [from (5)]

$$I'_m(t_1) = AI_{\mathrm{Na}}(t_1) + I_{\mathrm{K}}(t_1) \tag{5.13}$$

From (5.11) and (5.13) we can eliminate either I_{K} or I_{Na} to obtain the desired value of sodium and potassium ion components. These are

$$I_{\mathrm{Na}}(t_1) = \frac{I_m(t_1) - I'_m(t_1)}{1 - A} \tag{5.14}$$

and

$$I_{\mathrm{K}}(t_1) = \frac{AI_m(t_1) - I'_m(t_1)}{A - 1} \tag{5.15}$$

Hence the data obtained from the two experiments enable the two contributing function to be determined.

5.3. HODGKIN–HUXLEY CONDUCTANCE EQUATIONS

The data collected by Hodgkin and Huxley from their voltage clamp experiments on the squid axon were the basis of a quantitative model for the squid axon membrane behavior under both subthreshold and suprathreshold conditions. The critical components of the model are the equations for the conductances of sodium and potassium ions. The function of this section is to show how these conductance equations were obtained from their experimental data.

A remarkable aspect of the resulting equations is that they allow a membrane model to be constructed that not only reproduces the voltage clamp data itself but is capable of simulating new phenomena, such as the propagating action potential. Each ionic current is described by the product of a driving force with a conductance, as in the parallel-conductance model.

The driving force is the difference between the transmembrane potential and the ion's Nernst potential, and the conductance is a quantity that is determined experimentally. The formulation corresponds to the description given in (3.26) and (3.27) for the potassium and sodium ionic currents (chloride being neglected for the reasons given earlier). The same formulation has also been applied to evaluation of single-channel ion currents, as in (4.1).

The Hodgkin and Huxley scheme for evaluating the ionic conductivities, $g_{\mathrm{K}}(t)$ and $g_{\mathrm{Na}}(t)$ was as follows. Rearranging (3.26) and (3.27) gives

$$g_{\mathrm{K}}(t) = \frac{I_{\mathrm{K}}(t)}{(V_m - E_{\mathrm{K}})} \tag{5.16}$$

and

$$g_{\mathrm{Na}}(t) = \frac{I_{\mathrm{Na}}(t)}{(V_m - E_{\mathrm{Na}})} \tag{5.17}$$

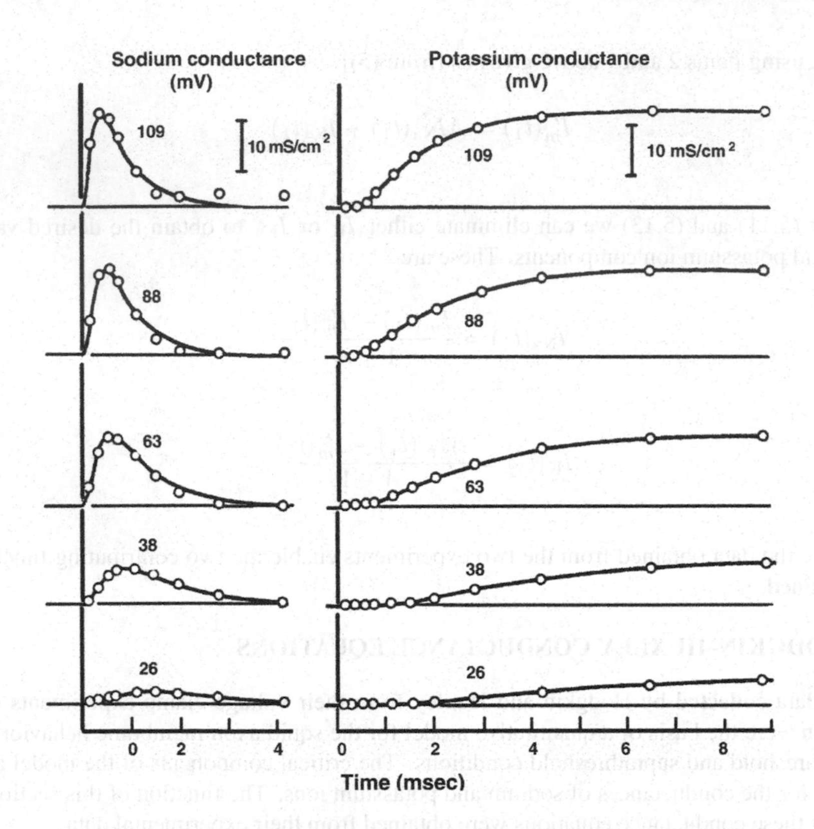

Figure 5.12. Conductance Changes Brought about by Clamped Depolarizations of Different Magnitudes. The circles represent values derived from the experimental measurements of ionic current, and the curves are drawn according to methods described in the text. The voltage clamp transmembrane potential values are in millivolts and are described relative to the resting value (i.e., v_m). From Hodgkin AL. 1958. Ionic movements and electrical activity in giant nerve fibers. *Proc R Soc* **148**:1–37. After Hodgkin AL, Huxley AF. 1952. A quantitative description of membrane current and its application to conduction and excitation in nerve. *J Physiol* **117**:500–544.

Because the denominators in (5.16) and (5.17) are constant during a voltage clamp, $g_K(t) \propto I_K(t)$ and $g_{Na}(t) \propto I_{Na}(t)$. A series of voltage-clamp experiments leading to a determination of $I_K(t)$ and $I_{Na}(t)$ by the methods of the last section is readily converted into families of potassium and sodium conductances. This procedure is illustrated in Figure 5.12.

5.3.1. Notation for Voltages and Currents

A major contribution of Hodgkin and Huxley was moving beyond their careful experimental studies into a general, theoretical model, described mathematically, relating membrane voltages and ionic current flows. That theory, as presented in the sections that follow, requires a recognition of two significant notational aspects in the descriptions of transmembrane voltages and currents, addressed here. These issues are not difficult but may be confusing if one does not realize the outlook that was adopted.

Notation for potentials and currents

Analyzing the time course of action potentials makes use of the introduction of an additional transmembrane potential variable. As before, the symbol V_m, using a capital V, will continue to be used for the transmembrane potential, measured as the potential just inside the membrane minus the potential just outside, that is, $V_m = \Phi_i - \Phi_e$.

But in the material that follows, the symbol v_m, using a lowercase v, will be used in some places to designate the difference in the transmembrane potential from its resting value. In other words, v_m will be defined by $v_m(t) = V_m(t) - V_m(\text{rest})$. Note that mathematically v_m differs from V_m only by a constant. This means that derivatives with respect to space or time of v_m are equal to the corresponding derivatives of V_m.

Why use v_m instead of V_m, when v_m seems to depend on a more complicated expression? To answer this question, consider the shape of an experimentally recorded action potential waveform such as that of Figure 5.4. The portion of the waveform showing the transmembrane potential at rest is easily identified. Deviations from that value, v_m, are easily measured as changes from that baseline.

In this regard v_m behaves like a "signal" in the engineering sense. In the absence of a signal (i.e., under resting conditions) v_m is appropriately zero. Thus, v_m in some cases better characterizes what is of most interest, namely, the magnitude and direction of changes in membrane voltage from its "natural" value at rest.

Just as it is useful to characterize the changes in the transmembrane potential from its zero or reference condition (i.e., $v_m = V_m - V_{\text{rest}}$), other time-varying potentials may be similarly described relative to their resting state. In particular, the value of the potential just inside or just outside the membrane is described by Φ_i or Φ_e; the change in either of these relative to their respective baseline value is designated ϕ_i or ϕ_e.

Flux J versus flow I

Another important notational issue involves fluxes versus flows. The careful reader will have noted the use of the symbol I_K (current) rather than J_K (flux) in equations such as (5.16), and will realize that the resulting conductance, g_K, seems to be a conductance value in units of Siemens, and not a conductivity. In fact, conductance g_K is more often expressed in Siemens/cm^2, or some other "per unit area" form.

Thus, for consistency with the notation used earlier in the text distinguishing flux versus flow, Eq. (5.16) should have been written with J_K instead of I_K, if a resulting g_K in Siemens per cm^2 is intended. An analogous concern arises for flows of other ions. However, in the presentation and use of Hodgkin–Huxley equations, a careful distinction in notation between flux versus flow normally is not made.

Thus, following the frequently (but not universally) accepted convention, in Eq. (5.16) we have used I_K for flux or flow. In many of the subsequent equations in this and the following chapters we have done the same.[5]

The notational shortcut of writing both flux and flow with an I has the advantage that many equations that otherwise would have to be written twice now can be written only once. Also, some computer code can be used in both cases without modification, as the difference will be signified by the choice of g_K. Finally, use of this convention is consistent with many other texts and references.

The price of this reduction in writing is that there is an increased possibility for confusion, that is, the reader is dependent on context (or perhaps the units given for g_K) to determine whether flow or flux is intended.

5.3.2. Mathematical model for potassium

A mathematical model was constructed by Hodgkin and Huxley to fit the data in Figure 5.12. The potassium conductance $g_K(t, v_m)$ was set equal to a fixed maximum value, $\bar{g}_K$, multiplied by n^4 $(0 < n < 1)$.

Potassium conductance equation

In a present-day interpretation (see Chapter 4), n^4 evaluates the fraction of open channels while $\bar{g}_K$ is the conductance when all channels are open. Their product evaluates the conductance as described in (4.24). Thus,

$$g_K(t, v_m) = \bar{g}_K n^4(t, v_m) \tag{5.18}$$

Hodgkin and Huxley assigned the power 4 to n in (5.18) because this choice gave the best fit to a large amount of potassium ion data. However, as discussed in Chapter 4, one can now give this choice a physical basis reflecting properties of the potassium protein structure. HH interpreted n as the probability of finding any one of four particles in the open state (this particle could be a subunit undergoing an "open" conformational change).

Assuming n to obey first-order kinetics, then just as explained in the material leading to (4.30), we have

$$\frac{dn(t, v_m)}{dt} = \alpha_n(v_m)(1 - n) - \beta_n(v_m)n \tag{5.19}$$

The rate constants α_n and β_n depend only on the transmembrane potential v_m. Accordingly, they are constant for a voltage-clamp experiment and permit (5.19) to be solved analytically. In this case we seek the solution of a first order differential equation with constant coefficients: the result is

$$n(t) = n_\infty - (n_\infty - n_0)e^{-t/\tau_n} \tag{5.20}$$

where

$$\tau_n = (\alpha_n + \beta_n)^{-1} \quad \text{and} \quad n_\infty = \alpha_n(\alpha_n + \beta_n)^{-1} \tag{5.21}$$

These are the same results as obtained in (4.36) and (4.37). The temporal variation for $n(t)$ arising from the first-order solution (5.20) has the correct form to fit the experimentally derived curves for $g_K(t)$ given in Figure 5.12.

One can combine the above equations, enabling one to rewrite (5.19) as

$$\frac{dn}{dt} = (n_\infty - n)/\tau_n \tag{5.22}$$

Potassium rates of change

To obtain an optimal fit of Eq. (5.22) to the measured data, for the *i*th voltage clamp v_{mi} one may adjust τ_n and n_∞. Note that since n_0 is the n_∞ of the rest period, just prior to clamp application, it is not at our disposal for curve fitting. Then, corresponding to these optimal values of $\tau_n(v_{mi})$ and $n_\infty(v_{mi})$, one can evaluate the corresponding rate constants for each (*i*th) case.

These rates are found by solving, simultaneously, the equations in (5.21) to yield

$$\alpha_n(v_{mi}) = \frac{n_\infty(v_{mi})}{\tau_n(v_{mi})} \quad \text{and} \quad \beta_n(v_{mi}) = \frac{[1 - n_\infty(v_{mi})]}{\tau_n(v_{mi})} \tag{5.23}$$

Here again the subscript *i* refers to a particular experiment whose voltage clamp is v_{mi}.

One can treat the resulting set $\alpha_n(v_{mi}), \beta_n(v_{mi})$ as samples of approximating analytic functions $\alpha(v_m), \beta(v_m)$. Hodgkin and Huxley, through curve fitting, described these (for potassium) to be

$$\alpha_n = \frac{0.01(10 - v_m)}{\left[\exp\left(\frac{10 - v_m}{10}\right) - 1\right]} \tag{5.24}$$

and

$$\beta_n = 0.125 \exp\left(\frac{-v_m}{80}\right) \tag{5.25}$$

where v_m is in mV and α, β are in msec^{-1}. Recall that $v_m = V_m - V_{\text{rest}}$.

5.3.3. Mathematical Model for Sodium

For sodium ionic currents the same overall approach is followed as described above for potassium, except that the sodium conductance is assumed to depend on the product of two parameters, *m* and *h*, where

$$g_{\text{Na}}(t, v_m) = \bar{g}_{\text{Na}} m^3(t, v_m) h(t, v_m) \tag{5.26}$$

In (5.26), $\bar{g}_{\text{Na}}$ is the maximum sodium conductance (*a constant*), *m* is an activation parameter ($0 < m < 1$), while *h* is an inactivation parameter ($0 < h < 1$).

We may interpret $m^3 h$ as the probability that a sodium channel is open. Hence, for a large population, $m^3 h$ is the fraction of the all-sodium channels that are open.

As with potassium, we can also assume that for the sodium protein structure to yield an open pore we require conformational changes in which each of four subunits are in an open position. For sodium three subunits have *m* as the probability of their being open, while the fourth is described by the probability *h* of being open.

Both parameters satisfy first-order differential equations similar to that for potassium's *n* variable, namely,

$$\frac{dm}{dt} = \alpha_m(1 - m) - \beta_m m \tag{5.27}$$

and

$$\frac{dh}{dt} = \alpha_h(1 - h) - \beta_h h \tag{5.28}$$

Maximum sodium conductivity

The value of $\bar{g}_{\mathrm{Na}}$ can be found from the measured Hodgkin and Huxley voltage clamp data from the asymptotically largest value of conductance obtained. It also corresponds to the situation where all-sodium channels are open. Thus, if there are N_{Na} sodium channels per unit area and each has a conductance of γ_{Na}, then $\bar{g}_{\mathrm{Na}} = N_{\mathrm{Na}}\gamma_{\mathrm{Na}}$.

Equations for m and h

Hodgkin and Huxley did not have knowledge of the structure and behavior of channels, except by inference. They were led to Eq. (5.26) by curve fitting, noting that their measured voltage clamp behavior of sodium conductance is second order (see Figure 5.12), and they achieved this by assigning *m* and *h* first-order behavior.

Equations (5.27) and (5.28) can be solved under voltage clamp conditions (where α and β are constants for each clamped value of v_m), giving

$$m(t) = m_\infty - (m_\infty - m_0)e^{-t/\tau_m} \tag{5.29}$$

and

$$h(t) = h_\infty - (h_\infty - h_0)e^{-t/\tau_h} \tag{5.30}$$

where

$$m_\infty = \frac{\alpha_m}{(\alpha_m + \beta_m)}, \qquad \tau_m = \frac{1}{(\alpha_m + \beta_m)} \tag{5.31}$$

and

$$h_\infty = \frac{\alpha_h}{(\alpha_h + \beta_h)}, \qquad \tau_h = \frac{1}{(\alpha_h + \beta_h)} \tag{5.32}$$

Assuming application of a transthreshold voltage clamp, *m* is seen to increase rapidly while *h* decreases slowly. The combination of m and h behavior, as expressed in the equation for sodium ion conductance (5.26), results in the expected second-order sodium conductance behavior (i.e., the conductance goes up and then comes down, even as the transmembrane potential remains constant.) At the same time, *m* and *h* are individually first order (meaning that they move from a starting to and an ending value).

Because the response to activation is for *m* to increase, it is called the *activating* parameter. Because the simultaneous *h* response is to *decrease*, it is described as an *inactivating* parameter.

Response of m and h to a stimulus

For $v_m > 30$ mV an action potential is certain to be elicited, and in the steady state that follows, $g_{\mathrm{Na}} \approx 0$. Consequently, since *m* is an increasing function, we deduce that $h_\infty \approx 0$ (complete inactivation). At rest we have seen that g_{Na} is relatively small, so $m_0 \approx 0$. Using the equation for sodium conductance (5.26) together with the asymptotic conditions on m and h with increasing time (5.31), and (5.32) gives

$$g_{\mathrm{Na}}(t) = \overline{g_{\mathrm{Na}}} m_\infty^3 h_0 \left(1 - e^{-t/\tau_m}\right)^3 e^{-t/\tau_h} \tag{5.33}$$

Examination of (5.33) shows a functional form (the product of a rising and decaying exponential) that is capable of matching the measured sodium behavior in Figure 5.12.

Time constants observed experimentally

For each voltage clamp v_m and corresponding experimental curve $g_{\text{Na}}(t, v_m)$, the values of τ_m, τ_h, m_∞ can be chosen in (5.33) so that it best fits the data. For a set of experimental voltage clamps v_{mi} one enumerates $\tau_m(v_{mi})$, $\tau_h(v_{mi})$, and $m_\infty(v_{mi})$. From these values one obtains the set of rate constants [rearrange (5.31) and (5.32)], namely,

$$\alpha_m(v_{mi}) = \frac{m_\infty(v_{mi})}{\tau_m(v_{mi})}, \quad \beta_m(v_{mi}) = \frac{1 - m_\infty(v_{mi})}{\tau_m(v_{mi})} \tag{5.34}$$

$$\alpha_h(v_{mi}) = \frac{h_\infty(v_{mi})}{\tau_h(v_{mi})}, \quad \beta_h(v_{mi}) = \frac{1 - h_\infty(v_{mi})}{\tau_h(v_{mi})} \tag{5.35}$$

Hodgkin and Huxley chose the following analytical expressions, which approximated their collection of data described in (5.34) and (5.35). These are

$$\alpha_m = \frac{0.1(25 - v_m)}{\exp\left[0.1(25 - v_m)\right] - 1} \quad \beta_m = 4\exp\left(-\frac{v_m}{18}\right) \tag{5.36}$$

and

$$\alpha_h = 0.07\exp\left(-\frac{v_m}{20}\right), \quad \beta_h = \left\{\exp\left[\frac{(30 - v_m)}{10}\right] + 1\right\}^{-1} \tag{5.37}$$

where, as before, v_m is in mV while α and β are in msec^{-1}. To obtain these results, valid also for $v_m < 30$ mV, an expression for h_∞ was necessary, and its derivation is described below.

Evaluating h_∞

In obtaining (5.36) and (5.37) it is necessary to first find $h_\infty(v_m)$ for all v_m (including $v_m < 30$ mV). Finding $h_\infty(v_m)$ was accomplished through a separate set of experiments.

First consider the plot of the h_∞ versus v_m in Figure 5.13. Note that for $v_m > 30$, $h_\infty \approx 0$, as assumed above. For normal resting conditions $h = 0.6$, while for hyperpolarizations of 30 mV or more $h = 1.0$, the maximum value. (This means that the largest action potentials are those elicited following such hyperpolarization.)

From the resting voltage just prior to the initiation of a voltage clamp, the h_∞ $(t = 0^-)$, can be found from Figure 5.13. But in view of (5.30), h cannot change in value instantly, and therefore $h_0(0^+) = h_\infty(0^-)$, which explains how h_0 is determined in (5.33).

An analytic expression that approximates the data in Figure 5.13 was devised by Hodgkin and Huxley:

$$h_\infty = \left\{1 + \exp\left[\frac{(v_m - v_{mh})}{7}\right]\right\}^{-1} \tag{5.38}$$

where v_{mh} is the value of v_m for $h_\infty = 0.5$ (in Figure 5.13, $v_{mh} = 2.5$ mV).

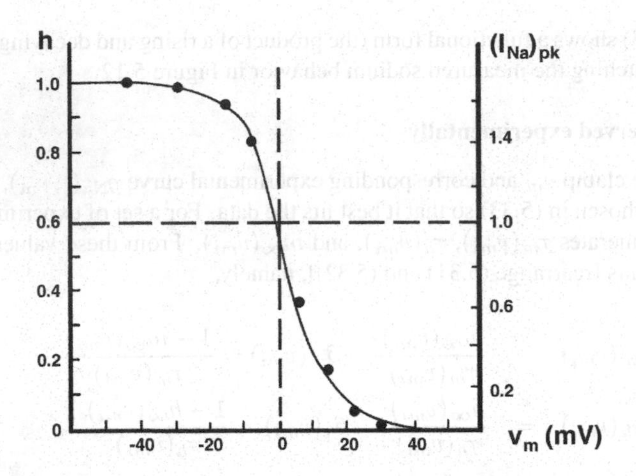

Figure 5.13. Sodium Inactivation Curve. The abscissa is the deviation from the resting potential (i.e., v_m). Dots are experimental points, and the smooth curve satisfies (5.38) for $v_{mh} = 2.5$ mV. The left vertical axis plots h as determined from these data, and the right vertical axis plots peak measured I_{Na} on a normalized scale (see text). From Hodgkin AL, Huxley AF. 1952. The dual effect of membrane potential on sodium conductance in the giant axon of *Loligo*. *J Physiol* **116**:497–506.

Two-step experiment to evaluate h_∞

A two-step experiment was performed by Hodgkin and Huxley [11] to evaluate h_∞. In step 1, described as the "conditioning period," a voltage clamp is established with a value v_c. (Reference to the conditioning period is indicated by the subscript c.) This clamp is maintained for a fixed time T_c that is large compared with τ_h. This long time duration ensures that for the conditioning step a steady-state value of h is reached, that is,

$$h(T_c) = (h_\infty)_c \tag{5.39}$$

At $t = T_c$ a suprathreshold clamp of fixed value v_t is applied. This second step was denoted the "test period" and is designated by the subscript t. Since $v_{mt} > 30$ mV the response is described by (5.33), which, with the present notation and $(h_0)_t$ as the initial h at the outset of the test period, becomes

$$g_{\text{Na}}(t) = \bar{g}_{\text{Na}} m_\infty^3 (h_0)_t (1 - e^{-t/\tau_m})^3 e^{-t/\tau_h} \tag{5.40}$$

Since, as we have noted, h cannot change discontinuously, the initial value of h in the test period is the final value of h in the previous conditioning period. Thus,

$$(h_0)_t = (h_\infty)_c \tag{5.41}$$

In (5.40) the values of m_∞, τ_m, and τ_h depend only on v_t, which is always chosen the same. Consequently, $g_{\text{Na}}(t) \propto (h_0)_t$.

In particular, because

$$I_{\text{Na}}(t) = g_{\text{Na}}(t)(V_t - E_{\text{Na}})$$

then

$$I_{\text{Na}}(t) \propto g_{\text{Na}}(t)$$

because V_t is fixed.

Thus, if the peak inward sodium current $(I_{\text{Na}})_{pk}$ is measured, then

$$h_\infty(v_c) \equiv (h_\infty)_c \propto (I_{\text{Na}})_{pk} \tag{5.42}$$

For the largest $(I_{\text{Na}})_{pk}$, namely $(I_{\text{Na}})_{mpk}$, we assign $h_\infty(v_c)$ the value of unity. Then all other values of h_∞ are given by

$$h_\infty(v_c) = \frac{(I_{\text{Na}})_{pk}}{(I_{\text{Na}})_{mpk}} \tag{5.43}$$

Noting (5.41), it is this $h_0(v_m)$ that is plotted in Figure 5.13.

Leakage current I_L

Recognizing that there are currents other than those of sodium and potassium ions, Hodgkin and Huxley introduced the *leakage current*, I_L, as a third component of the total membrane current. Leakage current takes into account, collectively, the currents of all ions other than those of potassium and sodium. Currents such as those of chloride ions, those of calcium ions, and those of any other charged particles moving through small holes in the membrane are a part of leakage.

In the Hodgkin–Huxley formalism, by analogy to potassium and sodium, the leakage current is written mathematically as

$$I_L = g_L(V_m - E_L) \tag{5.44}$$

In (5.44), E_L is set to a value that produces the expected or observed resting voltage of V_m, and g_L has a fixed value, rather than varying in time in the fashion of g_{K} or g_{Na}. Most of the time the leakage current has a small magnitude in comparison to the magnitudes of I_{K} or I_{Na}. Correspondingly, most of the time either g_{K} or g_{Na} is larger than g_L.

Even so, the values of g_L and E_L and the presence of leakage current more generally are not inconsequential. Leakage current is significant because its magnitude is comparable to that of other ionic currents during critical time periods when the other ionic currents also are small. Such time periods include times when V_m is near the resting state, and periods when sodium and potassium currents are similar in magnitude but opposite in sign, such as near the action potential's peak. Leakage currents also serve as a damping mechanism to dynamic changes of V_m.

5.4. SIMULATION OF MEMBRANE ACTION POTENTIALS

Once equations for describing the individual ionic currents are developed, they can be integrated into a model of the behavior of the membrane that finds membrane voltages and currents not only for voltage-clamp experiments but also for naturally evolving voltages. This section explains how that is done.

It could reasonably be expected that Hodgkin–Huxley ionic current equations would satisfactorily simulate any voltage-clamp experiment, because the model's parameters were chosen to fit the data from the voltage-clamp experiments that Hodgkin and Huxley performed. The critical question is whether the equations accurately predict the results for non-voltage-clamp situations, i.e., are the equations correct for naturally occurring action potentials? The answer is yes. This predictive power is why the HH formulation came to be considered a true formulation of how the membrane responds to many different situations and stimuli.

Because the use of the HH equations for membrane action potentials provides the foundation for answering many other questions, the next several sections are devoted to development of all the equations used, together with some results of using them.

Suppose one elicits an action potential for an axon in the chamber of Figure 5.7, but *without* the voltage clamp. Because the setup nevertheless demands axial uniformity, the entire membrane behaves synchronously. The action potential elicited therefore characterizes every patch of membrane. Under these circumstances, the action potential is known as a *membrane action potential*. Because V_m is a function of time, a membrane action potential is quite different from a voltage clamp. Nevertheless, as will be shown, the Hodgkin–Huxley equations are successful in simulating membrane action potentials.

5.4.1. Sum of Currents

Analytical evaluation of Hodgkin–Huxley membrane action potentials[6] begins with the assertion that the currents through the membrane follow the equation

$$I_m = I_K + I_{Na} + I_L + I_C \tag{5.45}$$

The picture associated with Eq. (5.45) is as follows: The total current through the membrane I_m arises from three ionic components and one capacitative one. The ionic currents are I_K, I_{Na}, and I_L. These currents reflect movement of potassium, sodium, and "leakage" ions through the membrane. The fourth term, I_C, is the current associated with charging or discharging the membrane capacitance.

The notation of (5.45) does not make clear whether the currents in the equation are constant or variable. In fact, at rest all the currents are constant. In contrast, during an action potential each current varies with time by a factor of 100 or more, with each current following a different time course. To make that explicit, I_m might be noted as $I_m(t)$, I_K as $I_K(t)$, and similarly for each one.

State variables

Questions involving evolution with time often are framed in terms of "state variables." State variables are the set of variables whose values, when known at a particular time, allow the other variables in the problem to be calculated for that same time. In the simulation of membrane action potentials using Hodgkin–Huxley formalism, the state variables are v_m, n, m, and h. (This statement means that all the other time-varying quantities, such as I_K and I_{Na}, can be found from the values of the state variables.)

The algorithm for time evolution focuses on state variables. That is, the algorithm must begin with an initial value for each one of them, and the algorithm must provide a method for finding a new value for each state variable at a later time. Most algorithms involve a time shift by a short time step, with many steps in succession required to calculate changes throughout a significant time interval. (Of course, the calculation also will use the values of other important quantities that do not change with time, e.g., membrane capacitance C.)

Whether a variable is or is not a state variable is a different issue from the variable's physical or physiological importance. Many fundamental quantities are nonetheless not state variables. For example, for a membrane action potential the sodium current, I_{Na}, is extremely important electrophysiology. Even so, I_{Na} is not a state variable, because I_{Na} can be computed from V_m, m, and h, together with constants such as E_{Na}. (If a different problem were to be considered, such as one with a long time duration where equilibrium potentials such as E_{Na} varied, that problem would required additional state variables, possibly concentrations.)

Tiny time steps

The independent variable for a membrane action potential is time. Though time varies continuously, for numerical analysis time is discretized into a sequence of particular time instants. Each time in the sequence is separated from the next by a time interval Δt.

An important consideration in finding a membrane action potential is the choice of a specific value of Δt. Normally HH simulations use Δt values in the range of 1 to 100 microseconds, with "small" time steps being those in the range of 1 to 10 msec.[7]

Suppose i is the index of listed times. An increase by one in this index corresponds to an advance in time by interval Δt. A satisfactory numerical algorithm must begin with the values of the state variables at time index i and then find the values of the state variables at time index $(i+1)$.

Voltage change for a time step

Equation (5.45) connects current components to the total current at a single instant of time. These currents may or may not be functions of time. Amazingly, for time variation, (5.45) can be turned inside out to become the central equation for determining changes in the membrane voltage for each time step.

Transforming (5.45) can be done because the capacitative current I_C is equal to $C_m dV_m/dt$, thus introducing time explicitly. Further, the time derivative dV_m/dt can be approximated as

$$\frac{dV_m}{dt} \approx \frac{\Delta Vm}{\Delta t} \tag{5.46}$$

While (5.46) is an approximation, it is an excellent approximation if Δt is sufficiently small.

After using (5.46) in Eq. (5.45), one can rearrange the result to be

$$\Delta V_m = \Delta t(I_m - I_{\mathrm{K}} + I_{\mathrm{Na}} + I_L)/C_m \tag{5.47}$$

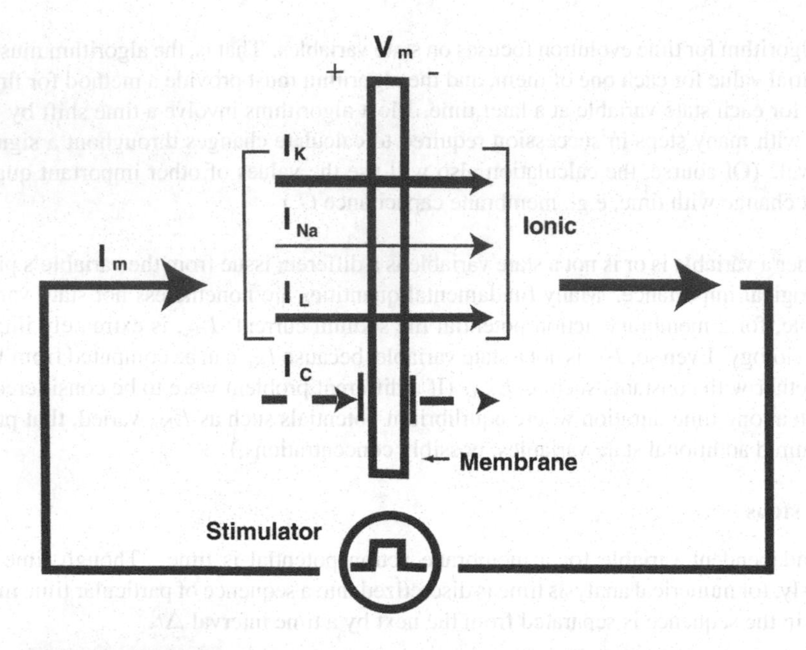

Figure 5.14. Membrane Voltage Change Due to Stimulus. The Figure shows a cartoon of a stimulator as it imposes total current I_m across a membrane (rectangle). The current from the stimulator divides into the components given by Eq. (5.47), including the ionic currents I_{K}, I_{Na}, and I_L. The remaining current is capacitive current I_C, shown as a dashed line. As I_C charges the membrane capacitance it modifies the transmembrane voltage V_m.

thus introducing ΔV_m and time increment Δt into what had seemed to be a static equation. In fact, Eq. (5.47) estimates the change in V_m, ΔV_m, that occurs when time advances by a short interval Δt (Figure 5.14).

A simulation program for finding $V_m(t)$ works by using (5.47) repeatedly. As an estimate of a plausible number of repetitions, note that a 1-second simulation period accomplished with a 1-microsecond time step corresponds to a million repetitions.

The picture of the meaning of Eq. (5.47) now is extended from that used with (5.45). In (5.47) one pictures the total current I_m as a known value of the total current, a current imposed on the membrane by its external environment, as shown in Figure 5.14. (At a particular time, the value of I_m might be zero, positive, or negative.) Equation (5.47) then shows how to get from I_m to one of the component currents, I_C, and from I_C to ΔV_m. That is, (5.47) shows how to connect the stimulus (or lack of one) to the transmembrane voltage change that follows.

Step by step, Eq. (5.47) says that to find the change in V_m, one begins with I_m and subtracts the ionic current $I_{\text{ion}} = I_{\mathrm{K}} + I_{\mathrm{Na}} + I_L$ (shown in brackets in Figure 5.14). The remaining current is the current of interest, $I_C = I_m - I_{\text{ion}}$ (shown by dashed lines in Figure 5.14). I_C is of interest because it is the current that modifies the membrane voltage by charging the membrane capacitance C. Thus I_C modifies V_m during time Δt by adding to the capacitance a charge $\Delta t(I_m - I_{\text{ion}})$.[8]

Examination of the terms of (5.47) makes clear that ΔV_m will sometimes be positive and at other times will be zero or negative. The variable polarity will occur because some currents on the right-hand side will usually be positive (I_K), others will usually be negative (I_{Na}), and some may be positive, negative, or zero (I_m, I_L).

Thus the sign of the result involves a summation of terms, some positive and some negative. Furthermore, the magnitude of each component current is likely to change markedly from one time to another, e.g., the stimulus current of step 1 varies markedly depending on whether, at a particular time, the stimulus is on or off. Thus the sign of V_m will sometimes be positive and at other times negative.

Changes in gating variables for a time step

A simulation program also must track changes in n, m, and h concurrently with tracking changes in V_m. Concurrent tracking of n, m, and h is necessary to allow I_K and I_{Na} to be evaluated at each time, as is required to find V_m at the following time, as shown in (5.47). For n, Eq. (5.19) gave its rate of change as

$$\frac{dn}{dt} = \alpha_n (1 - n) - \beta_n n \tag{5.48}$$

Again using the concept of (5.46) in (5.48), one can arrange the result to be

$$\Delta n = \Delta t \left[\alpha_n (1 - n) - \beta_n n\right] \tag{5.49}$$

Equation (5.48) is the result that is needed, as it shows how to find the change in n, Δn, from the values of the state variables present at a particular time. Analogous equations can be developed in a similar fashion for Δm and Δh and are given is a section below.

5.4.2. Algorithm for Advancing through Time

We now describe in more detail the series of steps needed to simulate the sequence of membrane events that occurs with the passage of time.

Starting values of the state variables

Living membrane exists continuously throughout its lifetime, so it has no fixed starting points or conditions in a fashion analogous to starting an automobile. (The fact that "time 0" is simply assigned arbitrarily to graphs is often frustrating to those new to the field, who want the time chosen to embody more physiological meaning than it actually has.) In the absence of reasons to do otherwise, however, starting conditions for each of the state variables normally are chosen as those that exist at rest.

For V_m, the starting condition is normally V_m at rest:[9]

$$V_m^o = V_r \approx -60 \text{ mV} \tag{5.50}$$

Recall that the stability of V_m at its chosen resting level may require a choice of E_L.

For n, m, and h the starting conditions are taken from Eqs. (5.21), (5.31), and (5.32):

$$n_o = n_\infty = \alpha_n/(\alpha_n + \beta_n) \tag{5.51}$$
$$m_o = m_\infty = \alpha_m/(\alpha_m + \beta_m)$$
$$h_o = h_\infty = \alpha_h/(\alpha_h + \beta_h)$$

where the α and β terms are evaluated using $v_m = 0$, corresponding to $V_m = V_r$.

In Figure 5.15 membrane voltage V_m and probabilities n, m, and h hold their initialized values through the period while $t < 0$. During this period one observes that $V_m = -60$, and marked variation from one probability to another, i.e., $n \approx 0.3$, $m \approx 0.5$, and $h \approx 0.6$.

During assignment of initial values, I_m must be assigned an initial value (such as $I_m^0 = 0$), consistent with the mathematical functions for I_m given below.

Time is the independent variable. Simulations often are started with t negative, e.g., $t = -100$ microseconds, so that a short baseline period corresponding to $t < 0$ is a part of the simulation record.[10]

Advancing the state variables step by step

The values of V_m, n, m, and h are moved forward step by step by repeatedly executing the steps that follow. Initially the starting set of values, which might be thought of as set $i = 0$, is used as a basis for finding set $i = 1$, the values of V_m, n, m, and h for time $t = \Delta t$.

Then the cycle is repeated. In general the cycle of steps begins with the state variable values for time index i to produce a set of values for time index $i + 1$. Such a cyclic process can be repeated over and over, thus allowing the simulation of indefinitely long time periods.

Advancing the state variables from time index i to time index $(i + 1)$ can be achieved by completing the following steps, in order:

1. Determine I_m, the total membrane current, for the interval Δt that extends from time (i) to time $(i + 1)$.

2. Estimate the change of V_m during Δt based on values at time i.

3. Estimate the change of n, m, and h during Δt based on values at time i.

4. Advance all the state variables from time (i) to time $(i + 1)$.

In the following sections we consider each of these steps, in turn.

1. Determine I_m For a space-clamped axon, the spatial uniformity ensures that no currents move along paths parallel to the membrane surface, so there can be no transmembrane currents created by adjacent segments of membrane. Thus the value of the total membrane current is equal to the stimulus that is applied, as in Figure 5.14.

For now assume that the stimulus consists of a depolarizing current $I_s(t)$ that begins at time $t = 0$ and lasts for duration T. Then the total membrane current is:

$$\begin{aligned} I_m &= 0 & t < 0 \\ I_m &= I_s & 0 <= t < T \\ I_m &= 0 & T \le t \le \infty \end{aligned} \tag{5.52}$$

In Figure 5.15 the top line shows a plot of I_s. One observes that I_m initially is zero, rises for a brief time when $t = 0$, and returns to zero shortly thereafter, as described in (5.52). A second stimulus pulse occurs at 10 msec.

In (5.52) note the values assigned to I_m at the time boundaries. For example, at time zero total membrane current I_m is set equal to I_s, not equal to zero. That is, I_m is assigned the value of I_s that will be present during the **following** time interval, Δt, because an assignment is being made at time index i involving the interval from i to $i + 1$. By the same reasoning, I_m becomes zero when $t = T$.

Often the above expressions for I_m are extended so that a stimulus train is created, with a new stimulus beginning periodically, e.g., every 20 milliseconds, rather than there being only a single stimulus. Also, I_s is often a constant (so the stimulus is rectangular), but nothing about the equation requires that to be so.

In general the stimulus might come from a natural source rather than one of artificial origin, in which case the time course of the stimulus likely would be longer and be a more complicated function of time.

2. Estimate ΔV_m. Using (5.47), one finds the incremental change of V_m during time step Δt to be:

$$\begin{aligned} \Delta V_m^i &= \frac{\Delta t}{C_m}[I_m^i - I_{\text{ion}}^i] \\ &= \frac{\Delta t}{C_m}[I_m^i - I_{\text{K}}^i - I_{\text{Na}}^i - I_L^i] \end{aligned} \tag{5.53}$$

where the individual ionic currents are known from (5.16), (5.17), (5.18), and (5.26) to be

$$\begin{aligned} I_{\text{K}}^i &= g_{\text{K}}^i(V_m^i - E_{\text{K}}) \\ I_{\text{Na}}^i &= g_{\text{Na}}^i(V_m^i - E_{\text{Na}}) \\ I_L^i &= g_L(V_m^i - E_L) \end{aligned} \tag{5.54}$$

where the variable conductivities are

$$\begin{aligned} g_{\text{K}}^i &= n_i^4(V_m^i - E_{\text{K}}) \\ g_{\text{Na}}^i &= m_i^3 h_i(V_m^i - E_{\text{Na}}) \end{aligned} \tag{5.55}$$

For computation, either the equations would have to appear in reverse order, i.e., (5.55), (5.54), (5.53), or, alternatively, the variables defined in later equations would have to have these definitions substituted in earlier ones. An advantage of keeping the steps distinct

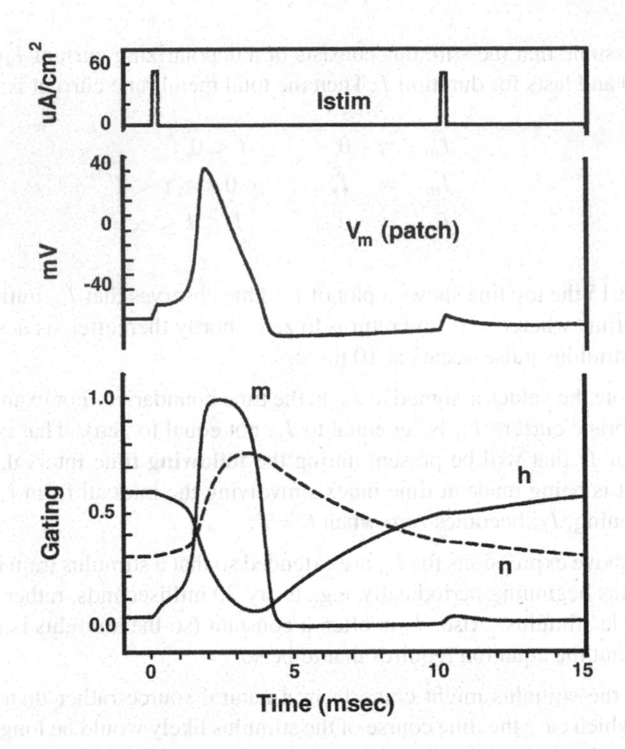

Figure 5.15. Computed Membrane Action Potential Using the Hodgkin–Huxley Equations. In addition to the temporal variation of $V_m(t)$, the gating variables temporal behavior [i.e., $m(t)$, $n(t)$, $h(t)$] are shown. In this simulation resting $v_m = -60$ mV, while the stimulus current starting at $t = 0$ is 53 μA/cm^2 for 0.2 msec. The temperature is 6.3°C.

is that oftentimes such values as the conductivities (the g values) become of interest in themselves, as well as serving as a means to determine V_m.

Note that many (but not all) of the quantities in Eqs. (5.53), (5.54), and (5.55) have a sub- or superscript i. The presence of the indexing i identifies each one as a quantity that has a value that changes from one time to another; quantities without an index hold constant values with time. The fact that all indices are i (rather than, say, mixed with $i+1$) signifies that the equations hold when all the quantities are for the **same** time instant.

In Figure 5.15 the second line of the Figure shows V_m plotted as it advances through many individual time steps. The lower lines of Figure 5.15 show the time course of m, h, and n. The changes in sodium and potassium conductivities and currents, associated with the passage of time and changes in V_m by the equations above, are shown in Figure 5.16, as the 3rd and 4th line of traces.

It is noteworthy that g_{Na} and g_K follow such a different time course, as is seen by comparing the solid and dashed lines in the figure, as these differences give rise to the observed time course of V_m. It also is remarkable that, at first glance, I_K and I_{Na} have wave shapes that are more or less identical except for opposite polarity. Of course, careful inspection of the Figure shows the I_{Na} wave shape to be slightly out of phase (and earlier than I_K).

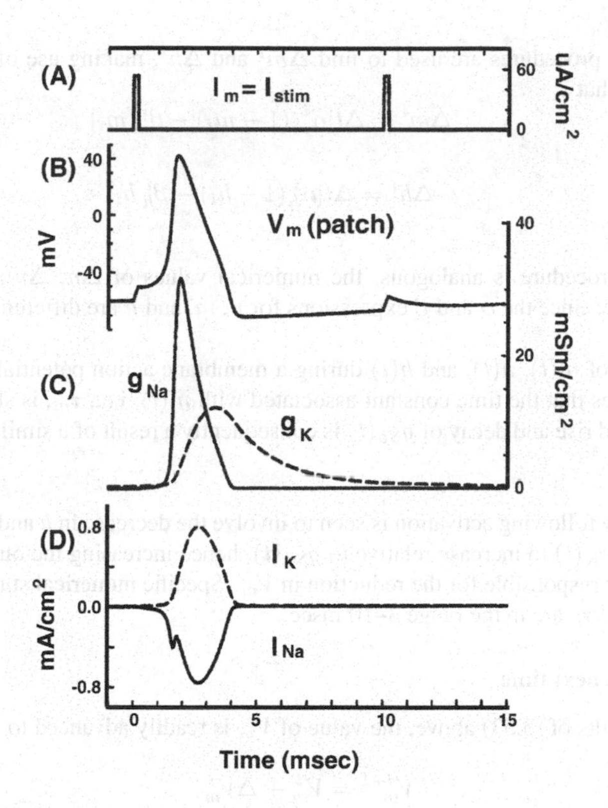

Figure 5.16. Membrane Potential and Currents following Stimulus. The stimulus is at the top (A). Below it are the calculated changes in membrane potential (B), sodium and potassium conductances (C), and sodium and potassium currents (D). All curves are for a squid giant axon membrane patch. The second stimulus is seen to elicit essentially no response even though it is of the same size and duration as the first (for which an action potential results, as is seen). It therefore identifies the condition as refractory. Since a larger stimulus would generate an action potential, this is a *relatively refractory period.* The stimulus amplitude is 53 $\mu A/cm^2$, and its duration is 0.2 msec. The second stimulus is similar in amplitude and duration and occurs after a delay of 15 msec. The resting potential is -60 mV while $T = 6.3°C$. Calculations were based on the Hodgkin–Huxley, equations.

In Figure 5.16 one notes the rapid rise and decay of $g_{Na}(t)$. In contrast, $g_K(t)$ has a delayed rise and more lasting elevation in magnitude. This behavior might have been anticipated as a result of what was learned from the voltage clamp measurements.

3. Estimate Δn, Δm, and Δh Estimation of the changes in state variables n, m, and h during interval Δt is accomplished following (5.49) as applied at time i:

$$\Delta n^i = \Delta t[\alpha_n^i(1 - n_i) - \beta_n^i n_i] \tag{5.56}$$

Evaluation of Eq. (5.56) is more complicated than Eq. (5.56) suggests. Several steps are required. First one uses v_m^i and Eqs. (5.24) and (5.25) for α_n and β_n to get the α_n^i and β_n^i numeric values for time i.[11] Second, one uses Eq. (5.56) to get Δn^i (the change in n for the interval Δt beginning at time i).

Analogous procedures are used to find Δm^i and Δh^i, making use of Eqs. (5.36) and (5.37), so that

$$\Delta m^i = \Delta t[\alpha_m^i(1 - m_i) - \beta_m^i m_i] \tag{5.57}$$

and

$$\Delta h^i = \Delta t[\alpha_h^i(1 - h_i) - \beta_h^i h_i] \tag{5.58}$$

Though the procedure is analogous, the numerical values of Δn, Δm, and Δh will be different, of course, since the α and β expressions for n, m, and h are different.[12]

The behavior of $m(t)$, $n(t)$, and $h(t)$ during a membrane action potential is shown in Figure 5.15. One notes that the time constant associated with $m(t)$, i.e., τ_m, is short relative to τ_n and τ_h.[13] The rapid rise and decay of $g_{\text{Na}}(t)$ is consequently a result of a similar time course for $m(t)$.

Early recovery following activation is seen to involve the decrease in *h* and the increase in *n*. This latter causes $g_K(t)$ to increase relative to $g_{\text{Na}}(t)$, hence increasing the outward component of current which is responsible for the reduction in V_m. Specific numerical study shows $\tau_m < 1$ msec, while τ_n and τ_h are in the range 3–10 msec.

4. Advance to the next time

Using the results of (5.53) above, the value of V_m is readily advanced to determine $V_m^{(i+1)}$ as

$$V_m^{i+1} = V_m^i + \Delta V_m^i \tag{5.59}$$

Similarly, the values of *n*, *m*, and *h* are readily advanced, using (5.56), (5.57), and (5.58) as

$$n^{i+1} = n^i + \Delta n^i \tag{5.60}$$

$$m^{i+1} = m^i + \Delta m^i \tag{5.61}$$

$$h^{i+1} = h^i + \Delta h^i \tag{5.62}$$

There are several reasons why it is advantageous to group Eqs. (5.59) through (5.62) as a separate 4th step, rather than commingling them with the computation of the various changes:

- First, the equations themselves rest on a stronger mathematical foundation, since all quantities in steps 1–3 are time coherent, i.e., are the values for the same time, time i, a condition that is part of their mathematical derivation.
- As a corollary, collecting the changes with time into a single region avoids inadvertently introducing small errors from time misalignment, or making the algorithm unintentionally sensitive to the ordering of steps 2 and 3.[14]
- Third, if one records (or prints) values during the computational cycle, printing done **prior to** step 4 will cause a coherent set of state variables and values derived from them (such as time, V_m, rate constants, currents) to be recorded, rather that having some quantities for one time and some for another.[15]

- Finally, positioning all the equations involving time transitions together allows passive or voltage-clamp simulations to be more readily incorporated as simulation alternatives (see below).

Units of calculation

In an algorithm for the simulation of HH transmembrane potentials, the use of a consistent set of units for currents and voltages does not happen naturally. Conflict arises because different parts of the calculation are done most naturally in different units. For example, specification of the stimulus current, I_s, is done most naturally in units of current, such as milliamperes, whereas quantities such as $\bar{g}_K$ are frequently read from reference tables in units of milliamperes per square centimeter, i.e., current per unit area. Inspection of the equations of steps 1 to 4 show them to remain valid either way, **so long as the units are consistent.**

The most common practice seems to be to use a "per-unit-area" formulation. Results found in those units are most readily compared to other results in the literature, i.e., membrane capacitance C becomes membrane capacitance per unit area C_m, which is known to be about 1 μF/cm^2, and no further conversion into the spatial dimensions of a particular preparation is required. In this regard, it is helpful to remember that under space-clamped conditions each membrane patch has an identical transmembrane voltage $V_m(t)$, so that the meaning of currents through the membrane are easily understood when they are expressed on a "per-unit-area" basis.

Time required to execute a simulation

By the year 2000, the power of desktop and laptop computers had grown to a level allowing one to routinely compute membrane action potentials in a minute or two, so that for membrane action potentials the time for execution of a simulation had largely ceased to be a limiting factor. For more complex spatially distributed simulations (considered mainly in later chapters of this text), time of execution continues to be a substantial consideration and often a limitation on what can be done.

Such situations as those requiring consecutive analysis of many sequential action potentials, or questions requiring analysis of action potential propagation in complicated anatomical structures such as the heart or brain, remain limited materially by limits on execution time.[16]

Passive and voltage-clamp simulations

Often one wishes to compare results of HH simulations of an active membrane to the sequence of transmembrane voltages and membrane currents that would result if the membrane were passive rather than active. (Here "passive" is used to mean that the membrane has constant conductivity, rather than having "active" sodium and potassium conductivities that change with time and membrane voltage.)[17]

The response of a passive (rather than active) membrane will be found by the algorithm above if parts of step 4, (5.60) through (5.62), are omitted. Conversely, the membrane can be voltage controlled by omitting or adjusting (5.59).

Other algorithms

The computational method outlined above is known as the explicit method, sometimes called the forward Euler method. A number of other approaches can be found in the literature.[18] Other algorithms usually are more complicated but also may provide more stability, accuracy, or computational speed.

5.4.3. Action Potential Characteristics

Action potentials have a number of special characteristics that are important and interesting in themselves. Several of these are detailed in this section.

Refractory periods

If an action potential is elicited, then a period ensues during which the membrane cannot be re-excited, the so-called *absolute refractory* state. After an interval it becomes possible to elicit an action potential, but it requires an abnormally high stimulus. This characterizes the *relative refractory condition*. An illustration is given in Figure 5.16, where the second stimulus, though equal in amplitude and duration to the first (which is transthreshold), fails to elicit an action potential.

Refractoriness can be understood mainly by the behavior of the inactivating parameter h. Following an action potential, h decreases to a very low value (see Figure 5.15). Consequently, even a very large stimulus elicits only a small sodium current, and this prevents re-excitation from occurring. Some time must elapse for h to recover to normal or near-normal values.

Time also is needed to permit n to decrease, because excitation requires bringing about the condition that $I_{\text{Na}} > I_{\text{K}}$. When that occurs, the net influx of cations increases V_m (algebraically). Increasing V_m initiates the regenerative process that characterizes the rising phase of the action potential (i.e., rising V_m creates a rising g_{Na}. In turn, the regenerative process elevates V_m still further, until a limit is reached).

Thus in Figure 5.15 it appears that the failure of the second stimulus to activate results from a depressed h (thus lowering I_{Na}) and also an elevated n (thus increasing I_{K}), as compared to values during the first stimulus.

Return to rest

Though there are refractory periods, it also is noteworthy that transmembrane voltage V_m and all the gating probabilities do return to their initial (resting) values gradually, as time elapses. In the simulation shown in Figure 5.15 it is seen that V_m and m are close to their resting values after about 6 milliseconds. Both n and h take longer, but are near their initial values by 15 milliseconds after the initial stimulus.

In Figure 5.15, note that V_m is *below* its resting value at 5 msec. That negativity occurs because n remains elevated so that the magnitude of I_{K} is greater than at rest, thus forcing V_m more negative. Ultimately the membrane returns to the same state as it was initially, where an action potential can be initiated by another stimulus of magnitude similar to the first.

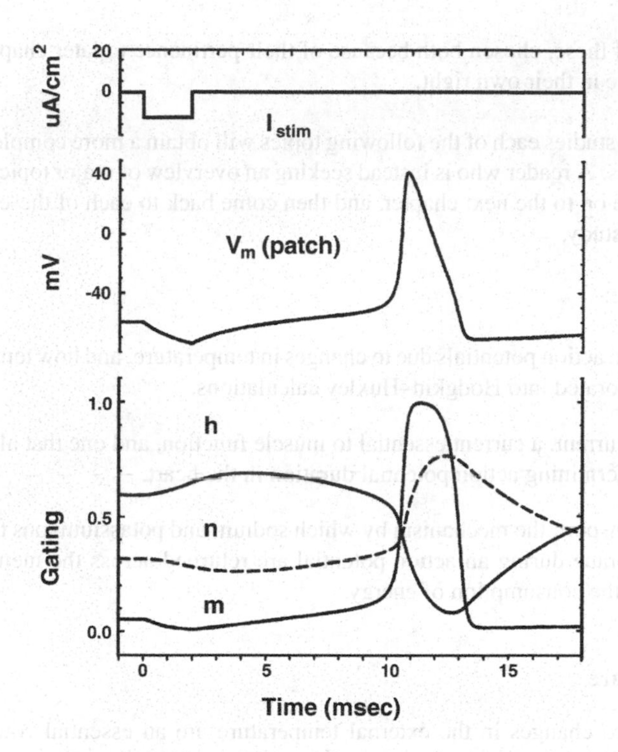

Figure 5.17. Anode Break Excitation. Computed values of m, n, h, and V_m from Hodgkin–Huxley equations. Space-clamped conditions. Values are computed during and after a hyperpolarizing pulse of duration 2 msec and magnitude 11.7 μA/cm^2, which starts at $t = 0$. The resting potential is -60 mV and the temperature is $T = 6.3°$C.

Anode break excitation

Figure 5.17 describes what happens after the termination of a 2-msec hyperpolarization and the sudden restoration of normal transmembrane potential. Just prior to release of the hyperpolarization, the value of h is elevated while m and n are reduced. However, m rapidly regains its normal value following restoration of normal V_m, since τ_m is relatively short.

The result, based on $\tau_m << \tau_n, \tau_h$, is that there is a depressed n, normal m, and elevated h. All three combine to promote $I_{\text{Na}} > I_{\text{K}}$. The consequence can be the initiation of excitation.

We shall see in the next chapter that with extracellular electrodes the membrane under the anode will be hyperpolarized during the stimulus. It is here that excitation can be initiated by the process just described [after a prolonged stimulation (hyperpolarization)]; this accounts for the name given it: *anode break* excitation.

5.5. BEYOND H-H MODELS

In the half-century since the advent of the Hodgkin–Huxley models, the quantitative analysis of electrophysiological events has advanced along multiple lines. The remainder of this chapter considers a few of these, chosen both because of their pertinence to later chapters, and because of their importance in their own right.

A reader who studies each of the following topics will obtain a more complete understanding of action potentials. A reader who is instead seeking an overview of major topics in bioelectricity may wish to move on to the next chapter, and then come back to each of these topics when it is required for later study.

The topics are:

- Changes in action potentials due to changes in temperature, and how temperature changes are incorporated into Hodgkin–Huxley calculations.
- Calcium current, a current essential to muscle function, and one that also plays a central role in determining action potential duration in the heart,
- Active transport, the mechanism by which sodium and potassium ions that diffuse across the membrane during an action potential are returned across the membrane, a process requiring the consumption of energy.

5.5.1. Temperature

Adaptations to changes in the external temperature are an essential component of living systems, as most physiological systems function at different rates at different temperatures. The expressions given by Hodgkin and Huxley for the α and β rate constants are for a temperature of 6.3°C, a natural temperature for a squid in seawater. Rate constants vary with temperature, as described here. Such a variation is consistent with experience and was introduced analytically in Chapter 3. As noted there, for equilibrium potentials, temperature is included in the equations for the Nernst potentials of sodium and potassium, e.g.,

$$E_K = -\frac{RT}{nF} \ln \frac{[K]_{in}}{[K]_{out}} \tag{5.63}$$

Definition of Q values

For rate constants, temperature changes are taken into account by adjusting the α and β values. Procedurally, one defines a parameter Q, also called Q_{10}, as

$$Q = 3^P \tag{5.64}$$

where

$$P = \frac{T - 6.3}{10} \tag{5.65}$$

In (5.65) the temperature must be specified in degrees Celsius. The reference temperature of 6.3°C comes from the temperature used by Hodgkin and Huxley for their original equations. That is, $Q = 1$ when the temperature is 6.3°C. With no Q present, the equations apply when the temperature is 6.3°C.

Temperature summary

The ratio of rate constants arising from a change of 10°C is described as its Q_{10}, and its value is normally assumed to be 3. The α and β in earlier equations can be converted from 6.3°C to any temperature T by multiplying by $Q = 3^{(T-6.3)/10}$ [e.g., $\alpha(T) = Q\alpha(6.3)$].

Use of Q values

The Q value affects the rates of change of n, m and h in the following way:

$$dn/dt = Q\alpha_n(1-n) - Q\beta_n n \tag{5.66}$$

$$dm/dt = Q\alpha_m(1-m) - Q\beta_m m \tag{5.67}$$

$$dh/dt = Q\alpha_h(1-h) - Q\beta_h h \tag{5.68}$$

That is, each α or β value at 6.3°C is replaced by Q times that value.

5.5.2. Calcium Currents

Intracellular and extracellular calcium concentrations are generally very small, as are calcium transmembrane currents. The study of calcium behavior before the patch clamp electrode was particularly difficult because sodium and potassium currents are so very much larger, making isolation of the calcium component very difficult. However, with a voltage clamp patch electrode it has been possible to obtain useful data in recent years.

A current–voltage curve for calcium is given in Figure 5.18. This was obtained using a patch pipette and an isolated bovine chromaffin cell. The potassium current was eliminated by using an intracellular potassium-free solution, which also included the potassium inhibitor Cs^+ and blocker TEA. The sodium current was blocked extracellularly with TTX.

The isolated calcium current is seen to be in the pA range, and roughly four orders of magnitude lower than peak sodium and potassium flux. A comparison with sodium also demonstrates the much larger depolarization required to elicit (Ca^{++}) activation (but, if the sodium action potential is inhibited, a true calcium action potential is obtained). For increasing depolarizations, the inward calcium current reaches a peak and then diminishes as the driving force ($E_{\text{Ca}} - V_m$) decreases.

As noted above, the intracellular free calcium concentration in excitable cells is very low, a typical value being 100 nM. Extracellular calcium may be four orders larger; the value in Figure 5.18 is 5 mM.

For the calcium single channel the ohmic behavior seen in sodium and potassium channels, and described in (3.26) and (3.27), cannot be expected. The reason is that at large depolarized potentials membrane flux is limited by the lack of availability of (intracellular) calcium ions. Because only a very small calcium efflux can take place, an inward-going rectification results.

The GHK current equation is an appropriate foundation for describing calcium currents [14]. Since calcium has valence $z_{\text{Ca}} = +2$, the calcium current, using (5.91), has the form

$$I_{\text{Ca}} = 4\frac{P_{\text{Ca}} V_m F^2}{RT} \frac{[\text{Ca}]_o - [\text{Ca}]_i e^{2V_m F/RT}}{1 - e^{2V_m F/RT}} \tag{5.69}$$

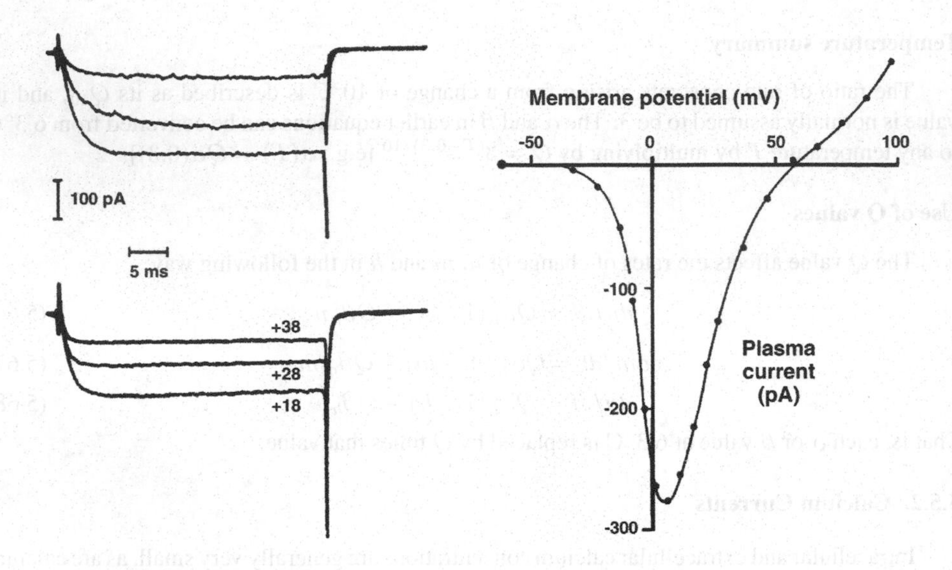

Figure 5.18. Current–Voltage Relations for plateau current amplitudes measured in bovine chromaffin cells. The cells contain CsCl, TEA, and EGTA and are bathed in a solution containing TTX and 5mM Ca. (These steps inhibit the otherwise overwhelming sodium and potassium currents.) From Fenwick EM, Marty A, Neher E. 1982. Sodium and calcium channels in bovine chromaffin cells. *J Physiol* **331**:599–635.

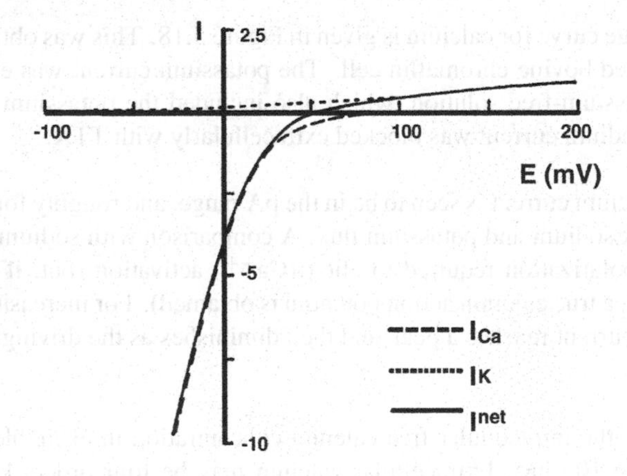

Figure 5.19. Theoretical *I–V* calcium Curve, as obtained from the GHK Eq. (5.69). The dashed line denotes the calculated values assuming $[\mathrm{Ca}]_i = 100$ nM and $[\mathrm{Ca}]_o = 2$ nM. Also plotted is the potassium current through the calcium channel based on (5.90), assuming $[\mathrm{K}]_i = 100$ nM, $[\mathrm{K}]_o = 2$ nM, and $P_{\mathrm{Ca}}/P_{\mathrm{K}} = 1000$. The solid curve is the total current. From Hille B. 1992. *Ionic currents*. Sunderland, MA: Sinauer Associates.

The single-channel *I–V* curve obtained from (5.69) is plotted in Figure 5.19 utilizing the above calcium concentrations. The curve shows the expected inward rectification.

The inward rectification is particularly large in view of the $[\mathrm{Ca}]_o/[\mathrm{Ca}]_i = 20000$ ratio and also that calcium is divalent. The calcium Nernst potential, given by $(58/2)\log_{10} 20000 = 124$

mV, is difficult to verify from the graph because of the small angle the curve makes with the zero-current axis.

Following Hille [8], we may also examine the effect of the presence of potassium, which while only slightly permeable in a calcium channel, can partially go through calcium channels anyway because of potassium's large concentration gradient. Assuming a relative permeability of $P_{Ca}/P_K = 1000$, the dotted curve in Figure 5.19 results for I_K. Also plotted is the net current $I_m = I_{Ca} + I_K$. This result shows the reversal potential at around 52 mV, less than half the calcium Nernst potential.

5.5.3. Active Transport

As we have noted, the electrical excitability of nerve and muscle depends on the ionic imbalance between intracellular and extracellular media. In view of the sodium influx both at rest and in an action potential, and in view of the potassium efflux under these same conditions, one would expect that after a while both intracellular and extracellular concentrations would reach a Donnan equilibrium (i.e., when the resulting concentrations generate equal Nernst potentials of all permeable ions resulting in equilibrium and an end to excitability).

This end would occur were it not for a process that transports these ions in the reverse direction. Since the above-described ion movements are directed down their electrochemical gradient, the reverse movement will require the expenditure of energy. For sodium, for example, transport out of the cell must overcome both the inward electric field and (inward) diffusional force.

Ion flow at rest and during an action potential is passive (down concentration gradients) and consequently the restoration of baseline conditions requires a similar movement of ions but necessarily against such gradients, hence requiring energy. The amount of energy must equal that consumed by current while at rest and during propagating action potentials.

This energy comes from an *active process*, an *ion pump*, which derives such energy from the hydrolysis of energy-rich ATP (adenosine triphosphate). The source of ATP is from the metabolism of the foods we consume. This represents an interesting chain of events, leading to the generation of electrical currents of the active cell.

Pump's characteristics

An experiment that illustrates the above remarks was performed by Hodgkin and Keynes [12] on the *Sepia* giant axon. The axon was first placed in a sodium-labeled potassium-free bath. The axon then was repetitively stimulated for a period of time.

The result was that the intracellular space became loaded with radioactive sodium. The axon then was transferred to a chamber, where it was perfused by normal seawater. The effluent carefully monitored for the $[Na^+]$-labeled ion (which is a measure of sodium efflux due to pump action).

As shown in Figure 5.20, in the first 100 min of measurement a log-linear efflux is detected. This log-linear relationship is explainable by a constant pump rate. With a constant rate, the

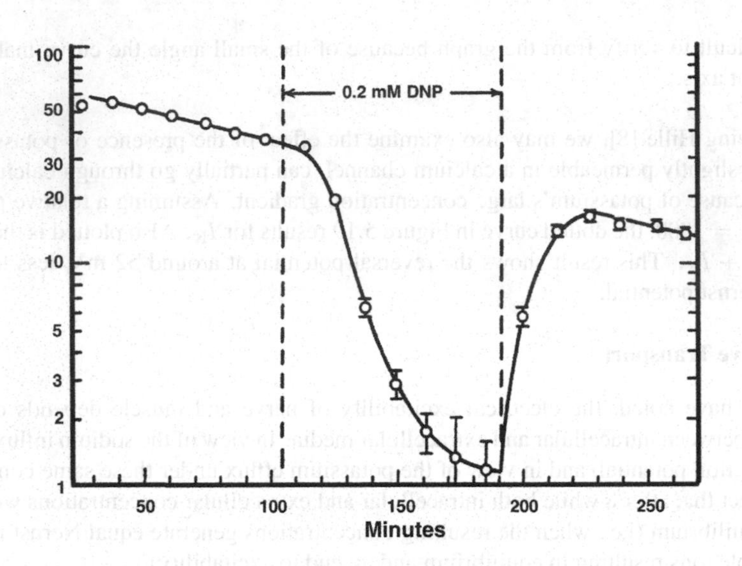

Figure 5.20. The effect of the metabolic inhibitor 2:4-dinitrophenol (DNP) on the efflux of radioactive sodium from a *Sepia* giant axon. From Hodgkin AL, Keynes RD. 1955. Active transport of cations in giant axons from *Sepia* and *Loligo*. *J Physiol* **128**:28–60.

efflux of labeled sodium is proportional to that present, a necessarily diminishing quantity. Stated mathematically,

$$-\frac{d[^{24}\mathrm{Na}^+]_i}{dt} = k[^{24}\mathrm{Na}^+]_i \tag{5.70}$$

so

$$[^{24}\mathrm{Na}^+]_i = A\exp(-kt). \tag{5.71}$$

where $[^{24}\mathrm{Na}^+]_i$ describes the intracellular concentration of the labeled sodium.

The addition of a metabolic inhibitor such as DNP or ouabain to the perfusate reduces the sodium efflux to a very small amount, as seen in Figure 5.20. This reduction confirms that the pump is metabolically driven and that the transport process is an active one.

Other experiments suggest that the rate of pumping is controlled by the intracellular (actual) sodium concentration. Furthermore, if the potassium is excluded from the extracellular medium, then the sodium efflux is reduced to one-third its normal value, suggesting that sodium extrusion is loosely coupled with potassium uptake. In fact, it had been thought at one time that for each sodium ion pumped out one potassium ion would be pumped in. If this were true, then there would be no net contribution to transmembrane current from the pump.

However, we now know that three sodium ions are extruded for two potassium ions taken up (a net current outflow) and consequently the active process contributes to the transmembrane current. Under these circumstances, we say that the pump is *electrogenic*.

Pump stoichiometry

A formal stoichiometric approach to pump behavior has been suggested by Chapman, Kootsey, and Johnson [3], namely,

$$\mathrm{ATP} + x[\mathrm{Na}]_i + y[\mathrm{K}]_o \longrightarrow \mathrm{ADP} + \mathrm{P}_i + x[\mathrm{Na}]_o + y[\mathrm{K}]_i \tag{5.72}$$

This expression describes the reduction of ATP to ADP + P_i in driving the process in an energy-consuming direction (i.e., the energy required for sodium efflux and potassium influx is provided by the energy derived from ATP).

As noted, $x = 3$, $y = 2$ appears to fit the experimental data, so that for each mole of ATP split, three moles of sodium are extruded, two moles of potassium are taken up, and a net efflux of one mole of cation occurs.

Pump included in steady-state model

The pump current can be included in a steady-state analysis. For example, it can be included in the parallel-conductance model. We continue to require that the total transmembrane current, I, under steady-state conditions be zero. However, the total current must now include the pump current, I_p, and hence,

$$I = I_{\mathrm{K}} + I_{\mathrm{Na}} + I_{\mathrm{Cl}} + I_p = 0 \tag{5.73}$$

Consequently, in place of (3.30) we have

$$g_{\mathrm{K}}(V_m - E_{\mathrm{K}}) + g_{\mathrm{Na}}(V_m - E_{\mathrm{Na}}) + g_{\mathrm{Cl}}(V_m - E_{\mathrm{Cl}}) = -I_p \tag{5.74}$$

Solving forV_m in (5.74) yields

$$V_m = \frac{g_{\mathrm{Na}}E_{\mathrm{Na}} + g_{\mathrm{K}}E_{\mathrm{K}} + g_{\mathrm{Cl}}E_{\mathrm{Cl}}}{g_{\mathrm{K}} + g_{\mathrm{Na}} + g_{\mathrm{Cl}}} - \frac{I_p}{g_{\mathrm{K}} + g_{\mathrm{Na}} + g_{\mathrm{Cl}}} \tag{5.75}$$

It is seen in (5.75) that the pump current contributes to the resting potential. In fact, since I_p (representing a net efflux of cation) is positive, Eq. (5.75) demonstrates (not surprisingly) that the pump causes an additional hyperpolarization of the membrane.

An application of (5.75) arises with fibers that are loaded with (additional) sodium by first being placed in a potassium-free medium at low temperature. Such a medium inhibits the pump and causes the accumulation of intracellular sodium through normal passive influx.

Placed subsequently in a normal extracellular medium, such fibers show resting potentials even more negative than the potassium Nernst potential, a result inexplicable from passive models alone. This condition can be explained as arising from an elevated I_p (due to elevated intracellular sodium concentration). Adrian and Slayman [1] obtained resting potentials 20 mV more negative than E_{K} in sodium-loaded muscle. Since the addition of ouabain (a metabolic inhibitor) to the extracellular medium is found to abolish this hyperpolarization, its metabolic origin was verified.

Under resting conditions the net passive (p) plus active (a) flux must be zero [as demanded in (5.73)]; however, this must also be true on an individual ion basis as well, since over time no change in intracellular or extracellular ionic composition can occur. Consequently,

$$p_{\mathrm{Na}} + a_{\mathrm{Na}} = 0 \tag{5.76}$$

and

$$p_{\mathrm{K}} + a_{\mathrm{K}} = 0 \tag{5.77}$$

where *p* is the passive and *a* the active flux of the subscripted ions; these are positive if outward, negative if inward. If the ratio of sodium to potassium ions exchanged by the pump is r (we have described $r = 1.5$), then more generally with $r = |a_{\mathrm{Na}}/a_{\mathrm{K}}|$, we have

$$r a_{\mathrm{K}} + a_{\mathrm{Na}} = 0 \tag{5.78}$$

Consequently, from (5.76) through (5.78) we get

$$r p_{\mathrm{K}} + p_{\mathrm{Na}} = 0 \tag{5.79}$$

From (5.78) and (5.79) it is apparent that the pumped sodium/potassium ratio must correspond to the passive ratio, a condition that must be true continuously while at rest and on the average under active conditions. If the GHK equation is used, and chloride is assumed essentially in equilibrium, then the resting condition (applied to the passive flux that is evaluated by this equation) requires

$$r I_{\mathrm{K}} + I_{\mathrm{Na}} = 0 \tag{5.80}$$

This result leads to

$$V_m = \frac{RT}{F} \ln \frac{r P_{\mathrm{K}}[\mathrm{K}]_o + P_{\mathrm{Na}}[\mathrm{Na}]_o}{r P_{\mathrm{k}}[\mathrm{K}]_i + P_{\mathrm{Na}}[\mathrm{Na}]_i} \tag{5.81}$$

This resulting equation for V_m may be seen as a replacement or improvement over (5.1), because it now takes ionic pumping into account. (Note that (5.81) reduces to (5.1) when $r = 1$].)

This expression (5.81) accounts for the resting potential lying closer to the potassium equilibrium potential (i.e., more negative) than otherwise expected. It fails to account for special situations (sodium loaded cells) where V_m is more negative than E_{K}. However, these instances appear to violate the assumption of a steady state (equilibrium) since the pump rate is abnormally high to compensate for abnormally high intracellular sodium.

5.6. APPENDIX: GHK CONSTANT-FIELD EQUATION

The GHK equation is an equation giving the amount of current across a membrane, I_{K}, as a function of the transmembrane voltage across the membrane V_m. The GHK equation is an alternative to the assumption by Hodgkin and Huxley that a linear approximation to the current–voltage relation, e.g., $I_{\mathrm{K}} = g_k(V_m - E_{\mathrm{K}})$, was sufficiently accurate.

Another aspect is that there are other ions than sodium, potassium, and chloride in the extracellular and intracellular spaces of excitable cells. For squid axon these other ions play a minor role and can be disregarded. In muscle and most nerve cells, however, *calcium* also is an important permeable ion.

5.6.1. Importance of Calcium Analysis

The calcium currents in physiological preparations are normally small. The small influx nonetheless does have important consequences. One such effect is the influence of the calcium ion on the gating properties of other ion channels. Calcium is also indispensable because of its role as an "intracellular messenger." By this term we refer to the action of calcium that results in muscle contraction. In other tissues it results in synaptic (chemical) release.

5.6.2. GHK Assumptions

Until now, we have assumed that the instantaneous current–voltage characteristic of an open ion channel is linear. Linearity, in addition to the independence principle, is the basis for the single-channel model, the parallel-conductance model, and the Hodgkin–Huxley model. Linear responses are, however, not quite right. Vertebrate Na and K channels in fact show a small rectification (meaning the plot of current-versus-voltage curves rather than falling along a straight line).

In other words, the relationship between current and voltage deviates from linearity. In this case a possibly better description is provided by the GHK current equation. This improvement is because, as we show below, the GHK current equation also shows an I–E curve that displays rectification.

Derivation of the GHK equations begins again with the Nernst–Planck equation (Chapter 3) and finds a description of current–voltage relationships. The goal is to do so without assuming linearity of current with voltage, as was done by Hodgkin and Huxley.

A further simplification that is used in deriving the GHK equations is to recognize that each membrane patch is essentially planar, in view of the very small membrane thickness. Thus a one-dimensional mathematical treatment is a good approximation, even when the macroscopic membrane shape is curved. One-dimensional variation allows variables to change as a function of distance across the membrane from intracellular to extracellular.

The difficulty that remains in following this plan, based on mathematical arguments only, is that the variation of C_p and also Φ within the membrane are unknown.

The assumption used by Goldman [17] to get around this difficulty was the following: Because the biological membrane is relatively thin, a plausible approximation to Φ within the membrane is to assume that it varies linearly. That is, the assumption is that the electric field is constant at each site within the membrane. With this assumption it becomes possible to integrate the Nernst–Planck equation across the membrane. Thus one can address questions such as how a change in electric potential across the membrane relates to changes in membrane permeability. In the GHK derivation it also was assumed (a) that steady-state conditions apply and (b) that variations in potential and concentrations within the membrane are transverse only.

These conditions are described in Figure 5.21 and in the following expression:

$$d\Phi/dx = [\Phi(d) - \Phi(0)]/d = -V_m/d \tag{5.82}$$

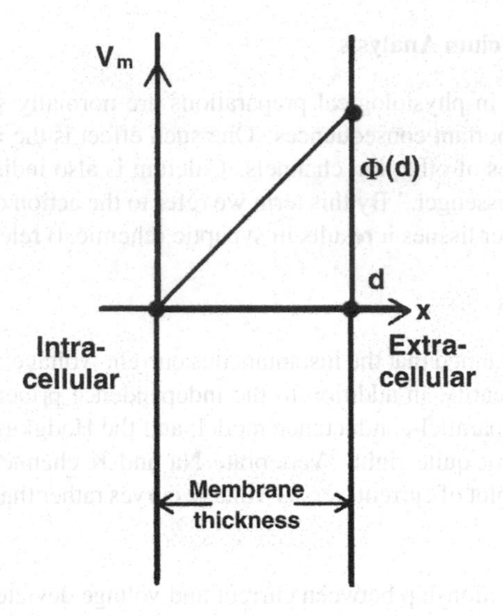

Figure 5.21. One-dimensional membrane model with linear variations of intramembrane potential for derivation of the constant-field equations of Goldman [17].

where V_m is the transmembrane potential and d is the membrane thickness, as shown in Figure 5.21. The polarity results from an association of $x = 0$ with the intracellular membrane edge and $x = d$ with the extracellular edge. For simplicity, we also restrict the following result to univalent cations (extension to multivalent ions is straightforward).[19]

5.6.3. Analysis for One Ion

Because $\nabla\Phi = d\Phi/dx$ and $\nabla C_p = dC_p/dx$ we obtain from (3.6)

$$j_p = -D_p \left[\frac{d[C_p]}{dx} + \frac{C_p F}{RT} \frac{d\Phi}{dx} \right] \tag{5.83}$$

as the flux of the pth ion per unit area. Specifically for the potassium ion, inserting the constant-field assumption expressed in (5.82) into (5.83) gives

$$\frac{d[C_K]}{dx} = -\frac{j_K}{D_K} + \frac{V_m F}{RTd}[C_K] \tag{5.84}$$

Rearranging (5.84) results in

$$\frac{d[C_K]}{-\frac{j_K}{D_K} + \frac{V_m F [C_K]}{RTd}} = dx \tag{5.85}$$

Equation (5.85) relates quantities at different x positions within the membrane. As such, it does not directly give the flux or flow across the membrane in terms of values across the membrane, or on its intracellular or extracellular edges. With the goal of finding the flow across the membrane from quantities known on the surface of the membrane, we integrate Eq. (5.85) across the membrane from the left ($x = 0$) to the right edge ($x = d$).

Performing the integration requires that one assume that the flow is at steady state. Under this condition, j_K is constant and has the same value at every x coordinate. (Otherwise, ionic concentrations would be changing at x positions where j_K changed.) Also, D_K is simply assumed to be constant within the membrane. The only variable quantity on the left-hand side of (5.85) is thus $C_K(x)$.

With these assumptions, integration with respect to x results in

$$\frac{RTd}{V_m F}\ln\left[\frac{\frac{V_m F}{RTd}[C_K]d - \frac{j_K}{D_K}}{\frac{V_m F}{RTd}[C_K]_0 - \frac{j_K}{D_K}}\right] = d \tag{5.86}$$

Equation (5.86) can be solved for j_K to yield

$$j_K = \frac{D_K V_m F}{RTd}\frac{[C_K]_d - [C_K]_0 e^{V_m F/RT}}{1 - e^{V_m F/RT}} \tag{5.87}$$

5.6.4. Boundaries at Membrane Surfaces

In Eq. (5.87) the concentration of potassium required is that within the membrane. This concentration within the membrane is, however, unknown, as the known potassium concentrations are those in surrounding intracellular and extracellular volumes.

The concentrations in these intracellular and extracellular spaces, however, do provide an important boundary condition for the concentrations within the membrane.

In other words, near the surfaces of the membrane the ionic concentrations within the membrane [i.e., those described in (5.87)] are related to those just outside the membrane, both on the intracellular and extracellular sides. They are linked by *partition coefficients* β. These partition coefficients are assumed to be identical at the two interfaces.

Consequently, if we denote edge 0 of the membrane to be in contact with the intracellular space of a cell and edge *d* in contact with the extracellular space (see Figure 5.21), then

$$[C_K]_d = \beta_K[\mathrm{K}]_{\mathrm{extra}} \quad \text{and} \quad [C_K]_0 = \beta_K[\mathrm{K}]_{\mathrm{intra}} \tag{5.88}$$

where [K] denotes the potassium concentration in the bulk media.

The electric current density due to potassium ion flow is J_K, which equals $F j_K$. We also can define potassium *permeability* P_K as

$$P_K \equiv D_K\,\beta_K/d \tag{5.89}$$

5.6.5. GHK equation for J

Using (5.88) and the definition (5.89) equation (5.87) becomes

$$J_K = \frac{P_K V_m F^2}{RT}\frac{[\mathrm{K}]_o - [\mathrm{K}]_i\, e^{V_m F/RT}}{1 - e^{V_m F/RT}} \tag{5.90}$$

Expression (5.90) is a major result. It is referred to as the Goldman–Hodgkin–Katz (GHK) current equation, in recognition of the contributions of these investigators [10]. If the ionic valence is not equal to +1, then a more general expression (for the pth ion) results, namely,

$$J_p = \frac{z_p^2 P_p V_m F^2}{RT} \frac{[p]_o - [p]_i \, e^{z_p V_m F/RT}}{1 - e^{z_p V_m F/RT}} \tag{5.91}$$

where, as usual, subscripts o and i refer to the extracellular and intracellular (bulk) media, and z_p is the valence.

5.6.6. Combined Flow of Several Ions

An expression similar to (5.90) arises for J_{Na} (replace K by Na), while for anions, such as J_{Cl}, only a slightly different expression results [substitute $z = -1$ in (5.91)], namely,

$$J_{\mathrm{Cl}} = \frac{P_{\mathrm{Cl}} V_m F^2}{RT} \frac{[\mathrm{Cl}]_i - [\mathrm{Cl}]_o \, e^{V_m F/RT}}{1 - e^{V_m F/RT}} \tag{5.92}$$

Our particular interest in potassium, sodium, and chloride currents arises since they are important components of the ion flux in most biological membranes. The total ionic current is the sum of the constituent ionic components and, assuming none in addition to K, Na, and Cl, we have

$$J = J_{\mathrm{K}} + J_{\mathrm{Na}} + J_{\mathrm{Cl}} \tag{5.93}$$

Substituting the expression for the indefinite ion p (5.91) into the term for each specific ion enables the summation (5.93) to become

$$J = \frac{V_m F^2 P_{\mathrm{K}}}{RT} \frac{w - y e^{V_m F/RT}}{1 - e^{V_m F/RT}} \tag{5.94}$$

where

$$w = [\mathrm{K}]_e + \frac{P_{\mathrm{Na}}}{P_{\mathrm{K}}}[\mathrm{Na}]_e + \frac{P_{\mathrm{Cl}}}{P_{\mathrm{K}}}[\mathrm{Cl}]_i \tag{5.95}$$

and

$$y = [\mathrm{K}]_i + \frac{P_{\mathrm{Na}}}{P_{\mathrm{K}}}[\mathrm{Na}]_i + \frac{P_{\mathrm{Cl}}}{P_{\mathrm{K}}}[\mathrm{Cl}]_e \tag{5.96}$$

5.6.7. GHK Resting Membrane Voltage

In a steady state, $\partial V_m/\partial t = 0$ and $J = 0$. (For a passive membrane, J is the total ionic flux, as in (5.93.) For $J = 0$, the GHK equations require that

$$(w - y e^{F V_m/RT}) = 0 \tag{5.97}$$

Hence one can solve the above equation for the resting transmembrane potential, V_{rest}, to get

$$e^{V_{\mathrm{rest}} F/RT} = w/y \tag{5.98}$$

or

$$V_{\mathrm{rest}} = \frac{RT}{F} \ln \frac{w}{y} \tag{5.99}$$

Applying (5.95) and (5.96) to (5.99) gives

$$V_{\text{rest}} = \frac{RT}{F} \ln \frac{P_{\text{K}}[\text{K}]_o + P_{\text{Na}}[\text{Na}]_o + P_{\text{Cl}}[\text{Cl}]_i}{P_{\text{K}}[\text{K}]_i + P_{\text{Na}}[\text{Na}]_i + P_{\text{Cl}}[\text{Cl}]_o} \tag{5.100}$$

Equation (5.100) is the GHK equation for resting transmembrane potential. It is a major result, as it allows the computation of an expected resting potential from a knowledge of membrane permeabilities. It applies specifically to an active membrane, for the case $J = 0$.

As with the parallel-conductance equation, the resting membrane potential is established by weighted contributions of the potassium, sodium, and chloride constituents. The weighting is described by permeabilities, in this case. While the permeabilities have the advantage of being defined in terms of basic physical parameters, in fact they are found experimentally.

5.7. NOTES

1. Some choices are available to the reader with the material in this chapter. A reader interested primarily in understanding the experimental and analytical methods used by Hodgkin and Huxley may wish to focus primarily on the first three sections. Conversely, a reader interested only in HH simulations may wish to move quickly to the fourth and fifth sections, referring back to earlier work only as needed. All readers may wish to use material from the sixth section selectively, depending on specific interest.
2. As compared to circuits designed by humans, the membrane's threshold response is more similar to that of a digital circuit, rather than the linear response of many analog systems.
3. Explain why the spatial action potential may (or may not) be a mirror image and changed in scale, compared to the temporal AP. Refer to Figure 8.2 to check your answer.
4. The Goldman-Hodgkin-Katz (GHK) equation is derived in the appendix of this chapter. Here this text quotes the GHK result and gives an explanation of its terms.
5. Note also that the distinction between flux and current used in this book so far is essentially their units, with current often in milliamperes and flux in milliamperes/cm^2. This association of the terms with these units is not used by everyone, i.e., one sees a number of variations in the units associated with the terms "current" and "flux" in the references and literature.
6. The Hodgkin and Huxley equations arise from measurements of transmembrane current from a very large number of channels. Consequently, they describe the space-averaged behavior of this ensemble. The HH model can be applied to any membrane element that has a number of channels large enough to allow a good statistical average. This condition is met in most experimental preparations and tissue models.
7. Making a choice of Δt involves issues of computational stability and accuracy, and also questions of what time resolution is needed in the result. In general more resolution and accuracy flow from smaller values of Δt, but shorter computation time and a smaller number of significant digits in intermediate results are benefits of larger values.
8. Recall that for a capacitor $\Delta V = \Delta Q/C$.
9. $V_r \approx -60mV$ for nerve. At rest it also is true that $V_m = (g_{\text{K}}E_{\text{K}} + g_{\text{Na}}E_{\text{Na}} + g_L E_L)/(g_{\text{K}} + g_{\text{Na}} + g_L)$ as shown in chapter 3.
10. Because calculations depend fundamentally on Δt rather than on absolute time t, the choice of time origin is arbitrary. Even so, certain conventions, such as applying the first stimulus at $t = 0$, often are followed.
11. Note that the HH equations for α and β require v_m as their argument, i.e., they are given by equations for deviation from baseline, rather than for V_m, the absolute transmembrane voltage. More recent membrane models usually define the functions with absolute voltage V_m as the argument.
12. There are several alternative plans for time-shifting values of n, m, and h that make use of known analytical results. For example, one such plan for n comes from (5.20). In this plan, first advance v_m. Then use v_m^{i+1} to obtain α_n^{i+1} and β_n^{i+1}. With these now obtain n_∞^{i+1} and τ_n^{i+1} using (5.21). An updated n_{i+1} is now obtained from n_i assuming a step change to v_m^{i+1} which is held for a time Δt.

$$n_{i+1} = n_\infty^{i+1} - (n_\infty^{i+1} - n_i)e^{-\Delta t/\tau_n^{i+1}}$$

Such a procedure is superior for simulating a voltage clamp, and might be superior otherwise when Δt is large. However, for short Δt, such a procedure is numerically sensitive and not an improvement, as can be seen from using a power series expansion of the e^{-u} term in the equation above.

13. That is, m changes quickly as V_m changes. Conversely, n and especially h respond, but more slowly.
14. Of course, one may wish to make the algorithm sensitive to the order of computation in steps 2 and 3, but if so one wants to do so in a planned and purposeful fashion rather than in some accidental manner.
15. Mixed times are especially pernicious when manually recalculating results to check for errors.
16. Historical footnote: Around 1950 Hodgkin and Huxley performed some of the first simulations of membrane action potentials. In the absence of present-day tools for high-speed numerical computation, the procedure used by Hodgkin and Huxley to simulate an action potential was the following. They assumed a uniformly propagating impulse, which enabled them to write an expression for the temporal behavior of the (propagating) action potential; the velocity is a parameter in this equation. A correct guess of the velocity was confirmed by a simulation that converged. More details of their simulation are given in Hodgkin AL, Huxley AF. 1952. A quantitative description of membrane current and its application to conduction and excitation in nerve. *J Physiol* **117**:500–544. However, by 1980 digital computer capability had advanced to the point where later investigators performed simulations of membrane action potentials routinely, although specialized facilities were required.
17. Such variations are of interest in their own right and also have value as a means of checking the calculations.
18. For example, Moore and Ramon (1974) were pathfinders; Beeler and Reuter adapted models to cardiac ventricular simulations (1977); Pollard, Hooke, and Henriquez (1992) show large-scale methods; Roth and Wikswo (1994) include the bidomain; and Cloherty, Dokos, and Lovel (2005) consider models from a more recent perspective. See the references at the end of this chapter.
19. In early models of the biological membrane it was viewed as analogous to an ion-exchange membrane (a homogeneous structure with uniformly distributed fixed charges). Both potential and permeable ion concentration was considered a function of a single transverse variable. A more recent view, the one largely taken in previous chapters, is that the membrane is predominantly an insulator pierced periodically by conducting channels that open and close. The latter are quasi-ohmic when open. The channel walls are lined with charges which add electrostatic forces to the electric field and diffusional forces. The constant-field model can be thought to apply, approximately, to all open ion channels. Hence, taking account of channel density, it applies to the macroscopic membrane itself.

5.8. REFERENCES

1. Adrian RH, Slayman CL. 1966. Membrane potential and conductance during transport of sodium, potassium, and rubidium in frog muscle. *J Physiol* **184**:970–1014.
2. Beeler GW, Reuter H. 1977. Reconstruction of the action potential of ventricular myocardial fibers. *J Physiol* **268**:177–210.
3. Chapman JB, Kootsey M, Johnson EA. 1979. Kinetic model for determining the consequences of electrogenic active-transport in cardiac muscle. *J Theor Biol* **80**:405–424.
4. Campbell DL, Giles WR, Hume JR, Doble D, Shibata EF. 1988. Reversal potential of the calcium current in bullfrog atrial myocytes. *J Physiol* **43**: 267–286.
5. Cole KS, Curtis HJ. 1939. Electrical impedance of the squid giant axon during activity. *J Gen Physiol* **22**: 649–670.
6. Cole KS, Marmont G. 1942. The effect of ionic environment upon the longitudinal impedance of the squid axon. *Fed Proc* **1**:15–16.
7. DiFrancesco D, Noble D. 1985. A model of cardiac activity incorporating ionic pumps and concentration charges. *Phil Trans R Soc London* **307**:353–398.
8. Hille B. 1992. *Ionic channels of excitable membranes*, 2d ed. Sunderland, MA: Sinauer Associates.
9. Hodgkin AL, Katz B. 1949. The effect of sodium ions on the electrical activity of the giant axon of the squid. *J Physiol* **108**:37–77.
10. Hodgkin AL, Huxley AF, Katz and B. 1952. Measurement of current–voltage relations in the membrane of the giant axon of *Loligo*. *J Physiol* **116**:424–448.
11. Hodgkin AL, Huxley AF. 1952. A quantitative description of membrane current and its application to conduction and excitation in nerve. *J Physiol* **117**:500–544.
12. Hodgkin AL, Keynes RD. 1955. Active transport of cations in giant axons from *Sepia* and *Loligo*. *J Physiol* **128**:28–60.
13. Keynes RD. 1951. The ionic measurements during nervous activity. *J Physiol* **114**:119–150.

14. Luo C-H, Rudy Y. 1994. A dynamic model of the cardiac ventricular action potential. *Circ Res* **74**:1071–1096.

15. Nastuk WL, Hodgkin AL. 1950. The electrical activity of single muscle fibers. *J Cell Comp Physiol* **35**:39–73.

16. Schoepfle G, Johns GC, Molnar GF. 1969. Simulated responses of depressed and hyper polarized medullated nerve fibers. *Am J Physiol* **216**:932–938.

17. Goldman DE. 1943. Potential, impedance, and rectification in membranes. *J Gen Physiol* **27**:37–60.

Additional References

Aidley DJ. 1978. *The physiology of excitable cells*. Cambridge: Cambridge UP.

Plonsey R. 1969. *Bioelectric phenomena*. New York: McGraw-Hill.

Randall JE. 1987. *Microcomputers and physiological simulation*, 2nd ed. New York: Raven Press.

Cronin J. 1987. *Mathematical aspects of Hodgkin–Huxley neural theory*. Cambridge: Cambridge UP.

Cloherty SL, Dokos S, Lovell NH. 2006. A comparison of 1D models of cardiac pacemaker heterogeneity. *IEEE Trans Biomed Eng* **53**:164–77.

Pollard AE, Hooke N, Henriquez CS. 1992. Cardiac propagation simulation. *Crit Rev Biomed Eng* **20**:342–359.

Moore JW, Ramon F. 1974. On numerical integration of the Hodgkin and Huxley equations for a membrane action potential. *J Theor Biol* **45**:249–273.

6

IMPULSE PROPAGATION

If a long thin fiber is initially depolarized at one end, the initial stimulus and initial depolarization at first has a limited extent. Thereafter *propagation* from the active (already depolarized) region to adjoining regions will take place.

This chapter considers "impulse propagation" (action potential conduction) in a single fiber. We note that the cylindrical fiber configuration is one that is found in nerve and striated muscle and is of interest in its own right. Furthermore, one can utilize the results from this specialized geometry in more complex and realistic preparations such as the nerve trunk, muscle bundle, and even for cardiac muscle.

It is helpful to keep in mind that broader usage of the word "propagation" has several distinct meanings. In one meaning, which does not apply to action potentials, propagation refers to an object that moves from place to place, such as a bowling ball rolling down the lane of a bowling alley. More abstractly, one thinks of a packet of energy moving from one place to another. In another meaning, which does apply to action potentials, propagation refers to a series of events where each event triggers one nearby, such as a wave of flame advancing through trees in a forest fire. In electrophysiology, "propagation" is used in the sense of the second meaning: each patch of excitable membrane initiates an action potential at an adjacent patch, which then creates its own sequence of action potential events, including initiating yet another patch.

The process by which excitation at one patch on a membrane initiates excitation at an adjacent patch is complex in that it involves multiple things happening at the same time. Each thing by itself is, however, relatively simple. Thus this chapter has a series of sections that consider each of the component parts. The challenge to the reader is to put these parts together, mentally, into a unified picture of propagation.

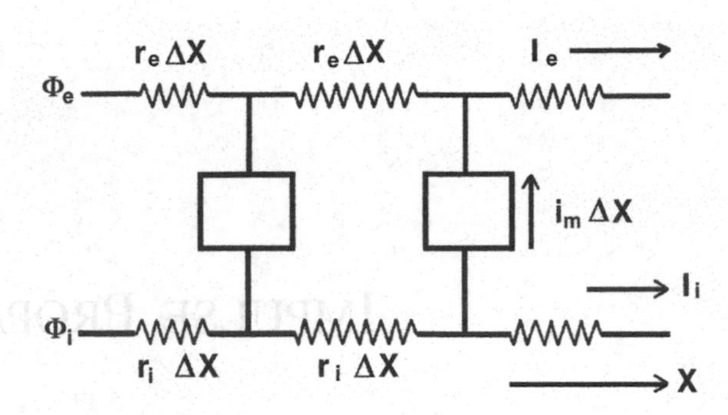

Figure 6.1. The Linear Core-Conductor Model for a single fiber lying in a restricted extracellular space. Longitudinal extracellular and intracellular currents are I_e and I_i, while extracellular and intracellular potentials are designated Φ_e and Φ_i respectively.

6.1. CORE-CONDUCTOR MODEL

The phrase core-conductor model refers broadly to a set of concepts used for analyzing problems existing mostly along one dimension, such as an excitable fiber. The phrase also refers to a set of fairly formal assumptions about the location and direction of allowed current flow.

The core-conductor model is closely tied to a collection of equations, called the *cable equations*, that show quantitatively how variables along and across the fiber relate one to another. The core-conductor model is extremely useful in electrophysiology, and it is the easiest framework for understanding propagation. Thus we consider each of the aspects of the core-conductor model, in turn.

6.1.1. Electrical Model for Single Excitable Fiber

An electrical model of a single fiber can be developed based on concepts from electric circuits. If the reasonable assumption of axial symmetry is made, then the resultant model is essentially one dimensional. For an excised fiber with a confined bounding fluid, extracellular currents flow in the longitudinal direction, except where they cross the membrane. Inside the confined intracellular region, current is again mostly one directional, along the fiber's axis.

The model depicted in Figure 6.1 reflects this basic expectation. In the figure, I_e is the longitudinal current in the extracellular region and I_i the longitudinal current in the intracellular region.[1]

The extracellular and intracellular resistance to flow is basically that of a cylindrically shaped resistance; it is portrayed in Figure 6.1.

In Figure 6.1 the potential along the extracellular path is designated Φ_e, while that along the intracellular path is Φ_i. For a graphical representation the structure is illustrated as a repetitive network of segment length Δx, but in fact $\Delta x \to 0$, and the analysis here is based on the continuum. The model illustrated in Figure 6.1 is known as the *linear core-conductor* model.

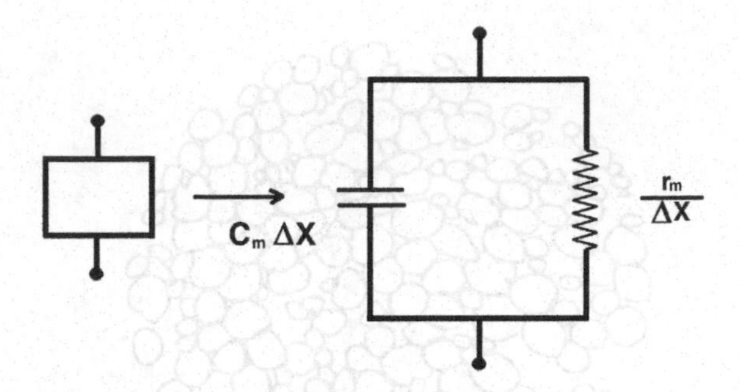

Figure 6.2. Electrical Representation of a Cylindrical Fiber Membrane element of length Δx. Under (linear) subthreshold conditions, r_m is constant.

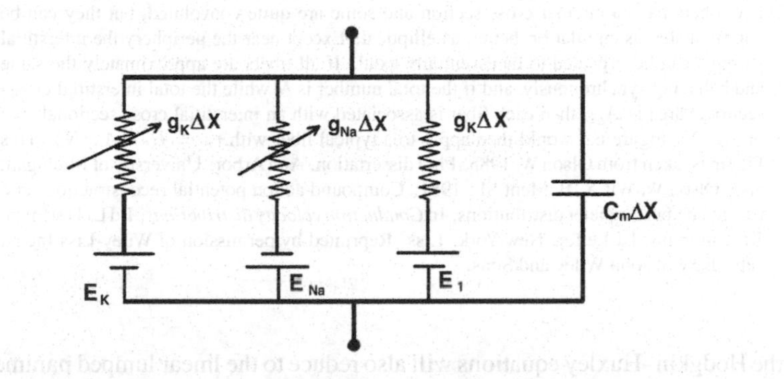

Figure 6.3. Electrical Representation of the Membrane for a fiber of length Δx under transthreshold conditions. The conductances $g_{\mathrm{K}}, g_{\mathrm{Na}}$ and g_l are found from the Hodgkin–Huxley equations and converted to units of S/cm for the linear core-conductor model.

The transmembrane electrical behavior of a fiber depends on the properties of the membrane. There are two distinct conditions of interest. The first is under subthreshold excitation when, as we have noted, each membrane segment behaves as a simple, passive, *RC* structure. The second is under transthreshold (or near-threshold) conditions when the membrane behavior is nonlinear and requires a description such as given by the Hodgkin–Huxley equations. The open box in Figure 6.1 is a symbol for either one of these two membrane conditions.

In Figure 6.2 we show, graphically, the model of a membrane element. Under subthreshold conditions, both R and C are nearly constant.

Figure 6.3 shows (the electrical representation under transthreshold conditions. In the latter Figure the schematic expression of the Hodgkin–Huxley model for ion current plus the capacitive current is given. The conductivities in Figure 6.3 are nonlinear and determined from the Hodgkin–Huxley equations ($g_{\mathrm{K}} = \overline{g}_{\mathrm{K}} n^4, g_{\mathrm{Na}} = \overline{g}_{\mathrm{Na}} m^3 h$, etc.). This representation is required for both transthreshold or near-threshold conditions (i.e., when the membrane behavior is nonlinear).

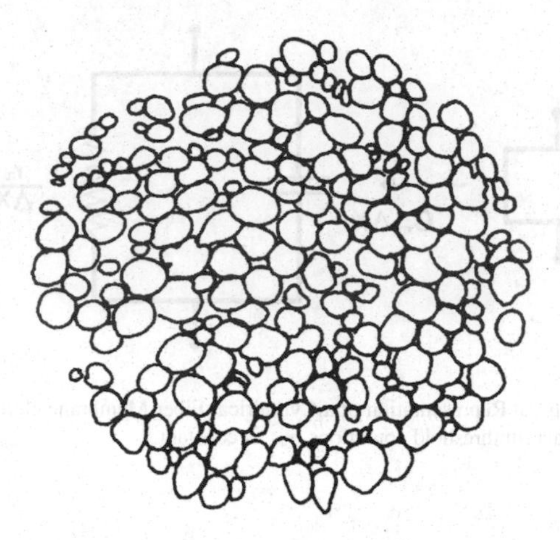

Figure 6.4. Photomicrograph of a Transverse Section of a Cat Saphenous Nerve Fascicle. Few fibers have a circular cross-section and some are quite convoluted, but they can be approximated as circular or, better, as elliptical. Except near the periphery the interstitial currents can be expected to be essentially axial. If all fibers are approximately the same and behaving synchronously, and if the total number is N while the total interstitial cross-sectional area is $A_{e'}$, then each fiber is associated with an interstitial cross-sectional area of A_e/N. Figure 6.1 would then apply to a typical fiber with $r_e = R_e/(A_e/N)$. This Figure is taken from Olson W. 1985. PhD dissertation. Ann Arbor: University of Michigan; also, Olson W, Wit X, BeMent SL. 1981. Compound action potential reconstructions and predicted fiber diameter distributions. In *Conduction velocity distributions*. Ed LJ Dorfman, KL Cummins, LJ Leifer. New York: Liss. Reprinted by permission of Wiley-Liss Inc., a subsidiary of John Wiley and Sons.

However, the Hodgkin–Huxley equations will also reduce to the linear lumped parameter network given in Figure 6.2 for subthreshold signals.

6.1.2. Core-Conductor Model Assumptions

The assumptions that underlie the linear core-conductor model are as follows:

1. Axial symmetry is assumed, that is, $\partial/\partial\xi = 0$ (where ξ is the azimuth angle). Thereby all field quantities are functions of r and x (cylindrical coordinates) at most. In fact, we usually assume that transmembrane and longitudinal currents as well as intracellular and extracellular potentials are functions only of the axial coordinate x (i.e., one dimensional). It is in this sense that the linear core-conductor model is *linear*.

2. Consistent with item 1, it is assumed that the external path carries axial current only. Thus the model may represent an excised fiber lying in the air but with a small film of extracellular fluid (appropriate in an in vitro study). But it would also describe a *typical fiber* in a large fiber bundle, such as illustrated in Figure 6.4. In this case extracellular (interstitial) current is also confined to the axial direction, except possibly for peripheral fibers.

   For a single fiber in an extracellular medium of considerable extent, sometimes one can set $r_e \approx 0$ in Figure 6.1, because the extracellular potentials are small. (Resistance r_e is low because there are wide pathways for parallel current.) In setting $r_e = 0$, one forgoes an ability to determine extracellular potential variations from this model, because they are set equal to zero at the outset. On the other hand, the linear core-conductor model may correctly evaluate all other (mainly intracellular) fields; the extracellular field can then be found subsequently using a field approach.

3. The internal conductive path is assumed to confine current to the axial direction alone. Since in general the fiber radius is many times smaller than fiber length, this approximation is normally very well satisfied.

4. For nerve and muscle under passive conditions the membrane is represented by a parallel combination of the leakage resistance r_m (Ωcm) and membrane capacitance c_m (μF/cm). Under active conditions, a constant r_m no longer suffices in determining the transmembrane ionic current because it is not a constant, and the Hodgkin–Huxley (or similar) formulation is required.

The core-conductor model and its mathematical children, the *cable equations*, apply imperfectly to any real situation. It is remarkable, however, how valuable they have been found to be, both in bioelectricity and in other problems (such as the trans-Atlantic telegraph cable). Their value arises because they capture essential relationships that are present between variables along and across the cable at each position along its length.

6.2. CABLE EQUATIONS

Application of Kirchhoff's laws (for electrical circuits) to the core-conductor model network leads to the *cable equations*, which are described in the sections that follow. These equations are the basic mathematical relationships used to study the electrical response of a uniform fiber to subthreshold and transthreshold stimuli.

The following sections are organized to answer three questions in succession:

1. How are axial currents related to the voltage across the membrane, V_m? The answers to this question provide the foundation required for answering the next two.

2. How does one find the membrane current, i_m, if one knows only V_m? Answering this question is critical to the sequence of steps in analyzing or simulating propagation.

3. Finally, if one knows the transmembrane potential V_m, how does one find the potentials ϕ_i and ϕ_e, the intracellular and extracellular potentials individually? Relating the potentials inside and outside to the transmembrane potentials is fundamental to understanding what occurs within and around fibers as a consequence of transmembrane voltage changes.

Essentially, these questions ask how all the other important variables can be found from V_m, if one knows, at a particular moment, V_m at all the positions up and down the length of the fiber. The questions are posed in this way because V_m often is the base variable known from measurement or computer simulation. If V_m is known, other values can usually be deduced.

6.2.1. Axial Currents

Axial currents from potentials

The decrease in potential per unit length along the intracellular (or interstitial) axial path equals the axial current times the resistance/length (i.e., the "IR drop") according to Ohm's law. Consequently,

$$\frac{\partial \Phi_e}{\partial x} = -I_e r_e \tag{6.1}$$

and

$$\frac{\partial \Phi_i}{\partial x} = -I_i r_i \tag{6.2}$$

In (6.1) and (6.2) I_e and I_i are the axial extracellular and intracellular current, and r_e and r_i are the intracellular and extracellular axial resistances per unit length. The axial variable is x. The minus signs (the lack of which are the source of many errors) arise because we define positive longitudinal currents to be flowing in the positive x direction. The potential must decrease with increasing x for current to flow in the positive x direction (because current flows from a higher to a lower potential region).

Axial current linked to membrane currents, stimulation

Intracellular Axial. If current leaves the intracellular space by crossing the membrane, then the longitudinal intracellular current will show an axial decrease while a positive transmembrane current will also be seen. The loss of longitudinal current (per unit length) must precisely equal the transmembrane current (per unit length), because the total current must be conserved. This conclusion is a simple application of the conservation-of-current principle of Kirchhoff's current law.

Expressed mathematically, where i_m is the transmembrane current per unit length, we have

$$\frac{\partial I_i}{\partial x} = -i_m \tag{6.3}$$

Thus in Figure 6.5 I_i increases at b because i_m is negative (inward), and I_i decreases at d because i_m is positive (outward).

Extracellular Axial. The extracellular longitudinal current may increase with increasing axial distance x either from the arrival of current that crosses the membrane (transmembrane current, i_m) or due to the introduction of a stimulus current from outside the preparation through inserted electrodes.

For convenience, the stimulus current is expressed as a current per unit length, i_p, where i_p is positive for current entering the extracellular space via *polarizing* electrodes. These units then correspond to the same units used for i_m. Taking both i_p and i_m into account, we have

$$\frac{\partial I_e}{\partial x} = i_m + i_p \tag{6.4}$$

Thus in Figure 6.5 I_e increases at a, because some of the stimulus current remains on the extracellular side, that is, in the Figure i_m is negative, but i_p is even more positive.

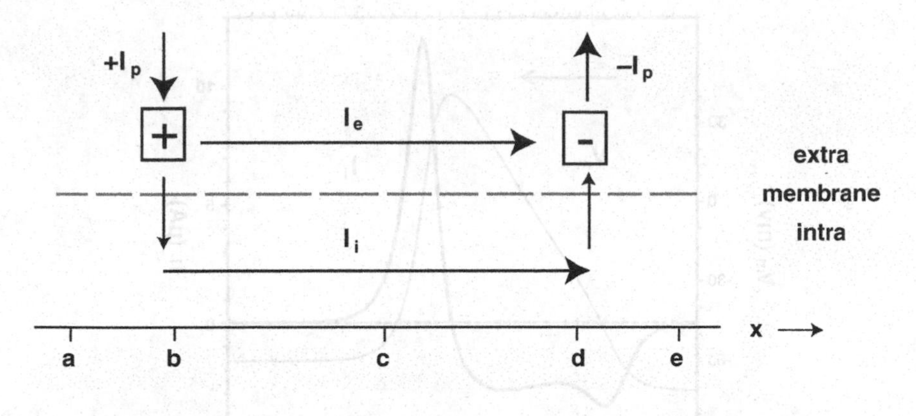

Figure 6.5. Current Pathways Cartoon. A membrane (dotted line) separates the extracellular space (above) from intracellular space (below). Letters a through e at the bottom identify particular x coordinates. Stimulus current I_p enters the extracellular space through an extracellular current source at b, and leaves the extracellular space through an extracellular current sink at d. As drawn, the stimulus current divides into an intracellular component I_i and an extracellular component I_e at b, then returns at d. A negative transmembrane current (negative because inward) exists at b and a positive transmembrane current exists at d. Total current I is zero at a and e but equal to I_p at c. For purposes of discussing an example, it is convenient to assign $I_p = 3$ and $I_i = 1$, so $I_e = 2$. Thus the membrane current is -1 at b and $+1$ at d.

It is also possible to insert polarizing electrodes into the intracellular space via microelectrodes, in which case i_p must be included in (6.3) and not (6.4).

Since i_p describes applied current *density*, current from a *point* electrode will require a delta function description. A single electrode at the origin inserting 1 mA will be described by $i_p = 1.0\delta(x)$ mA; δ is a unit delta function, namely, $\delta(x) = 0$ $(x \neq 0)$, $\delta(x) = \infty$ $(x = 0)$, and $\int \delta(x)dx = 1$ (assuming the limits include the origin).

Total Axial Current. The total axial current refers to the sum of the currents in the intracellular and extracellular regions at each point along the fiber. Analytically, use of the total axial current is particularly helpful in determining the effects of stimuli, where one knows the total current applied but not always its division across the membrane. Suppose I is defined as

$$I = I_i + I_e \tag{6.5}$$

Then from (6.3) and (6.4) one has

$$\frac{\partial I}{\partial x} = i_m + (-i_m + i_p) = i_p \tag{6.6}$$

Thus in Figure 6.5 $I = 0$ at a and e, but $I = I_i + I_e = 3$ at c. (An interesting exercise is to apply each of Eqs. (6.3)–(6.5) to each of the x coordinates a to e in Figure 6.5.)

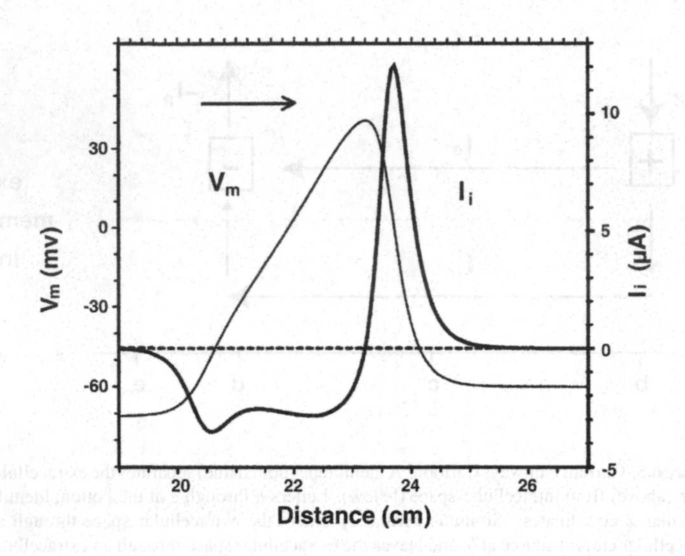

Figure 6.6. Transmembrane Potential and Axial Current from simulation of propagation on squid axon of diameter 600 μm at $T = 6.3°$C. Hodgkin–Huxley membrane parameters and equations are utilized. $R_i = 30\ \Omega$cm and $R_e = 20\ \Omega$cm. The Figure describes the spatial behavior of the transmembrane potential (scale on the left) and the intracellular axial current (scale on the right). The lines overlap so that the time relationships are evident. V_m is given in mV and I_i in μA.

Axial currents linked to transmembrane potentials

Since, by definition, $V_m \equiv \Phi_i - \Phi_e$, we have

$$\frac{\partial V_m}{\partial x} = \frac{\partial \Phi_i}{\partial x} - \frac{\partial \Phi_e}{\partial x} = -r_i I_i + r_e I_e = -r_i I_i + r_e(I - I_i) \tag{6.7}$$

where (6.1) and (6.2) have been employed.

Now, by simplifying (6.2), we obtain

$$\frac{\partial V_m}{\partial x} = -(r_i + r_e)I_i + I r_e \tag{6.8}$$

Rearranging this equation (6.8) algebraically, one obtains

$$I_i = \frac{-1}{(r_i + r_e)} \left[\frac{\partial V_m}{\partial x} - I r_e \right] \tag{6.9}$$

An example of the relationship of axial current and V_m is seen in Figure 6.6, where V_m and I_i are plotted on a common horizontal axis. As expected from expression (6.9), axial current I_i is most positive in the Figure near $x = 24$ cm, where the slope of the V_m waveform is its most negative.

In the simulation plotted in the figure, the time shown is after the stimulus is over; hence $I = 0$, so that (6.9) shows that I_i is proportional to the negative of the derivative of V_m. As the action potential moves to the right (in the figure, toward positive x), one sees that the largest intracellular current is also to the right. The direction of the axial current is to the right, toward the part of the fiber not yet depolarized.

6.2.2. Membrane Currents

With the core-conductor model it is easy to find membrane current at any point along the fiber from $V_m(x)$ or from $\phi_i(x)$. The following sections consider these cases in turn.

Membrane current from V_m

The equations developed above to answer the first question now provide us a base of reference information that can be immediately exploited to answer a critical question: How can the membrane current at a particular point along the fiber be found, if one knows V_m along the fiber?

If (6.8) is differentiated with respect to x, then

$$\frac{\partial^2 V_m}{\partial x^2} = -(r_i + r_e)\frac{\partial I_i}{\partial x} + r_e\frac{\partial I}{\partial x} \tag{6.10}$$

Substituting for the derivatives of I_i (6.3) and I (6.6), we have

$$\frac{\partial^2 V_m}{\partial x^2} = (r_i + r_e)i_m + r_e i_p \tag{6.11}$$

Equation (6.11) is valid under the core-conductor assumptions made earlier. Here again we note that the use of either V_m or v_m is correct, since both will have the same spatial derivative.

Algebraic rearrangement of (6.11) gives the expression for the membrane current as

$$i_m = \frac{1}{(r_i + r_e)}\left(\frac{\partial^2 V_m}{\partial x^2} - r_e i_p\right) \tag{6.12}$$

Equation (6.12) shows that the membrane current, i_m, is proportional to the second spatial derivative of the transmembrane potential V_m, if $i_p = 0$. The proportionality involves both the intracellular and extracellular quantities, r_i and r_e. These relationships are illustrated in Figure 6.7, where one sees that the outward (positive) peak of transmembrane current i_m is at the leading edge of the action potential, as it moves to the right. The peak inward current occurs during the upstroke of v_m. Two later and much smaller peaks of inward and outward current occur later in the action potential.

Membrane current from Φ_i

Consider now the equation for i_m that results from differentiating Eq. (6.2) with respect to x and then using (6.3) for $\partial I_i/\partial x$. The result is

$$i_m = \frac{1}{r_i}\frac{\partial^2 \phi_i}{\partial x^2} \tag{6.13}$$

Equation (6.13) is valid whether or not the assumption of axial extracellular current [necessary for (6.1), and hence (6.23), to be true] is satisfied. Here i_m is found solely from intracellular quantities.

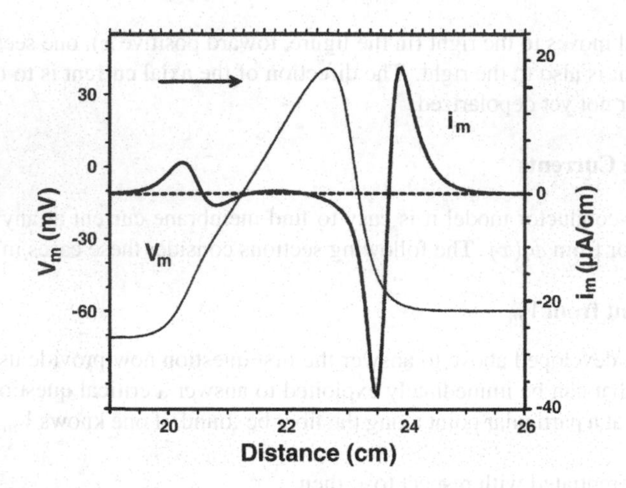

Figure 6.7. Propagating Action Potential and Transmembrane Current. The data come from a simulation of propagation on squid axon of diameter 600 μm at $T = 6.3°$C. The parameter values follow those of Hodgkin–Huxley, whose equations are utilized.

Note that up to this point no assumptions have been made about whether the membrane is subthreshold or transthreshold and few regarding the presence or absence of stimulating currents. The expressions reflect the linear (resistive) character of the intracellular and extracellular regions. The expressions, in other words, describe the intrinsic electrical properties of fibers (aside from their membranes).

6.2.3. Potentials ϕ_i and ϕ_e from V_m

We now pick up again from the answer to question 1 to pursue a different line of reasoning, that of finding ϕ_i and ϕ_e if one knows $V_m(x)$

From (6.8), the equation for the spatial derivative of V_m, and using (6.2), which gives the derivative of ϕ with respect to x, one obtains

$$\frac{\partial V_m}{\partial x} = \frac{(r_i + r_e)}{r_i}\frac{\partial \Phi_i}{\partial x} + Ir_e \tag{6.14}$$

Rearranging (6.14) gives

$$\frac{\partial \Phi_i}{\partial x} = \frac{r_i}{r_i + r_e}\frac{\partial V_m}{\partial x} - \frac{r_i r_e}{r_i + r_e}I \tag{6.15}$$

because

$$\frac{\partial \Phi_e}{\partial x} = \frac{\partial \Phi_i}{\partial x} - \frac{\partial V_m}{\partial x} \tag{6.16}$$

Then substituting (6.15) into (6.16) results in

$$\frac{\partial \Phi_e}{\partial x} = \left(\frac{r_i}{r_i + r_e}\frac{\partial V_m}{\partial x} - \frac{r_i r_e}{r_i + r_e}I\right) - \frac{\partial V_m}{\partial x} \tag{6.17}$$

and therefore

$$\frac{\partial \Phi_e}{\partial x} = \frac{r_e}{r_i + r_e}\frac{\partial V_m}{\partial x} - \frac{r_i r_e}{r_i + r_e} I \tag{6.18}$$

If it is assumed that all applied currents lie in a finite region near the coordinate origin, then at $x = \infty$ the membrane may be assumed to be in the resting state and hence $v_m = 0$.[2]

We also choose as a reference $\phi_e(\infty, t) = 0$, so that $\phi_i(\infty, t) = 0$. If (6.18) is now integrated from an arbitrary value of x to ∞, then

$$\phi_i(\infty, t) - \phi_i(x, t) = \frac{r_i}{r_i + r_e}[v_m(\infty, t) - v_m(x, t)] - \frac{r_i r_e}{r_i + r_e}\int_x^\infty I(x)dx \tag{6.19}$$

and, recalling boundary conditions at infinity, we have

$$\phi_i(x, t) = \frac{r_i}{r_i + r_e} v_m(x, t) + \frac{r_i r_e}{r_i + r_e}\int_x^\infty I(x)dx \tag{6.20}$$

In (6.20), the first term on the right corresponds to the homogeneous solution of (6.15) (i.e., with $I = 0$), while the second term is the particular solution. Similarly, from (6.18) integration yields

$$\phi_e(\infty, t) - \phi_e(x, t) = -\frac{r_e}{r_i + r_e}[v_m(\infty, t) - v_m(x, t)] - \frac{r_i r_e}{r_i + r_e}\int_x^\infty I(x)dx \tag{6.21}$$

for which

$$\phi_e(x, t) = -\frac{r_e}{r_i + r_e} v_m(x, t) + \frac{r_i r_e}{r_i + r_e}\int_x^\infty I(x)dx \tag{6.22}$$

Consider a propagating action potential, aside from some short initial period of stimulation. Under this circumstance,

- $I = 0$ and the integrals in (6.20) and (6.22) drop out.
- If all i_p lie in the region $x \leq 0$, say, then for $x \geq 0$ it is again true that $I = 0$ and the integrals drop out (for $x \geq 0$).
- For these conditions ϕ_i, ϕ_e, and v_m are linked by voltage-divider expressions.

The voltage-divider expressions are

$$\phi_i = \frac{r_e}{r_i + r_e} v_m \tag{6.23}$$

and

$$\phi_e = -\frac{r_i}{r_i + r_e} v_m \tag{6.24}$$

Note that (6.23) and (6.24) satisfy the requirement that $v_m = \phi_i - \phi_e$.

6.3. PROPAGATION

When excitation is initiated at a site on an excitable fiber, the changes at the site of initiation include a rapid increase in sodium permeability, an influx of sodium ions, and a change in transmembrane potential to a voltage approaching the sodium Nernst potential. These changes are those of a membrane patch undergoing an action potential, as described in the preceding chapter.

That is not all that happens, as it is observed that the action potential at the first patch somehow initiates excitation at the sites nearby, and then those patches at sites at still greater distances, until a chain-reaction process has occurred, and all adjacent excitable membranes have undergone the action potential sequence. The question arises naturally as to how this sequence happens, that is, why does any portion of the fiber become excited, other than the first? One might argue that, after all, excitation was initiated only in one place.

6.3.1. Local Circuit

To understand why a single patch on a continuous, excitable and unmyelinated fiber cannot be active alone, consider the currents that flow in the vicinity of the leading edge of an action potential. As shown in Figure 6.6, there is an intense intracellular axial current in the midst of the action potential upstroke.

In Figure 6.6 the influx of sodium associated with the rising phase of the action potential crosses the membrane and then flows in the direction of propagation. The current in the direction of propagation is responsible for depolarizing the axon ahead of the excitation wave (i.e., the spike of axial current in the Figure is positive, so its direction is $+x$, or to the right in the figure.) At a particular site, there is temporally increasing depolarization ahead of the action potential, which eventually results in activation of the membrane at that site. Thereby the region of excitation shifts, and propagation of the activation wave moves to the right.

Corresponding to the intense axial current over a short length of the axis, there are intense transmembrane currents. As shown in Figure 6.7, there is an intense outward transmembrane currents in the region where the transmembrane potential begins to rise, and an intense inward current midway through the upstroke. This inward current, largely due to an influx of sodium ions, powers the overall excitation process.

A cartoon depicting the currents flowing during activation is shown in Figure 6.8. In the figure, propagation is moving from left to right. The band in the middle is the membrane, with the region above the extracellular volume and the region below the intracellular volume. Multiple current pathways during the sodium influx of activation are illustrated in the figure, crossing the membrane toward the intracellular side around point A. Some sodium influx near A is offset by potassium efflux and is ignored. Also not depicted is the constant resting potential and associated membrane charges. One might consider these to be superimposed on the time-varying component under consideration.

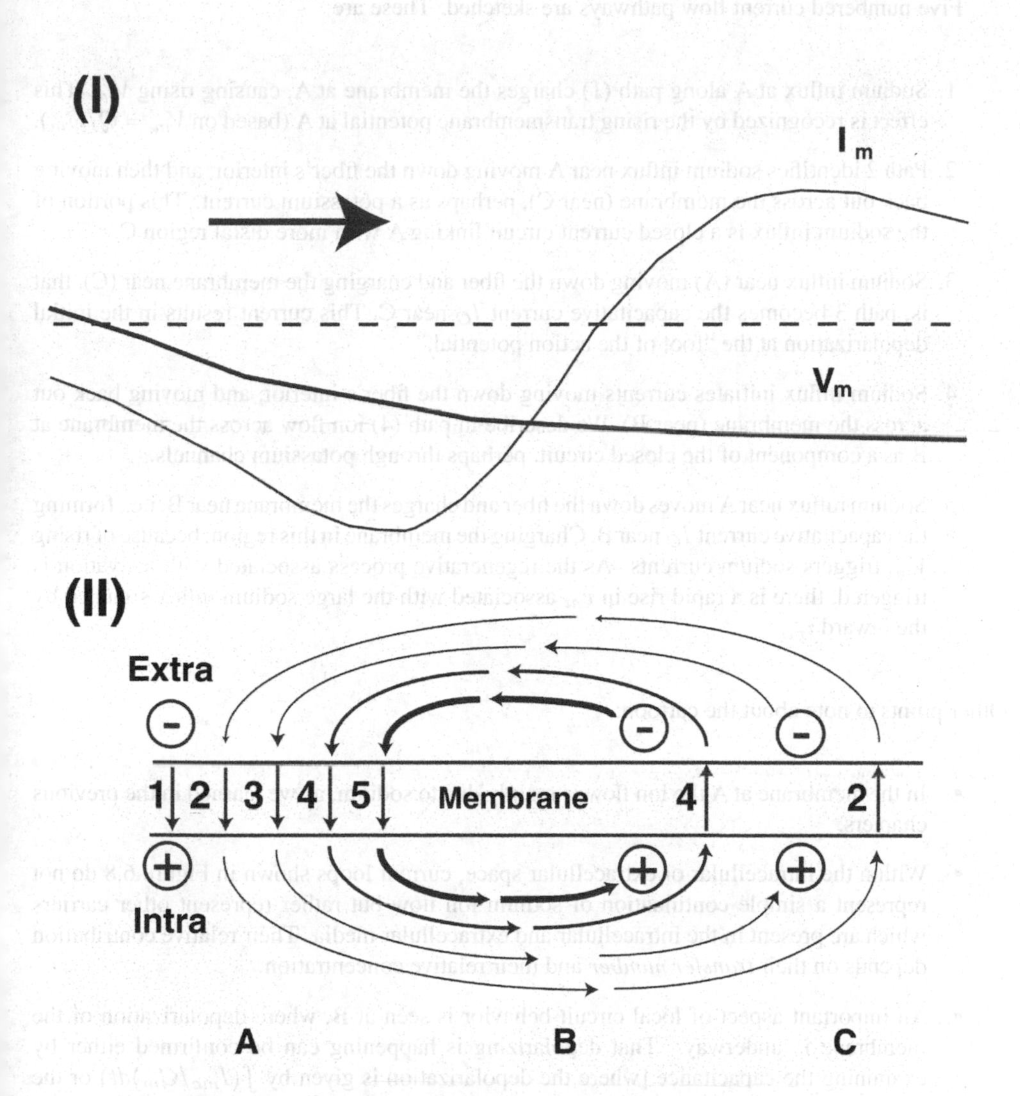

Figure 6.8. Local Circuit Pattern. Panel I (top): The arrow indicates the direction of excitation. Sketches from Figure 6.7 of a segment of the V_m and I_m curves also are present. Panel II (bottom) is a sketch of possible current pathways. The two horizontal lines represent the fiber's membrane, with the extracellular volume above and the intracellular volume below. Five possible current flow pathways are depicted (and discussed in the text). Letters A, B, and C along the axis at bottom identify three major regions as A, sodium influx; B, transition; C, potassium outflux. Along the fiber at the moment depicted, currents of propagation are inward at A, due to the inward movement of sodium ions. At B there is an outward movement of potassium ions. At C there is in effect an outward capacitative current associated with axial current down the longitudinal pathways. Horizontal locations A, B, and C drawn in panel II show currents that correspond approximately to the transmembrane potentials and currents at the same horizontal position.

Five numbered current flow pathways are sketched. These are

1. Sodium influx at A along path (1) charges the membrane at A, causing rising V_m. This effect is recognized by the rising transmembrane potential at A (based on $V_m = Q/C_m$).

2. Path 2 identifies sodium influx near A moving down the fiber's interior, and then moving back out across the membrane (near C), perhaps as a potassium current. This portion of the sodium influx is a closed current circuit linking A with more distal region C.

3. Sodium influx near (A) moving down the fiber and charging the membrane near (C), that is, path 3 becomes the capacitative current I_C near C. This current results in the initial depolarization at the "foot of the action potential."

4. Sodium influx initiates currents moving down the fiber's interior, and moving back out across the membrane (near B). We describe in path (4) ion flow across the membrane at B as a component of the closed circuit, perhaps through potassium channels.

5. Sodium influx near A moves down the fiber and charges the membrane near B, i.e., forming the capacitative current I_C near B. Charging the membrane in this region, because of rising V_m, triggers sodium currents. As the regenerative process associated with activation is triggered, there is a rapid rise in v_m associated with the large sodium influx signaled by the inward i_m.

Other points to note about the cartoon:

- In the membrane at A the ion flow is mainly due to sodium, as we learned in the previous chapters.

- Within the intracellular or extracellular space, current loops shown in Figure 6.8 do not represent a simple continuation of sodium ion flow but rather represent other carriers which are present in the intracellular and extracellular media. Their relative contribution depends on their *transfer number* and their relative concentration.

- An important aspect of local circuit behavior is seen at B, where depolarization of the membrane is underway. That depolarizing is happening can be confirmed either by examining the capacitance [where the depolarization is given by $\int (I_{mc}/C_m)dt)$ or the resistance (where there is an $I_{mr}R_m$ drop of the same magnitude, where $I_m = I_{mr} + I_{mc}$). Eventually the depolarization at B will cause the transmembrane potential to reach threshold, at which time conditions at B become those shown in Figure 6.8 for site A (i.e., site B will then become the point of high sodium influx). Such a shift occurs continuously with uniform propagation.

- During the early phase of the action potential at C, when the membrane can be characterized as passive and linear, it is an outward membrane current that accounts for the depolarization.

- The cartoon is illustrative but not comprehensive. It includes artistically drawn attributions of the dominant effects. In fact, a mixture of current components of all aforementioned kinds are present at A, B, and C.

As an aside, we note that when the active region of the membrane is at a distance from a resting patch, an *outward* membrane current at the patch depolarizes it slightly. The current flow at the patch arises solely from the electric circuit currents from the active region some distance away. In contrast, when propagation results in the patch itself becoming an active site, an *inward* current depolarizes it further. In the latter case it is essentially diffusional forces which are responsible for the depolarizing current. Circuit theory does not recognize this non-electrical generator, making the matter *seem* paradoxical when forced into a circuits perspective.[3]

In summary, the local currents are a consequence of differences of potential between the portion of the fiber where the action potential is occurring and the adjacent region. These currents are known as *action or local circuit* currents. They create an unstable transmembrane potential pattern, because the currents flowing forward from the excitation wave into the adjacent unexcited regions have the effect of starting an action potential there too. The overall result is that action potentials are created in sequence down the fiber, with each active region on the fiber exciting the next unexcited region nearby.

6.3.2. Mathematics of Propagating Action Potentials

We now consider propagation again, this time from a mathematical perspective. Propagation requires an action potential at one site to initiate an action potential at an adjacent site. To understand this process, it is helpful to begin again with the membrane current equation:

$$I_m(x,t) = I_{\text{ion}}(x,t) + I_C(x,t) \tag{6.25}$$

The membrane current equation (6.25) is written here with the functional dependence (x, t) included explicitly, so as to emphasize that the equation must hold at each site along the fiber, and at each time in the action potential's evolution. In that sense, the equation might be thought of as identifying a large set of relationships, of which each particular x and t is one example.

As for a patch, conversion of the seemingly static relationship of (6.25) into an equation for temporal variation requires recognition that the capacitative current involves the equation for the V_m time derivative, i.e.,

$$I_C(x,t) = C_m \frac{\partial V_m(x,t)}{\partial x} \tag{6.26}$$

so that

$$\frac{\partial V_m(x,t)}{\partial t} = \frac{1}{C_m}(I_m(x,t) - I_{\text{ion}}(x,t)) \tag{6.27}$$

The above equation (6.27) is integrated with respect to time, usually numerically, to find the temporal evolution of V_m. [Units for Eqs. (6.26) and (6.27): If C_m is in Farads, then I_m is in Amperes. Alternatively, if the units of C_m are F/cm^2 then I_m and I_{ion} must be A/cm^2. The latter are chosen more frequently, so as to have units compatible with HH and most other membrane models.] Performing an integration of equation (6.27), whether analytic or numerical, requires an expression for each of the two currents on the equation's right-hand side, the ionic current I_{ion}, and the total current I_m. Such expressions are given in the sections that follow.

So far this sequence of equations follows the pattern of development that was used for an action potential for a single patch, as was done in Chapter 5. The formal inclusion of the (x, t)

dependence identifies the fact that action potentials can vary from one site to another, but it does not show how one site connects to another, or that they are linked in any way.

Why therefore is there propagation, rather than simply individual action potentials at individual sites along the fiber? In other words, how does any spatial linkage or temporal sequence become a part of the mathematics? To resolve this question, each of the two currents in (6.27) must be examined carefully.

Membrane ionic current

The ionic current arises because transmembrane current density is related to the transmembrane potential V_m through intrinsic membrane properties.

The particular relationship depends on the species and type of tissue, and is exemplified by the membrane equations that were given by Hodgkin–Huxley, as described. Thus from that work we have the ionic current equation

$$\begin{aligned} I_{\text{ion}}(x,t) &= g_{\text{K}}(x,t)\,(V_m(x,t) - E_{\text{K}}) \\ &+ g_{\text{Na}}(x,t)\,(V_m(x,t) - E_{\text{Na}}) \\ &+ g_L\,(V_m(x,t) - E_L) \end{aligned} \tag{6.28}$$

In equation (6.28) the dependence of ionic current on location and time is noted explicitly by the inclusion of the dependence (x, t) after those values that change in these respects. Note that the transmembrane potential V_m is used to determine in all three ionic components to determine the total ionic current. The transmembrane potential is a function of (x, t), i.e., in general it will have a different value at each time and at each position at that time. Conductances g_{K} and g_{Na} also are functions of space and time because they depend on the underlying probabilities n, m, and h.

The ionic current equation (6.28) does not, however, give any particular insight into the mechanisms of propagation. That is, I_{ion} depends only on the transmembrane voltage and the conductances at its own site. Thereby, there is nothing in the ionic current equation that links one point on the fiber with any other.

Total membrane current for fiber

The mechanism of linkage, and thereby of propagation, only becomes evident when one examines the total current term $I_m(x, t)$ in the equation for the V_m time derivation (6.27). Recall from (6.12) above that

$$i_m = \frac{1}{(r_i + r_e)}\left(\frac{\partial^2 V_m}{\partial x^2} - r_e i_p\right) \tag{6.29}$$

Recall (from Chapter 2) that the transmembrane current per unit area, I_m, is related to i_m (the current per unit length) through the cylindrical geometry, i.e.,

$$I_m = i_m/(2\pi\, a) \tag{6.30}$$

Thus we get the membrane current equation

$$I_m = \frac{1}{(2\pi a)(r_i + r_e)}\left(\frac{\partial^2 V_m}{\partial x^2} - r_e i_p\right) \tag{6.31}$$

This equation contains within its terms the explanation for linkage and propagation, as discussed below.

Alternative forms for the membrane current equation

There are several alternative forms and special cases of the membrane current equation (6.31) that frequently appear. These include:

Transmembrane stimulus. In this general form for I_m, an explicit term I_s is added to account for a transmembrane stimulus, in the fashion of Chapter 5, giving

$$I_m = \frac{1}{(2\pi a)(r_i + r_e)}\left(\frac{\partial^2 V_m}{\partial x^2} - r_e i_p\right) + I_s \tag{6.32}$$

Membrane current if no stimulus. A condition associated with a propagating action potential on a single fiber is that, generally, $i_p = 0$ and $I_s = 0$. That is, once propagation is initiated propagation continues even in the absence of any external stimulus. Furthermore, essentially from the equations for defining I (6.5) and differentiating it (6.6), $I_i = -I_e$. Under these conditions the equation for the second derivative of V_m (6.31) specializes to

$$I_m = \frac{1}{(2\pi a)(r_i + r_e)}\left(\frac{\partial^2 V_m}{\partial x^2}\right) \tag{6.33}$$

Membrane current if high extracellular conductance. Frequently, the fiber is considered to lie in an extensive extracellular medium where one can assume $r_e \approx 0$ (e.g., the Hodgkin–Huxley experimental chamber for study of squid axon permits this approximation).

In addition, the conversions of Chapter 2 can be used to express resistance per unit length, r_i, in terms of intracellular resistivity, R_i (Ωcm). With these changes (6.33) converts to

$$I_m = \frac{a}{2R_i}\left(\frac{\partial^2 V_m}{\partial x^2}\right) \tag{6.34}$$

Discussion of membrane current equations

The membrane current equation (6.31) relates density of current crossing the membrane (also called the transmembrane current density) to the second spatial derivative of transmembrane potential. Note in particular that the *spatial* derivative of V_m appears in (6.31), and that the spatial derivative depends on the relationship of V_m at x with V_m at other points in its neighborhood.

Said differently, even if V_m were to remain constant at a particular x, a change in V_m at points in its neighborhood will change the spatial derivative of V_m at x.[4]

From a more physical or physiological perspective, the relationship expressed in (6.31) is tied to the fact that in a fiber currents can flow axially. Regions are linked by currents coming inward across the membrane at one place, and moving outward at another. Thus if one looks again at regions A and B in Figure 6.8, one sees membrane currents flowing inward across the membrane at A, axially down interior of the fiber, and then across the membrane outward in region B. That is, the membrane current going out at B depends on the membrane current coming in at A.

Equation (6.31) also ties transmembrane currents to the curvature of $V_m(x)$. The association of these transmembrane currents with the transmembrane voltage was seen in Figure 6.7, where the peaks of i_m occur at the points of peak curvature (extrema of 2nd spatial derivative) of the curve for $V_m(x)$.

Note that the relationship of transmembrane voltage to membrane currents arises from a condition imposed by the structure of the conducting region (of the load seen by the membrane), and is thereby embodied in the mathematics of the membrane current equation (6.31). It is this dependence of membrane current at one x on the transmembrane voltages in the surrounding neighborhood that is the cause of propagation.

In other words, the mechanism that allows an action potential at one place to trigger an action potential at a site nearby is the mechanism of axial currents from one place to another. The result is that current exits the membrane at places different from those where it enters, leading to excitation at a different membrane location.

6.3.3. Numerical Solutions for Propagating Action Potentials

A numerical solution for propagating action potentials is built on the mathematics above and then furthered by a series of additional steps. The necessary steps are:

- Division of the continuous variable time into a list of discrete times separated by interval Δt, and division of the length of the fiber into a discrete set of spatial segments of length Δx.
- Presentation of the key mathematical equations of propagation in a form that corresponds to the discretization of time and space.
- Development of a step-by-step procedure for the numerical solution of the discretized variables.

We consider each of these items in turn.

Division into discrete steps in space and time

First we divide the fiber into a sequence of discrete axial elements with membrane crossings at the ends, as illustrated in Figure 6.9. The division also establishes sequential points (or *nodes*) spaced at intervals Δx as sites of membrane crossings. The nodes and crossings start at one end of the fiber and continue to the other.

The significance of the division is that one now thinks of most quantities that vary with axial position, such as $V_m(x)$, as having a tabulated value at each node, and changing in some simple fashion (such as linearly) in between nodes, so that the values at the nodes are a sufficient representation of what is happening. In this fashion the description of the actual continuous function is replaced by a list of values. With enough nodes, there is not much difference in

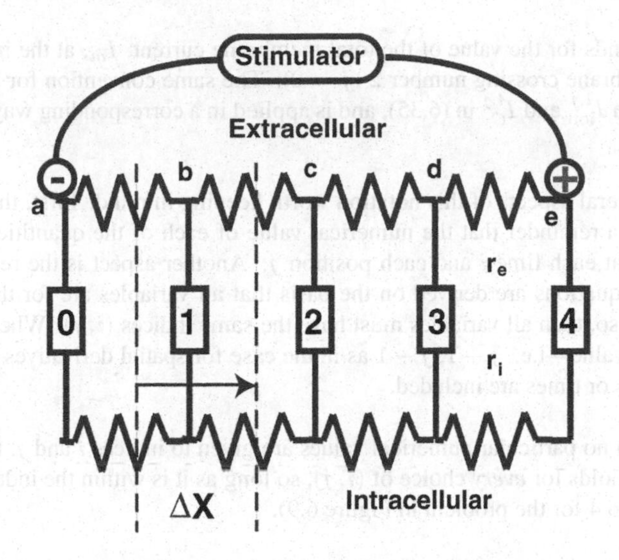

Figure 6.9. Fiber Model for Propagation. The fiber is represented by a network of electrical components, as in Figure 6.1. The continuous fiber lies along the x axis. There is a resistive extracellular path (along top line) with resistance per length r_e, a resistive intracellular path (along bottom line) with resistance per length r_i, and a discrete set of membrane crossings.

practice between one view and the other. However, because of the large changes occurring in short distances within an action potential, a good choice of node spacing is not always apparent, and the best choice may vary depending on the goal of the analysis.

We also choose a temporal discretization, Δt.[5]

An example of a discretized fiber is given in Figure 6.9. Here there are only five membrane crossings, separated by axial distance Δx and numbered 0 to 4.[6] A stimulator is connected to the extracellular nodes outside of crossings 0 and 4. The stimulator has its current sink at a and current source at e. Each membrane crossing represents a segment of the fiber surrounding the crossing, as suggested by the dotted lines around crossing 1. The width of each band is Δx, except for the bands at the ends, which have a width half this much, because membrane is present on only one side of the node.

Propagation equations discretized

The equation for the total current (6.25), given in discretized form, is

$$I_m^{i,j} = I_{\text{ion}}^{i,j} + I_C^{i,j} \tag{6.35}$$

In (6.35) i is present as an index of time. One might think of i as the line number for a list of the times to be evaluated. The j is present as an index of spatial position, so in Figure 6.9 $j = 0$ signifies the leftmost membrane crossing, $j = 1$ the next to the right, up to $j = 4$ as the crossing on the right.

Thus $I_m^{0,2}$ stands for the value of the total membrane current, I_m, at the initial time instant ($i = 0$), for membrane crossing number 2 ($j = 2$). The same convention for indices i and j is also followed with $I_{\text{ion}}^{i,j}$ and $I_C^{i,j}$ in (6.35), and is applied in a corresponding way in the equations below.

There are several aspects of this notation worth keeping in mind. First, the presence of the indices serves as a reminder that the numerical value of each of the quantities so indexed has a different value at each time i and each position j. Another aspect is the relative timing and location: Often equations are derived on the basis that all variables are for the *same* time and *same* position. If so, then all variables must have the same indices (i, j). When there is present a different index value—i.e., $j + 1$, $j - 1$ as in the case for spatial derivatives—one knows that different positions or times are included.

Finally, when no particular numerical values are given to indices i and j, the implication is that the equation holds for *every* choice of (i, j), so long as it is within the index's range (e.g., j in the range of 0 to 4 for the problem in Figure 6.9).

The equation for capacitative current (6.26) becomes, in discretized form,

$$I_C^{i,j} = C_m \frac{\partial V_m^{i,j}}{\partial x} \approx \frac{C_m \Delta V_m^{i,j}}{\Delta t} \tag{6.36}$$

and the discretized equation for the capacitative current (6.36) can be rearranged and substituted into the equation for the total membrane current (6.38).

A central result follows: The equation for the change in V_m during interval Δt is:

$$\Delta V_m^{i,j} = \frac{\Delta t}{C_m}(I_m^{i,j} - I_{\text{ion}}^{i,j}) \tag{6.37}$$

Note that in these last two equations, the change in voltage ΔV_m and change in time Δt were treated as ordinary variables. Thus Δt was moved from one side of the expression to the other in the discretized capacitative current (6.36) and in the equation for the change in V_m, where Δt appears on the equation's right-hand side.

The numerical solution will proceed by evaluation of the equation for the change in V_m, that is, solutions will be found by using (6.37) repeatedly. The evaluation is, at each time step, first done for each of the spatial locations along the fiber. Then time is stepped forward, and another set of ΔV values is found.

To find the total membrane current term of (6.37), I_m, one begins with Eq. (6.31), converting each term to discretized form, with the result:

$$I_m^{i,j} = \frac{1}{(2\pi a)(r_i + r_e)}\left(\frac{\Delta^2 V_m^{i,j}}{\Delta x^2} - r_e i_p^{i,j}\right) \tag{6.38}$$

A special issue arises with the second spatial derivative of V_m. How is the approximation found? The second derivative is the spatial rate of change of the first derivative, so one finds it useful first to find approximations the first spatial derivative. An approximation to the first spatial derivative of V_m at the midpoint of the interval to the right of position j (position jp) is

$$\frac{\Delta V_m{}^{(i,jp)}}{\Delta x} \approx \frac{(V_m^{i,j+1} - V_m^{i,j})}{\Delta x} \tag{6.39}$$

A similar approximation to the first derivative at the midpoint of the interval to the left of j (position jn) is

$$\frac{\Delta V_m{}^{(i,jn)}}{\Delta x} \approx \frac{(V_m^{i,j} - V_m^{i,j-1})}{\Delta x} \tag{6.40}$$

With the first derivatives available, it is straightforward to express the second spatial derivative of V_m at position j as the approximate spatial rate of change between them, or

$$\frac{\Delta^2 V_m^{i,j}}{\Delta x^2} \approx \left(\frac{\Delta V_m^{(i,jp)}}{\Delta x} - \frac{\Delta V_m^{(i,jn)}}{\Delta x} \right) / \Delta x \tag{6.41}$$

$$\approx \frac{V_m^{i,j+1} - 2V_m^{i,j} + V_m^{i,j-1}}{\Delta x^2} \tag{6.42}$$

That is, (6.41) estimates the second spatial derivative of V_m at time i and position j. Geometrically, this derivative quantifies the curvature of V_m, its change with distance of the slope.

Sequence of operations

0. Initialize. One must first assign values to structural constants (a, R_i, C_m), to equilibrium potentials (E_K, E_{Na}, E_L), to maximum conductances ($\bar{g}_K$, $\bar{g}_{Na}$, g_L), to magnitude and timing parameters for the stimulus, and perhaps to other constants. Further, one must assign initial values to all the state variables.

Although the evolution of transmembrane events depends only on Δt, normally a absolute time variable t must be initialized, because some parameters (such as stimulus on/off) usually depend on absolute time. Time often is kept as an integer number, in microseconds (or tenths of microseconds) so that there is no ambiguity in the edges of the timing of stimuli, and often is initialized at a negative time (such as –100 microseconds), so that a first stimulus can be delivered at $t = 0$, with the record showing unequivocally the rise from baseline due to the stimulus, and so that there is no accumulated round-off in t after, say, a million iterations.

In an HH fiber, an initial value will be needed for V_m and for channel probabilities n, m, and h. In contrast to the situation for a patch simulation as described in Chapter 5, for propagation these state variables must be assigned values at every position along the fiber, i.e., throughout the range of index j. If the membrane is to begin at rest, then channel probabilities n, m, and h are initialized by assigning each of them the value at $t = \infty$ (5.21), (5.31), and (5.32). Specifically, for all j,

$$\begin{aligned} n_{0,j} &= n_\infty = \alpha_n/(\alpha_n + \beta_n) \approx 0.3 \\ m_{0,j} &= m_\infty = \alpha_m/(\alpha_m + \beta_m) \approx 0.05 \\ h_{0,j} &= h_\infty = \alpha_h/(\alpha_h + \beta_h) \approx 0.6 \end{aligned} \tag{6.43}$$

where the α and β terms are evaluated using $v_m = 0$, corresponding to $V_m = V_r$.

Additionally, for a start from rest V_m must be initialized to its resting value, at all nodes, so for all j:

$$V_m^{0,j} = V_r \approx -60 \text{ mV} \tag{6.44}$$

Recall that the stability of V_m at its chosen resting level may require a choice of E_L. At rest it also is true that $V_m = (g_K E_K + g_{Na} E_{Na} + g_L E_L)/(g_K + g_{Na} + g_L)$, as shown in Chapter 3, so for V_r and E_L to be consistent at rest, E_L must be

$$E_L = [(g_K + g_{Na} + g_L)V_r - (g_K E_K + g_{Na} E_{Na})]/g_L \tag{6.45}$$

1. Find $I_m^{i,j}$. That is, for time i, find $I_m^{i,j}$, the total current, at all positions j. So doing can be accomplished by inserting the results for the second spatial derivative of V_m (6.41) back into the equation for I_m (6.38) to reach

$$I_m^{i,j} = \frac{1}{(2\pi a)(r_i + r_e)}\left(\frac{V_m^{i,j-1} - 2V_m^{i,j} + V_m^{i,j+1}}{\partial x^2} - r_e i_p^{i,j}\right) + I_s^{i,j} \tag{6.46}$$

From the perspective of understanding the mechanism of propagation, the equation for the total membrane current (6.46) is critical, for it is the *only* step in the algorithm where a result at central position j depends on values of variables at adjacent positions $j-1$ and $j+1$. That is, the total membrane current of a central segment is determined by the relationship of the transmembrane potential at the central segment to that of its neighbors, not the status of the central segment alone.

The complexity of this step often is underestimated, for there are special cases galore. First, there are special cases at the ends of the fiber, where the 2nd derivative expression cannot apply (e.g., at node 0 there is no $V_m^{i,j-1}$. Then there are special cases for both types of stimuli, both in time and, for i_p, in position. For stimuli, transitions are important, and normally the stimulus must be considered "on" if it is to have effect between time index i and $i+1$.[7]

2. Find $I_{ion}^{i,j}$. That is, for time i, find $I_{ion}^{i,j}$, the sum of ionic currents, at all positions j.

$$I_{ion} = g_K^{i,j}(V_m^{i,j} - E_K) + g_{Na}^{i,j}(V_m^{i,j} - E_{Na}) + g_L(V_m^{i,j} - E_L) \tag{6.47}$$

where

$$g_K^{i,j} = \bar{g}_K n_{ij}^4 \tag{6.48}$$

$$g_{Na}^{i,j} = \bar{g}_{Na} m_{ij}^3 h_{ij} \tag{6.49}$$

Note that I_{ion} is an addition of terms frequently having varying signs, where the net residual often has a smaller magnitude than that of any individual term. Thereby, addition with many significant digits is required, lest the result lose any significance, numerically.

3. Estimate $\Delta V_m^{i,j}$. That is, using data for time i, estimate $\Delta V_m^{i,j}$, the estimated V_m change between time i and time $i+1$, for all positions j:

$$\Delta V_m^{i,j} = \frac{\Delta t}{C_m}(I_m^{i,j} - I_{ion}^{i,j}) \quad \text{for all } j \tag{6.50}$$

The summation intrinsic to this computation again is numerically sensitive and requires high precision, as the sum may require adding numbers that are large in magnitude but having different signs. The summation is especially sensitive because the I_{ion} term is itself a sensitive summation. Further, in many situations (such as a stimulus near threshold), the physiologically meaningful result is computed as the accumulation over time of many small changes.

4. Estimate Δn, Δm, Δh. That is, using data for time i, estimate $\Delta n^{i,j}$, $\Delta m^{i,j}$, and $\Delta h^{i,j}$, the estimates of n, m, and h changes from time i to time $i+1$. In mathematical form, these estimates are:

$$\Delta n^{i,j} = \Delta t[\alpha_n^{i,j}(1 - n^{i,j}) - \beta_n^{i,j}\, n^{i,j}] \tag{6.51}$$

$$\Delta m^{i,j} = \Delta t[\alpha_m^{i,j}(1 - m^{i,j}) - \beta_m^{i,j}\, n^{i,j}] \tag{6.52}$$

and

$$\Delta h^{i,j} = \Delta t[\alpha_h^{i,j}(1 - h^{i,j}) - \beta_h^{i,j}\, h^{i,j}] \tag{6.53}$$

Note that every position along the fiber will have its own set of changes for n, m, and h, and that, correspondingly, each position will have different α and β rates.[8]

5. Advance time by one step. Use the data from time i, as found in the preceding steps, to compute values for all the state variables for time $i+1$:

$$V_m^{i+1,j} = V_m^{i,j} + \Delta V_m^{i,j} \quad \text{for all } j \tag{6.54}$$

The values of n, m, and h are readily advanced by

$$n^{i+1,j} = n^{i,j} + \Delta n^{i,j} \tag{6.55}$$

$$m^{i+1,j} = m^{i,j} + \Delta m^{i,j} \tag{6.56}$$

$$h^{i+1,j} = h^{i,j} + \Delta h^{i,j} \tag{6.57}$$

This sequence of computational steps advances V_m and n, m, and h by one time step. The process can be repeated through successive time steps by returning to step 1 and repeating steps 1 through 5. This method is called an explicit method (or sometimes Euler's method) and has the advantages of being fast and relatively simple. The Euler method also serves as a point of reference for many other methods, such as predictor-corrector algorithms.[9]

6.3.4. Stability and the Mesh Ratio

It is well known that algorithms such as the one above contain the potential for numerical disaster. The disaster occurs when the computed results begin to oscillate wildly and nonsensically over a series of time steps, with values taking on greater and greater magnitudes, until the computation fails.

Because such disasters have occurred many times in the simulation of propagation, such questions also have been studied specifically in the context of the Hodgkin–Huxley equations. As a consequence, a literature is available that includes descriptions of the properties of many algorithms. Most are more complicated than the one above, but thereby have advantageous properties[10].

One result of numerical analysis as well as practical experience is the identification of the *mesh ratio*[11] as an important numerical indicator. The mesh ratio is

$$\text{mesh ratio} = \frac{\Delta t}{r_i c_m \Delta x^2} \tag{6.58}$$

Numerical procedures for finding solutions for V_m in time and space remain stable, when values for Δt and Δx are selected that make the mesh ratio small, as compared to one. Conversely, solutions may become unstable when the mesh ratio becomes greater than one.

The origin of the mesh ratio can be seen from the following example. Suppose we picture a simple fiber represented by the grid shown in Figure 6.9, extended to be 10 segments long (with 11 nodes 0 to 10). The conductivity of the extracellular region is high compared to that of the intracellular region. Further, assume there is no stimulus, and assume the membrane supports no ionic currents.

Consider a situation where V_m at time 0 ($i = 0$) is zero everywhere, except at node 5 ($j = 5$), the center of the extended fiber. At node 5, $V_m^{0,5} = 1$. Units are unspecified but the problem occurs even if the "1" is only a small deviation, e.g., 1×10^{-6} Volts. Of course, what one would expect to happen in a physical environment with these characteristics is that the perturbation at the one point spreads out and dies away, because there is no mechanism to keep it in place or increase it.

To compare that intuitive result to the outcome of the propagation algorithm, we use the equation for ΔV_m, from the given starting conditions. In particular, at the time i we have for site j from (6.50)

$$\Delta V_m^{i,j} = \frac{\Delta t}{C_m}[I_m^{i,j} - I_{\text{ion}}^{i,j}] \tag{6.59}$$

and for I_m from (6.46) we have

$$I_m^{i,j} = \frac{1}{(2\pi a)(r_i + r_e)}\left[\frac{V_m^{i,j-1} - 2V_m^{i,j} + V_m^{i,j+1}}{\Delta x^2} - r_e i_p^{i,j}\right] + I_s^{i,j} \tag{6.60}$$

We now substitute equation (6.60) in (6.59). To simplify the analysis (and to show that the result does not depend on special cases), we make the stimulus current $I_s = 0$, the ionic current $I_{\text{ion}} = 0$, and the extracellular resistance $r_e = 0$. The result is

$$\Delta V_m^{i,j} = \frac{\Delta t}{C_m}\left(\frac{1}{2\pi a r_i}\left[\frac{V_m^{i,j-1} - 2V_m^{i,j} + V_m^{i,j+1}}{\Delta x^2}\right]\right) \tag{6.61}$$

From (2.57) we know that $C_m = c_m/(2\pi a)$. With that substitution in (6.61) and some factoring, (6.61) becomes

$$\Delta V_m^{i,j} = \left(\frac{\Delta t}{r_i c_m \Delta x^2}\right)[V_m^{i,j-1} - 2V_m^{i,j} + V_m^{i,j+1}] \tag{6.62}$$

Note that the term in the parentheses is the mesh ratio, and note that the mesh ratio always has a value that is greater than zero.

Table 6.1. Mesh Ratio and Changes in V_m.

node	time 0	1	2	3	4
0	0	0	0	0	0
1	0	0	0	0	16
2	0	0	0	8	-128
3	0	0	4	-48	448
4	0	2	-16	120	-896
5	1	-4	24	-160	1120
6	0	2	-16	120	-896
7	0	0	4	-48	448
8	0	0	0	8	-128
9	0	0	0	0	16
10	0	0	0	0	0

Each table entry is a value of V_m for a node (rows) and time (columns). Entries were calculated with (6.62), with a mesh ratio of 2. The calculation was unstable, so the lone nonzero value of V_m at time zero (node 5) produced V_m entries at later times that showed increasing magnitudes.

Inserting values from the example stated above, one can see the consequences. For the first step $V_m = 1$ at node 5 and $V_m = 0$ at all other nodes. That is, at node 5 the first time step produces a change in V_m at position 5 of [from (6.62)]

$$\Delta V_m^{0,5} = \left(\frac{\Delta t}{r_i c_m \Delta x^2}\right)[-2V_m^{0,5}]. \tag{6.63}$$

Equation (6.63) is unsettling, even as it shows explicitly the significance of the mesh ratio. In other words, (6.63) shows that the magnitude of V_m may not get smaller, as it should, because the change will be proportional to the mesh ratio. In fact, inspection of (6.63) shows that the magnitude of V_m will actually increase if the mesh ratio is greater than one.

Furthermore, the increase may continue from the first time step to the second, and on and on thereafter. Thus V_m at $j = 5$ may grow without bound.[12]

Such a result is shown in Table 6.1, where the "time 0" column gives initial values at each node, showing all zeros except for an initial entry of 1 at node 5. The table follows with values through four time steps (successive columns).

At time 1, one can see that V_m at node 5 has a larger magnitude (4) and the opposite sign (now negative). Additionally, at time 1 the entries for V_m at nodes 4 and 6 have become nonzero (value of 2). At time 2, V_m at node 5 has grown to 24 (the previous magnitude times 6); at time 2 there are 5 nonzero values. This process continues, so that by time 4, V_m at node 5 has grown to a value of 1,120. Moreover, nonzero values extend from sites 1 to 9, with alternating signs.

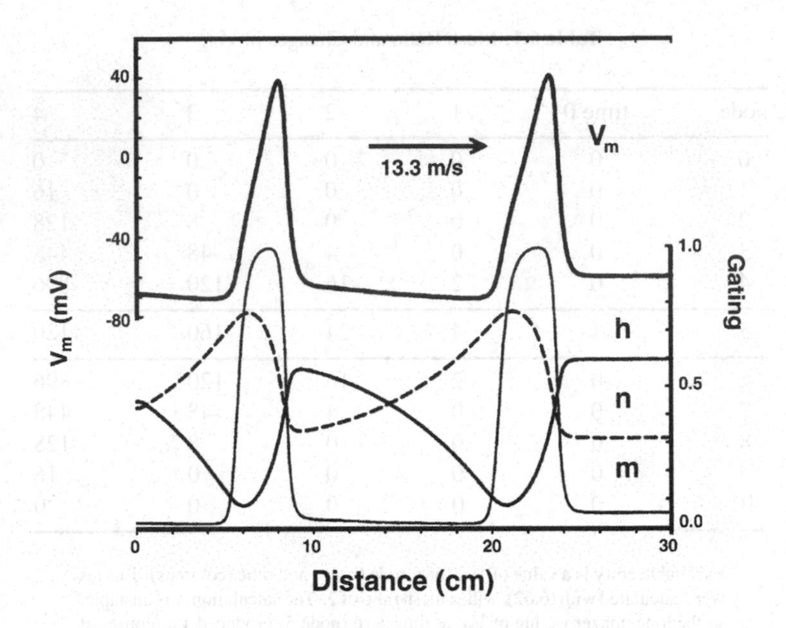

Figure 6.10. Simulation of Propagation on squid axon of diameter 600 μm at $T = 6.3°$C. Hodgkin–Huxley membrane parameters and equations are utilized. The Figure describes the behavior of *gating variables* and transmembrane potential as functions of the axial coordinate. The velocity of propagation is 13.3 m/sec. Following Hodgkin–Huxley, $R_i =$ 30 and $R_e = 20\ \Omega$cm.

Mesh ratios often are increased into the unstable range inadvertently. For example, an initial, approximate solution may be done with a large Δx so as to test a computer program with faster execution and smaller output files. If the solution then is refined to get better spatial resolution by shrinking Δx by a factor of 4, then the mesh ratio *increases* by a factor of 16. The increased mesh ratio easily can move the solution from stability to instability.

The result may be answers that are wrong, in that they are a product of an unstable algorithm, not the underlying electrophysiology. The remedy for the problem is to make Δt smaller, so that the mesh ratio remains well below one,[13] even though decreasing Δt will cause a corresponding increase in computation time.

6.3.5. Simulated HH Propagating Action Potentials

The propagation of two successive action potentials is illustrated in Figure 6.10. These waveforms result from a simulation of action potential propagation on a uniform fiber with squid axon parameters based on the Hodgkin–Huxley equations, the Hodgkin–Huxley description of the squid axon, and at 6.3 °C.

In this Figure the values of the gating variables m, n, and h from the initiation to the termination of the action potential are shown. Because the Figure is plotting spatial variation, it gives a different perspective than that obtained with a temporal plot for a membrane patch, which was depicted in Chapter 5. Spatially, the variable m may be observed to rise and fall in spatial synchrony with the rise and fall of v_m, characterizing the activation process. In contrast, recovery of gating variables n and h trails well behind, and in this case extends over nearly 10 cm.

In this example, the second action potential is elicited 10 msec after the first, and is possible because the gating variables are by that time nearly back to resting conditions. An interesting detail is that because m, n, and h are not quite back to their initial resting values, the second action potential is slightly different in morphology.

The velocity of propagation of 13.3 m/sec. This velocity is less than the value 18.8 m/sec obtained by Hodgkin and Huxley, and arises because their velocity was found for a higher temperature, 18.5 °C, in contrast to Figure 6.10, which is at 6.3°C.

6.3.6. Velocity Constraint for Uniform Propagation

For uniform propagation, the space–time behavior of $V_m(x,t)$ must satisfy the wave equation, namely,

$$V_m(x,t) = V_m(t - \frac{x}{\theta}) \tag{6.64}$$

where θ is the velocity of propagation. Using (6.64) and the chain rule gives

$$\frac{\partial V_m}{\partial x} = (-1/\theta)\frac{\partial V_m}{dt} \tag{6.65}$$

Applying the chain rule a second time results in

$$\frac{\partial^2 V_m}{\partial x^2} = (1/\theta^2)\frac{\partial^2 V_m}{dt^2} \tag{6.66}$$

Consequently, (6.34) can be written

$$I_m = \frac{a}{2R_i\theta^2}\frac{\partial^2 V_m}{\partial t^2} \tag{6.67}$$

Hodgkin and Huxley used this relationship to solve for the action potential waveform by equating it to the ionic current expression, Eq. (6.67). For a uniformly propagating wave (where $I_0 = 0$, because there is no stimulus during uniform propagation) we have

$$\begin{aligned}\frac{a}{2R_i\theta^2}\frac{\partial^2 V_m}{\partial t^2} &= C\frac{\partial V_m}{\partial t} + g_K(V_m - E_K)\\ &+ g_{Na}(V_m - E_{Na}) + g_L(V_m - E_L)\end{aligned} \tag{6.68}$$

Numerically, the procedure followed by Hodgkin and Huxley was to guess at θ and then step through the solution as a function of the single time variable, t. For incorrect θ the solution was found to diverge, but with the correct θ the V_m went through an appropriate action potential and then returned toward the resting values.

Their procedures have now been replaced by procedures more suitable to digital computer methods. With these methods, solutions are found as functions of both space and time and for the full action potential's time course, for structures of moderate complexity.

Velocity and diameter

Inspection of (6.68) shows that an important result can be deduced without having to solve the equation explicitly. Note that any solution to (6.68) for $V_m(t)$ will continue to be a solution if

$$\frac{a}{2R_i\theta^2} = \text{constant} = \frac{1}{K} \tag{6.69}$$

so long as the membrane properties (reflected in the behavior of g_K, g_{Na}, and g_L) remain unchanged.

Consequently, for unchanging membrane properties, velocity θ relates to radius a by

$$\theta = \sqrt{\frac{aK}{2R_i}} \tag{6.70}$$

One pair of experimental values of θ and α are required for its evaluation of K, a value that is constant only within a particular context. That is, K depends on the species, type of membrane, and temperature. R_i is tied to the composition of the intracellular volume.

Although the change in velocity θ associated with changes in K or R_i can be predicted by (6.70), often K and R_i are constant in any one nerve bundle. Thus the significance of (6.70) is most often that it then ties velocity to fiber diameter when other values do not change.

Squid nerve example

Such predictions are reasonably well confirmed by particular experimental results. For example, where d is the fiber diameter in μm, for squid axon at 18.3°C it was reported by Hodgkin and Huxley that the velocity roughly followed the following equation:[14]

$$\theta = \sqrt{d} \quad m/\text{sec} \tag{6.71}$$

One can see that for this particular membrane and temperature the fraction $K/2R_i \approx 1$, in the units chosen. Similar empirical coefficients can be developed for other kinds of tissues and temperatures.

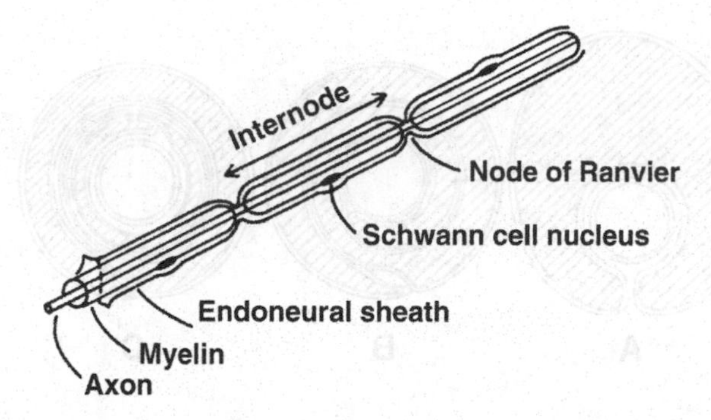

Figure 6.11. Diagram Showing the Structure of a Myelinated Nerve Fiber. Reprinted with permission from Aidley DJ. 1978. *The physiology of excitable cells*. Cambridge: Cambridge UP.

6.4. PROPAGATION IN MYELINATED NERVE FIBERS

6.4.1. Myelin Sheath

Most nerve fibers are myelinated, that is, they are coated with an essentially lipid material except at periodic points of exposure. An illustration of a myelinated fiber is given in Figure 6.11. The gaps in the myelin are called *nodes of Ranvier*, and these are regularly spaced with internodal distances ranging from 1 to 2 mm. (As a rough empirical rule the internodal length equals $100d$, where d is the fiber diameter in μm.)

Study of the myelin sheath in a given internode shows that it is made up of a single (Schwann) cell that has wrapped around the axon many times. This process is described in Figure 6.12, and one notes that after the growth process is completed almost all axoplasm has been squeezed out, leaving only layer after layer of plasma membrane. The myelin is, indeed, lipid in the same sense that the plasma membrane is predominantly lipid.

Some very rough data on the internodal leakage and capacitive properties in comparison with that of a single plasma membrane are given in Table 6.2. Since the myelin sheath is composed of layers of cell membranes, its leakage resistance should be the sum of the membrane leakage resistances and its capacitance the membrane capacitance divided by the number of layers. The data in Table 6.2 are consistent with this view if one assumes that the myelin contained 100 membrane layers.

Table 6.2. Electrical Properties of Myelin Sheath and Cell

	Specific leakage resistance	Specific capacitance
	(Ω cm^2)	(F/cm^2)
Myelin sheath	10^5	10^{-8}
Cell membrane	10^3	10^{-6}

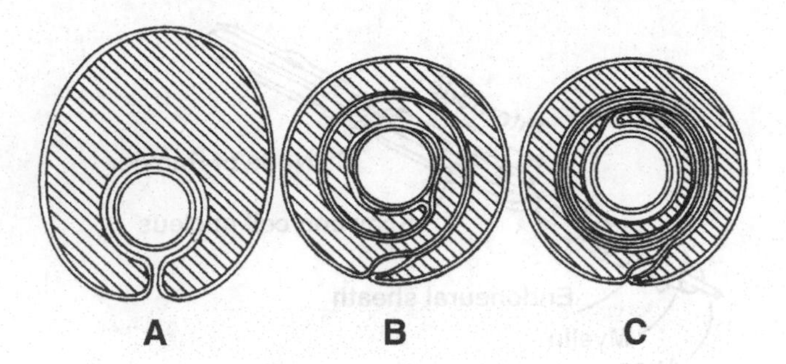

Figure 6.12. The Development of the Myelin Sheath by Vertebrate Schwann Cells in the Sequence. $A \to B \to C$. Reprinted with permission from Robertson JV. 1960. The molecular structure and contact relationships of the cell membrane. *Prog Biophys* **10**:343–417. Copyright ©1960, Pergamon Journals Ltd.

6.4.2. Saltatory Propagation

Because of the relatively high myelin leakage resistance, little transmembrane current leaves the cell in the internodal region. Instead, transmembrane currents are confined essentially to the node (of Ranvier). An electrical description of the node is therefore of special interest; its subthreshold properties are described by a specific resistance and capacitance of

$$R_m = 20\Omega\,\text{cm}^2 \quad C_m = 3\mu F/cm^2 \tag{6.72}$$

The Frankenhaeuser–Huxley model [1] for frog nerve provides these values as well as those under active conditions. (The F-H equations are described in Chapter 12.) The gap width itself is on the order of 1 μm.

More recently, it has been found that sodium channels are mainly confined to the nodes while potassium channels appear to be located mainly in the internodal region. The high concentration of sodium channels at the node contributes to an intensive inward current when excited: a current that is needed to depolarize the long internode and following node (currents needed to charge the associated capacitances) [2].

Figure 6.13 reproduces the Tasaki–Takeuchi experiment [5]. A special chamber was prepared that permits the division of extracellular space into three isolated compartments. With this, one can force any ionic current that normally flows between the central and outer compartments to pass through a resistor; the voltage measured across the resistance is, consequently, a measure of the transmembrane current in the central compartment.

With the fiber arranged so that there is no node in the central compartment, one sees only a capacitively coupled voltage (Figure 6.13A). But if the central region contains a node, then a clearly measured transmembrane current arises (Figure 6.13B). The experiment clearly supports the conclusion that transmembrane current is essentially confined to the nodes.

Propagation in a myelinated fiber proceeds from an active node to the next adjoining node by virtue of local circuit currents. Since there is little current loss in the internodal space,

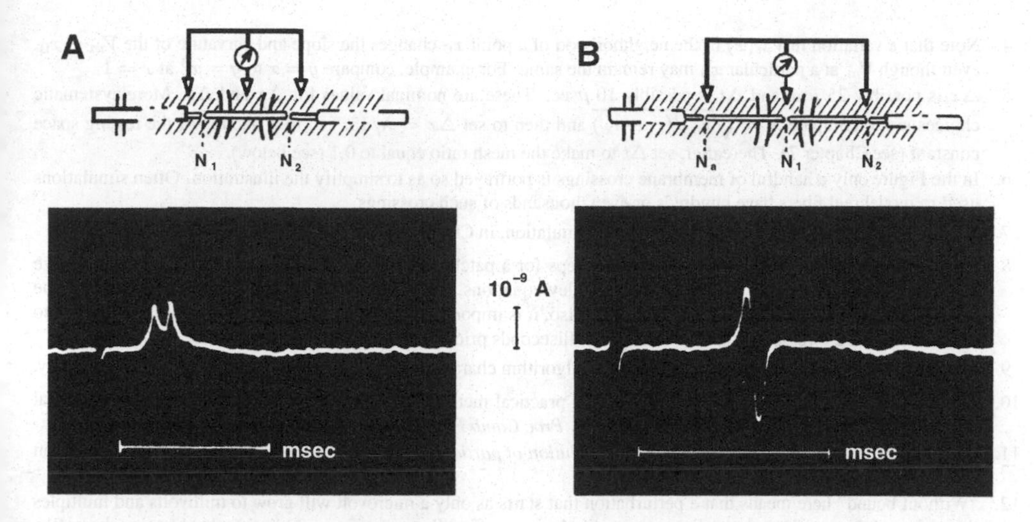

Figure 6.13. The Radial Currents in a short length of a myelinated fiber during the passage of an action potential. The upper part of the Figure describes the recording arrangement and the lower part the recorded currents. (A) The tracing shows the membrane current when the middle pool of Ringer does not contain a node. (B) The membrane current is measured when the middle pool contains a node. Nodes are identified as N_0, N_1, N_2. The three extracellular compartments are insulated from each other by the indicated insulating diaphragm. From Tasaki I, Takeuchi T. 1942. Weitere studien über den aktionsstrom der narkhaltiger nervefaser und über die elektrosaltatorische übertragung des nervenimpulses. *Pflugers Arch Ges Physiol* **245**:764–782.

activation can more easily influence a more distal excitable membrane than with an equivalent unmyelinated fiber. One consequence is a higher velocity for myelinated fibers. In contrast to (6.71), an empirical expression for myelinated fibers at frog temperature is

$$\theta = 6d \quad \text{m/sec} \qquad (d \text{ in } \mu\text{m}) \tag{6.73}$$

Since activation spreads from node to node (jumping from one to the next), propagation is said to be *saltatory* (from *saltare*—to leap or dance).

Considerably more information about myelinated nerves and the electrodes and protocols used to control them is presented in Chapter 12, Functional Electrical Stimulation (FES).

6.5. NOTES

1. In some places we use the subscript *o* (outside) to signify that a parameter is applied to the extracellular space. In this chapter we use the subscript *e* instead. There is no uniform convention regarding this notation. In many situations there is an extracellular space within the organ itself (frequently called the interstitial space). Sometimes the symbol *e* is used to designate the latter space and *o* to designate the (passive) space surrounding the organ.
2. From Chapter 5, recall that lowercase v_m, ϕ_i, and ϕ_e are used to signify changes from the resting values of V_m, Φ_i, and Φ_e. Note, however, that because resting values are constant $\partial V_m/\partial x = \partial v_m/\partial x$, $\partial \Phi_i/\partial x = \partial \phi_i/\partial x$, and $\partial \Phi_e/\partial x = \partial \phi_e/\partial x$. In the equations that follow, all potentials are expressed in terms of their values relative to rest, in contrast to the immediate previous treatment.
3. An entertaining and instructive discussion of this matter is given by Jewett [3].

4. Note that a variation in $V_m(x)$ in the neighborhood of a point x_0 changes the slope and curvature of the V_m at x_0, even though V_m at a particular x_0 may remain the same. For example, compare $y = x$ to $y = x^3$ at $x = 1$.
5. Δx is possibly 25 μm, and Δt is possibly 10 μsec. These are nominal values for Δx and Δt. More systematic choices are to compute $\lambda = \sqrt{r_m/(r_i + r_e)}$ and then to set $\Delta x = \lambda/15$, i.e., to a fraction of the resting space constant (see Chapter 7). Thereafter, set Δt to make the mesh ratio equal to 0.1 (see below).
6. In the Figure only a handful of membrane crossings is portrayed so as to simplify the illustration. Often simulations used to model real fibers have hundreds or even thousands of such crossings.
7. See discussion of stimulus timing for the patch simulation, in Chapter 5.
8. Also, as noted in the discussion of the related steps for a patch simulation in Chapter 5, it is important to realize that the α and β values are, in the Hodgkin–Huxley equations, functions of $v_m = V_m - V_r$, the deviation of the transmembrane voltage from rest, in millivolts. Also, it is important to recall that α and β have units of msec^{-1}, so that Δt must be expressed in (or converted to) milliseconds prior to their multiplication.
9. In general, other methods offer improvements in algorithm characteristics in exchange for an increase in complexity.
10. For example, see Crank J, Nicholson P. 1947. A practical method for numerical evaluation of solutions of partial differential equations of the heat-conduction type. *Proc Camb Phil Soc* 43:50–67
11. For example, see Smith GD. 1985. *Numerical solution of partial differential equations*, 3rd ed. Oxford: Clarendon Press.
12. "Without bound" here means that a perturbation that starts as only a microvolt will grow to millivolts and multiples thereof, until the membrane's nonlinear properties begin to affect (but not necessarily limit) what happens thereafter.
13. An alternative is a different algorithm, as this kind of instability arises from the way this method extrapolates from the current time instant to a future one.
14. This equation is unlikely to be a good approximation for other temperatures or types of membrane.

6.6. REFERENCES

1. Frankenhaeuser B, Huxley A. 1964. The action potential in the myelinated nerve fiber of *Xenopuslaevis* as computed on the basis of voltage clamp data. *J Physiol* **171**:302–315.
2. Hille B. 2001. *Ionic channels of excitable membranes*, 3rd ed. Sunderland, MA: Sinauer Associates.
3. Jewett D. 1984. The case of the missing outward current. *The Physiologist* **27**:437–439.
4. Randall E. 1987. *Microcomputers and physiological simulation*, 2nd ed. New York: Raven Press.
5. Tasaki I, Takeuchi T. 1942. Weitere studien über den aktionsstrom der narkhaltiger nervefaser und über die elektrosaltatorische übertragung des nervenimpulses. *Pflugers Arch Ges Physiol* **245**:764–782.

7

ELECTRICAL STIMULATION OF EXCITABLE TISSUE

In designing systems for stimulation, a qualitative understanding together with mathematical descriptions of responses to stimulation are essential. The response of excitable cells to naturally occurring or artificial stimuli is a subject of great importance in understanding natural function of nerve and muscle, because most stimuli are produced by the natural system itself. Both electric and magnetic field stimulation are used in research investigations and in clinical diagnosis, therapy, and rehabilitation. This chapter focuses primarily on responses to electrical stimuli, which are more frequent, and examines several biological preparations.

The core idea of stimulation is the following: A current, arising from an external stimulator or natural source, is introduced into a cell or its neighborhood. The current creates transmembrane voltage in nearby membrane. The membrane responds passively (i.e., with constant membrane resistance), so long as the voltage produced is below a threshold level. When the threshold level is reached, the membrane responds with an action potential, or some other active response.

From the perspective of the analysis of the effects of stimulation, critical issues revolve around what strength and time duration of a stimulus is required for the stimulus to cause the tissue to move through its initial, passive state to reach the threshold level for active response.[1] The answers depend, as one would expect, on a number of variables, importantly including the geometry of the tissue being stimulated, its electrical characteristics, and the location of the stimulus electrodes.

Analysis of stimuli focuses especially on mathematical relationships between the current applied as the stimulus and the resulting transmembrane potentials. Such knowledge, when quantitative, then allows one to draw quantitative conclusions about the strength and duration of stimuli that will result in transmembrane potentials above the threshold level in new or future situations, as well as those already explored experimentally.[2]

The initial sections of this chapter provide the simple mathematics giving the transmembrane voltages produced by a constant current stimulus, in a *spherical cell.* These current–voltage equations then are manipulated to produce strength–duration curves. A few real cells can be idealized as spherical, and the idealization is useful and interesting because, in a spherical cell, the response to stimulus depends only on time.

In other words, the spherical cell has a geometrical uniformity that avoids the additional complexity of spatial variation. Thereby the results for a spherical cell serve as a relatively simple beginning point and reference for more complicated cellular structures. As might be expected, more complicated cell structures lead to a correspondingly more complicated space–time behavior. That is, though idealized and relatively simple, the spherical cell analysis shows most clearly many of the fundamental concepts of stimulation, and also introduces most of the terminology used in stimulation.

The main part of the chapter considers *fibers*. Here fibers are idealized as having cylindrical geometry. Initially, the mathematical expressions relating currents to voltages along the fiber are established. Using these relationships in one spatial dimension, we then evaluate a stimulus just outside the membrane, initially just the steady-state response. Thereafter, the time evolution response (also called the transient response) for an intracellular electrode is found.

With one-dimensional analysis completed, the chapter moves on to field stimulation, a three-dimensional situation as the stimulus electrodes may be moved away from the fiber surface. With field stimulation, we examine both subthreshold and transthreshold situations. Fiber simulations under transthreshold conditions evaluate circumstances where stimuli may lead to propagating action potentials. Such simulations permit an evaluation of the classical concept of threshold, revealing conditions where it is not dependable.

Most fibers evaluated in this chapter are assumed to be infinitely long. That obviously is an approximation, as often a real fiber is much longer than the region affected by a stimulus. To examine it more carefully, however, in the final section, we examine the differences in behavior of a fiber that has a finite rather than infinite length.

7.1. SPHERICAL CELL STIMULATION

We begin with the study of the spherical cell, as illustrated in Figure 7.1. While the spherical cell's shape is a poor model for most biological cells, the simplicity of its electrical behavior makes it of interest. It is interesting because one can analyze the cell's response to a stimulus in a thorough way, taking into account all the central factors. Since the same central factors are present for a much broader set of cell shapes and circumstances, the response of the spherical cell serves as a guide to those also.

An analysis of the response of a spherical cell to an intracellular *subthreshold* stimulating current shows that the intracellular region is isopotential, to a good approximation. If one pictures the cell placed within an extensive extracellular region, then the extracellular volume also will be virtually isopotential. Consequently, all points on the cell membrane elements will have very nearly the same transmembrane potential. (The transmembrane potential has to be uniform because all intracellular potentials are nearly the same, and all extracellular potentials are nearly the same).

Consequently, the response of any patch on the cell's membrane will be the same as any other patch, and the entire membrane will behave synchronously.

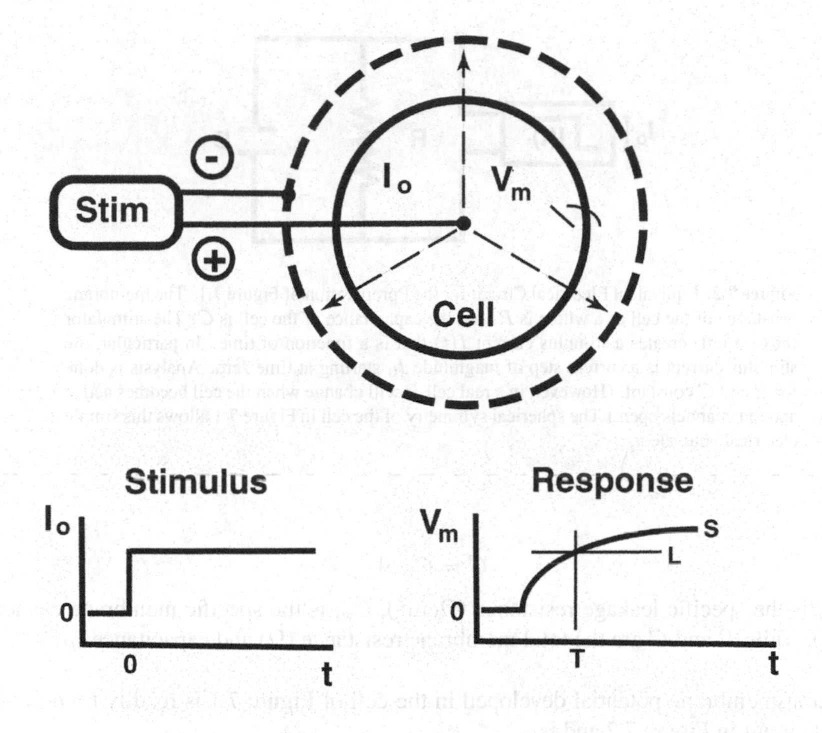

Figure 7.1. Top: A stimulator (left) applies a current I_0 to the center of a spherical cell. Current flows symmetrically outward (arrows) through the membrane (solid circle). Current is collected symmetrically at the periphery of the surrounding extracellular bath (dashed sphere). Bottom: A current step of magnitude I_o is applied (lower left) by the stimulator between the intracellular and extracellular electrodes. The stimulus current continues indefinitely during time t. The current produces a rising transmembrane voltage, v_m (solid curve), that does not have the step waveform of I_0. Even though the stimulus current I_0 continues on, the rise of v_m approaches limiting level $v_m = S$. Level S is called the "strength" of the stimulus. Of particular interest is the time T required to reach a "threshold" voltage level $V_T = L$ (short lines crossing v_m curve at lower right). The v_m curve is sketched as the response if membrane resistance R_m is constant. Furthermore, the concept of this simplified view of stimulation is that R_m will change abruptly once v_m reaches threshold voltage level L, as an active membrane response will occur thereafter.

7.1.1. Spherical Cell's Response to a Current Step

What is the response of an spherical cell to the application of a stimulating subthreshold current step? The arrangement is depicted in Figure 7.1.

Because the intracellular and extracellular regions are essentially isopotential, all membrane elements are electrically in parallel. Thus the entire cell in Figure 7.1 can be represented by a single lumped-RC circuit, and both R and C will be constant under subthreshold conditions.

The corresponding electrical circuit is illustrated in Figure 7.2, where, for a membrane surface area A, we have from (2.57)

$$R = \frac{R_m}{A} \tag{7.1}$$

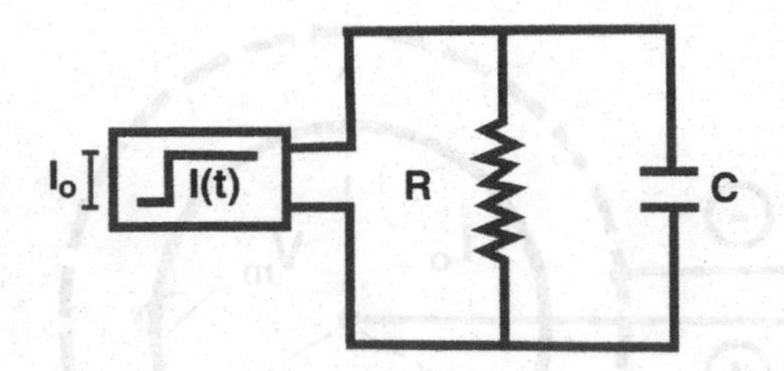

Figure 7.2. Equivalent Electrical Circuit for the Ppreparation of Figure 7.1. The membrane resistance of the cell as a whole is R, and the capacitance of the cell is C. The stimulator (box on left) creates a stimulus current $I(t)$ that is a function of time. In particular, the stimulus current is a current step of magnitude I_0 starting at time zero. Analysis is done for R and C constant. (However, in a real cell R will change when the cell becomes active and ion channels open.) The spherical symmetry of the cell in Figure 7.1 allows this simple electrical equivalent.

and (2.54)

$$C = C_m A \tag{7.2}$$

Here, R_m is the specific leakage resistance (Ωcm^2), C_m is the specific membrane capacitance ($\mu\text{F/cm}^2$), while R and C are the total membrane resistance (Ω) and capacitance (μF).[3]

The transmembrane potential developed in the cell of Figure 7.1 is readily found from the equivalent circuit in Figure 7.2 and is

$$v_m = I_0 R(1 - e^{-t/\tau}) \tag{7.3}$$

Rewriting (7.3) for a stimulus just strong enough and long enough to reach a threshold voltage level V_T with stimulus duration T, we have

$$V_T = S(1 - e^{-T/\tau}) \tag{7.4}$$

In Eq. (7.4) time constant $\tau = R_m C_m = RC$ and stimulus strength $S = I_0 R$. Note that parameter S is the steady-state voltage approached by v_m as $t \to \infty$. The quantity S can be thought of as a measure of the depolarizing strength of the applied stimulus current I_0; in fact, it is the maximum depolarization that can be produced passively by I_0. We also note that the time constant τ is independent of A (the cell area). Finally, in (7.3) we use v_m (rather than V_m) since the quantity of interest is the change in the transmembrane potential caused by the stimulus, relative to its baseline.

7.1.2. Strength–Duration

It is well known experimentally and theoretically that as stimulus strength S is increased, a shorter stimulus duration T is needed to reach a particular transmembrane voltage. To examine the correspondence mathematically, suppose that the transmembrane voltage threshold needed for initiate activation is fixed at $v_m = V_T$,[4] and a stimulus strength S greater than V_T is used. The consequence by (7.4) will be that membrane voltage V_T will be reached with a shorter stimulus duration, T, than $T \to \infty$.

What stimulus duration T is necessary? Rearranging (7.4) to isolate the term containing T, one gets

$$e^{T/\tau} = \frac{1}{(1 - V_T/S)} \tag{7.5}$$

By taking the natural log of (7.5), one can find either T or τ if the other parameters are known. Thus, where log is the natural logarithm,

$$T = \tau \log(\frac{1}{(1 - V_T/S)}) = \tau \log(\frac{S}{(S - V_T/S)}) \tag{7.6}$$

A more subtle use of (7.5) occurs when one wishes to find τ from two pairs of values of S and T. In this case one can solve for τ by writing (7.5) twice, and taking the ratio before taking the log.

Weiss–Lapicque equation

Rearranging (7.4) in a different way, one sees that the relationship between stimulus strength S and threshold voltage V_T can be written as

$$S = V_T/(1 - e^{-T/\tau}) \tag{7.7}$$

Division on both sides of (7.7) by the membrane resistance R leads to

$$I_{\text{th}} = \frac{I_R}{(1 - e^{-T/\tau})} \tag{7.8}$$

Eq. (7.8) often is called the Weiss–Lapicque equation.[5] There is a specialized terminology used in connection with this equation, as discussed in the next section.

Rheobase

In (7.8) I_R is named the *rheobase*, while I_{th} is the minimum current required to reach threshold with stimulus duration T.

From (7.8) one sees that the rheobase, I_R, is the minimum stimulus intensity that still produces a threshold value of transmembrane voltage, as the stimulation duration grows long (conceptually, as $T \to \infty$). V_T is the strength at rheobase, or *rheobase voltage*. The colorful terminology of *rheobase* and *chronaxie* was introduced by Lapicque [2].[6]

A plot of S versus T for fixed V_T is given in Figure 7.3. The curve depicts the *strength–duration* relationship for a threshold stimulus. The curve shows an exponential decay to the rheobase voltage, and divides all strength–duration combinations into two groups. Those in region A produce transmembrane voltages that exceed threshold. Combinations precisely on the line $V_T = L$ produce transmembrane voltages exactly equal to threshold. Strength–duration combinations in region B produce transmembrane voltages less than threshold. Of these, the

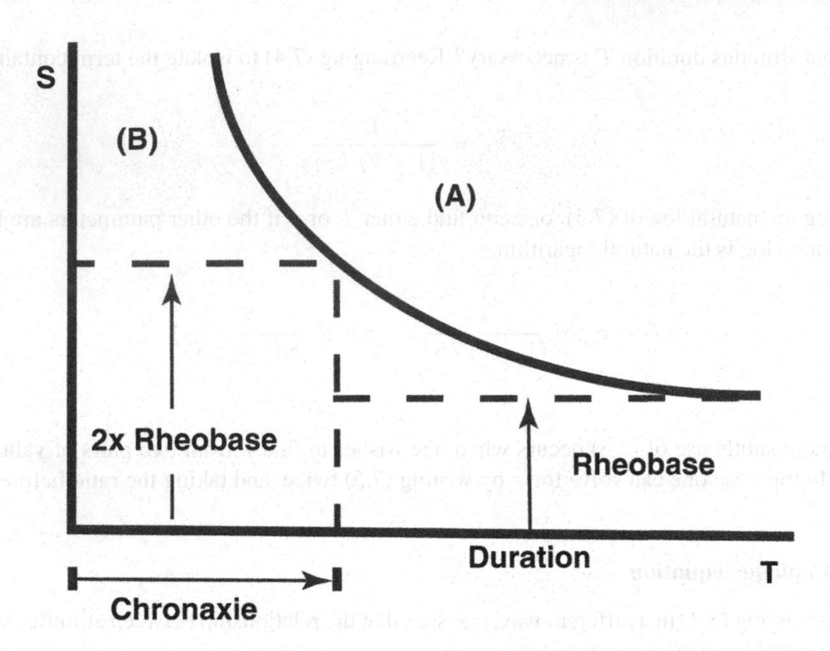

Figure 7.3. Strength–Duration Curve. Line $V_T = L$ shows the combinations of stimulus strength S (on the vertical axis) and stimulus duration T (on the horizontal axis) that are just sufficient to reach the threshold level. Combinations on side A of line L are above threshold and may lead to action potentials, while combinations on side B are below threshold. *Rheobase* is the value of stimulus current that is just sufficient to reach L with a long stimulus duration T. *Chronaxie* is the stimulus duration required to reach L if the stimulus current is set to twice rheobase.

graph makes clear that stimuli with a strength less than the rheobase voltage will never reach threshold, whatever their duration.

Chronaxie

The pulse duration when the stimulus strength S is *twice* rheobase is called *chronaxie.* From (7.7) chronaxie, T_c, can be found analytically, since at chronaxie $S = 2V_T$. Multiplying through by the term in parentheses, we have

$$V_T = 2V_T(1 - e^{-T_c/\tau}) \tag{7.9}$$

Equation (7.9) can be simplified to

$$e^{-T_c/\tau} = 1/2 \tag{7.10}$$

so after inverting, taking the natural log, and solving for T_c one has

$$T_c = \tau \ln 2 = 0.693\tau \tag{7.11}$$

Chronaxie is significant as a practical time period required to reach the threshold voltage when using a practical stimulus strength. In a comparison of different membranes or the same membrane under different conditions, chronaxie provides a nominal measure of excitability.

7.1.3. Stimulus Theory vs Experimental Findings

When the previous analysis is compared to experimental studies the results are modestly in accord, both qualitatively and quantitatively. The agreement is good enough to be useful in providing a qualitative understanding of the way experimental results change, as one or more experimental variables change. For example, the theory provides a guide to understanding why a greater stimulus current can create an action potential despite a shorter stimulus duration, or understanding why a sufficiently low stimulus current never creates an action potential, whatever the stimulus duration.

Even so, significant differences between the simple spherical-cell theory and experimental findings also are evident. Some reasons for such differences are as follows:

1. We assumed that the network in Figure 7.2 was valid up to threshold transmembrane potentials, while from Figure 5.6 we know that linearity holds up to 50% of threshold (if that much). Beyond 50% the assumption is at best a weak approximation.

2. The spherical cell stimulated with an intracellular electrode is a special case. In general, stimulating electrodes are extracellular and produce a response which depends on electrode location as well as the cell geometry. These parameters all affect the distribution and extent to which various membrane elements are depolarized, the conditions that ensue following termination of the stimulus, and hence the outcome regarding the initiation of an action potential. An example will be considered toward the end of this chapter. Some improvements in the model have been suggested based on a time-varying threshold, but even this possibility seems sensitive to the specific geometry and stimulus waveform.

3. A fixed threshold fails to account for its increase with time when the stimulus duration is comparable to the time constant of the inactivation parameter h (i.e., τ_h). The effect is described quantitatively by the Hodgkin–Huxley model based on the change in h with depolarizing or hyperpolarizing stimuli. This phenomena is known as *accommodation* and conflicts with the idea of a fixed threshold. For example, if the stimulus waveform were a ramp that reaches "threshold" after a time delay comparable to τ_h, then a diminished value of h at that point would require a yet higher stimulus. The "threshold," in other words, is now elevated. A slowly rising ramp could be chasing an ever elusive threshold and excitation fail to be elicited even though very high values of voltage are reached.

4. For stimuli with durations that are short (less than the sodium activation time constant τ_m), stimulation will grow more difficult, in that threshold v_m will rise (a fact noted by Lapicque [2]). Because the regenerative activation process will not be initiated at termination of the stimulus, even the transmembrane voltage that would be threshold, if the stimulus was longer, the stimulus may fail to produce a response. In this situation one must investigate whether the effective *RC* membrane can retain an adequate depolarizing voltage following the brief stimulus to continue opening sodium channels to the point that activation occurs. This question will be considered later on in this chapter with an example using an active membrane.

The above reasons are not a rationale for discarding the theory. Rather, they simply say that the theory has to be used with recognition that it is an approximation.

7.2. STIMULATION OF FIBERS

In the preceding section we considered the subthreshold response of a spherical cell, where all parts of the cell membrane were affected or changing in the same way, all the time. Now we examine the response to stimulation of a *fiber*.

At first we examine the behavior of the fiber under subthreshold (electrotonic) conditions, as was the case for the sphere. In fibers we expect subthreshold behavior that is similar to that of spherical cells in some respects, but we also expect that there will be some major differences.

One kind of difference occurs because of the length of fibers. Events at different sites along the fiber will occur at different times, because of the capacitance in the fiber's membrane. A second kind of difference is the corollary of the first: Adjoining segments of the fiber often are responding to a stimulus to different degrees and thus have differing transmembrane voltages, with the result that there are currents flowing within and along the fiber. Finally, fibers are evaluated using stimuli placed in different locations, which may be inside or outside the membrane, or distant from the whole fiber.

All of these aspects of fiber stimulation may occur in real fibers. Because of their number and complexity, addressing these aspects requires a number of the sections that follow.

When the excursion in transmembrane voltage is sufficiently small, the corresponding membrane current can be found from a passive admittance. Such subthreshold conditions are referred to as linear or *electrotonic*. For nerve (and approximately for muscle), the membrane can then be characterized electrically with a parallel *RC* network with constant values of *R* and *C*. This passive description is in contrast with the nonlinear behavior beyond threshold, where the potassium and sodium conductances are no longer independent of v_m.

An examination of membrane properties under linear (subthreshold) conditions is important, since these are frequently present in clinical and experimental studies. Furthermore, in the case of a propagating action potential, regions ahead of the activation site, where critical depolarization is taking place (e.g., region C in Figure 6.5), will be subthreshold during a critical initial interval. In addition, in the design of a stimulator, the membrane may often be considered as linear up to the point of activation.

7.2.1. Fiber Equations

It is immensely valuable in subsequent sections (and in analyzing fibers in general) to have available some basic equations for relationships among voltages and currents at points along the fiber. Thus we develop some of those here. They are of interest in their own right but will prove to be essential starting points in later sections.

Under subthreshold conditions, we have

$$i_m = \frac{v_m}{r_m} + c_m \frac{dv_m}{dt} \tag{7.12}$$

where r_m is the membrane resistance times unit length (Ωcm), c_m is the capacitance per unit length (μF/cm), and i_m is the transmembrane current per unit length (mA/cm). (At transmembrane voltages above threshold (7.12) still applies, but it is less useful because r_m must be treated as a variable.)

An interesting and useful result can be found from (7.12) if one recalls from the cable equations (6.11) that

$$\frac{\partial^2 v_m}{\partial x^2} = (r_i + r_e)i_m + r_e i_p \tag{7.13}$$

Substituting (7.12) into (7.13) gives

$$\lambda^2 \frac{\partial^2 v_m}{\partial x^2} - \tau \frac{\partial v_m}{\partial t} - v_m = r_e \lambda^2 i_p \tag{7.14}$$

where we have defined the following normalizing parameters:

$$\begin{aligned} \lambda &= \sqrt{\frac{r_m}{r_i + r_e}} \\ \tau &= r_m c_m \end{aligned} \tag{7.15}$$

For steady-state conditions ($\partial/\partial t = 0$), Eq. (7.14) simplifies to

$$\lambda^2 \frac{d^2 v_m}{dx^2} - v_m = r_e \lambda^2 i_p \tag{7.16}$$

When the stimulus current is zero ($i_p = 0$), Eq. (7.16) becomes simply

$$\lambda^2 \frac{d^2 v_m}{dx^2} - v_m = 0 \tag{7.17}$$

[which is also the homogeneous form of equation of (7.16)]. The solution of (7.17) is

$$v_m = Ae^{-x/\lambda} + Be^{x/\lambda} \tag{7.18}$$

where *A* and *B* are arbitrary constants. Rather than introducing the stimulating current i_p explicitly in (7.16) to obtain the particular solution, we can, instead, impose boundary conditions on the solution for the region where $i_p = 0$, namely, $|x| > 0$. But this solution is that given by (7.18). The boundary conditions at $x = 0$ and $x = \infty$ will serve to evaluate the constants *A* and *B* in (7.18). This approach is illustrated in the following sections.

7.2.2. Space and Time Constants

In the previous section we introduced the constants λ and τ (7.15). These quantities are referred to as the space (or length) constant and time constant of a fiber, respectively. Both are important parameters that characterize the response of a fiber to applied stimuli.

Under steady-state conditions λ is the distance over which the voltage and current change by the factor *e*, as identified in Eq. (7.18). For spherical-like cells only, τ is the time for the transient response to a current step to differ from its steady-state magnitude by the factor $1/e$, as seen in (7.4). For a fiber, we will presently show that τ is a measure of the time it takes for the transient response to a current step to reach a particular fraction of its steady-state value, where the fraction depends on the distance from the site to the point of stimulation.

Constants λ and τ are important because, frequently, they can be measured directly. Furthermore, λ and τ have a consistent meaning for many different fiber structures, so they may be used for characterization and comparison.

For circular cylindrical axons with constant membrane properties and with $r_e \approx 0$,

$$\lambda = \sqrt{\frac{r_m}{r_i + r_e}} \approx \sqrt{\frac{r_m}{r_i}}. \tag{7.19}$$

(The condition $r_e \approx 0$ applies when the extracellular space is large.) Converting r_i to R_i and r_m to R_m, by using (2.55) and (2.52), gives

$$\lambda = \sqrt{\frac{R_m/2\pi a}{R_i/\pi a^2}} \tag{7.20}$$

When simplified this equation becomes

$$\lambda = \sqrt{\frac{aR_m}{2R_i}} \tag{7.21}$$

where *a* is the fiber radius. Note that λ varies directly as the square root of fiber radius.

7.3. FIBER STIMULATION

The stimulus currents to be discussed are introduced into a biological preparation with the goal of changing the transmembrane voltage. In most situations, the electrode or electrodes through which the current is injected are outside the target fiber(s).

If injecting current extracellularly changes the transmembrane potential, by how much does it do so? And where? The following material examines these questions in an idealized geometry, but one that nonetheless includes the essential elements needed for insight into a experimental and clinical situations.

7.3.1. Extracellular Stimulus, Steady-State Response

Suppose a single small electrode is placed in the bounded extracellular space just outside a cylindrical fiber, while a pair of electrodes to remove the current lie extracellularly at $\pm\infty$. Suppose the fiber is at rest, infinitely long, the location of the proximal electrode identifies the coordinate origin ($x = 0$), and the fiber structure satisfies the assumptions of the core-conductor model. Note that this arrangement imposes symmetry between positive and negative regions. Also, we expect that a portion of the injected current will enter and flow within the intracellular space of the fiber; it will be constrained to flow longitudinally along the *x* axis.

With the above arrangement, where will current go? As a first guess it might seem that the injected current would remain in the extracellular space until removed by the distal electrodes.

This would be the case if the membrane was a perfect insulator. But for biological membranes it is reasonable to expect current to cross the membrane, particularly if the fiber is long (since the effective leakage resistance can become very low given an adequate axial distance).

With increasing x, this inflowing transmembrane current builds up the total intracellular current, I_i, while the extracellular current decreases by an equal amount. An equilibrium is reached for large enough x, where $r_i I_i = r_e I_e$. At this point the spatial rate of decreasing voltage is the same in both intracellular and extracellular space so, from a Kirchhoff loop, the transmembrane voltage and hence transmembrane current is zero. (Thus for $x \to \infty$ there is no further change in either I_e or I_i.)

In summary, one can expect the transmembrane current to be greater in the region near the stimulus site and to diminish to essentially zero at sites more distant from the stimulus. In the limited region where the stimulus produces a transmembrane current it must also produce an associated transmembrane potential. Thus we conclude somewhat intuitively that stimuli from extracellular electrodes can be used to create hyperpolarizing or depolarizing potentials over an extent of fiber near the stimulus electrode.

We now move to examine these expectations quantitatively. The current entering the preparation from the electrode can be idealized *as a spatial* delta-function source, that is,

$$i_p = I_0 \delta(x) \tag{7.22}$$

where $\delta(x)$ is a unit delta function.

The definition of $\delta(x)$ has three parts:

$$\begin{aligned} \delta(x) &= 0 \quad \text{for } x \neq 0 \\ \delta(0) &= \infty \\ \int_{-\infty}^{\infty} \delta(x)dx &= 1 \end{aligned} \tag{7.23}$$

Note that the delta function is zero except at the origin, where it is infinite, but its integral is finite (equal to unity) provided the interval of integration includes the origin.

From (7.22) and (7.23) we can identify I_0 as the total applied current while $i_p(x)$ is the current density (current per unit length); the latter is zero except at the origin, where it is infinite. If the delta-function source is used in the equations governing v_m under subthreshold and steady-state conditions, we have from Eq. (7.16)

$$\lambda^2 \frac{d^2 v_m}{dx^2} - v_m = r_e \lambda^2 I_0 \delta(x) \tag{7.24}$$

Now we seek the solution to the differential equation in (7.24). A good strategy is to first find the solution to the corresponding homogeneous equation, as that solution will apply to all points other than $x = 0$. Then, with that solution viewed as a boundary-value problem, we evaluate undetermined coefficients through the boundary conditions at the origin (which result from the introduction of the applied current at this point). We will follow that strategy in the following section.

Boundary conditions around the stimulus site

To establish the boundary condition at the stimulus site, the origin, suppose Eq. (7.24) is integrated from $x = 0^-$ to $x = 0^+$, i.e., from just to the left of the origin to just to the right of it. The result is

$$\lambda^2 \frac{dv_m}{dx}\bigg|_{x=0^+} - \lambda^2 \frac{dv_m}{dx}\bigg|_{x=0^-} - [v_m(0^+) - v_m(0^-)]\Delta x = r_e \lambda^2 I_0 \tag{7.25}$$

where $\Delta x = 0^+ - |0^-|$. As distance Δx approaches zero, the middle term goes to zero, since (on physical grounds, at least) v_m is continuous. Note that the term on the right-hand side no longer contains the δ function (whose integral was replaced by unity).

Rewriting (7.25) we obtain

$$\lambda^2 \left(\frac{dv_m}{dx}\bigg|_{x=0^+} - \frac{dv_m}{dx}\bigg|_{x=0^-} \right) = r_e \lambda^2 I_0 \tag{7.26}$$

and we note that $\partial v_m / \partial x$ is discontinuous at $x = 0$. The discontinuity, furthermore, is proportional to the strength of the current source I_0.

We will use this result below and evaluate derivatives near the stimulus site, to get the boundary condition needed there.

The homogeneous solution at steady state

For sites along the fiber away from the origin there are no applied currents, so the homogeneous equation (7.17) applies, namely,

$$\lambda^2 \frac{d^2 v_m}{dx^2} - v_m = 0 \tag{7.27}$$

Equation (7.27) has the solution

$$v_m(x) = Ae^{-x/\lambda} + Be^{x/\lambda} \tag{7.28}$$

Thus one sees that v_m at all points along the fiber can be found from (7.28) once values for constants A and B are determined from the boundary conditions.

We now consider the appropriate choices of constants A and B. The choices must satisfy the conditions imposed by the source at $x = 0$ and also the requirements when $|x| \to \infty$.

The necessary outcomes are summarized in Table 7.1. The choice of $A = 0$ for $x < 0$ and $B = 0$ for $x > 0$ is necessary because the solution for v_m caused by applying a finite current I_0 must go to zero as the distance from the stimulus becomes large. Because v_m must be symmetric about the origin, there being no physical difference between the positive x side versus the negative side, it is also concluded in Table 7.1 that both A and B are equal to the same constant, C.

Table 7.1. Boundary Conditions

x range	A	B
$x < 0$	0	C
$x > 0$	C	0

These choices of A and B are required for the transmembrane potential to decline to zero far from the stimulus site.

Imposing these conditions results in equal but opposite axial currents at symmetric points about the origin, an outcome that is consistent with the symmetry. Thus Eq. (7.28) can be written as

$$\begin{aligned} v_m(x) &= Ce^{x/\lambda} \quad x \leq 0 \\ v_m(x) &= Ce^{-x/\lambda} \quad x \geq 0 \end{aligned} \tag{7.29}$$

A more compact form of (7.29) is

$$v_m(x) = Ce^{-|x|/\lambda} \tag{7.30}$$

Imposing the boundary condition at the origin

The coefficient C in (7.30) can now be found, since the solution must also satisfy (7.26). To impose this boundary condition at the origin dv_m/dx is first evaluated from Eq. (7.29). The result is

$$\begin{aligned} \frac{dv_m}{dx} &= \frac{C}{\lambda}e^{x/\lambda} \quad x < 0 \\ \frac{dv_m}{dx} &= -\frac{C}{\lambda}e^{x/\lambda} \quad x > 0 \end{aligned} \tag{7.31}$$

Substituting (7.31) into (7.26) gives

$$\left(-\frac{C}{\lambda}e^{-x/\lambda}\Big|_{x=0^+} - \frac{C}{\lambda}e^{x/\lambda}\Big|_{x=0^-} \right) = r_e I_0 \tag{7.32}$$

The solution for C from (7.32) is

$$C = -\frac{r_e \lambda I_0}{2} \tag{7.33}$$

The steady-state solution

Using the value of C obtained in (7.33) and substituting into (7.30) gives the desired solution, namely,

$$v_m = -\frac{r_e \lambda I_0}{2} e^{-|x|/\lambda} \tag{7.34}$$

Inspection of Eq. (7.34) provides a quantitative response to the questions and speculations posed at the beginning of this section. These are summarized below.

1. The stimulus clearly affects the transmembrane potential, since v_m is nonzero for all values of x.

2. The effect of the stimulus varies markedly with x. The largest change in transmembrane potential occurs at the site of the stimulus, where $x = 0$. As one moves away from the stimulus site, v_m decreases exponentially, falling by a factor of e every length λ.

3. For a given stimulus current I_0, the magnitude of the change in transmembrane potential increases as extracellular resistance r_e increases.

4. Note from the sign of (7.34) that a *positive* current injected at the origin leads to a more *negative* transmembrane potential. That is, membrane under an *anode* becomes *hyperpolarized* as a result of current influx into the intracellular region.

5. Note that the space constant λ may also be regarded as a measure of the distance from a source (at the origin) to which the disturbance in v_m essentially extends.

7.3.2. Intracellular Stimulus, Time-Varying Response

We now turn our attention to an investigation of the *temporal transient* behavior under the same stimulus condition, rather than the *steady-state* response evaluated above.

Determining transient behavior requires a solution to the general expression of (7.14). As before, we first seek a solution to the homogeneous equation and introduce the applied current through a boundary condition at the origin. We consider an unbounded extracellular medium and assume that the stimulus current is introduced *intracellularly*.

This geometry permits introducing the simplification that $r_e \approx 0$. The resulting equation is

$$\lambda^2 \frac{\partial^2 v_m}{\partial x^2} - \tau \frac{\partial v_m}{\partial t} - v_m = 0 \tag{7.35}$$

The space constant λ and time constant τ are as defined in (7.15). (If the applied current is introduced extracellularly, the solution obtained here can be converted to this condition, as described in a later section.)

A simplified notation results from introducing the normalized spatial and temporal variables (X,T), defined by

$$X = \frac{x}{\lambda} \quad \text{and} \quad T = \frac{t}{\tau} \tag{7.36}$$

Hence (7.35) becomes

$$\frac{\partial^2 v_m}{\partial X^2} - \frac{\partial v_m}{\partial T} - v_m = 0. \tag{7.37}$$

We seek the transmembrane potential, v_m, arising from the introduction of a current step at the origin.

Reduction to one variable by the Laplace transform method

We proceed by taking Laplace transforms with respect to T of each term in (7.37). The Laplace transform of $\partial v_m/\partial T$ is $s\overline{v}_m - v_m(0,X)$, where the overbar indicates a Laplace transform.

The initial condition of v_m at $t = 0$, namely $v_m(0,X)$, is assumed to be zero. It is initially zero because we consider the response of a resting cable to an applied current that starts at $t \geq 0$. Consequently, we get

$$\frac{\partial^2 \overline{v}_m}{\partial X^2} - (s+1)\overline{v}_m = 0 \tag{7.38}$$

The advantage of introducing the Laplace transform is that the partial differential equation (7.37) in X and T has been converted into an ordinary differential equation in X (7.38).

The solution to (7.38) is

$$\overline{v}_m = Ae^{-X\sqrt{s+1}} + Be^{X\sqrt{s+1}} \tag{7.39}$$

Because $\overline{v}_m$ cannot increase without bound for $x \to \infty$, $B = 0$ (for the infinite cable). Thus

$$\overline{v}_m = Ae^{-X\sqrt{s+1}}, \quad X \geq 0 \tag{7.40}$$

The boundary condition at the origin

At $x = 0$, the site of introduction of the current I_0 into the intracellular space, because of symmetry, $I_0/2$ flows into the positive x region and $I_0/2$ into the region $x < 0$.

This applied current as a function of time is in the form of a step that we designate $u(t)$, the unit step function. This function is described by $u(t) = 0$ for $t \leq 0$ and $u(t) = 1$ for $t \geq 0$. There is a discontinuity at $t = 0$. Applying Ohm's law in the intracellular space at $x = 0$, we have

$$\left.\frac{\partial \Phi_i}{\partial x}\right|_{x=0} = -\frac{I_0 u(t) r_i}{2}. \tag{7.41}$$

For the extracellular space at $x = 0$ there is no longitudinal current (it begins to appear when $x > 0$), so

$$\left.\frac{\partial \Phi_e}{\partial x}\right|_{x=0} = 0 \tag{7.42}$$

If we subtract (7.42) from (7.41) and then note from (7.36) that $\partial/\partial x = (1/\lambda)\partial/\partial X$,[7] we get

$$\left.\frac{\partial v_m}{\partial X}\right|_{x=0} = -\frac{I_0 u(t) r_i \lambda}{2} \tag{7.43}$$

Taking the Laplace transform of both sides of (7.43), where the Laplace transform of $u(t)$ is included as $1/s$, gives

$$\left.\frac{\partial \overline{v}_m}{\partial X}\right|_{x=0} = -\frac{I_0 r_i \lambda}{2s} \tag{7.44}$$

We also can evaluate the left-hand side of (7.44) from (7.40). So doing gives

$$\begin{aligned}\left.\frac{\partial \overline{v}_m}{\partial X}\right|_{X=0} &= -A\left[(s+1)^{1/2}e^{-X(s+1)^{1/2}}\right]_{X=0} \\ &= -A\sqrt{s+1}\end{aligned} \tag{7.45}$$

One obtains an equation for *A* by equating (7.45) and (7.44). This yields

$$A = \frac{r_i \lambda I_0}{2s\sqrt{s+1}} \tag{7.46}$$

Substituting (7.46) into (7.40) gives $\overline{v}_m$ as a function of *s*, namely,

$$\overline{v}_m = \frac{I_0 r_i \lambda}{2s\sqrt{s+1}} e^{-X\sqrt{s+1}}, \quad X > 0 \tag{7.47}$$

Time-varying response to stimulus

The desired solution for the time-varying response is found by taking the inverse transform of (7.47). Finding the inverse transform is most readily accomplished by consulting a table of Laplace transforms,[8] which demonstrates that

$$\begin{aligned} v_m(X,T) &= \frac{r_i \lambda I_0}{4}\left\{e^{-X}\left[1 - \operatorname{erf}\left(\frac{X}{2\sqrt{T}} - \sqrt{T}\right)\right]\right. \\ &\quad \left. - \; e^{X}\left[1 - \operatorname{erf}\left(\frac{X}{2\sqrt{T}} + \sqrt{T}\right)\right]\right\} \end{aligned} \tag{7.48}$$

This result is for an infinite cable, based on the introduction of I_0 at $X = 0$, and describes conditions for $x > 0$ (those for $x < 0$ can be found by symmetry). One can also replace x by $|x|$, which gives the expected symmetry and an expression valid for all x. On restoring the original coordinates x and t, (7.48) becomes

$$\begin{aligned} v_m(x,t) = \frac{r_i \lambda I_0}{4} &\left\{ e^{-|x|/\lambda}\left[1 - \operatorname{erf}\left(\frac{|x|}{2\lambda}\frac{\sqrt{\tau}}{t} - \frac{\sqrt{t}}{\tau}\right)\right]\right. \\ &\left. - e^{|x|/\lambda}\left[1 - \operatorname{erf}\left(\frac{|x|}{2\lambda}\frac{\sqrt{\tau}}{t} + \frac{\sqrt{t}}{\tau}\right)\right]\right\} \end{aligned} \tag{7.49}$$

In (7.47) and (7.48), erf is the *error function* defined by

$$\operatorname{erf}(y) = \frac{2}{\sqrt{\pi}}\int_0^y e^{-z^2}\,dz \tag{7.50}$$

Note that erf $(\infty) = 1$ and erf $(-\infty) = -1$. The result in (7.49) tacitly assumes sinks of strength $-I_0/2$ at $x = \pm\infty$.[9]

7.3.3. Examination of Temporal Response

For a given value of time the spatial behavior is exponential-like but not exponential. For $t >> \tau$ (i.e., for the temporal condition approaching the steady state), $v_m(x)$ tends toward a

Table 7.2. Temporal Morphology at Different Values of x

x	Steady-state fraction at $t = \tau$
0	0.843
λ	0.632
2λ	0.372
3λ	0.157
4λ	0.0453
5λ	0.00862

Stimulus: current step at $x = 0$.

true exponential in x, as shown in (7.34), and also as obtained from (7.49). The presence of the membrane leakage resistance is responsible for a continuous decrement of v_m with increasing x while λ describes the rate of this effect.

In the temporal behavior of $v_m(x,t)$ given by Eq. (7.49), τ characterizes this behavior. Thus as noted, when $t > \tau$ the response rapidly approaches steady-state values. Figure 7.4 plots families of curves derived from Eq. (7.49), which expresses the above ideas graphically. These results show that time is required to reach steady state owing to the presence of membrane capacitance and resistance, and this membrane time constant is a measure of that time. Further, the response is spatially confined to a region near the site of the stimulus and λ is a measure of its extent.

For a fixed x, the temporal behavior is not a true exponential; its shape is not readily apparent by inspection of (7.49). If we determine from (7.49) the fraction of steady-state amplitude reached at $t = \tau$ as a function of x, the data in Table 7.2 are obtained. The rapid decrease in value seen in Table 7.2 also reflects a temporal waveform that is not exponential. Only at $x = \lambda$ does the magnitude of the fraction of steady-state amplitude reached at $t = \tau$ equal that obtained with an exponential waveform (i.e., $1 - 1/e$).

7.4. AXIAL CURRENT TRANSIENT

Questions: How much axial current does the stimulus generate? Does the axial current start quickly? Where? Current is injected intracellularly, so does it all flow down the intracellular volume, or does some go outside? Is the current flow pattern quickly established, or does it take a long time to reach equilibrium?

Stimuli often are used to manipulate the actions of excitable tissue, so understanding a fiber's response to stimuli as a function of the magnitude or position of the stimulus site has a natural interest and utility. As an example of the mathematical results developed to this point, let us evaluate the response of a semi-infinite fiber with a bounded extracellular space to such a stimulus. In particular, let us consider the application of an intracellular current step, of magnitude I_0, at the coordinate origin ($x = 0$) at $t = 0$. (To simplify the consideration, we assume that the remote electrode is at $+\infty$.)

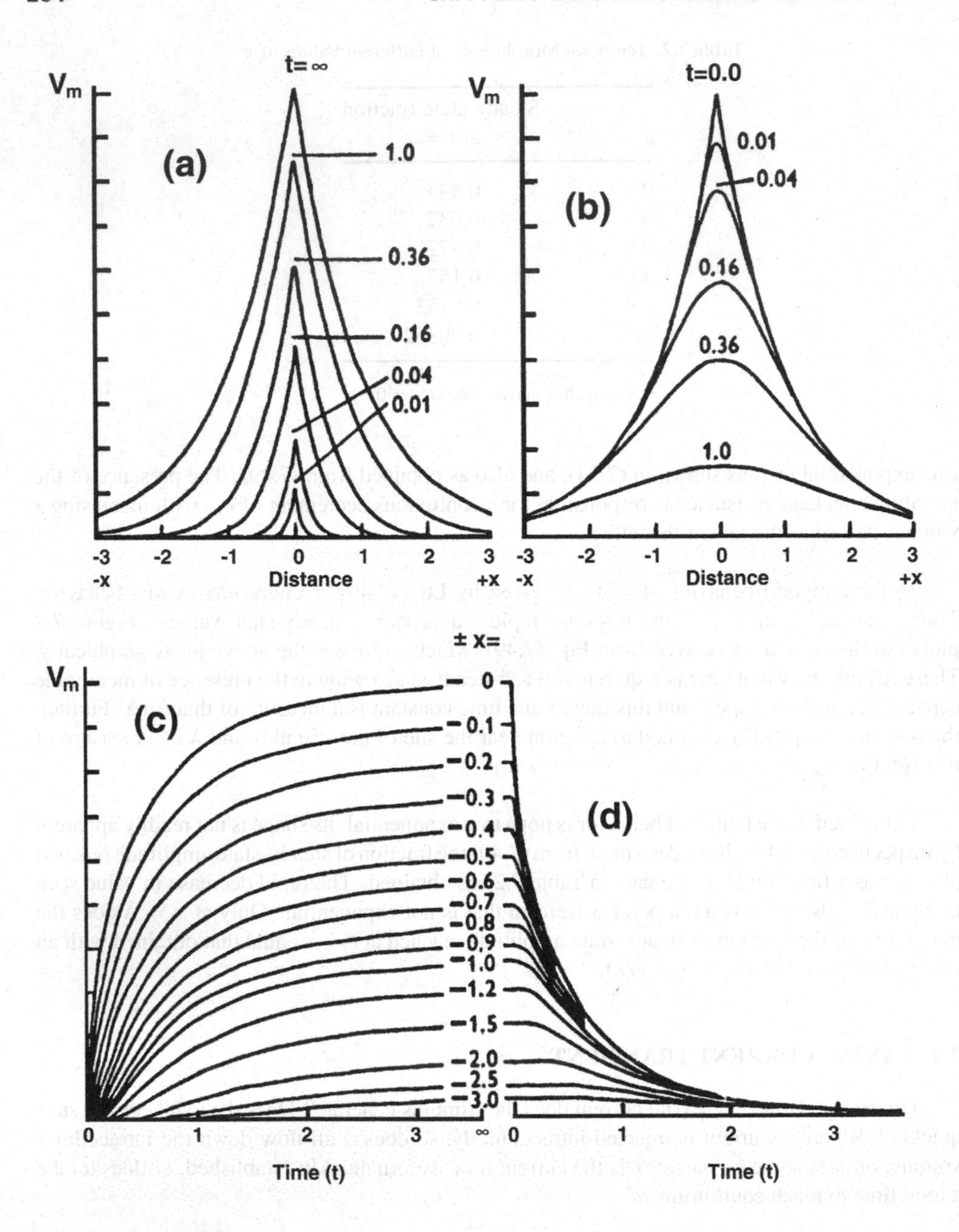

Figure 7.4. Theoretical Distribution of Potential Difference across a passive nerve membrane in response to onset (a and c) and cessation (b and d) of a constant current applied intracellularly at the point $x = 0$. (a) and (b) show the spatial distribution of potential difference at different times, and (c) and (d) show the time course of the potential at different distances along the axon. Time (t) is in time constants, τ, and distance (x) is in space constant, λ. From Aidley DJ. 1978. *The physiology of excitable cells*. Cambridge: Cambridge UP. After Hodgkin AL, Rushton WAH. 1946. The electrical constants of a crustacean nerve fiber. *Proc R Soc London, Ser B* **133**:444–479. Reprinted with permission of Cambridge University Press.

To find quantitative answers to the questions, first we examine the given parameters and observe that

- At $t = 0$, $v_m = 0$ everywhere since the membrane capacitances have yet to receive any charge from the applied current.
- Because $v_m \equiv 0$ signifies a short-circuited membrane, the applied current, at $t = 0$, divides instantaneously between intracellular and extracellular space in inverse proportion to the axial resistances [i.e., $I_i = (r_e/(r_i + r_e))I_0$ and $I_e = (r_i/(r_i + r_e))I_0$].
- Because extracellular space is assumed bounded r_e is not negligibly small.
- Now for $x >> \lambda$ and at steady state, (7.34) describes $v_m \approx 0$, so $r_i I_i = r_e I_e$.
- Also, because $I_i + I_e = I_0$, then $I_i = r_e I_0/(r_i + r_e)$, and $I_e = r_i I_e/(r_i + r_e)$, hence approximating their initial values.

This close approximation to the initial values suggests that at large x there is a transient of negligible magnitude. It further suggests that the axial current response to a step is essentially instantaneous. We will examine this hypothesis quantitatively by deriving and evaluating an expression for the axial intracellular steady-state current. That is,

$$I_i = \frac{r_e}{r_i + r_e} I_0 \quad \text{when} \quad x \to \infty$$

so at a more proximal site (smaller x) I_i will be greater than this limiting value. It will be greater by an amount equal to the total outflow of transmembrane current between x and infinity.

That is, for finite values of x

$$I_i(x) = \frac{r_e}{r_i + r_e} I_0 + \int_x^{\infty} i_m dx \tag{7.51}$$

Equation (7.34) gives the steady-state v_m for a current I_0 applied extracellularly at the origin of an infinite fiber; $I_0/2$ is removed at $\pm\infty$. In view of the intracellular–extracellular symmetry, we obtain a similar expression for a current applied intracellularly by interchanging subscripts i and e; also, there is a change in sign (since transmembrane current is oppositely directed).

Using this reasoning, and since we are now taking I_0 to be the total current in the positive x direction rather than $I_0/2$ in (7.34) (since this is a semi-infinite cable), we get

$$v_m = r_i \lambda I_0 e^{-|x|/\lambda} \tag{7.52}$$

At steady state the transmembrane current is entirely through r_m (i.e., there is no capacitive current), so $i_m = v_m/r_m$. With this relationship and using (7.52), Eq. (7.51) becomes

$$I_i(x) = \frac{r_e}{r_i + r_e} I_0 + \int_x^{\infty} \frac{r_i \lambda I_0}{r_m} e^{-x/\lambda} dx \tag{7.53}$$

where $0 < x < \infty$. Performing the integration and simplifying the results leads to the expression

$$I_i(x) = \frac{r_e}{r_i + r_e} I_0(1 + \alpha e^{-x/\lambda}) \tag{7.54}$$

where $\alpha = r_i/r_e$. Note that $I_i(0) = I_0$, while $I_i(\infty) = (r_e/(r_i + r_e))I_0$, as expected. For muscle bundles and for cardiac tissue, experimental data support $\alpha = 1$ for estimates of I_i.

It also is informative to use Table 7.2 to examine issues related to the temporal response:

- For values of x equal to 2λ or less, the transient amplitude becomes large ($> 13\%$). For this x range, as shown in Table 7.2 the transient time is on the order of τ.
- At $x = 5\lambda$ we have the result that $I_i(5\lambda)$ changes its relative magnitude from $t = 0$ to $t = \tau$ by only 0.862%. From (7.49), we can determine that the time required to achieve 65% of steady state (an effective time constant) is roughly 3τ, hence fairly long. Nevertheless, since the change is so small, so that the time required to achieve it may not matter.
- For $x = 10\lambda$ achieving steady state will take much longer ($\approx 5\tau$), and the change in magnitude during this transient will be even smaller (0.005%). Under many circumstances these changes are insignificant.

7.5. FIELD STIMULUS OF AN INDIVIDUAL FIBER

In this section we examine the subthreshold membrane response of a single fiber of infinite length. The fiber is assumed to be lying in an unbounded conducting medium. The stimulus field arises from an *external* point current source. We picture our goals as follows:

- The site of the stimulus may be away from the fiber, so an expression is to be derived for the induced transmembrane potential given the source–fiber distance, *h*. Also known is the current magnitude, I_0, and fiber and medium properties.
- We wish to find an expression that will permit an examination of the relationship between induced transmembrane potential and the stimulating source field.
- The analysis and thus the result depend on the assumption of linearity (subthreshold conditions) and that $a/h << 1$, where *a* is the fiber radius.[10]

In the subsequent sections the same physical arrangement is considered, but under transthreshold stimulus levels. This additional analysis permits an examination of *threshold* and a determination of its constancy as various parameters are changed.

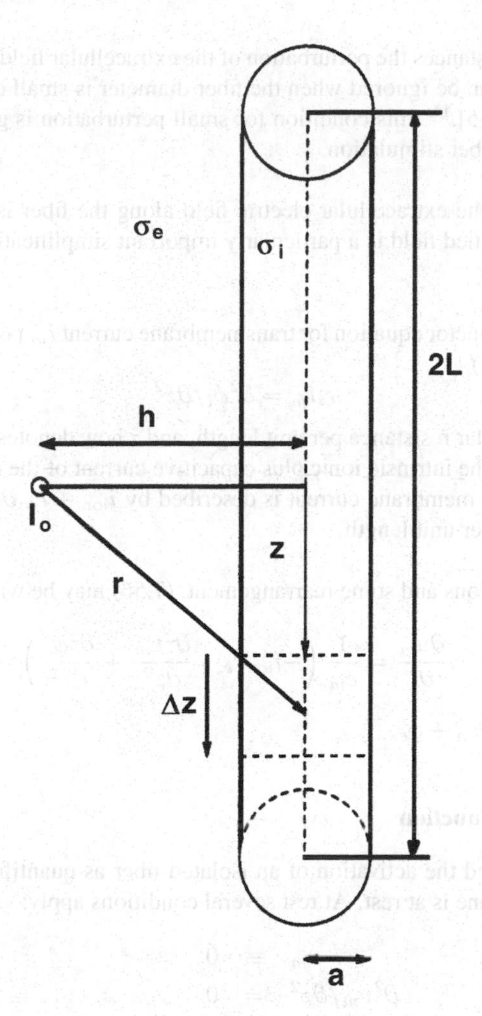

Figure 7.5. Geometry of Source and Fiber. A single current point source of magnitude I_0 is placed at a distance h from a circular cylindrical fiber of length $2L$. The extracellular region is unbounded, uniform, and has a conductivity σ_e. The fiber radius is a and its intracellular conductivity is σ_i. The fiber's centerline lies along the coordinate z axis. The length is divided into elements Δz for numerical calculations.

7.5.1. The Electric Field from a Point Source

Let us consider the response of an unmyelinated fiber lying in an unbounded conducting medium due to an applied electric field of a point current source. The field, ϕ_a, has the form described in (2.21):

$$\phi_a = I_0/(4\pi\sigma_e r) \tag{7.55}$$

In (7.55) I_0 is the current strength, and σ_e the conductivity of the medium, where the extracellular space being designated with subscript e, and r is the distance from the source to an arbitrary field point. A description of the geometry is given in Figure 7.5.

Under many circumstances the perturbation of the extracellular field resulting from the presence of the fiber itself can be ignored when the fiber diameter is small compared to its distance from the (point) source [5].[11] This condition for small perturbation is generally satisfied under the usual conditions of fiber stimulation.

As a consequence, the extracellular electric field along the fiber is essentially the applied field. Use of the unmodified field is a particularly important simplification for the evaluation of fiber excitation.

The linear core-conductor equation for transmembrane current i_m per unit length of unmyelinated fiber is, from (6.13),

$$r_i i_m = \partial^2 \phi_i / \partial z^2 \tag{7.56}$$

where r_i is the intracellular resistance per unit length, and z now denotes the axial variable. This current must also equal the intrinsic ionic plus capacitive current of the membrane, as discussed in obtaining (6.31). The membrane current is described by $i_{\text{ion}} + c_m \partial v_m / \partial t$, where c_m is the membrane capacitance per unit length.

With these substitutions and some rearrangement, (7.56) may be written as

$$r_i \frac{\partial v_m}{\partial t} = \frac{1}{c_m} \left(-i_{\text{ion}} r_i + \frac{\partial^2 v_m}{\partial z^2} + \frac{\partial^2 \phi_e}{\partial z^2} \right) \tag{7.57}$$

where ϕ_i is replaced by $v_m + \phi_e$.

7.5.2. The Activating Function

Rattay [7] considered the activation of an isolated fiber as quantified by (7.57). He noted that at $t = 0$ the membrane is at rest. At rest several conditions apply:

$$\begin{aligned} v_m &\equiv 0 \\ \partial^2 v_m / \partial z^2 &= 0 \\ i_{\text{ion}} &= v_m / r_m = 0. \end{aligned} \tag{7.58}$$

Using these resting conditions with (7.57), he established that, initially upon application of the stimulus,

$$r_i \partial v_m / \partial t = (1/c_m) \partial^2 \phi_e / \partial z^2 \tag{7.59}$$

Initial change follows activating function

Rattay argues that (7.59) provides a foundation for the following conclusions: Where activation may occur corresponds to the region where $\partial^2 \phi_e / \partial z^2$ is positive, since having $\partial^2 \phi_e / \partial z^2$ positive will make $\partial v_m / \partial t > 0$ initially. Conversely, the region that will hyperpolarize (i.e., where $\partial v_m / \partial t < 0$ for small t) is where $\partial^2 \phi_e / \partial z^2$ is negative, according to (7.57), and this region will not initiate activation. Because of the role played by

$$A(x) = \partial^2 \phi_e / \partial z^2 \tag{7.60}$$

Rattay named the function the *activating function*.

For *subthreshold* linear conditions the ionic current may also be evaluated by v_m/r_m, where r_m is the fiber membrane resistance per unit length. Under steady-state conditions ($\partial v_m/\partial t = 0$) and with $i_{\text{ion}} = v_m/r_m$, (7.57) becomes

$$\frac{\partial^2 v_m}{\partial z^2} - \frac{v_m}{\lambda^2} = -\frac{\partial^2 \phi_e}{\partial z^2} \tag{7.61}$$

where, of course, $\lambda = \sqrt{r_m/r_i}$. Since the axial applied electric field, E_z, is the negative z derivative of ϕ_e, Eq. (7.61) can also be written as

$$\frac{\partial^2 v_m}{\partial z^2} - \frac{v_m}{\lambda^2} = \frac{\partial Ez}{\partial z} \tag{7.62}$$

Equation (7.62) describes the effect of the applied field on the target fiber through the solution for v_m, the induced transmembrane potential. The axial derivative of E_z is seen as the applied or "forcing" function in the differential equation for v_m. For a fiber of infinite length, the response as described by $v_m(z)$ should correspond, more or less, to the applied function.

Thus the peak depolarization, of particular interest in clinical design, could be expected to be located where $\partial E_z/\partial z$ attains its maximum values. To the extent that such a correspondence is true, the activating function is a valuable tool since a possibly complex solution for the actual v_m is avoided.

The activating function is only the beginning

The activating function is not the solution for $V_m(t)$, but only its initial rate of change. $V_m(t)$ changes over time during and after the stimulus interval, and it is clear that the v_m arising from $\partial E_z/\partial z$ depends in some way on the entire function E_z, not just the location and magnitude of its initial values or its peak values.

Furthermore, the v_m response can be expected to depend in some way on the fiber properties, as perhaps described simply by the parameter λ. In addition, for finite fibers, boundary conditions must be introduced into the solution of (7.62) and the boundary conditions may have an important influence on the morphology of v_m.

7.5.3. The V_m Response over Time

From a formal point of view the activating function has the role of an applied function in the differential equation (7.62). While the form of $v_m(x)$ may evolve to become similar to that of the forcing function, another possibility is that $v_m(x)$ will not be the same in important respects.

Consequently, more mathematical results are needed to know what the stimulus does, quantitatively. Thus we now proceed to find a solution for v_m. When that solution is obtained, there will be the opportunity to compare it to the activating function to see what looks the same, and what looks different.

The following mathematical development will show that the transmembrane potential response, over time, can be found by means of a convolution. The convolution shows the interaction between the effects of the external field on the fiber, and the response of the fiber to that field.

In particular, we are considering the case where the stimulus is sufficiently small, so that the membrane behavior is linear and may therefore be described by a parallel resistance and capacitance (r_m in Ωcm and c_m in μF/cm). Equation (7.62) may then be written as

$$r_i c_m \frac{\partial v_m}{\partial t} + r_i \frac{v_m}{r_m} - \frac{\partial^2 v_m}{\partial z^2} = \frac{\partial^2 \phi_e}{\partial z^2} u(t) \tag{7.63}$$

where $i_{\text{ion}} = v_m / r_m$ and $u(t)$ is a unit step (included here to signify that the activating function is switched on at $t = 0$).

Initial rate of change of V_m

Examining (7.63), one sees that if a stimulus is initiated when the fiber is at rest, then both v_m and $\partial^2 v_m / \partial z^2$ are zero. Thus, rearranging (7.63), one has an equation for the initial rate of change of v_m along the fiber as

$$r_i c_m \frac{\partial v_m}{\partial t} = \frac{\partial^2 \phi_e}{\partial z^2} u(t) \tag{7.64}$$

That is, $\partial v_m / \partial t$ is proportional to the activating function, as noted above. The proportionality coefficient is determined by characteristics of the fiber, specifically its time constant $r_i c_m$.

Transformations to find V_m response

We will find it useful here to have the fiber response to a unit *intracellular* point source. Thus, using (7.63) we seek the solution of

$$\lambda^2 \frac{\partial^2 v_m}{\partial z^2} - \tau \frac{\partial v_m}{\partial t} - v_m = -r_i \lambda^2 \delta(z) u(t) \tag{7.65}$$

where $\delta(x)$ is a unit Dirac delta function (7.23). We see that the desired solution to (7.65) is the solution found for (7.35).

We assume that the extracellular medium being unbounded supports the assumption that $r_e \approx 0$.[12] An examination of (7.65) is facilitated by introducing normalized variables defined by

$$X = \frac{z}{\lambda} \quad \text{and} \quad T = \frac{t}{\tau} \tag{7.66}$$

where $\tau = r_m c_m$ and $\lambda = \sqrt{r_m / r_i}$. Substituting (7.66) into (7.65) results in

$$\frac{\partial^2 v_m}{\partial X^2} - \frac{\partial v_m}{\partial T} - v_m = -r_i \lambda^2 \delta(z) u(t) \tag{7.67}$$

which is essentially (7.37), except that in (7.67) the stimulus current is included explicitly.

The solution to (7.67) is given in (7.48). Since we seek a unit impulse response which divides into the positive and negative z directions, we require $I_0 = 1$ in (7.48). Hence

$$G(X,T) = \frac{r_i \lambda}{4} \left\{ e^{-X} \left[1 - \text{erf}\left(\frac{X}{2\sqrt{T}} - \sqrt{T} \right) \right] - e^{X} \left[1 - \text{erf}\left(\frac{X}{2\sqrt{T}} + \sqrt{T} \right) \right] \right\} \tag{7.68}$$

Function G thus gives the transmembrane voltage produced by a unit current at position X and time T.[13]

Transmembrane potential's response to stimulus

We now wish to extend the results found for a delta function source to find results from a distributed source. Applying (7.67) to (7.63) yields

$$\frac{\partial^2 v_m}{\partial X^2} - \frac{\partial v_m}{\partial T} - v_m = -\lambda^2 \frac{\partial^2 \phi_e}{\partial X^2} u(t) \tag{7.69}$$

Comparison of (7.69) and (7.67) shows that they differ only in the forcing function (the terms on the right). Additionally, the one equation (7.67) provides the impulse response while the other equation (7.69) is for a continuous forcing function (namely, $\partial^2 \phi_e / \partial X^2$).

Moreover, the system is linear, since we are restricting consideration to passive membrane. Thus the solution to (7.69) is the convolution of $\partial^2 \phi_e / \partial X^2$ with the fiber impulse response.

If we take into account the additional factor r_i in (7.67) as well as the normalized coordinates, we obtain

$$v_m = \frac{\lambda}{r_i} \int_{-\infty}^{\infty} f(\xi) G(X - \xi, T) d\xi \tag{7.70}$$

where ξ is a dummy variable for X and $f(\xi) = \partial^2 \phi_e / \partial z^2 |_{z=\xi}$. The coefficient λ in (7.70) arises from the change in variable, where $dz = \lambda d\xi$.[14]

The interpretation of Eq. (7.70) is a follows. Function $f(\xi)$ comes from the field stimulus, as it affects the fiber. Function $G(X - \xi)$ is the fiber's response to a stimulus given at one point along the fiber. The convolution integrates (in effect, it adds up) the fiber's responses to the stimuli created along the fiber by the external field.

Equation (7.70) is an application of linear systems theory, where the output to an arbitrary input is expressed in terms of the impulse response (system function). A similar expression appropriate for myelinated fiber stimulation was derived by Warman *et al* [10].

Note that any stimulus may be considered a sequence of impulse functions.

7.5.4. Isolated Single Fiber and a Point Current Source

The problem at hand is described in Figure 7.5, but with $L = \infty$. In other words, the point current source is located at a distance h from the unmyelinated fiber of infinite extent. The foot of the perpendicular determines the origin of the coordinate system. The desired solution is given formally by Eq. (7.70).

Again we consider the subthreshold response. This response can be evaluated using the Fourier transform. Now the Fourier transform of a convolution is the product of the Fourier transform of each convolving function.

Thus if F denotes the Fourier transform and F^{-1} the inverse transform, then (7.70) can be expressed as

$$v_m(X,T) = \frac{\lambda}{r_i} F^{-1}[F[f(X)]F[G(X,T)]] \tag{7.71}$$

The applied field arising from the point current source, (7.55), is simply given by

$$\phi_e(z) = \frac{I_0}{4\pi\sigma_e\sqrt{h^2+z^2}} \tag{7.72}$$

The spatial second derivative of Φ_e, in the direction of the fiber's axis, is

$$f(z) = \frac{\partial^2 \Phi_e}{\partial z^2} = \frac{I_0}{4\pi\sigma_e} \frac{2z^2 - h^2}{(h^2+z^2)^{5/2}} \tag{7.73}$$

Consequently, where $H = h/\lambda$,

$$f(X) = \frac{I_0}{4\pi\sigma_e\lambda^3} \frac{2X^2 - H^2}{(H^2+X^2)^{5/2}} \tag{7.74}$$

Accordingly, substituting (7.68) and (7.74) into (7.71) gives [5]

$$\begin{aligned} v_m(X,T) &= \frac{\lambda}{r_i} F^{-1} \left\{ \frac{I_0}{16\pi\sigma_e\lambda} \times F\left[\frac{2X^2 - H^2}{(H^2+X^2)^{5/2}}\right] \times \right. \\ & F\left[e^{-X}\left(1 - \operatorname{erf}\left(\frac{X}{2\sqrt{T}} - \sqrt{T}\right)\right) \right. \\ & \left.\left. - e^{X}\left(1 - \operatorname{erf}\left(\frac{X}{2\sqrt{T}} + \sqrt{T}\right)\right)\right]\right\} \end{aligned} \tag{7.75}$$

7.5.5. Activation Function's Prediction versus Response

With the results above, we now can compare the activating function with the actual membrane response. It is helpful to visualize the results. To this end, the transmembrane potential created by (7.75) for one example is plotted in Figure 7.6 at three times following the onset of the stimulus. The example has a point cathodal stimulus at distance h from a fiber, with details given in the Figure caption.

At the first of the three times plotted, 0.01 msec after the start of the stimulus, the wave shape of the activating function is a good approximation to that of the transmembrane potential. Recall that the impulse response, $G(X,T)$, for small T, is approximately a delta function (see Figure 7.4). The consequence is that the early transmembrane potential response has a shape like that of the activating function, consistent with the convolution equation (7.70). Thus for a very short stimulus (short in comparison to the time constant), the activating function is a good predictor of the resulting transmembrane potential and correctly shows the regions of depolarization and hyperpolarization produced by the stimulus.

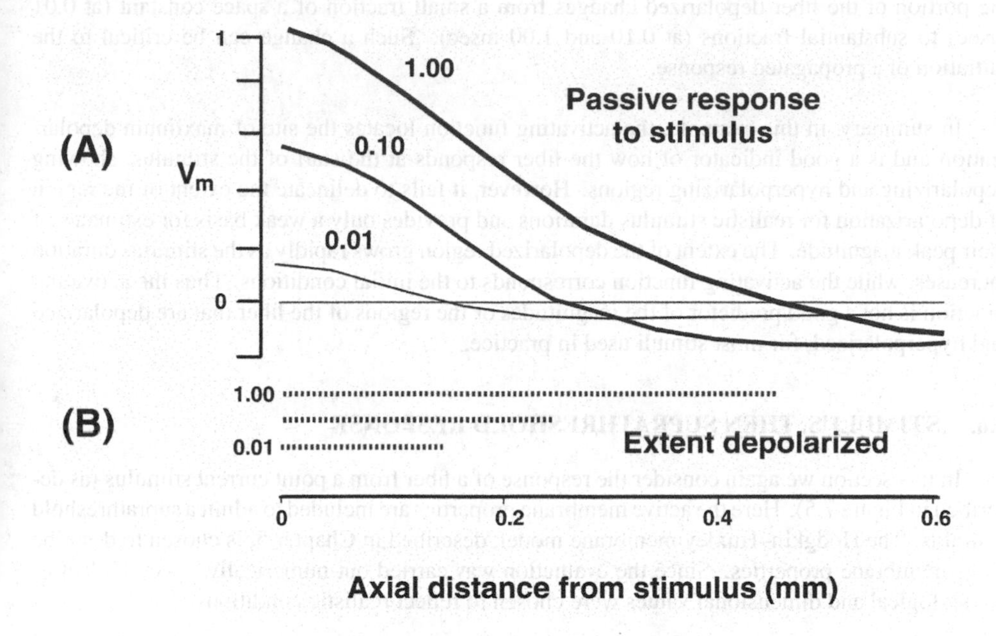

Figure 7.6. Time Evolution of the Induced Transmembrane Voltage along a Fiber. The response is that from a field stimulus, and the three lines shown the response at three times following stimulus onset, i.e., $V_m(x)$ at 0.01, 0.10, and 1.00 msec after the start of a stimulus. The point current stimulus is at distance $h = 0.02$ cm from a fiber described in Figure 7.5. [Other parameters are $\lambda = 0.86$ mm. $\tau = 1.5$ msec, $\sigma_e = 33.3$ mS/cm, and $I_0 = -0.44$ mA.] In the top panel (A), the vertical axis plots V_m on a normalized scale to facilitate comparison of the plots. (A value of 1.0 on the normalized scale is approximately 30 mV.) In the lower panel (B), three horizontal bars show the extent of the depolarized region. The horizontal axis (bottom) applies to both (A) and (B). Distances along the horizontal axis are given in millimeters from the point directly under stimulus and thus also are approximately the distance in space constants. (Only one half of the spatial response is shown, because the two sides are symmetric.) The activating function has a wave shape similar to that of the 0.01 msec curve. Adapted from Plonsey R, Barr RC. 1995. Electric field stimulation of excitable tissue. *IEEE Trans Biomed Eng* **42**:329–336. Copyright ©1995, IEEE.

It would be convenient if what happened at later times were simple multiples of the result for 0.01 msec. However, such is not the case. Thus the limitations of making estimates using the activating function are seen when examining the true response as the stimulus grows longer, as shown in the plots for 0.10 and 1.00 msec. The result of a longer stimulus is that the transmembrane potential grows larger in magnitude, but not linearly. Thus the peak v_m for the 0.10-msec plot is roughly three times that of the plot for 0.10 msec, and the plot for 1.00 msec is roughly two times that of 0.10 msec, even though the stimulus has 10× the duration. Finally, the peak amplitude never increases much beyond that for 1.00 msec (for this example), even for much longer stimuli.

The transmembrane potential response also grows wider. As the stimulus gets longer, the extent of the fiber that is depolarized by the stimulus grows larger, as seen explicitly in Figure 7.6B, where the extent of the depolarized region is identified by a horizontal line for each time. Further, the portion of the fiber depolarized changes from a small fraction of a space constant (at 0.01 msec) to substantial fractions (at 0.10 and 1.00 msec). Such a change can be critical to the initiation of a propagated response.

In summary, in this example, the activating function locates the site of maximum depolarization and is a good indicator of how the fiber responds at the start of the stimulus, showing depolarizing and hyperpolarizing regions. However, it fails to delineate the extent of the region of depolarization for realistic stimulus durations and provides only a weak basis for estimates of their peak magnitude. The extent of the depolarized region grows rapidly as the stimulus duration increases, while the activating function corresponds to the initial conditions. Thus the activating function is not a good predictor of the magnitudes or the regions of the fiber that are depolarized and hyperpolarized, for most stimuli used in practice.

7.6. STIMULUS, THEN SUPRATHRESHOLD RESPONSE

In this section we again consider the response of a fiber from a point current stimulus (as described in Figure 7.5). Here the active membrane properties are included to admit a suprathreshold stimulus. The Hodgkin–Huxley membrane model, described in Chapter 5, is chosen to describe these membrane properties. Since the evaluation was carried out numerically, specific electrophysiological and dimensional values were chosen to reflect realistic conditions.

7.6.1. Numerical Methods for Finding V_m

As in the previous section, the fiber and source geometry is specified in Figure 7.5. The fiber is assumed to be circular cylindrical with radius $a = 0.002$ cm. A stimulus current of magnitude I_0 is located a distance h from the fiber. The duration of the stimulus is denoted by t_d and is varied, as is h.

The stimulus threshold was determined for various combinations of h and t_d by repeated trials. Threshold was judged by the presence or absence of a propagating action potential at sites several space constants, λ, from the site of excitation. The threshold stimulus was such that a 10% increase resulted in propagation. (It was the largest stimulus for which propagation did *not* result.)

7.6.2. Results of Space-Clamped and Field Stimulation

Space-Clamped Threshold. For a reference, threshold was determined for a transmembrane stimulus under space-clamped conditions. A space clamp can be achieved by considering an axially uniform transmembrane potential stimulus. The result is plotted in Figure 7.7 (inset) [1]. One sees that with a space clamp the threshold is nearly independent of stimulus duration and requires 7 mV of depolarization.

Field Stimulation. An examination of Figure 7.7 shows that for field stimulation, the threshold voltage is no longer independent of stimulus duration, in general. The degree of deviation from the space-clamped result is seen to depend on both the stimulus–fiber distance h and the stimulus duration t_d.

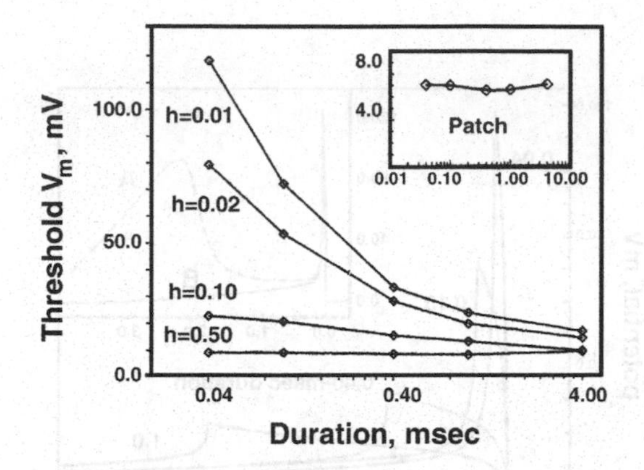

Figure 7.7. Threshold Values of v_m versus Stimulus Duration for a Point Stimulus. (**Inset**: results for patch geometry for comparison.) The transmembrane voltage at the end of the stimulus is shown for a stimulus condition that is just below threshold. Patch data are for the condition of no spatial variation. All potentials shown are relative to a baseline of −57 mV. **Outer**: Each curve is for a different source–fiber distance as shown (h given in cm). Results shown are for $z = 0$, the shortest fiber–stimulus distance. Membrane properties are: $E_K = -72.1$ mV, $E_{\mathrm{Na}} = 52.4$ mV, $\bar{g}_{\mathrm{Na}} = 120$ mS/cm^2, $\bar{g}_K = 36$ mS/cm^2, $g_l = 0.30$ mS/cm^2. Fiber properties are: $R_m = 0.148\ \Omega\mathrm{cm}^2$, $\lambda = 0.086$ cm, $C_m = 1.0\ \mu$F/cm^2. From Barr RC, Plonsey R. 1995. Threshold variability in fibers with field stimulation of excitable membranes. *IEEE Trans Biomed Eng* **42**:1185–1191. Copyright ©1995, IEEE.

The highest thresholds were measured for the smallest duration and smallest value of h. Thus for $t_d = 0.04$ msec and $h = 0.01$ cm, a threshold value of 118 mV was obtained. In contrast, for long stimulus duration and large source–fiber distance, results were obtained that are similar to those for the patch. For example, when $h = 0.5$ cm, the threshold value of 8 mV corresponds to stimulus durations of 0.04–4.0 msec. (One could have anticipated such a result, since the axial variation of the applied field is increasingly uniform, approaching space-clamped conditions, for increasing h.)

Temporal Transmembrane Potential Waveforms. The temporal response following a just subthreshold stimulus is given in Figure 7.8 and is helpful in interpreting all results shown in Figure 7.7. For the shortest durations of 0.01 msec, we see that a transmembrane potential of 118 mV marks the threshold voltage. This elevated voltage is required to maintain a large enough voltage following the termination of the stimulus to open sufficient sodium channels, since the activation gate time constant τ_m is several tenths of a millisecond.

The effect of stimulus decay based on the membrane time constant is a contributing factor in Figure 7.7. But a second contributing factor affecting the membrane decay depends on the source–field distance. To understand this effect, a plot of the spatial transmembrane potential $v_m(z)$ is given in Figure 7.9. Here, we note that the central depolarized region is flanked by hyperpolarized regions.

Thus the depolarized membrane decay is also accelerated by longitudinal current flow into the hyperpolarized regions. This current will be enhanced for smaller values of h, which reduces the distance to the peak hyperpolarized position (in Figure 7.9 it is at around 0.15 cm).

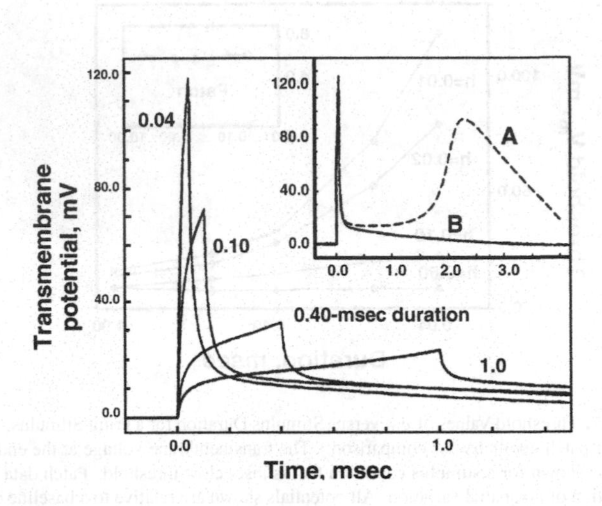

Figure 7.8. Transmembrane Potential as a Function of Time for stimuli that are just below and just above threshold. **Inset**: Curve A is for a just transthreshold stimulus and B for a just subthreshold stimulus. The source–fiber distance is 0.01 cm. Stimulus magnitudes were 1.40 (A) and 1.30 mA (B). **Outer**: Temporal responses for just subthreshold stimuli for stimulus duration as shown (in msec). From Barr RC, Plonsey R. 1995. Threshold variability in fibers with field stimulation of excitable membranes. *IEEE Trans Biomed Eng* **42**:1185–1191. Copyright ©1995 IEEE.

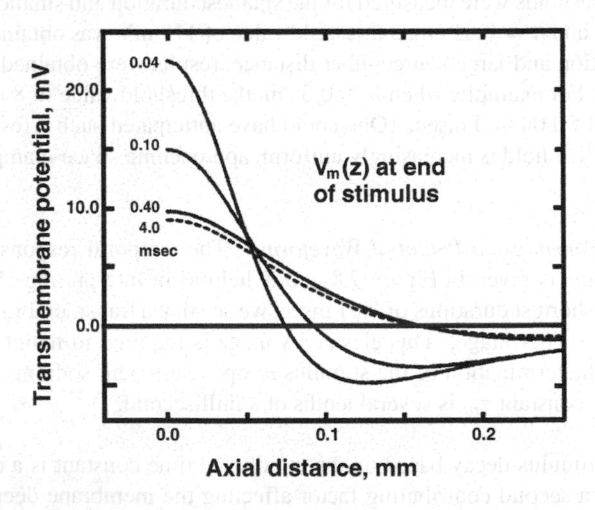

Figure 7.9. Spatial Distribution of Transmembrane Potential, $v_m(z)$, at the End of the Stimulus. Each curve is labeled with the duration of the stimulus. The source–fiber distance is 0.10 cm. In each case the stimulus magnitude is for a just subthreshold response. From Barr RC, Plonsey R. 1995. Threshold variability in fibers with field stimulation of excitable membranes. *IEEE Trans Biomed Eng* **42**:1185–1191. Copyright ©1995 IEEE.

This factor explains the very large threshold requirements for small values of h, seen in Figure 7.7. For $h = 0.5$ cm, the depolarized region is so broad that the behavior is similar to that shown for the patch (membrane decay is entirely due to the *RC* component alone).

7.6.3. Comments on the Concept of Threshold

For the design of a practical stimulator, it is highly desirable to specify a target threshold that can be relied on to achieve fiber activation. Specifying a target threshold allows one to use linear analysis to estimate stimulus parameters. Other factors that arise in a practical stimulator design are introduced in Chapter 12.

What is clear here is that the actual threshold value that exists at the end of just a transthreshold stimulus may range from 7 to 118 mV, depending on the stimulus duration and the distance from the stimulus electrode to the fiber. If a value of $h \approx 0.5$ cm or $h \approx 0.1$ cm and $t_d \approx 0.5$ msec is consistent with other design criteria, then a fixed threshold of around 8 mV can be assumed for an HH membrane. Otherwise, an elevated threshold value must be initially assumed in a linear treatment. In every case, a nonlinear membrane analysis is eventually desired to be followed by appropriate animal and human measurements.

7.7. FIBER INPUT IMPEDANCE

Many questions about the electrical properties of fibers can be framed in terms of the *input impedance*. For example, the effects of cable length are examined in the section below by comparing the input impedance for realistically short lengths with that of infinite lengths.

The input impedance, Z_0, is defined to be

$$Z_0 = v_m / I_i \tag{7.76}$$

and is evaluated at the point where the stimulus is applied. This evaluation requires both polarizing electrodes to be at the origin with one in the intracellular and the other in the extracellular space; the subthreshold applied current, I_i, and the resulting voltage, v_m, enter (7.76) to evaluate the input impedance. Note that the transmembrane voltage appearing in (7.76) is compared to the *longitudinal* intracellular current I_i.

We note first that the assumed stimulus satisfies the condition under which (7.76) is derived, namely, that $i_p = 0$ for $0 < x < \infty$. Consequently, using (7.30), we have

$$v_m = Ce^{-|x|/\lambda} \tag{7.77}$$

Assuming $r_e = 0$ permits (6.9) to be expressed as

$$I_i = -\frac{1}{r_i}\frac{\partial V_m}{\partial x} = -\frac{1}{r_i}\frac{\partial v_m}{\partial x} \tag{7.78}$$

Substituting (7.77) in (7.78) results in[15]

$$I_i = \frac{C}{r_i \lambda} e^{-|x|/\lambda} \tag{7.79}$$

Since $\lambda \approx \sqrt{r_m/r_i}$ when $r_e \approx 0$,

$$I_i = \frac{C}{\sqrt{r_m r_i}} e^{-|x|/\lambda} \tag{7.80}$$

and Z_0 (at $x = 0^+$) is given by

$$Z_0 = \frac{Ce^{-|x|/\lambda}}{\frac{C}{\sqrt{r_i r_m}} e^{-|x|/\lambda}} \tag{7.81}$$

or

$$Z_0 = \sqrt{r_i r_m} \tag{7.82}$$

So, for an infinitely long cable with $r_e \approx 0$, the input impedance is the square root of the product of membrane and intracellular resistance.

7.7.1. Cables of Finite Length

Much of the above analysis has been based on the assumption of an infinitely long cable. Of course, no cables are infinitely long. In this section, the consequences of this discrepancy are examined. Specifically, the differences in the steady state are compared for cables of finite and infinite lengths.

The overall strategy used here is based on the cable input impedance. We have seen that for an infinitely long cable, Z_0 is $\sqrt{r_m r_i}$. Now we consider the input impedance Z_{in} of a cable of arbitrary length, L, terminated by an arbitrary impedance, Z_L.

For the specific case of a fiber of length L terminated in a short circuit ($Z_L = 0$), the input impedance, Z_{in}, will be of interest. This is because the extent to which Z_{in} corresponds to Z_0 provides a quantitative measure of the extent to which the finite cable approaches the input behavior of the infinite length cable.

There are a number of important applications. One arises in an examination of the behavior of a network of neurons, such as found in the central nervous system. This is shown to depend in part on the impedance behavior of short fibers (neurons). Interest in neural networks is not limited to neurophysiologists but to those working on artificial neural networks as computer processors. Further material on both topics is given in [3, 6, 11].

7.7.2. Finding Z_{in} for a Finite Length Cable

Consider an axon in an extensive extracellular medium ($r_e \approx 0$), of finite length ($x = L$), and terminated with an arbitrary load impedance Z_L. Assume an input voltage to the cable of $v_m = v_0$ applied at $x = 0$. For $x > 0$, $i_p = 0$, so the homogeneous form of (7.16) applies, namely, (7.17) or

$$\lambda^2 \frac{\partial^2 v_m}{\partial x^2} - v_m = 0 \tag{7.83}$$

The solution of (7.83) has already been given as (7.18)

$$v_m(x) = Ae^{-x/\lambda} + Be^{x/\lambda} \tag{7.84}$$

Note that since the cable is finite in length we can no longer set $B = 0$ based on the boundary condition at infinity. Now the relationship between v_m and I_i is available from the cable equations. Since $i_p = 0$, except at the origin, (7.78) is valid, and we rewrite it here for convenience as

$$I_i = \frac{1}{r_i}\frac{\partial v_m}{\partial x} \tag{7.85}$$

Substituting (7.84) into (7.85) and evaluating I_i gives

$$I_i(x) = \frac{1}{Z_0}(Ae^{-x/\lambda} - Be^{x/\lambda}) \tag{7.86}$$

where $Z_0 = \sqrt{r_m r_i}$ from (7.82).

At $x = 0$ we have $Z_{\text{in}} = V(0)/I(0)$, so from (7.84) and (7.86) Z_{in} is given by

$$Z_{\text{in}} = Z_0\left(\frac{A+B}{A \ B}\right) \tag{7.87}$$

For a cable of infinite extent, we must set $B = 0$ to avoid a potential that grows indefinitely, and consequently from (7.87), $Z_{\text{in}} = Z_0$, which corresponds to earlier results (i.e., the input impedance of an infinite cable is Z_0).

For cables of finite length and arbitrary termination the input impedance requires the evaluation of A and B in (7.87). This evaluation is facilitated by an evaluation of a factor involving the terminal impedance known as the *reflection coefficient*. In the next section we define the reflection coefficient and show how it introduces the boundary condition at the load located at $x = L$.

Reflection coefficient: Now at $x = L$, $Z_L = V_m(L)/I_i(L)$, so dividing (7.84) by (7.86) for $x = L$ gives

$$Z_L = Z_0\left(\frac{Ae^{-L/\lambda} + Be^{L/\lambda}}{Ae^{-L/\lambda} - Be^{L/\lambda}}\right) \tag{7.88}$$

We define the *reflection coefficient*, Γ, as

$$\Gamma = \frac{Z_L + Z_0}{Z_L - Z_0} \tag{7.89}$$

Substitution of (7.88) in (7.89) and simplification of the resulting expression yields the following relationships:

$$\Gamma = \frac{Ae^{-L/\lambda}}{Be^{L/\lambda}} \tag{7.90}$$

$$Z_L = \left[\frac{\frac{Ae^{-L/\lambda}}{Be^{L/\lambda}} + 1}{\frac{Ae^{-L/\lambda}}{Be^{L/\lambda}} - 1}\right] Z_0 = Z_0\left(\frac{\Gamma + 1}{\Gamma - 1}\right) \tag{7.91}$$

Table 7.3. Normalized Input Impedance of Finite Length Cable

L/λ	Z_{in}/Z_0
0.1	10.0
0.5	2.16
1	1.31
2	1.04
3	1.01

By substituting (7.90) into (7.87) we obtain an expression for Z_{in} in terms of Γ:

$$Z_{in} = Z_0 \left(\frac{\Gamma e^{2L/\lambda} + 1}{\Gamma e^{2L/\lambda} - 1} \right) \tag{7.92}$$

The name "reflection coefficient" comes about from similar definitions used in the study of traveling electromagnetic waves, where the wave may by reflected in whole or in part from discontinuities in a cable, such as at its termination.

For example, when $Z_L = Z_0$ the termination is equivalent to an infinite cable and consequently the finite cable itself behaves as the proximal element of an infinite cable. In this case nothing will be "reflected," of course. From (7.89), a termination of $Z_L = Z_0$ results in $\Gamma = \infty$. In contrast, if $Z_L = 0$ (short circuit) or $Z_L = \infty$ (open circuit), then $\Gamma = \pm 1$, and the termination introduces a maximum discontinuity (everything "reflected"). This outcome is recognized in (7.92) with both $\Gamma = \pm 1$ and small L/λ.

While the present nomenclature has been utilized due to a superficial analogy with EM waves, the physical situation is, of course, quite different.[16]

7.7.3. Z_{in} for an Open Circuit Termination

A finite cable with a sealed end can be regarded as a cable that ends in an open circuit. That is, $Z_L = \infty$ and $\Gamma = 1$. For a cable of length L with such a termination, we have from (7.92)

$$Z_{in} = Z_0 \left(\frac{e^{2L/\lambda} + 1}{e^{2L/\lambda} - 1} \right) = Z_0 \coth \left(\frac{L}{\lambda} \right) \tag{7.93}$$

Equation (7.92) confirms that $Z_{in} = Z_0$ when $\Gamma = \infty$, while when $\Gamma = \pm 1$, Z_{in} depends on L/λ [e.g., $Z_{in} = Z_0 \tanh(L/\lambda)$ for $\Gamma = -1$]; further details are found in the next section.

Table 7.3 shows the result of evaluating (7.93) numerically to find Z_{in}/Z_0. It indicates that for short cables, defined by $L < \lambda$, there are substantial deviations in behavior from that of an infinite cable. On the other hand, Table 7.3 also shows that as L increases beyond λ, the input rapidly becomes indistinguishable from that of an infinite cable. In particular, the input impedance is within 1% of Z_0 if L is 3λ or more.

7.8. MAGNETIC FIELD STIMULATION

For an applied time-varying magnetic field, Faraday's law describes an induced (free-space) electric field, namely,

$$\oint \overline{E} \cdot \overline{dl} = -\frac{d}{dt} \int_s \mu_0 \overline{H} \cdot \overline{dS} \tag{7.94}$$

where μ_0 is the permeability (normally taken as free space) and H the magnetic field. Induced secondary sources at conductive discontinuities (in particular, the torso–air interface) must be included in a realistic evaluation of $\overline{H}$ and hence the induced electric field at human nerve/muscle from applied external magnetic fields. The electric field at nerve/muscle fibers, once obtained, follows all principles described in this chapter. An overview of magnetic field applications can be found in Stuchly [9], and some theoretical considerations in Plonsey [4].

7.9. NOTES

1. The concept of a voltage threshold as a fixed point of sudden transition to an active response, a classical conceptual starting point, does not hold up consistently when examined in detail.
2. It is worth noting that the behavior of the tissue after it crosses threshold and becomes active (which often means once the sodium channels open) is not usually a focus of the analysis of stimuli. That is because most of the time the active response depends primarily on the tissue's intrinsic membrane response and cellular structure rather than on external stimuli. However, active response also is affected if the external stimulus is large enough and long enough, e.g., in cardiac defibrillation.
3. Often R_m and C_m are used directly, and I_o is converted to Amperes/cm^2.
4. Later in the chapter we examine critically the classical notion that activation is automatically achieved once the transmembrane potential reaches a critical transmembrane voltage.
5. One sees Eq. (7.8), sometimes with alternative variable names, used to relate threshold current and rheobase in many contexts. For example, in Chapter 12 the same equation is used in connection with functional electrical stimulation.
6. Lapicque gave the equation $i = \alpha/(1 - e^{-\beta t})$, where i was intensity, t was duration, and α and β were two constants.[2]
7. If the current had been applied extracellularly, then $\partial\Phi_i/\partial x = 0$ at $x = 0$ and $\partial\Phi_e/\partial x = -I_0 u(t) r_e/2$ at $x = 0$. In this case we would replace (7.43) by $\partial v_m/\partial X = I_0 u(t) r_e \lambda/2$ (i.e., replace $-I_0 r_i$ by $I_0 r_e$). From symmetry, we interchange subscripts i and e and, in addition, change the current sign to reflect it being oppositely directed. These expressions, however, assume a limited extracellular space where currents are essentially axial and a one-dimensional Ohm's law applies.
8. One can use Eq. 30 in Appendix V of Carslaw HS, Jaeger JC. 1959. *Conduction of heat in solids*. Oxford: Oxford UP.
9. If the current were introduced into (a bounded) extracellular rather than intracellular space, then the coefficient on the right-hand side of (7.49) would equal $-r_e \lambda I_0/4$, and one can confirm that this expression reduces to (7.34) when $t \to \infty$. Note that (7.49) applies even if $r_e = 0$ for a fiber in an unbounded extracellular region. For an extracellular applied current, it is required that $r_e \neq 0$ to invoke the aforementioned symmetry.
10. The importance of the subthreshold case is the following: even if the goal of the stimulation is to bring the fiber above threshold, it must pass through a subthreshold state first. Thus the results are broadly applicable.
11. The assumption of a thin fiber and relatively large source–fiber distance assures that the azimuthal potential variation is relatively small compared with the axial variation. For the azimuthal potential behavior, the fiber roughly doubles the values of the applied field at the nerve periphery; these potential variations are relatively small for a thin fiber at a large distance from the source, compared to variations along the axis. The secondary field arising from the axial variations can be shown to be negligible [6].
12. The stimulating current, I_0, is relatively large in order that the field it generates in the *extracellular* volume conductor, given by (7.55), can induce subthreshold or suprathreshold depolarization of the target fiber. When considering the response of the fiber itself to an *intracellular* unit impulse current, its behavior is little affected by the relatively small extracellular field when the extracellular medium is unbounded. The linear core-conductor equation therefore describes the intracellular fields correctly with $r_e = 0$; the large extracellular point source and its field are zero in this situation.
13. Exercise: what are the units for G?

14. Exercise: give the units for each quantity on the right side of (7.70) and show that they combine to give the units needed on the left side.
15. The reason why Eq. (7.79) differs from (7.54) might not be apparent. However, this arises because in deriving (7.54) we assumed an intracellular applied current I_0 located at $x = 0$ with the removal of this current at infinity (whether it is removed intracellularly or extracellularly will have no effect in the region $0 < x < L$ so long as L is finite). In (7.79), the electrode pair carrying I_0 into the intracellular space and out of the extracellular region are both at $x = 0$.
16. Propagation of microwave energy along cables or waveguides results from the injection of energy at the proximal end; this energy diminishes with distance due to losses. For nerve/muscle, only a trigger to initiate a propagating action potential is assumed at the proximal end; the energy is derived and expended all along the fiber and there is no attenuation. An analogy to the biological case (regarding energy) is the behavior of a fuse, except that the fuse can be used only once.

7.10. REFERENCES

1. Barr RC, Plonsey R. 1995. Threshold variability in fibers with field stimulation of excitable membranes. *IEEE Trans Biomed Eng* **42**:1185–1191.
2. Lapique L. 1909. Definition experimentale de l'excitation. *C R Hebd Seanas Acad Sci* **67**:280–283.
3. Partridge LD, Partridge DL. 1993. *The nervous system, its function and integration with the world.* Cambridge: MIT Press.
4. Plonsey R. 1981. Generation of magnetic fields by the human body. In *Biomagnetism*. Ed SN Erné, HD Halbohm, H Lübbig. Berlin: de Gruyter.
5. Plonsey R, Barr RC. 1995. Electric field stimulation of excitable tissue. *IEEE Trans Biomed Eng* **42**:329–336.
6. Rall W. 1977. Core conductor theory and cable properties of neurons. In *Handbook of physiology*, Section 1: *The nervous system*, Vol. 1: *Cellular biology of neurons*, pp. 39–97. Bethesda, MD: American Physiological Society.
7. Rattay F. 1987. Ways to approximate current-distance relations for electrically stimulated fibers. *J Theor Biol* **125**:339–349.
8. Rattay F. 1990. *Electrical nerve stimulation*. New York: Springer.
9. Stuchly MA. 1990. Applications of time-varying magnetic fields in medicine. *CRC Rev Biomed Eng* **18**:89–124.
10. Warman EN, Grill WM, Durand D. 1992. Modeling the effects of electric fields on nerve fibers: Determination of excitation thresholds. *IEEE Trans Biomed Eng* 39:1244–1254.
11. Zahner DA, Micheli-Tzanakou E. 1995. Artificial neural networks: definitions, methods, applications. In *Biomedical engineering handbook.* Ed JD Bronzino. Boca Raton, FL: CRC Press.

8

EXTRACELLULAR FIELDS

8.1. BACKGROUND

Even body surface electrodes detect the small currents that flow as part of membrane activation deep within the torso volume. The electrocardiogram is a familiar such example: the sources of these body surface potentials are the combined action currents of many cardiac cells.

The electroencephalogram arises from the brain, and the electromyogram and electroneurogram arise from other organs. In each case the electric potential field in the surrounding volume conductor arises from nerve or muscle sources within the region. These sources are generated by cellular membrane activity.[1] Where do the extracellular wave shapes come from? How can one extracellular waveform have a different wave shape from another nearby? How do they relate to the transmembrane potential waveforms? These are the questions this chapter seeks to answer.

One knows that the extracellular potentials must arise from the transmembrane currents and therefore must be linked to the transmembrane potentials, but the basis of this relationship is not apparent from inspection of the temporal waveforms. Thus the focus of this chapter is to develop the relationship between the intracellular and transmembrane events that satisfies the governing equations.

An ongoing research goal is to understand the details of the origins of the measured signals through both qualitative and quantitative examination of the signal generation, and to do so in enough detail that one can formulate clinical interpretations of the timing, wave shapes, and amplitudes of specific features of the measurements. The goal of this chapter is a more limited one: to develop and describe mathematical relationships that link the cellular action potential with the volume conductor fields (action current fields) associated with them, and thus to provide a basis for understanding the origin of the extracellular wave shapes such as those shown in Figure 8.1.

To this end, the first section of the chapter focuses on providing a framework of ideas that can serve as background knowledge from which more quantitative analysis can take hold. This background includes comparing temporal to spatial transmembrane potential relationships. The

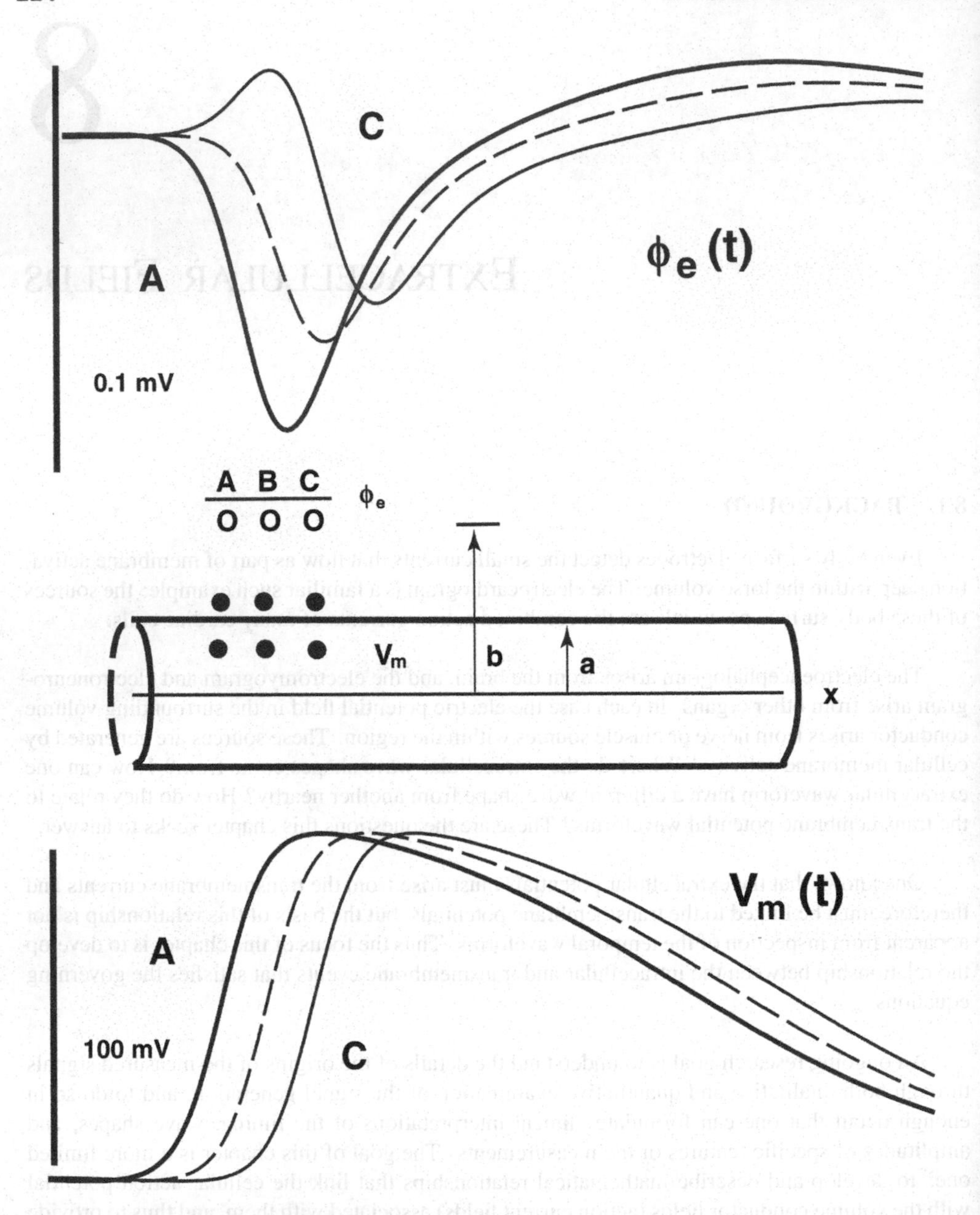

Figure 8.1. Intracellular and Extracellular Temporal Waveforms. A sketch of a short section of a long cylindrical fiber is shown in the middle panel. Transmembrane electrodes (closed circles) and extracellular electrodes (open circles) are drawn at three points along the axis, with their respective columns labeled A–C. Transmembrane potentials $V_m(t)$ for each column are shown in the panel below. Extracellular waveforms $\Phi_e(t)$ are shown in the panel at the top. (These are unipolar waveforms, i.e., potential with respect to a distant reference.) The vertical bars (top, bottom) give a voltage calibration. Each trace has a 10-ms duration. Locations A, B, and C are at $x = 0, 1, 2$ mm, respectively. Radius $a = 0.1$ mm and distance $b = 1.0$ mm. These waveforms are based on a computer simulation that uses a mathematically defined template function for $V_m(t)$. $R_i = 100$ and $R_e = 30$ Ωcm.

section also introduces and explains some terminology used, so as to provide a language and general background about the origin of extracellular observations.

A qualitative explanation is insufficient, however, so the next major portion of the chapter provides a quantitative explanation for the origin of extracellular fields from single fibers. This framework is one that is often referenced because it offers a well developed basis for understanding extracellular measurements quantitatively. (Extracellular fields from planar excitation waves, often used for cardiac excitation, are evaluated in the next chapter.)

Here we restrict attention to a single cylindrical fiber. This cylindrical shape is ubiquitous in nerve and muscle, and adaptation of the results to other shapes is often possible. This section includes discussion of how potentials are found from membrane currents, how the membrane currents are found if they are at first unknown, and how the membrane currents can be grouped together ("lumped") for simplicity. The mathematical formulation of fiber fields has relative simplicity and thus intuitive appeal.

The final major portion of the chapter lets go of the single-fiber cylindrical-fiber geometry and starts the analysis again from scratch. Here the goal is to examine the extracellular potentials generated by a single cell of arbitrary shape. On its face, this problem is a narrow one of interest only in an experimental or research context.

This impression is deceiving. In fact, because an arbitrary shape is used, the solution can be applied to many different situations (including gaining a better understanding of waveforms from cylindrical fibers). The more general approach also can be extended to multicellular organs such as the heart (as is done in Chapter 9). That is, while the approach is more abstract, this portion of the chapter produces results that embody fewer assumptions and that can be applied to more situations.

8.1.1. Opportunities

Because all excitable tissues lie within a volume conductor, extracellular currents extend throughout the entire conducting space, diminishing in amplitude with distance from the source.

Thus the opportunity provided by extracellular fields is that potentials can be sampled outside cells, or even at a distance, without damage to the electrically active cells, or signal perturbation resulting from electrode entry into the tissue of the organ. Noninvasive measurements from the surface of the thorax, extremities, and head were among the first electrophysiological measurements historically and remain the most numerous and valuable in clinical electrophysiology today. The temporal signal so obtained corresponds to the electrophysiological behavior of an organ and therefore contains information of clinical value.

The challenge is in the interpretation of the extracellular measurement. Extracellular potential waveforms are weaker in magnitude and thus have features more easily obscured by measurement noise, and also are more variable in wave shape.

Consider Figure 8.1, which shows the temporal transmembrane potential waveforms $V_m(t)$ and extracellular waveforms $\Phi_e(t)$ around a cylindrical conductor. As seen in the bottom panel, the transmembrane potential waveforms all have similar wave shapes. Conversely, the three extracellular waveforms at three positions 1 mm from the fiber's axis have markedly different wave shapes from each other and from the transmembrane potential waveforms beneath them. (Note the lack of correlation between the shape of the extracellular waveform in column A, above, and that of the transmembrane potential waveform in column A, below, and similarly for columns B and C). Additionally, the peak-to-peak magnitudes of these extracellular waveforms are much smaller, roughly a thousand times smaller than those of the transmembrane potentials. Moreover, the duration of the major deflections of the extracellular waveforms is shorter.

The predominant mode of interpretation of clinical waveforms is the accumulation of a large number of measurements in a standardized way from patients with known conditions, and then the retrospective classification of waveforms into categories, using both the intuition of the investigators or more formal statistical procedures.

Wave shape interpretation based on understanding of the mechanism of origin, coupled with quantitative evaluation of amplitudes, is nonetheless of great interest. As knowledge of the mechanisms of origin for particular situations increases, the approach is gradually changing to a more mechanistic mode of interpretation.

8.1.2. Spatial Rather than Temporal

To understand qualitatively (as well as quantitatively) how extracellular waveforms come to be what they are, it is necessary to model the underlying electrophysiological process. The model, while simplified from the actual tissue structure, must take into account the core elements of the process: the tissue membrane properties, geometry, and the various electrical parameters (e.g., volume conductor impedances, inhomogeneities, and anisotropies).

The temporal waveforms of Figure 8.1, when redrawn as spatial distributions (Figure 8.2, panel A), show some unexpected additional waveforms.[2] Inspecting the series of $V_m(x)$ distributions and the cartoons of Figure 8.2, one observes the following:

- As seen in Figure 8.2A, excitation begins at the center of the fiber (at $x = 0$). This conclusion is reached by looking at the progression of $V_m(x)$ patterns for $t = 8, 6, 4, 2$ and extrapolating backward to time $t = 0$.
- As seen in Figure 8.2A in the line for 2 milliseconds, a series of wave shapes for $V_m(x)$ have patterns that do not recur at later times.
- As seen in Figure 8.2B, the initial deflection seen spatially evolves into two distinct excitation waves, one on the positive x side and the other on the negative x side.
- Once the excitation waves in opposite directions are separated, each one, considered separately, propagates uniformly. As a pair, however, they do not propagate uniformly, because the velocities on the left and right sides have opposite sign.
- Once separated, the spatial action potentials are mirror images. Note that the one on the left looks like $V_m(t)$ in that the fast upstroke is on the left, while the one on the right has a wave shape that is reversed, in that the fast upstroke is on the right.

To understand the origin of extracellular waveforms, a key first step is to describe the underlying transmembrane potentials as distributions in space, which of course change with time,

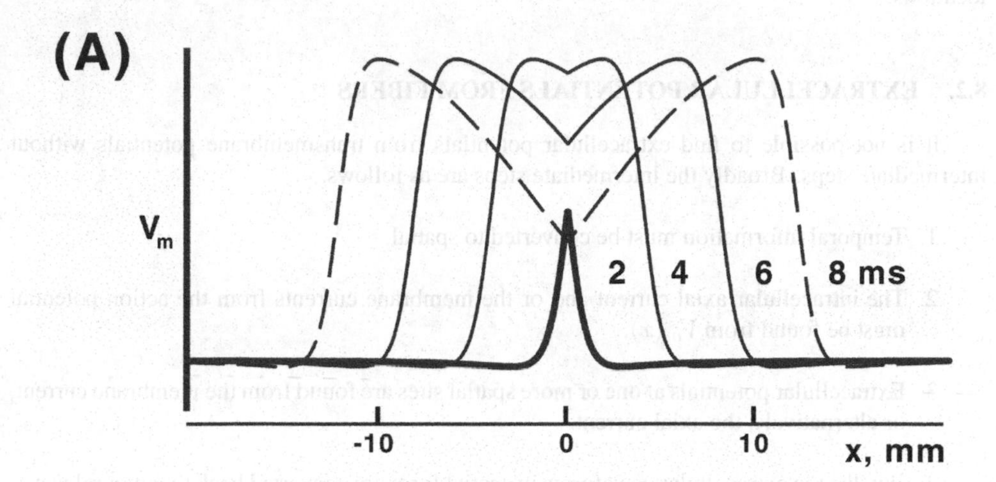

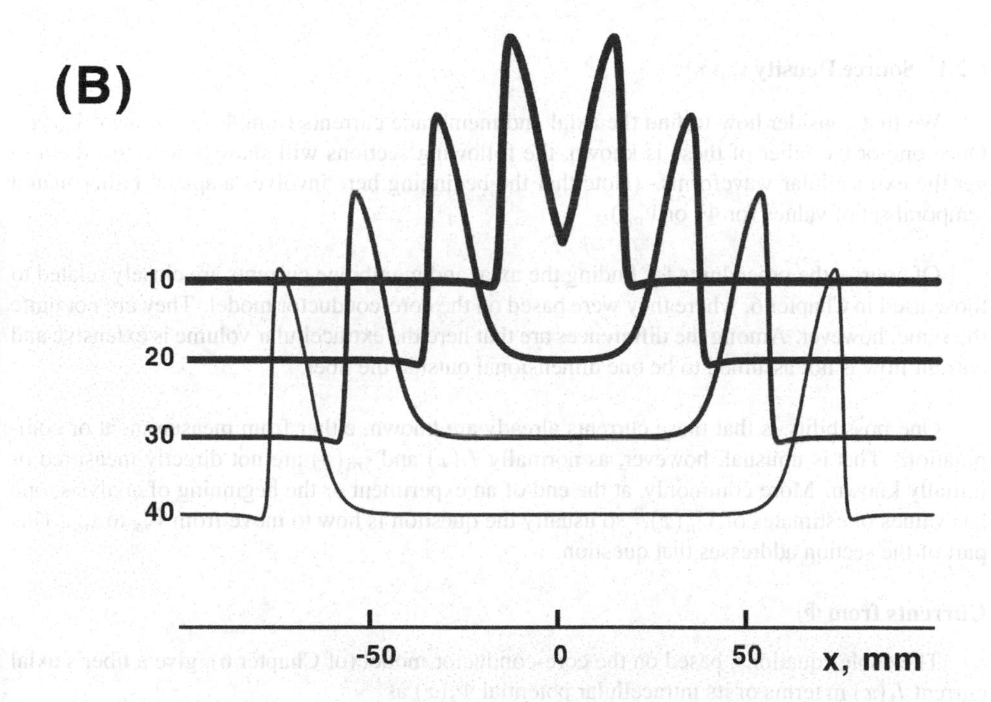

Figure 8.2. Spatial Transmembrane Potentials. Panel A shows the transmembrane potential as a function of distance along the fiber, for early times after excitation. Panel B shows $V_m(x)$ for several later times, given by the number beside each trace. For illustration, the different traces are displaced vertically. Note the multiplicity of wave shapes present in the $V_m(x)$ traces, as compared to $V_m(t)$, which is shown in Figure 8.1.

rather than describing the transmembrane potentials as a set of voltages versus time at various locations.

8.2. EXTRACELLULAR POTENTIALS FROM FIBERS

It is not possible to find extracellular potentials from transmembrane potentials without intermediate steps. Broadly the intermediate steps are as follows.

1. Temporal information must be converted to spatial.

2. The intracellular axial current and or the membrane currents from the action potential must be found from $V_m(x)$.

3. Extracellular potentials at one or more spatial sites are found from the membrane current, or alternatively, the axial current.

4. Finally, the extracellular waveforms in spatial form are converted back to temporal waveforms, if that is the desired final form.

8.2.1. Source Density $i_m(x)$

We first consider how to find the axial and membrane currents from $\Phi_i(x)$ or from $V_m(x)$. Once one or the other of these is known, the following sections will show how to use them to get the extracellular waveforms. (Note that the beginning here involves a spatial rather than a temporal set of values for Φ_i or V_m.)

Of course, the procedures for finding the axial and membrane currents are closely related to those used in Chapter 6, where they were based on the core-conductor model. They are not quite the same, however. Among the differences are that here the extracellular volume is extensive and current flow is not assumed to be one dimensional outside the fiber.

One possibility is that these currents already are known, either from measurement or computation. That is unusual, however, as normally $I_i(x)$ and $i_m(x)$ are not directly measured or initially known. More commonly, at the end of an experiment or the beginning of analysis, one has values or estimates of $V_m(x)$,[3] so usually the question is how to move from V_m to i_m. This part of the section addresses that question.

Currents from Φ_i

The cable equations, based on the core-conductor model (of Chapter 6), give a fiber's axial current $I_i(x)$ in terms of its intracellular potential $\Phi_i(x)$ as

$$I_i = -\frac{1}{r_i}\frac{\partial \Phi_i}{\partial x} \tag{8.1}$$

This relationship continues to hold here even though the extracellular volume now is extensive, rather than one dimensional. Similarly, if one knows the axial current one can determine the membrane current by

$$i_m = -\frac{\partial I_i}{\partial x} \tag{8.2}$$

As shown in chapter 6, substitution of (8.2) into (8.1) gives

$$i_m = \frac{1}{r_i}\frac{\partial^2 \Phi_i}{\partial x^2} \tag{8.3}$$

For circular cylindrical axons the axial intracellular resistance, r_i, per unit length, assuming uniform axial current density, is

$$r_i = \frac{R_i}{\pi a^2} \tag{8.4}$$

where a is the axon radius and R_i is the resistivity of the axoplasm (Ωcm). Accordingly, (8.3) becomes

$$i_m = \frac{\pi a^2}{R_i}\frac{\partial^2 \Phi_i}{\partial x^2} = \pi a^2 \sigma_i \frac{\partial^2 \Phi_i}{\partial x^2} \tag{8.5}$$

In (8.5), $\sigma_i = 1/R_i$ is the conductance per centimeter (S/cm) of the axoplasm.

Thus, if one knows the intracellular potential Φ_i as a function of distance along the fiber, then the computation of $i_m(x)$ is done by finding the second spatial derivative, and using (8.5). The resulting $i_m(x)$ then is used to find $\Phi_e(x)$ at one or more field points, as described below.[4]

Currents from V_m

Often, however, one does not know $\Phi_i(x)$ but instead has $V_m(x)$. If there is a simultaneous set of values for $\Phi_e(x)$ just outside the fiber membrane, then one can use the definition $V_m(x) = \Phi_i(x) - \Phi_e(x)$ to get $\Phi_i(x)$. Then one can use (8.5) above. That is an unusual situation, however, so usually one resorts to the following argument.

We noted when discussing the core-conductor model that if the extracellular space were extensive, $r_e << r_i$, and we could choose $r_e \approx 0$. If $r_e \approx 0$, then $\Phi_e << \Phi_i$ and $\Phi_e \approx 0$. Consequently, $V_m = \Phi_i - \Phi_e \approx \Phi_i$.[5] Using this approximation, one has

$$i_m = \frac{1}{r_i}\frac{\partial^2 \Phi_i}{\partial x^2} \approx \frac{1}{r_i}\frac{\partial^2 V_m}{\partial x^2} \tag{8.6}$$

In the above expressions, recall that $\partial V_m/\partial x = \partial v_m/\partial x$ and $\partial \Phi_i/\partial x = \partial \phi_i/\partial x$.

Using the definition of r_i as done above (8.4), one gets

$$I_i = -\frac{\pi a^2}{R_i}\frac{\partial \Phi_i}{\partial x} = -\pi a^2 \sigma_i \frac{\partial V_m}{\partial x} \tag{8.7}$$

and

$$i_m = \frac{\pi a^2}{R_i}\frac{\partial^2 V_m}{\partial x^2} = \pi a^2 \sigma_i \frac{\partial^2 V_m}{\partial x^2} \tag{8.8}$$

As an example, consider Figure 8.3, which gives a transmembrane potential and the intracellular and membrane currents as determined using equations (8.7) and (8.8). Specifically,

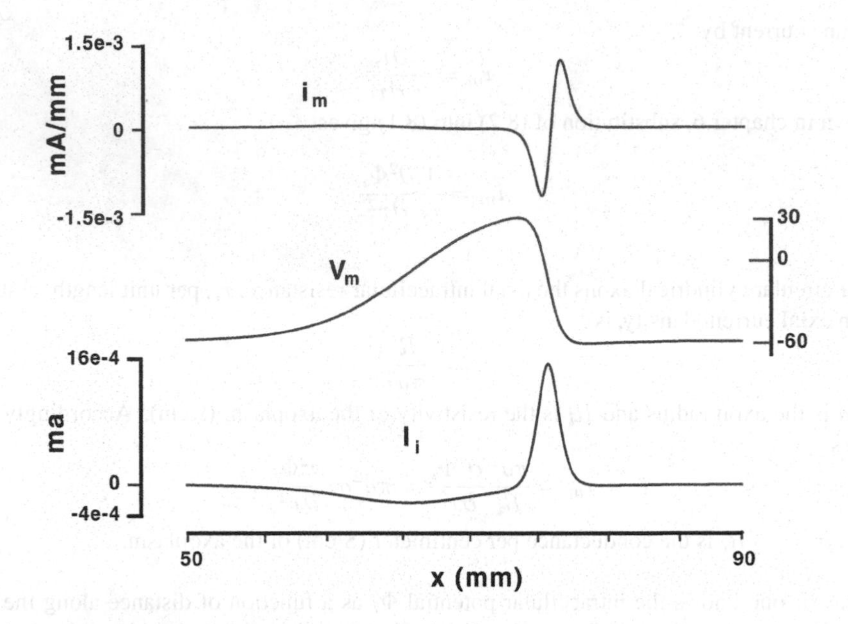

Figure 8.3. Transmembrane Potential V_m, Intracellular Current I_i, and Transmembrane Current i_m. The Figure shows a transmembrane potential (middle). Other traces give the transmembrane (top) and intracellular axial (bottom) currents, as determined from the transmembrane potential using Eqs. (8.7) and (8.8).

A. The top panel (A) gives the transmembrane current. As the negative of the spatial derivative of I_i, it has markedly positive and negative deflections on the leading and trailing edges of the I_i waveform.

B. The middle panel (B) gives the transmembrane potential along the fiber, as specified by a mathematical template function.

C. The lower panel (C) gives the intracellular axial current. Note that it has a sharp upward spike in the region of the action potential's upstroke.

It seems contradictory that here we assume that the extracellular resistance is insignificant, while in the following sections a nonzero extracellular resistance is essential for the creation of extracellular waveforms. To resolve the paradox one remembers the context. Here the extracellular resistance per length is assumed small *relative to that of the intracellular space.* That can be so if the extracellular resistivity is of significant value. One aspect is that the extracellular resistance per length is the extracellular resistivity *divided by* by the cross-section for extracellular current flow. In an extensive extracellular volume, that cross-section is large.

8.2.2. Action Currents

Action currents are those associated with action potentials, and their propagation. Drawings of action current loops provide a formative mental picture of the origin of extracellular potentials.

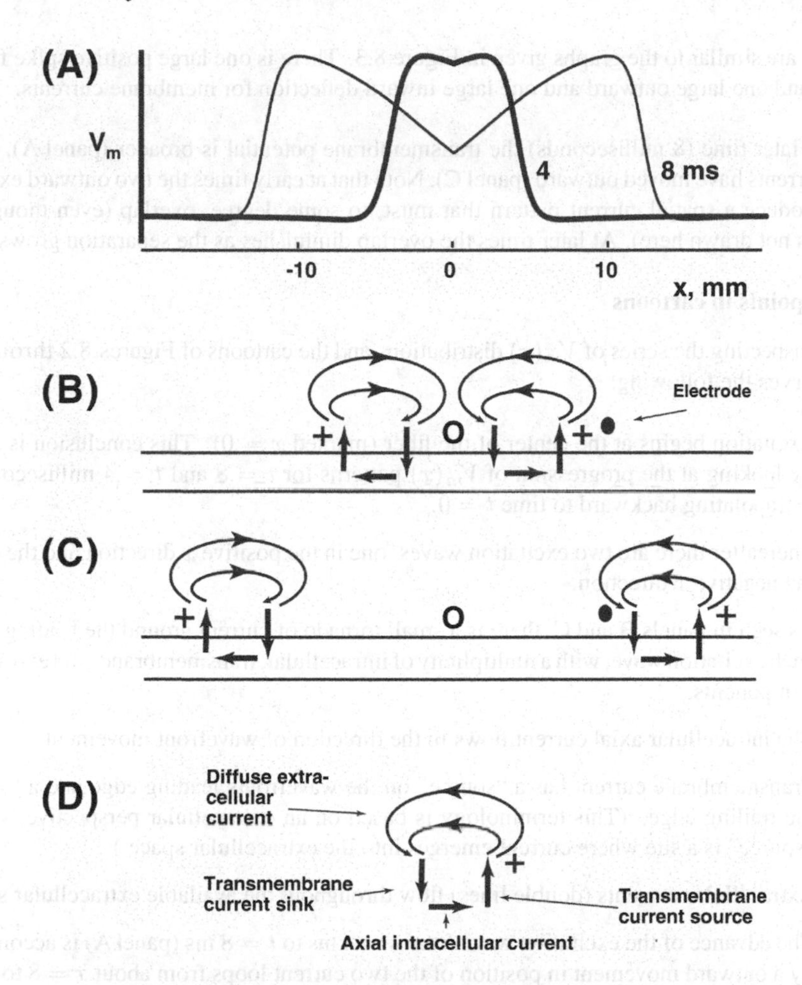

Figure 8.4. Action Current Cartoon. Panel A shows the transmembrane potential as a function of distance along the fiber for 4 and 8 milliseconds. Excitation began at $x = 0$ and spread from there in both directions. Panel B shows a cartoon of the current flow at 4 msec, and Panel C for 8 msec. Labels identifying elements of the current loops of panels B and C are given in panel D. The open and closed dots in B and C are hypothetical electrode positions. In B–D, for purposes of illustration the source–sink distance is widened, compared to that implied by the upstrokes of panel A.

These depictions are helpful even when they are only qualitative, as such pictures can be refined as individual elements become quantitative through mathematical analysis.

Here we start first with a qualitative depiction. The cartoon of Figure 8.4 redraws (in panel A) the curves of the spatial distribution of transmembrane potentials, for times of 4 and 8 milliseconds. Cartoon drawings of the action currents at 4 milliseconds are given in panel B. Two current patterns are drawn, each associated with one of the two excitation waves progressing outward from $x = 0$. (Each component of the current path is identified in panel D.) Note that the elements of these drawings are similar to the graphs given in Figure 8.3. There is one large positive spike for axial current, and one large outward and one large inward deflection for membrane currents.

At a later time (8 milliseconds) the transmembrane potential is broader (panel A), and the action currents have moved outward (panel C). Note that at early times the two outward excitation waves produce a spatial current pattern that must, to some degree, overlap (even though such overlap is not drawn here). At later times the overlap diminishes as the separation grows larger.

Notable points in cartoons

By inspecting the series of $V_m(x)$ distributions and the cartoons of Figures 8.2 through 8.4, one observes the following:

- Excitation begins at the center of the fiber (marked $x = 0$). This conclusion is reached by looking at the progression of $V_m(x)$ patterns for $t = 8$ and $t = 4$ milliseconds and extrapolating backward to time $t = 0$.
- Thereafter there are two excitation waves, one in the positive x direction and the other in the negative x direction.
- As seen in panels B and C, there is a small tornado of current around the leading edge of each excitation wave, with a multiplicity of intracellular, transmembrane, and extracellular components.
- The intracellular axial current flows in the direction of wavefront movement.
- Transmembrane current has a "source" on the wavefronts leading edge and a "sink" on the trailing edge. (This terminology is based on an extracellular perspective, so that a "source" is a site where current emerges into the extracellular space.)
- Extracellular currents (double lines) flow throughout the available extracellular space.
- The advance of the excitation wave from $t = 4$ ms to $t = 8$ ms (panel A) is accompanied by a outward movement in position of the two current loops from about $x = 8$ to $x = 16$ mm (panels B to C).
- The electrode near the center (open circle) is always closer to sinks than sources, so its potential is always negative. (Compare to trace A of Figure 8.1.)
- The electrode away from the center (solid circle) initially is closest to a source (panel B) and then closest to a sink (panel C), so one expects its potential to change from initially positive to later negative. (Compare to trace C of Figure 8.1.)

Cartoon limitations

One draws the conclusions listed above from the cartoons rather cautiously, as they lack any rigorous quantitative basis, i.e., they are just drawings. For example, comparison with the quantitative graphs of Figure 8.3 makes it clear that the cartoons are not quite right on the following points:

- Intracellular current flows most intensively in one region, as drawn, but also flows in the opposite direction in a different site, not drawn.

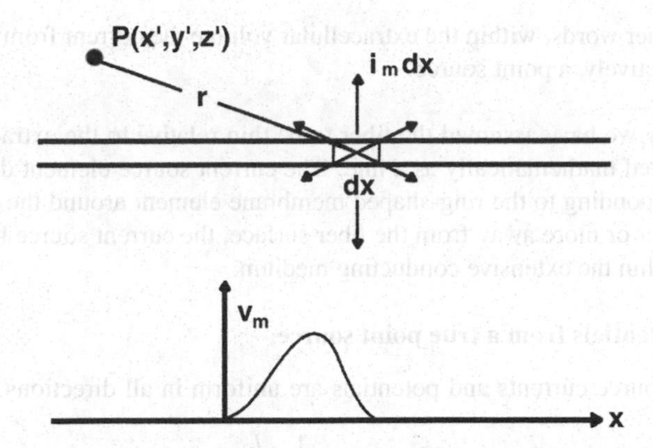

Figure 8.5. Element of a Fiber. (a) An action potential is propagating on a fiber in the positive x direction. The fiber is divided into mathematical segments, and one such segment is drawn. The fiber lies in a uniform extracellular medium of conductivity σ_e that is infinite in extent. The field arising from the action currents at an arbitrary field point P is desired. The coordinates of P are (x', y', z'). The source element is at (x, y, z). Shown is the current emerging from the fiber element dx (magnitude $i_m dx$). (b) The monophasic action potential $V_m(x)$.

- Transmembrane current exists outside of the one source and one sink drawn for each excitation wave.
- It is unexplained how the sources and sinks translate into specific extracellular currents or how their magnitudes diminish with increased distance from the fiber axis.
- No provision was available for understanding overlapping effects from two or more excitation waves.

For these reasons and others, it is essential to focus and strengthen the analysis of action currents and extracellular potentials through a more quantitative approach, as is done in the following section.

8.2.3. Quantitative Formulation of Extracellular Potentials

A cylindrical fiber carrying an action potential propagating in the x direction produces potentials throughout the surrounding medium. Here we assume the fiber is lying in an extensive conducting medium, so the geometry is shown in Figure 8.5a. A sketch of a monophasic action potential in the fiber is drawn in Figure 8.5b.

For analysis the fiber is divided conceptually into small elements along its length. Fiber element dx, identified in Figure 8.5, lies within the region occupied by an action potential. Out of this differential fiber element a current emerges into the extracellular region. The amplitude of this current from the element is the transmembrane current per unit length, i_m, times the length dx (i.e., $i_m dx$). From the perspective of the large extracellular region, the transmembrane current emerges from a very small spatial region into an effectively unbounded space (except for the fiber itself). In other words, within the extracellular volume the current from element dx creates potentials as, effectively, a point source.

For simplicity we have assumed the fiber to be thin relative to the extracellular volume, so that it can be treated mathematically as a line. The current source element described here has a ring shape, corresponding to the ring-shaped membrane element around the fiber. Even so, at a distance a diameter or more away from the fiber surface, the current source behaves virtually as a point source within the extensive conducting medium.

Extracellular potentials from a true point source

For a point source currents and potentials are uniform in all directions. Recall from (2.8) that

$$\Phi_e = \frac{1}{4\pi\sigma_e}\frac{I_0}{r} \tag{8.9}$$

where r is the distance from the point source to the field point, σ_e is the extracellular conductivity, and I_0 is the source's amplitude. The currents from a fiber obey related equations.

Extracellular potentials from a distributed sources

In the case of a cylindrical fiber, the analog to the point source I_0 is the transmembrane current $i_m\,dx$. The transmembrane current is, of course, not confined to a point, but instead varies along the fiber. However, that distribution can be considered to be the same, in the limit, as a distribution of point sources. Therefore the contribution of $i_m\,dx$ to Φ_e can be rewritten, by analogy to (8.9), as

$$d\Phi_e = \frac{1}{4\pi\sigma_e}\frac{i_m}{r}\,dx \tag{8.10}$$

Finding the potential requires an integration over the full length of the fiber, as shown in (8.11),

$$\Phi_e(P) = \frac{1}{4\pi\sigma_e}\int_L \frac{i_m dx}{r} \tag{8.11}$$

In writing (8.9), (8.10) and (8.11) one recognizes the following points:

- The expression $i_m dx$ is used instead of I_0, since $i_m dx$ is the current emerging from one segment of the fiber, as shown in Figure 8.5.
- Current sources are considered to lie on the fiber axis, even though currents emerge from rings around the fiber.
- The expression (8.10) is a differential contribution to the extracellular potential $d\Phi_e$. To get Φ_e, an integral over the length of the fiber is required, so that all segments along the fiber that have nonzero membrane current are taken into account.
- In (8.11), r is the distance from each element of current along the fiber $i_m dx$ to the field point P, the location for which Φ_e is to be determined.
- Length L, over which the integration is performed, must be long enough that all membrane currents are contained within that length.

- Different extracellular potentials Φ_e will be present at different field points P because each field point will lead to a different set of values of variable r, even though $i_m(x)$ is unchanged. That is, different extracellular locations will give different weights to the same set of transmembrane currents (sources) along the fiber's length.

Putting (8.8) into (8.11) gives the desired expression for Φ_e from V_m as

$$\Phi_e = \frac{a^2 \sigma_i}{4\sigma_e} \int \frac{\partial^2 V_m/\partial x^2}{r} dx \tag{8.12}$$

Equation (8.12) is used widely as it depends on V_m, which often is known. The equation may be used to find potentials at field points that may be chosen to be either close to or far from the active fiber. The same equation can be used to find the potential at only one position, or repeatedly to find potentials for each location within a family of positions.

By using (8.12) twice, one can find the voltage between two field points. Often such a voltage is needed to predict or confirm a measurement. For example, since linearity will apply in the volume conductor outside the fiber, the voltage V_{ab} between two field points a and b will be

$$V_{ab} = \Phi_e(a) - \Phi_e(b) \tag{8.13}$$

where $\Phi_e(a)$ is found using (8.12) and the geometrical coordinates for field point a, with an analogous procedure for b.

As a matter of terminology, when position a is close to an electrophysiologically active membrane and b is relatively far away, the voltage so measured often is called *unipolar*, while if a and b are close together (e.g., within a millimeter), the measurement often is called *bipolar*. (Bipolar recordings are approximations of a measurement of the spatial derivative of the potentials.) This terminology can be misleading, as two electrodes are used both for unipolar and bipolar recordings.

Transfer function in convolution form

The geometrical mathematics can be confusing because one often thinks of a set of field points along the fiber's direction (variable x), some distance away. At the same time each field point involves summing contributions from points on the axis of the fiber for its whole length (another variable x).

To keep things straight, a more formal writing of variable r is helpful, where these variables are separated into x and x'. Specifically, if the element $i_m dx$ is located at the coordinate (x, y, z), and if the point at which the potential field is desired is located at (x', y', z'), then

$$r = \sqrt{(x - x')^2 + (y - y')^2 + (z - z')^2} \tag{8.14}$$

Often the coordinate origin is placed on the fiber axis so that points on the axis have $y = z = 0$. Then with r as written above, integrating along the axis of the fiber involves varying x (with $y, z = 0$), keeping everything else constant. Conversely, moving the field point along the direction of the x axis involves varying x', while keeping y', z' constant (but not both zero).

Writing r in this way, one has

$$\Phi_e(x',y',z') = \frac{1}{4\pi\sigma_e}\int_L \frac{i_m(x)dx}{\sqrt{(x-x')^2+(y')^2+(z')^2}} \tag{8.15}$$

The integral in (8.15) extends over the interval in x occupied by the action potential; also, as discussed above, we have set $y = z = 0$ in (8.15).

The equation for Φ_e (8.15) can be rewritten, in the form of a *convolution*, as

$$\Phi_e(x',y',z') = \frac{1}{4\pi\sigma_e}\int_L H(x-x')i_m(x)dx \tag{8.16}$$

where $H(x-x')$ is defined by

$$H(x-x') = \frac{1}{\sqrt{(x-x')^2+(y')^2+(z')^2}} \tag{8.17}$$

In convolution form equation (8.17) for $\Phi_e(x)$ has $H(x-x')$ is its *kernel*. Convolution form has the advantage of making separating the terms for the sources $i_m dx$ from the terms linking the sources to the field points $H(x-x')$.[6] In engineering, function H often is referred to as the *transfer function*.

Examples of the quantities shown in the convolution equation for the extracellular potential (8.16) are given in Figure 8.6. In the figure:

A. Panel A plots the function H for several field points. The solid line is for a field point at $x = -10$ millimeters, while the two dotted lines are for $x = 0$ (center) and $x = 10$, respectively. One notes that the same wave shape is present in each of these H plots, but shifted in space so that the peak occurs at the x coordinate of the field point.

B. Panel B shows i_m for 4 msec, and in a second trace shows the extracellular potential waveform $\phi_e(x)$ computed from $i_m(x)$ and H. In panel A one notes that the H wave shape has a significant breadth, implying that the extracellular potential at $x = 0$ will be a weighted sum of membrane currents around $x = 0$, not just those at $x = 0$ precisely. That expectation is seen demonstrated in panel B, where the sharp deflections and return to the baseline near $x = 0$ (center) on the i_m waveform become more rounded. Note that the trace for $\phi_e(x)$ does not return to zero at $x = 0$, even though the i_m curve does return to zero there.

C. Panel C of Figure 8.6 shows waveforms for $i_m(x)$ and $\phi_e(x)$ for 8 ms. At this later time (as compared to panel B) the two excitation waves are separated. While the separation is evident from $\phi_e(x)$, the waveform in notably smoother than is $i_m(x)$. A more favorable way of viewing $\phi_e(x)$ on this trace is that its observation shows not only what is happening underneath the field point but also the time course of propagation as it approaches or moves away. A spatially broader response often is advantageous in following the movement of excitation waves.

The major conclusion from this Figure is that the wave shape of $\phi_e(x)$ follows that of $i_m(x)$. However, the $\phi_e(x)$ wave shape is smoother, and it reflects membrane currents some distance away (laterally), as well as the membrane current at the position directly beneath the field point.

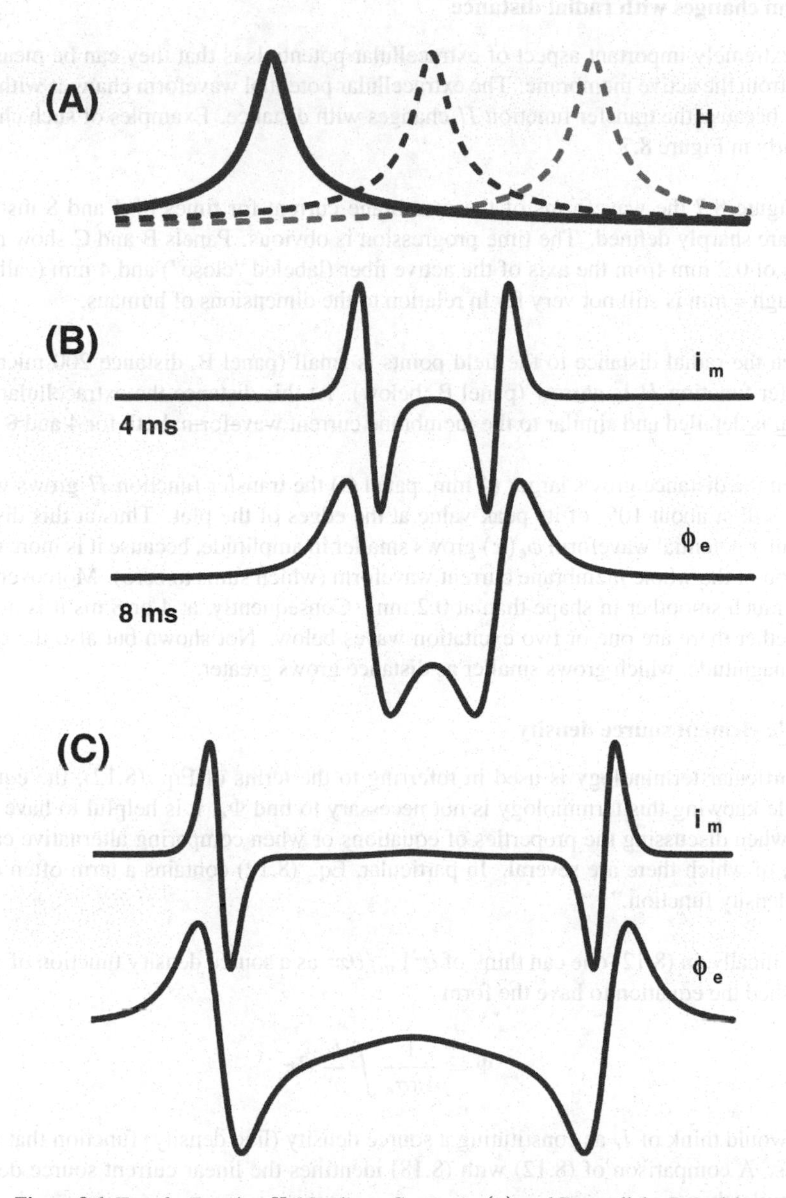

Figure 8.6. Transfer Function H, Membrane Current $i_m(x)$, and Extracellular Potentials $\Phi_e(x)$. Panel A: Plots of the transfer function H. Transfer function $H(x - x')$ is given for three values of x'. The solid line is for $x' = -10$ mm, while the two dashed lines are for $x' = 0$ (centered) and $x' = 10$ (on right). Panel B: Membrane current $i_m(x)$ at 4 milliseconds and extracellular potential $\Phi_e(x)$ along a line a distance of 1 mm from the fiber axis. Panel C: Membrane current $i_m(x)$ at 8 milliseconds and extracellular potential $\Phi_e(x)$ along a line a distance of 1 mm from the fiber axis. The extracellular potential distribution $\Phi_e(x)$ for 4 ms (thick line) and 8 ms (thin line). The 4-ms potential function comes from the convolution of H with i_m for 4 ms, and similarly for 8 ms.

Waveform changes with radial distance

An extremely important aspect of extracellular potentials is that they can be measured at a distance from the active membrane. The extracellular potential waveform changes with distance, however, because the transfer function H changes with distance. Examples of such changes are seen clearly in Figure 8.7.

In Figure 8.7 the waveforms of the membrane current for times of 4 and 8 ms (panel A, bottom) are sharply defined. The time progression is obvious. Panels B and C show results for distances of 0.2 mm from the axis of the active fiber (labeled "close") and 4 mm (called "far"), even though 4 mm is still not very far in relation to the dimensions of humans.

When the radial distance to the field points is small (panel B, distance 200 micrometers), the transfer function H is narrow (panel B, below). At this distance the extracellular potential waveform is detailed and similar to the membrane current waveform, both for 4 and 8 ms.

When the distance grows larger (4 mm, panel C) the transfer function H grows wider, and in fact is still at about 10% of its peak value at the edges of the plot. Thus at this distance the extracellular potential waveform $\phi_e(x)$ grows smaller in amplitude, because it is more nearly the summation of the whole membrane current waveform (which sums to zero). Moreover, $\phi_e(x)$ at 4 mm is much smoother in shape than at 0.2 mm. Consequently, at 4 or 8 ms it is not entirely clear whether there are one or two excitation waves below. Not shown but also the case is the relative magnitude, which grows smaller as distance grows greater.

Monopole element source density

A particular terminology is used in referring to the terms of Eq. (8.12), the equation for Φ_e. While knowing this terminology is not necessary to find Φ_e, it is helpful to have the terms in mind when discussing the properties of equations or when comparing alternative calculation methods, of which there are several. In particular, Eq. (8.12) contains a term often called the "source density function."

Specifically, in (8.12) one can think of $\partial^2 V_m/\partial x^2$ as a source density function of x. That is, if we wished the equation to have the form

$$\Phi = \frac{1}{4\pi\sigma_e} \int \frac{I_\ell}{r}\, dx \tag{8.18}$$

then we would think of I_ℓ as constituting a source density (line density) function that lies along the x axis. A comparison of (8.12) with (8.18) identifies the linear current source density. I_ℓ, more completely as

$$I_\ell = \pi a^2 \sigma_i \frac{\partial^2 V_m}{\partial x^2} \tag{8.19}$$

where the dimension of I_l is current per unit length. Sometimes one loosely refers to the source density as $\partial^2 V_m/\partial x^2$, ignoring the (constant) coefficient $\pi a^2 \sigma_i$, but the coefficient must be included in any quantitative evaluation.

An element of the source defined here is called a *monopole* source. A characteristic of monopole sources is that they have $1/r$ dependence in the equations for Φ_e. This kind of source

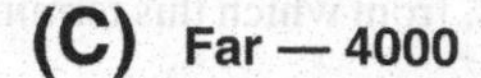

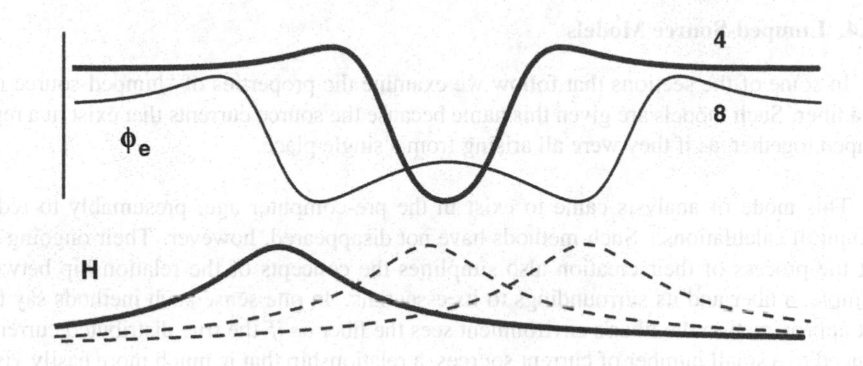

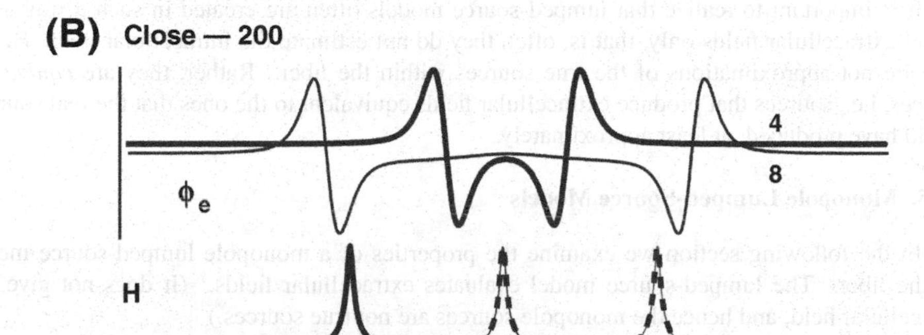

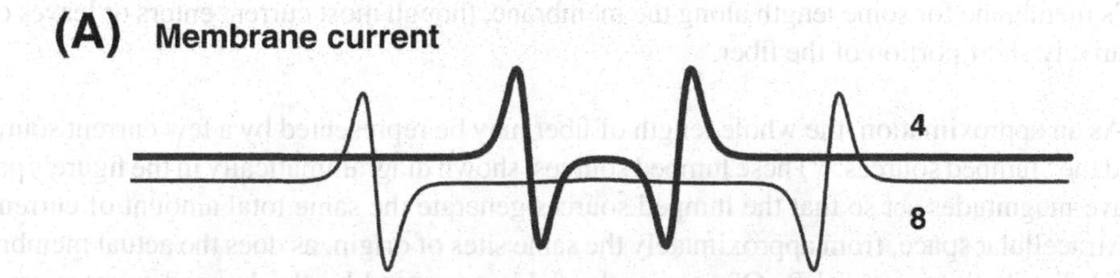

Figure 8.7. Changes with Distance of Extracellular Waveforms. Panel A shows the transmembrane current waveform. Panel B shows data for a distance of 200 micrometers from the fiber axis. Data plotted is H for positions –10, 0, and 10 mm, and $\phi_e(x)$. Panel C shows the same data at a distance of 4000 micrometers (4 mm). In all panels, the small numbers 4 and 8 are to identify curves for 4 milliseconds and 8 milliseconds after excitation begins at the center. (For illustration, the 8-ms plot is slightly displaced downward.)

distribution is sometimes called a *single* source density in contrast to the dipole (or double) source density which we discuss in a later section. There are many analogies between these sources and those with similar mathematical forms in the study of electrostatics, from which this terminology comes.

8.2.4. Lumped-Source Models

In some of the sections that follow we examine the properties of "lumped-source models" for a fiber. Such models are given this name because the source currents that exist in a region are lumped together, as if they were all arising from a single place.

This mode of analysis came to exist in the pre-computer age, presumably to reduce the amount of calculations.[7] Such methods have not disappeared, however. Their ongoing merit is that the process of their creation also simplifies the concepts of the relationship between, for example, a fiber and its surroundings to its essentials. In one sense such methods say that to a first approximation the fiber's environment sees the fiber *as if* the true distributed currents were reduced to a small number of current sources, a relationship that is much more easily visualized and understood, as well as computed with fewer steps.

It is important to realize that lumped-source models often are created in such a way as to model extracellular fields only, that is, often they do not estimate the intracellular field. Hence they are not approximations of the true sources within the fiber. Rather, they are *equivalent sources*, i.e., sources that produce extracellular fields equivalent to the ones that the real sources would have produced, at least approximately.

8.2.5. Monopole Lumped-Source Models

In the following section we examine the properties of a monopole lumped-source model for the fiber. The lumped-source model evaluates extracellular fields. (It does not give the intracellular field, and hence the monopole sources are not true sources.)

For example, suppose there is an action potential in a fiber, as shown in Figure 8.2A, where transmembrane current is leaving the inside of the fiber and entering the extracellular space, as shown in the figure's panel B. Inspection of the Figure shows that current enters and leaves the fiber's membrane for some length along the membrane, though most current enters or leaves over a relatively short portion of the fiber.

As an approximation, the whole length of fiber may be represented by a few current sources, called the "lumped sources." These lumped sources, shown diagrammatically in the figure's panel C, have magnitudes set so that the lumped sources generate the same total amount of current in the extracellular space, from approximately the same sites of origin, as does the actual membrane current distribution of panel B. Of course, the fields generated by the lumped sources are not exactly the same as those of the true distributed sources, except perhaps asymptotically at large source–field distances.

The question arises as to how to determine the magnitudes and positions of each of the lumped sources, so that they create, to a good approximation, the same extracellular waveform as did the original $i_m(x)$ distribution. There are several ways to do this process. One way is given in the paragraphs that follow. This way involves two steps: (1) approximate the actual action potential by a triangular shaped waveform, and then (2) find the exact sources that would arise from the waveform with this altered shape.

Triangular action potential approximation

In Figure 8.8D the action potential of panel A is redrawn. As redrawn, the original wave shape of $V_m(x)$ is approximated by a triangular action potential. The triangular action potential has an activation slope that corresponds to the peak slope of the depolarizing phase of V_m and has a recovery slope that corresponds to the peak slope in the recovery phase of V_m. (For the figure, the triangle was formed by drawing straight lines over the original action potential that had slopes the same as the maximum slope in the depolarization and repolarization phases.)

Once triangularized, the width of the activating phase of the triangular action potential is $w_a = x_1 - x_2$, and the width of the recovery phase is $w_r = x_2 - x_3$ (panel D). As a consequence, the slope of the triangularized waveform during depolarization is $A_{\max} = V_{\text{pp}}/w_a$, and the slope during recovery is $B_{\max} = V_{\text{pp}}/w_r$. Note that V_{pp} is the peak-to-peak magnitude of the triangularized action potential, not the value of the transmembrane potential at the peak.[8]

Three lumped sources

A quantitative basis for the lumped monopole approximation may be developed by first recalling (8.12), the expression for the extracellular potential from an action potential, which was

$$\Phi_e = \frac{a^2 \sigma_i}{4\sigma_e} \int_L \frac{\partial^2 V_m / \partial x^2}{r} \, dx \tag{8.20}$$

Note that, once triangularized, the second derivative is zero outside of the corner points. Thus the question arises as to how to evaluate the integral across the corners. Consider in particular corner 3, which has x_a to the left of this corner and $\partial^2 V_m / \partial x^2 = 0$ at x_a. With x_b to the right of corner 3, then at x_b $\partial^2 V_m / \partial x^2$ again equals zero.

We then can find the magnitude of the source at corner 3 by integrating from x_a to x_b:

$$M = \pi a^2 \sigma_i \int_{x_a}^{x_b} \frac{\partial^2 V_m}{\partial x^2} dx = \pi a^2 \sigma_i \left(\left. \frac{\partial V_m}{\partial x} \right|_{x_b} - \left. \frac{\partial V_m}{\partial x} \right|_{x_a} \right) \tag{8.21}$$

Aside from the coefficient, the result is simply the difference between the first derivatives evaluated at the ends of the interval.

In the triangular waveform, the derivatives are readily available from the peak-to-peak transmembrane voltage, V_{pp}, together with spatial widths w_a and w_r of the activation and repolarizing phases, as shown in Figure 8.8. Because the derivatives are used in several combinations, it is convenient to define the maximum slope during activation, $A_{\max}$, as

$$A_{\max} = V_{\text{pp}}/w_a \tag{8.22}$$

and the maximum slope during repolarization as

$$B_{\max} = V_{\text{pp}}/w_r \tag{8.23}$$

Note that $A_{\max}$ and $B_{\max}$ are defined in a way that makes both of them unsigned quantities.

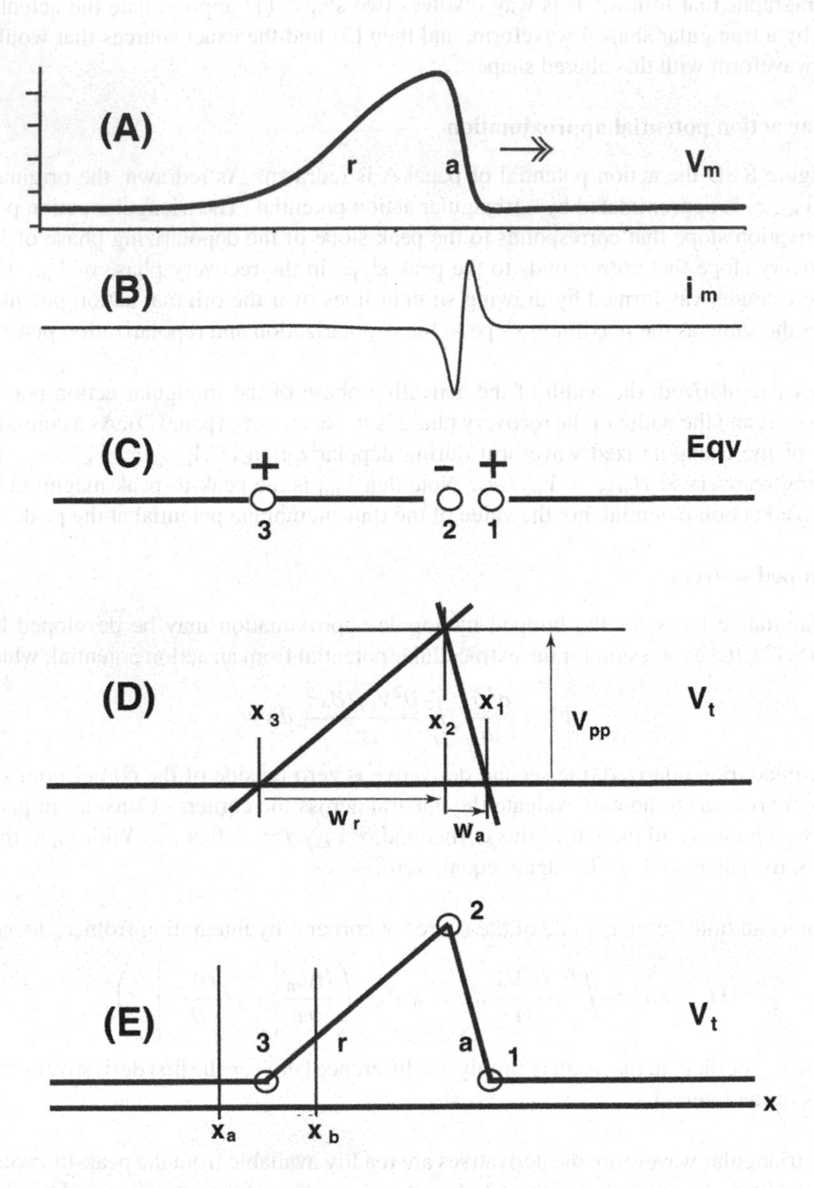

Figure 8.8. Monopole Sources. (A) Monophasic action potential $V_m(x)$. The activation and recovery phases are identified with the small letters a and r. (B) Membrane current $i_m(x)$. (C) Lumped equivalent monopole sources. (D) Triangularized action potential. The sides of the triangle have a slope equal to the maximum slope of V_m of panel A. Widths w_a and w_r are for the activation and recovery phases, spatially. (E) Triangular action potential with encircled sites where there is a slope change. Sites 1 to 3 lead to monopole sources M_1 to M_3. The activation phase of V_t has slope A_{max} and the recovery phase has slope B_{max}.

Since the second derivative is triphasic, there are three lumped sources, and, in particular, the central region is negative and is flanked by positive side regions. We designate the total monopole strength in the aforementioned regions M_1, M_2, M_3, corresponding to points 1, 2, and 3 on the figure.

Making use of the same procedure used for M_3 above, and making use of the definitions of A_{max} and B_{max}, we can find the whole set lumped sources as

$$M_1 = \pi a^2 \sigma_i A_{max} \tag{8.24}$$

$$M_2 = -\pi a^2 \sigma_i (B_{max} + A_{max}) \tag{8.25}$$

$$M_3 = \pi a^2 \sigma_i B_{max} \tag{8.26}$$

Thus the single-layer sources for the triangular action potential consist of three discrete monopoles with magnitude and locations as follows:

$$\pi a^2 \sigma_i A_{max}$$

at the "activation" vertex,

$$\pi a^2 \sigma_i B_{max}$$

at the "recovery" vertex, and

$$-\pi a^2 \sigma_i (A_{max} + B_{max})$$

at the foot of the altitude from the triangle's peak. These monopoles constitute exact equivalent sources for the triangle. Consequently, they are also approximations to the sources of an actual action potential for which the triangle is an approximation.

The above source model is sometimes referred to as a *tripole* model because it consists of the three monopoles M_1, M_2, M_3. It is often cited as a practical approximation to the true sources associated with a propagating action potential in nerve and skeletal muscle.

The field generated by the three point sources is the sum of fields from the three sources, namely,

$$\Phi_p = \frac{a^2 \sigma_i}{4\sigma_e} \left(\frac{M_1}{r_1} + \frac{M_2}{r_2} + \frac{M_3}{r_3} \right) \tag{8.27}$$

In terms of the slopes of the triangularized action potential, the equation is

$$\Phi_p = \frac{a^2 \sigma_i}{4\sigma_e} \left(\frac{A_{max}}{r_1} - \frac{A_{max} + B_{max}}{r_2} + \frac{B_{max}}{r_3} \right) \tag{8.28}$$

where the distance from the field point to each point source of magnitude A_{max}, $(A_{max} + B_{max})$, and B_{max} is r_1, r_2, and r_3, respectively.

Test of approximations

If the field point is at a distance r_1 from the midpoint of this interval that is large compared to $(x_b - x_a)$, then the mean-value theorem permits the approximation of

$$\Phi_e^1 = \frac{a^2\sigma_i}{4\sigma_e} \int_{x_a}^{x_b} \frac{\partial^2 V_m}{\partial x^2} \frac{1}{r_1} dx \tag{8.29}$$

by the equation with r factored out from under the integral sign, so that

$$\Phi_e^1 \approx \frac{a^2\sigma_i}{4\sigma_e} \frac{1}{r_1} \int_{x_a}^{x_b} \frac{\partial^2 V_m}{\partial x^2} dx \tag{8.30}$$

A comparison of these results can provide a quantitative test of the degree of approximation in the lumped model. One sees from inspection of the equation that the degree of approximation is likely to be poor for a field point close to the membrane surface, but excellent for field points distant from the membrane surface, because that will determine the variability of r over the interval from x_a to x_b.

8.2.6. The Dipole Formulation

A wonderful thing happens in this subsection of the chapter. The analysis presented to this point shows how extracellular potentials can be found from a knowledge of membrane currents. With the magic of mathematics, that formulation is here transformed into another formulation, one not at all obvious from the work done so far, and one that has a completely different physical interpretation. To wit, earlier we had

$$\Phi_e = \frac{a^2\sigma_i}{4\sigma_e} \int \frac{\partial^2 V_m/\partial x^2}{r} dx \tag{8.31}$$

If we write (8.31) in the form

$$\Phi_e = \frac{a^2\sigma_i}{4\sigma_e} \int_{-\infty}^{\infty} \frac{\partial}{\partial x}\left(\frac{\partial V_m}{\partial x}\right) \frac{1}{r} dx \tag{8.32}$$

then we can integrate by parts using the standard formula

$$\int_L u dv = \Big|_L uv - \int_L v\, du \tag{8.33}$$

with the assignments of $v = \partial V_m/\partial x$ and $u = 1/r$).

The result is an alternate expression for Φ_e, namely,

$$\Phi_e = \frac{a^2\sigma_i}{4\sigma_e} \left\{ \left[\frac{\partial V_m}{\partial x} \frac{1}{r} \right]_{-\infty}^{\infty} - \int_{-\infty}^{\infty} \frac{\partial V_m}{\partial x} \frac{d(1/r)}{dx} dx \right\} \tag{8.34}$$

In (8.34) the integrated part drops out so long as the action potential is not at the ends of the fiber, because far to the left (i.e., $x \to -\infty$) and far to the right (i.e., $x \to \infty$) the membrane is at rest and $\partial V_m/\partial x \equiv 0$. Additionally, in the term on the right, it is necessary to integrate over only the region where $\partial V_m/\partial x$ is nonzero. Thus

$$\Phi_e = \frac{a^2}{4}\frac{\sigma_i}{\sigma_e}\int_L \left[-\frac{\partial V_m}{\partial x}\right]\frac{d(1/r)}{dx}dx \tag{8.35}$$

where length L on the final integral is only long enough to include the region where $V_m \neq 0$.

The directional derivative was shown in Chapter 1 to be

$$\frac{d(1/r)}{dx} = \overline{a}_x \cdot \nabla\left(\frac{1}{r}\right) \tag{8.36}$$

Substitution of the directional-derivative relationship (8.36) enables (8.35) to be rewritten as

$$\Phi_e = \frac{a^2\sigma_i}{4\sigma_e}\int\left[-\frac{\partial V_m}{\partial x}\overline{a}_x\right]\cdot\left[\nabla\left(\frac{1}{r}\right)\right]dx \tag{8.37}$$

Dipole interpretation

The integral in (8.37) can be given a physical interpretation. Recall that a dipole in the x direction ($\overline{p} = p\,\overline{a}_x$) generates a field in the surrounding uniform conducting medium as described in Chapter 2, resulting in

$$\Phi_d = \frac{1}{4\pi\sigma_e}\overline{p}\cdot\nabla\left(\frac{1}{r}\right) = \frac{1}{4\pi\sigma_e}p\,\overline{a}_x\cdot\nabla\left(\frac{1}{r}\right) \tag{8.38}$$

We now wish to identify, in Eq. (8.37) for Φ_e, the terms that correspond to the dipole moment, p, in Eq. (8.38). To this end, note that in Eq. (8.37) an element of fiber, dx, contributes to the total potential an amount

$$d\Phi_e = \frac{a^2\sigma_i}{4\sigma_e}\left(-\frac{\partial V_m}{\partial x}\right)\overline{a}_x\cdot\nabla\left(\frac{1}{r}\right)dx \tag{8.39}$$

so that after multiplying and dividing by π and rearranging slightly, one has

$$d\Phi_e = \frac{1}{4\pi\sigma_e}\left(-\pi a^2\sigma_i\frac{\partial V_m}{\partial x}\right)\overline{a}_x\cdot\nabla\left(\frac{1}{r}\right)dx \tag{8.40}$$

A comparison of (8.38) with (8.40) permits the identification of

$$-\pi a^2\sigma_i\,\partial V_m/\partial x\,\overline{a}_x$$

as the dipole element p. More precisely, because the equation is for $d\Phi_e$, the identified term is an axial dipole element.

Dipole source density

Rather than the generic p, a distinct notation for a linear axial dipole source density function proves convenient, so we define $\overline{\tau}_\ell$ as

$$\overline{\tau}_\ell \equiv -\pi a^2 \sigma_i \frac{\partial V_m}{\partial x} \overline{a}_x = I_i \overline{a}_x \tag{8.41}$$

where $\overline{\tau}_\ell$ has the dimensions of current (current times length per unit length) and is oriented in the x direction.[9]

Making use of the earlier Eq. (8.7) for axial current I_i, one has

$$\overline{\tau}_\ell = I_i \overline{a}_x \tag{8.42}$$

showing that the axial dipole density is proportional to the axial current.[10]

Thus we can write an equation for Φ_e in terms of dipole sources as

$$\Phi_e = \frac{1}{4\pi\sigma_e} \int \overline{\tau}_\ell \cdot \left[\nabla \left(\frac{1}{r}\right)\right] dx \tag{8.43}$$

or

$$\Phi_e = \frac{1}{4\pi\sigma_e} \int I_i \overline{a}_x \cdot \left[\nabla \left(\frac{1}{r}\right)\right] dx \tag{8.44}$$

Dipoles throughout cross-section

In the derivations above we assumed that the fiber had a negligible cross-sectional area, i.e., the sources are essentially concentrated on the axis as line sources. A more rigorous treatment will show that with respect to external fields the sources in (8.37) fill the fiber cross-section, so that for field points very close to the fiber in this distribution would have to be taken into account. In fact, it would be seen that the source is uniform through the cross-section. Since the cross-section is πa^2, then from (8.41) we would deduce that a more general source specification is

$$\overline{\tau}_v = -\sigma_i \frac{\partial V_m}{\partial x} \overline{a}_x \tag{8.45}$$

where $\overline{\tau}_v$ is a volume dipole density. Each axial element of the fiber therefore represents a double-layer disk of source. Plonsey [4] gives further details, and this subject is explored more below in the section on potentials from a single cell.

Lumped dipole source model

The propagating action potential of a nerve fiber is given in Figure 8.3 along with its axial and membrane currents. As noted earlier, the action potential propagates axially at a uniform velocity, so that waveforms in Figure 8.9 all satisfy the functional form of a traveling wave $f(t-x/\theta)$, where θ is the velocity of propagation. Since the spatial behavior of $V_m(x)$ is illustrated in Figure 8.9, with activation on the "right" and recovery to the "left," the wave necessarily propagates in the positive x direction. As discussed in a previous section, the equivalent double layer source density

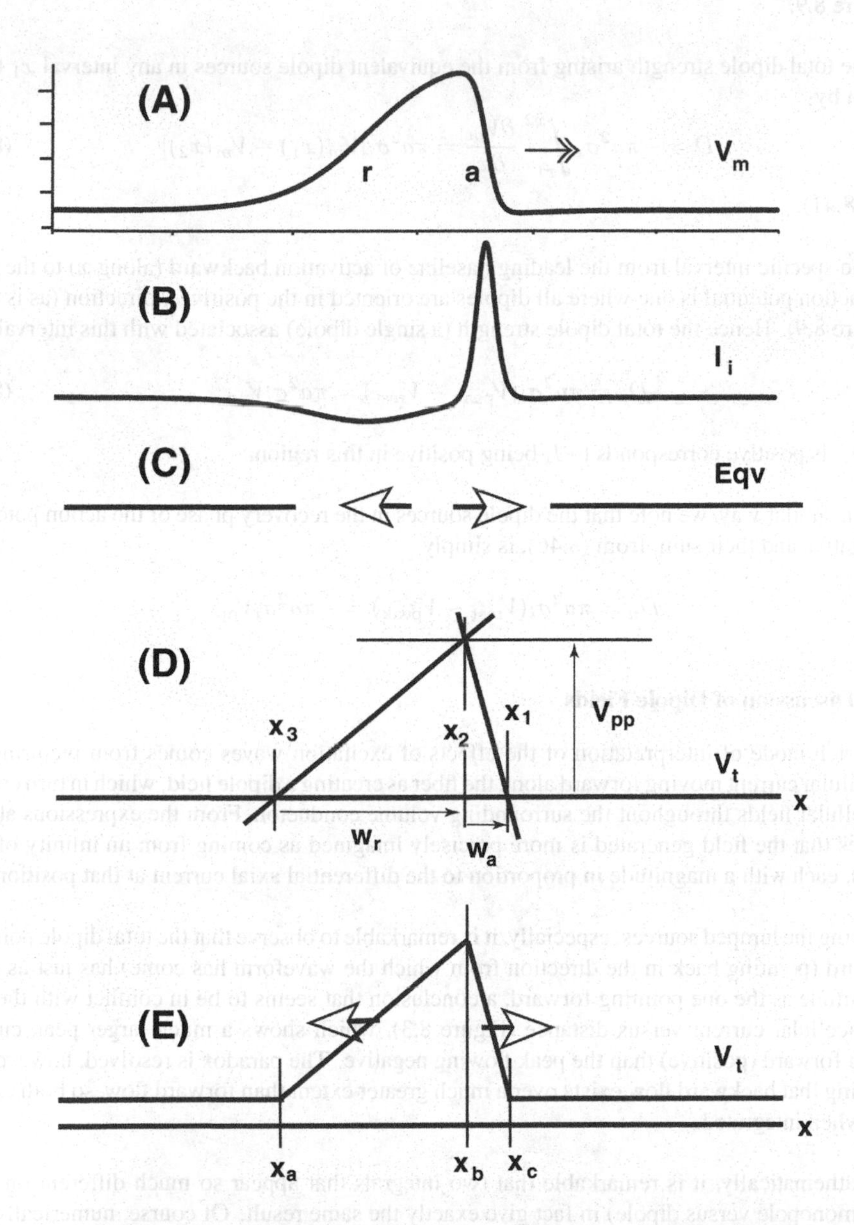

Figure 8.9. Dipole Sources. The action potential $V_m(x)$ and its first spatial derivative are given along with the approximating triangularized action potential $V_t(x)$ and axial current I_i, as found from the spatial derivative of V_m. Depolarization and repolarization spatial widths w_d and w_r are for the approximating triangular action potential.

is proportional to I_i [see Eq. (8.41)]. These functions and a sketch of the dipole sources are given in Figure 8.9.

The total dipole strength arising from the equivalent dipole sources in any interval x_1 to x_2 is given by

$$D = -\pi a^2 \sigma_i \int_{x1}^{x2} \frac{\partial V_m}{\partial x} = \pi a^2 \sigma_i [V_m(x_1) - V_m(x_2)] \tag{8.46}$$

using (8.41).

The specific interval from the leading baseline of activation backward (along x) to the peak of the action potential is one where all dipoles are oriented in the positive x direction (as is clear in Figure 8.9). Hence the total dipole strength (a single dipole) associated with this interval is

$$D_p = \pi a^2 \sigma_i (V_{\text{peak}} - V_{\text{rest}}) = \pi a^2 \sigma_i V_{\text{pp}} \tag{8.47}$$

That D_P is positive corresponds to I_i being positive in this region.

In a similar way, we note that the dipole sources in the recovery phase of the action potential are negative and their sum, from (8.46), is simply

$$D_n = \pi a^2 \sigma_i (V_{\text{rest}} - V_{\text{peak}}) = -\pi a^2 \sigma_i V_{\text{pp}} \tag{8.48}$$

8.2.7. Discussion of Dipole Fields

A rich mode of interpretation of the effects of excitation waves comes from picturing the intracellular current moving forward along the fiber as creating a dipole field, which in turn creates extracellular fields throughout the surrounding volume conductor. From the expressions above, one sees that the field generated is more precisely imagined as coming from an infinity of tiny dipoles, each with a magnitude in proportion to the differential axial current at that position.

Using the lumped sources, especially, it is remarkable to observe that the total dipole pointing backward (pointing back in the direction from which the waveform has come) has just as great a magnitude as the one pointing forward, a conclusion that seems to be in conflict with the plot of intracellular current versus distance (Figure 8.3), which shows a much larger peak current flowing forward (positive) than the peak flowing negative. The paradox is resolved, however, by observing that backward flow exists over a much greater extent than forward flow, so both can be equal when integrated.

Mathematically, it is remarkable that two integrals that appear so much different on their faces (monopole versus dipole) in fact give exactly the same result. Of course, numerically that might not be so, since the sequence of calculations is not the same, but outside of extraordinary cases the results are very close.

8.2.8. Dipole Asymptotic Field Configuration

This section discusses the characteristics generated by dipole sources as the distance from the sources gets large, in comparison to the spatial extent of the action potential. The spatial extent of action potentials varies markedly, because the variations in duration and velocity vary according to species, location, and the presence or absence of myelination. Nonetheless, both skeletal muscle and nerve examples show that sometimes the entire spatial extent of the upward deflection of an action potential is accommodated simultaneously along an excitable fiber.

For instance, for skeletal muscle, there might be an action potential 2 msec in duration and a velocity of 3 m/sec, and therefore an action potential extent of 0.6 cm.[11] Many muscles are this long, or longer. In the last section we saw that the source associated with a spatial action potential can be described, approximately, by two equal but oppositely directed dipoles, and at large distances the volume conductor field is therefore that of a quadrupole. As discussed below, quadrupole fields vary with distance as $O(1/r^3)$.

However, there are situations where the field is essentially dipolar and characterized by a dependence on r of $O(1/r^2)$, and this dipole component will tend to dominate any simultaneous quadrupole component that is also present. For example, dipole fields dominate in the case when examining cardiac muscle because, in distinction to nerve and skeletal muscle, the cardiac action potential has a duration of 200–300 msec, so that normally the "activation dipole" and the "recovery dipole" do not exist on a cardiac fiber (or in the whole heart) at the same time. Thus in the chapter on cardiac electrophysiology we shall see that activation and recovery sources and their fields are treated separately because they are temporally separate.

Even for skeletal muscle and nerve, conditions arise that introduce "asymmetry" to the leading and trailing dipole. For example, when a propagating action potential reaches a termination (e.g., when an action potential on a motoneuron reaches the neuromuscular junction), the leading dipole will fade away and, for a short time, the trailing dipole alone will remain. In this interval the source would be described as dipolar. Propagation along curved fibers could also introduce asymmetry, since the leading and trailing dipoles are no longer collinear. A discussion with experimental illustrations can be found in Deupree and Jewett [2].

Another consideration regarding an expected dipole or quadrupole field concerns practical source–field distances. We have seen that a quadrupole far field requires that the dipole separation to source–field distance ratio be small. Using the skeletal muscle parameters above, the dipole separation would be, say, 0.3 cm, and hence the far field, if at a distance which is, say, ten times larger, would start at $\rho = 3$ cm. But if measurements are made within a muscle itself or even at the surface of an extremity, the dipole fields will not have fully canceled and the dominant dipole contribution would still be observed.

8.2.9. Quadrupole Source Density

Many readers will want to skip over this material on quadrupoles on first reading, as many people consider it to be a theoretical embellishment that moves too far beyond intuition or application to justify the time required for its understanding. Nonetheless, we include a quadrupole presentation here as quadrupole concepts are straightforward extensions of concepts developed already. For a reader willing and able to deal with a little more mathematical complexity, quadrupole concepts offer the reward of a different and more unified understanding of the nature of the extracellular fields of nerves and muscle, and how they change with distance from their sources.

Our discussion of quadrupolesbegins by noting that equation (8.35) may be integrated by parts once again. The result is

$$\Phi_e = -\frac{a^2\sigma_i}{4\sigma_e}\Bigg|_{-\infty}^{\infty} V_m \frac{d(1/r)}{dx} + \int V_m \frac{d^2(1/r)}{dx^2} dx. \tag{8.49}$$

The integrated part drops out, since the functions and their derivatives are zero outside the region occupied by the action potential which is of finite extent. Thus

$$\Phi_e = \frac{a^2\sigma_i}{4\sigma_e} \int V_m \frac{d^2(1/r)}{dx^2} dx \tag{8.50}$$

This expression can be interpreted if we first consider the field from two unit dipoles oriented along x but in opposite directions and separated by a differential amount. If they were both placed at the origin, they would cancel and produce no field. But if the one pointing in the positive x direction (labeled as plus) is then moved a small distance dx along $+x$, there will be a residual field (incomplete cancellation).

This field is simply the change in the positive dipole field arising from its displacement by dx. The change can be expressed mathematically as the directional derivative of the dipole field expression with respect to source coordinate x times dx. The result is

$$\Phi_q = \frac{\partial \Phi_d}{\partial x} dx = \frac{1}{4\pi\sigma_e} \frac{\partial}{\partial x}\left(\frac{d(1/r)}{dx}\right) dx = \frac{1}{4\pi\sigma_e} \frac{d^2(1/r)}{dx^2} dx \tag{8.51}$$

The field, Φ_q, is a *quadrupole field* arising from the equal and opposite axial dipoles. The quadrupole magnitude is the product of the dipole strength times the separation. It should be evaluated in the limit that the dipole magnitude becomes infinite as the separation goes to zero such that the product remains finite; this would place the quadrupole source at the origin.

If we let $q(x)$ represent an axial quadrupole source density, then from (8.50) and (8.51)

$$q(x) = \pi a^2 \sigma_i V_m(x) \tag{8.52}$$

Because $V_m(x)$ is ordinarily monophasic, then from (8.50) we recognize that Φ_e is a summation of elementary quadrupole contributions whose coefficient, $V_m(x)dx$, is all of one sign. For increasing values of r, the direction from each quadrupole element to the field point will be increasingly similar and we will have, asymptotically, a lumped quadrupole source whose magnitude is $\pi a^2 \sigma_i \int V_m(x) dx$.

Lumped quadrupolar source

In (8.47) and (8.48) we showed how the positive and negative dipole distributions can each be approximated by a single (lumped) dipole. From (8.47) and (8.48), we note that each equivalent dipole is of equal magnitude but has opposite signs (oppositely oriented).

If each lumped dipole were located at the "center of gravity" of its respective distributed dipole moment densities, then their fields would best approximate the true field (the field of the distributed source). Intuitively, one expects this lumped approximation to improve for field points at increasing distances from the active region, since the distance from each source element to the field becomes essentially equal and the spatial distribution of dipole source elements no longer influences the summation.

The above source description of two equal and opposite axial dipoles corresponds to the axial quadrupole that we described earlier. The quadrupole strength is given by the product of dipole magnitude and the separation of the two component dipoles. The lumped dipole approximates the distributed source, an approximation that improves for increasing source–field distance. Consequently, the lumped quadrupole description is also an approximation that is improved asymptotically for increasing the source–field distance; it not only depends on the extent of each dipole distribution but also on the separation of the two distributions.

Suppose the waveform of an action potential were rectangular with a magnitude equal to the peak value of V_m, as shown in Figure 8.8. Then the (exact) distributed source would consist of a negatively oriented lumped dipole at the activation site and a positively oriented dipole (of equal magnitude) at the recovery site. That these dipoles are discrete is verified by using (8.42), which shows the density function to be zero everywhere except at the two discontinuities in V_m, where they are infinite.

Accordingly, the source is discrete. While the density function is infinite the total source is finite (the function is integrable). In fact, the lumped dipole magnitudes are evaluated in (8.47) and (8.48). Thus an approximate lumped dipole representation for Figure 8.8 can also be regarded as an exact solution for an approximating action potential of rectangular shape (with initiation of activation and recovery corresponding to the "center of gravity" position noted above).

Quadrupole approximation

The extracellular potential at large distances from a fiber carrying an action potential is, asymptotically, the stereotyped waveform corresponding to an axial quadrupole source. This conclusion is reached in several ways, including recognition that the source is approximately that of two opposed dipoles of equal magnitude.

The dipoles are separated by a finite distance, so if the ratio of dipole separation to source–field distance is small a quadrupole field will result, because dipole cancellation is almost complete. Thus for field points at distances which are large compared to dipole separation, the resulting field is essentially that of a quadrupole.

An expression for generating a true (mathematical) quadrupole field was derived in (8.51). Also, an algebraic expression for an axial quadrupole can be found by summing expressions for each of the two contributing dipole fields. We choose the axial dipoles to lie along x equidistant from the origin by $x_\Delta = dx/2$.

The geometry is described in Figure 8.10. We define p through $\rho^2 = y^2+z^2$ and consequently, if the field point is at (x, y, z), then its radial distance from the coordinate origin is

$$r = \sqrt{\rho^2 + x^2} \tag{8.53}$$

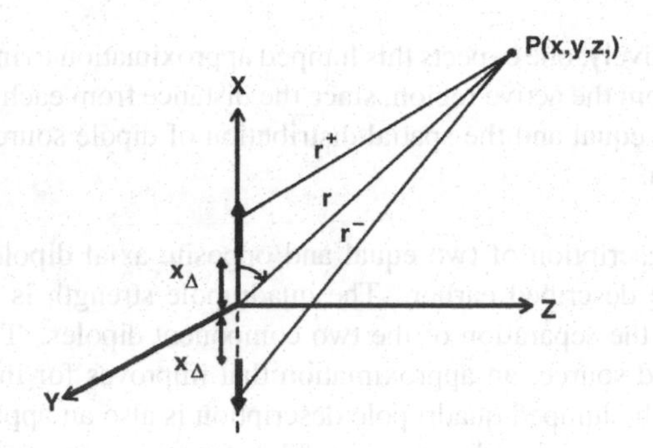

Figure 8.10. Two equal and opposite dipoles in the x direction are located symmetrical about the origin. Their separation, $2x_\Delta = dx$, is very small compared to the distance to the field point. The polar angle with the x axis is θ.

We require

$$\frac{\partial^2(1/r)}{\partial x^2} = \left(\frac{3x^2 - r^2}{r^5}\right) \tag{8.54}$$

and, using (8.51), evaluate the quadrupole field at point P:

$$\Phi_p = \frac{1dx}{4\pi\sigma_e}\left(\frac{3\cos^2\theta - 1}{r^3}\right) = \frac{\mathcal{Q}_d}{4\pi\sigma_e}\left(\frac{3\cos^2\theta - 1}{r^3}\right) \tag{8.55}$$

In (8.55), the quadrupole strength, $\mathcal{Q}_d$, equals the dipole strength (here unity) and separation (dx) while $\theta = \arccos(x/r)$. This waveform can be fully explored in the xz plane in view of rotational symmetry about the x axis; it is seen to be symmetrical and triphasic as described through (8.55). The field has a negative peak at $x = 0$ and positive peaks at $x = \pm\sqrt{1.5}\,z$.

Note, therefore, that the separation of positive peaks increases linearly with the radial (in cylindrical coordinates) distance of the field point. Using these values one can confirm that the ratio of positive to negative peak amplitudes is 0.8.

Using the triangle action potential approximation, each lumped dipole strength equals $(v_{tm})_{\max}$. Since the quadrupole strength depends on both the dipole magnitude and dipole separation, and taking the lumped dipoles at the center of their distribution, the results equals

$$\mathcal{Q} = V_{\text{pp}}\left(\frac{D_1 + D_2}{2}\right) \tag{8.56}$$

8.3. POTENTIALS FROM A CELL

The material presented so far in this chapter has focused on finding the extracellular potentials from a single cylindrical fiber in a large volume conductor. In this section we "erase the board" and start again. This time the goal is to examine the potentials, especially the extracellular potentials, generated by a single active cell. The cell may have any cell shape or size.

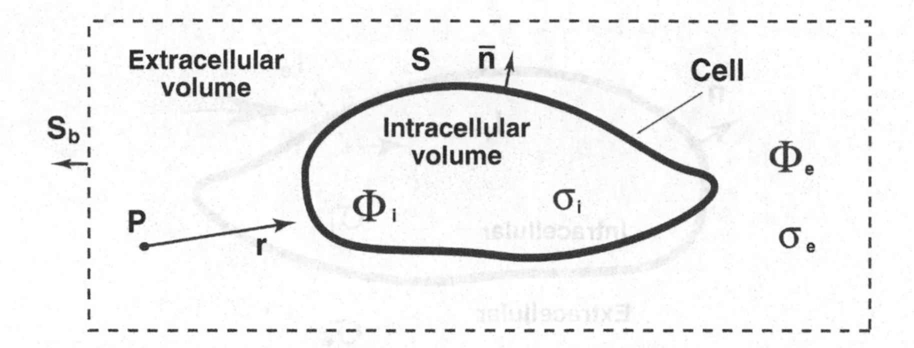

Figure 8.11. An Excitable Cell of Arbitrary Shape. The intracellular and extracellular regions are assumed to have constant conductivities σ_i and σ_e, respectively. The potentials inside and outside the cell are Φ_i and Φ_e, which vary with position and over time. The cell membrane (solid line) is treated as an interface (i.e., as a boundary but having zero thickness). The extracellular volume extends out to bounding surface S_b.

Such a cell is sketched in Figure 8.11. Here the cell has an irregular shape, with intracellular conductivity σ_i and extracellular conductivity σ_e. The extracellular volume is large, ultimately limited by bounding surface S_b, which here is assumed to be a long distance away. The goal is to find the potential generated by the cell at point P, a point outside the cell but within the extracellular volume. In the cell an action potential may have been evoked and may be propagating over the membrane surface, S. Consequently, V_m may vary from one point to another over the cell surface.

8.3.1. The Membrane as Primary Source

The cell's membrane is thin in comparison to other dimensions of the cell (a membrane thickness of roughly 5 nanometers as compared to a diameter of 10,000 nanometers or more). We shall consider the membrane to be of zero thickness and of high resistance, as a volume of small magnitude. Not much current flows within the membrane, but at some points along the cell surface there is intracellular current crossing the cell membrane into the extracellular volume, or vice versa.

Continuity of normal current

Currents associated with such a crossing are pictured in Figure 8.12. As shown diagrammatically in the figure, conservation of current requires that the component of the intracellular current normal (perpendicular) to the membrane surface on the intracellular side be equal to the component to the current normal to the membrane surface on the extracellular side, i.e.,

$$\overline{I}_i \cdot \overline{n} = \overline{I}_e \cdot \overline{n} \tag{8.57}$$

where $\overline{n}$ is a unit vector that is normal to the point on the membrane surface, and $\overline{I}_i$ and $\overline{I}_e$ are intracellular and extracellular current vectors. Note that $\overline{n}$ is the same unit vector, pointing outward, on both the left and right side of the equation.

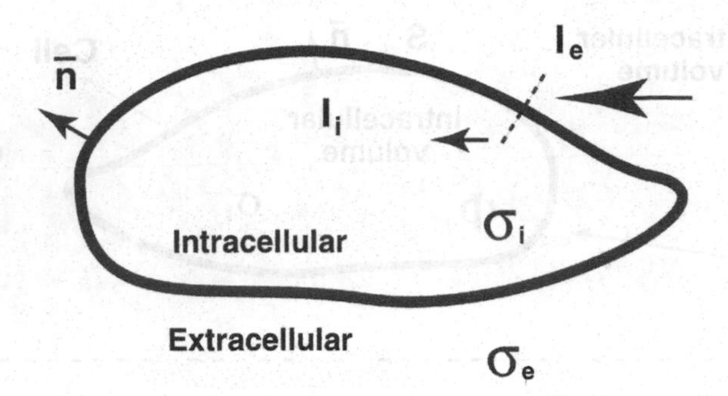

Figure 8.12. Current Crossing the Membrane. The intracellular and extracellular regions are assumed uniform with conductivities σ_i and σ_e, respectively. The cell membrane (solid line) is treated as an interface (i.e., it has zero thickness). By convention, unit vector $\overline{n}$ is normal to the surface and points outward. Only a portion of I_e (its normal component) crosses the membrane to become I_i.

To express this condition in terms of potentials, one recalls that the current is proportional to the gradient of the potential, so that the condition becomes

$$\sigma_i \nabla \Phi_i \cdot \overline{n} = \sigma_e \nabla \Phi_e \cdot \overline{n} \tag{8.58}$$

where Φ_i and Φ_e are the intracellular and extracellular potentials, respectively just inside and outside the membrane, while n is the outward surface normal.

Nonzero V_m

By definition of V_m, we have a second membrane boundary condition, namely,

$$V_m = \Phi_i - \Phi_e = V_m \neq 0 \tag{8.59}$$

That is, V_m is not zero in general, though it varies with position on the membrane, across a physiological range, and with time. Most of the time the transmembrane potential is not zero, i.e., usually

$$v_m = \phi_i - \phi_e = V_m \neq 0 \tag{8.60}$$

That is, we are not assuming that the voltage variables have their baseline values.

Primary sources

In examining whether or not Laplace's equation holds in different locations one realizes that in some locations Laplace's equation holds, but not others. In particular

- In the intracellular volume, there are many small charged ions that move under the influence of the electric fields within the cell, but have no net movement otherwise. Thus Laplace's equation holds in this region.
- In the extracellular volume, there again are many small charged ions that move under the influence of extracellular fields, but have no net movement otherwise. Extracellular ionic

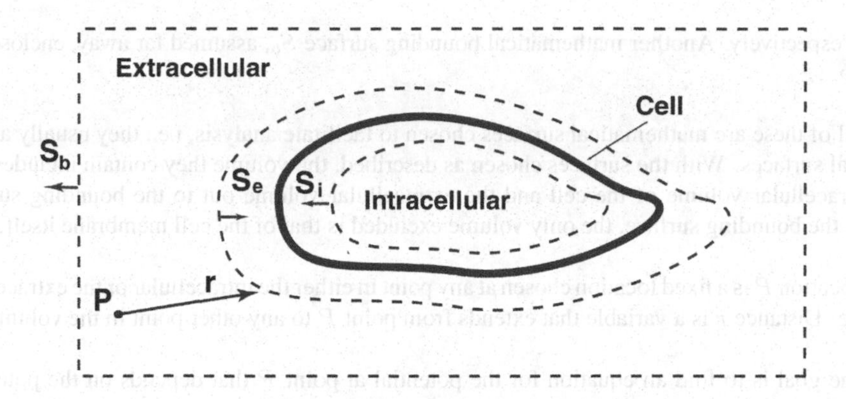

Figure 8.13. Green's theorem surfaces around cell.

concentrations are different than those of the intracellular volume, so the conductivity is different, but we assume the conductivities can be assumed constant in each region.

- Within the membrane, charges may move forcefully because of concentration differences across the membrane, as well as in response to electric fields across the membrane. Because the concentration differences play a major role in charge movement, and concentration differences are a non-electrical effect, Laplace's equation cannot be expected to hold in this region. Rather, the diffusion-based effects drive the currents throughout the extracellular and intracellular volumes, so they are the *primary sources* for the electrical currents throughout the intracellular and extracellular volumes.

8.3.2. Solution for the Cell's Potential

We now wish to find the extracellular potential from the cell by exploiting the power of Green's second identity as applied to this problem's geometry. Recall from Chapter 1 that Green's second identity is

$$\int_U (a\nabla^2 b - b\nabla^2 a)\, du = \oint_S (a\nabla b - b\nabla a) \cdot d\vec{S} \tag{8.61}$$

In (8.61) note that the symbol U is used for volume, rather than the conventional V, so as to avoid confusion with voltages.

The power of (8.61) is that one can choose any scalar functions of position for a and b (excepting only a few pathological cases) and (8.61) still holds.

Another remarkable aspect is that the surface S surrounding the volume can be quite complicated. In particular, we choose a surface that has three parts, as shown in Figure 8.13.

Mathematical surfaces are placed just inside and just outside the physical cell surface, as shown by the dotted lines labeled S_i and S_e. A gap is shown between the cell surface (heavier solid line) and mathematical surfaces S_i and S_e, but the gap is for the purpose of illustration only, as S_i and S_e are intended to be adjacent to the cell surface on the intracellular and extracellular sides, respectively. Another mathematical bounding surface S_b, assumed far away, encloses the volume.

All of these are mathematical surfaces chosen to facilitate analysis, i.e., they usually are not physical surfaces. With the surfaces chosen as described, the volume they contain includes both the intracellular volume of the cell and the extracellular volume out to the bounding surface. Within the bounding surface, the only volume excluded is that of the cell membrane itself.

Location P is a fixed location chosen at any point in either the intracellular or the extracellular volume. Distance r is a variable that extends from point P to any other point in the volume.

The goal is to find an equation for the potential at point P that depends on the potentials across the cell surface, the conductivities, and the geometry. With the advantage of having done this problem before we make some informed choices for a and b, namely,

$$a = 1/r \qquad b = \sigma\Phi \tag{8.62}$$

In (8.62) distance r extends from point P to any surface point, and Φ is the electric potential. Note that conductivity σ is a variable that takes on the value σ_i for points inside the cell or σ_e for points outside.

We also get to choose the surface enclosing the volume. We choose as the surface the combination of bounding surface S_b and surfaces S_i and S_e just inside and outside a cell membrane, respectively. Thus the associated volume is all the volume inside the surrounding boundary, excluding only the cell membrane itself.

Thus, using the assignments (8.62) in Green's identity (8.61), and defining as $S = S_b + S_i + S_e$, we have

$$\int_U \left[\frac{1}{r}\nabla^2(\sigma\Phi) - \sigma\Phi\nabla^2(\frac{1}{r})\right] du = \oint_S \left[\frac{1}{r}\nabla(\sigma\Phi) - \sigma\Phi\nabla(\frac{1}{r})\right] \cdot \vec{dS} \tag{8.63}$$

We will now consider the individual terms that are present in (8.63) and see that many of them are special in one way or another. Consider first the terms on the left hand side of the equation. First,

$$\int_U \left[\frac{1}{r}\nabla^2(\sigma\Phi)\right] du = 0 \tag{8.64}$$

This integral equals zero because $\nabla^2(\sigma\Phi) = \sigma\nabla^2(\Phi) = 0$ at all points in both the intracellular and extracellular volumes. Conductivity σ can be factored because it is a constant within either the intracellular volume or the extracellular volume, and there are no differential elements du where σ makes a transition.

Second, consider the other term on the left hand side of (8.63). This term is nonzero at only a single point, namely,

$$\int_U -\sigma\Phi\nabla^2(\frac{1}{r})\, du = 4\pi\sigma_e\Phi_e(P) \tag{8.65}$$

That is, $\nabla^2(1/r) = 0$ everywhere except at point P, where r becomes zero. There the integral equals -4π and $\sigma = \sigma_i$ or $\sigma = \sigma_e$ according to whether P is intracellular or extracellular (as discussed in chapter 1). Also, at this point variable $\Phi = \Phi_e(P)$. Thus one reaches the result shown.

Now consider the terms on the right-hand side of (8.63). First, we assume that

$$\oint_{S_b} \left[\frac{1}{r} \nabla(\sigma\Phi) - \sigma\Phi\nabla(\frac{1}{r}) \right] \cdot \vec{dS} = 0 \tag{8.66}$$

because both Φ and its gradient become smaller with distance, and in both cases there is an increasing divisor as r grows larger. That is, we assume the bounding surface is too far away to have any effect on the result. It is simpler to make this assumption, which is a good one for many situations, but a solution still can be found with a closed boundary.

Second, consider the terms involving the gradient of potential, as integrated over S_i and S_e:

$$\oint_{S_i+S_e} \left[\frac{1}{r} \nabla(\sigma\Phi) \right] \cdot \vec{dS} = 0 \tag{8.67}$$

These integrals sum to zero because of the continuity of current condition given in the section above. In this regard, note that every point on the cell surface is, geometrically, virtually a point on surfaces S_i and S_e. The same current passes through both surfaces, so at each point their sum is zero, because the surface vectors point in opposite directions.

Thus the remaining terms result in

$$4\pi\sigma\Phi_e(P) = \int_{S_i} -\sigma_i\Phi_i\nabla(\frac{1}{r}) \cdot \vec{dS} + \int_{S_e} -\sigma_e\Phi_e\nabla(\frac{1}{r}) \cdot \vec{dS}_e \tag{8.68}$$

Solution as a surface integral

Using the relationship $d\vec{S}_e = -d\vec{S}_i$, these two integrals can be merged, with the resulting surface simply referred to as S, so that

$$\Phi_e(P) = \frac{1}{4\pi\sigma} \oint_S (\sigma_e\Phi_e - \sigma_i\Phi_i)\nabla(\frac{1}{r}) \cdot \vec{dS} \tag{8.69}$$

where σ is equal to σ_i or σ_e, depending on the location of point P.

Equation (8.69) is the result we were trying to obtain, as it gives the extracellular potential at an arbitrary point outside the cell in terms of the potentials just inside and just outside the cell, and the related conductivities.

Thus the potential at P from a cell with boundary S also can be written as

$$\Phi_e(P) = \frac{1}{4\pi\sigma} \oint_S (\sigma_i\Phi_i - \sigma_e\Phi_e) \frac{\vec{a}_r \cdot \vec{a}_n}{r^2} dS \tag{8.70}$$

Equation (8.70) applies to any shape cell and identifies the source as a double layer of strength $(\sigma_i\Phi_i - \sigma_e\Phi_e)\overline{a}_n$ in the outward normal direction.

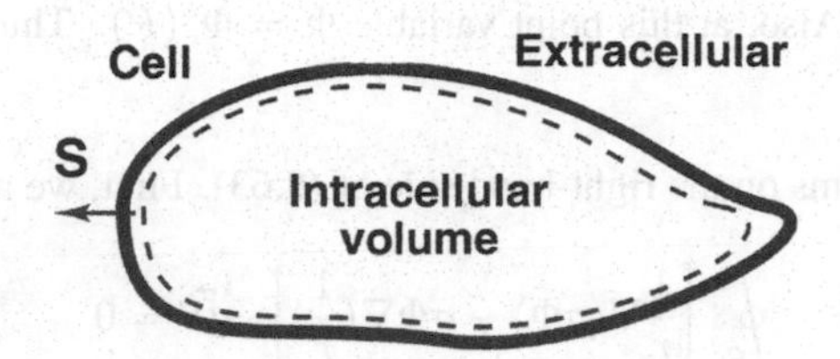

Figure 8.14. Gauss Surface around Cell. The dotted lines just inside the cell membrane draw the Gaussian surface. The small separation between the Gaussian surface and the cell membrane (dark solid line) is simply for illustration. An element of this surface is $\vec{dS}$, an outwardly pointing vector. Volume U is the intracellular volume of the cell, and the volume within the Gaussian surface.

It differs from the equivalent sources discussed earlier in this chapter in that it generates fields intracellularly as well as extracellularly. Because of the variable σ coefficient, it is also an equivalent source. There are many fewer approximations in its derivation, essentially that the membrane thickness is ignored. If (8.93) is applied to a circular cylindrical cell, making several approximations would lead to the earlier expressions [5].

The equation for $\Phi_e(P)$ also can be expressed in terms of the solid angle as

$$\Phi_e(P) = \frac{1}{4\pi\sigma} \oint_S (\sigma_i \Phi_i - \sigma_e \Phi_e)\, d\Omega \tag{8.71}$$

where $d\Omega$ is the increment of solid angle of the surface, as seen from location P.

Solution as a volume integral

Remarkably, the solution for the potential at P can be transformed from the form of (8.69), a surface integral, into a volume integral that has an interesting interpretation. To do this requires, first, defining the surface function $(\sigma_i \Phi_i - \sigma_e \Phi_e)$ on the cylindrical volume.

We assume that $(\sigma_i \Phi_i - \sigma_e \Phi_e)$ is a function of z only.

Assuming such axial symmetry one can take $(\sigma_i \Phi_i - \sigma_e \Phi_e)$ as uniform through any cross-section, and otherwise a function only of z within the fiber. This, of course, maintains its correct value at the surface. Then the transformation that we want is straightforward and as follows.

We define vector function $\overline{A}$ as giving the excitation along z:

$$\overline{A} = (\sigma_e \Phi_e - \sigma_i \Phi_i)\nabla(\frac{1}{r}) \tag{8.72}$$

If cell excitation is along a single axis, such as the x axis, we can extend the definition through the cell volume by giving the function the same value in the interior as on its surface, e.g., constant within cross-sectional planes. (See also details in Plonsey [4].)

With this extension, (8.69) can be rewritten as

$$\Phi_e(P) = \frac{1}{4\pi\sigma_e}\oint_S \overline{A}\cdot d\vec{S} \tag{8.73}$$

Recall that when surface S surrounds volume u then by Gauss's Theorem (Chapter 1)

$$\oint_S \overline{A}\cdot d\vec{S} = \int_u \nabla\cdot\overline{A}\,du \tag{8.74}$$

Thus,

$$\Phi_e(P) = \frac{1}{4\pi\sigma_e}\int_u \nabla\cdot\overline{A}\,du \tag{8.75}$$

Recall that for a scalar multiplier b and vector function $\overline{B}$, a vector identity (Chapter 1) is

$$\nabla\cdot(b\vec{B}) = \vec{B}\cdot\nabla b + b\nabla\cdot\vec{B} \tag{8.76}$$

To subdivide $\overline{A}$ as defined in (8.72) according to (8.76), one makes the assignments

$$b \equiv (\sigma_e\Phi_e - \sigma_i\Phi_i) \tag{8.77}$$

and

$$\vec{B} \equiv \nabla(\frac{1}{r}) \tag{8.78}$$

Thus, making use of Eqs. (8.75) through these definitions, we have

$$\Phi_e(P) = \frac{1}{4\pi\sigma_e}\int_v \left\{\nabla(\frac{1}{r})\cdot\nabla(\sigma_e\Phi_e - \sigma_i\Phi_i) + (\sigma_e\Phi_e - \sigma_i\Phi_i)\nabla^2(\frac{1}{r})]\right\} du \tag{8.79}$$

In the equation for potentials (8.79), $\nabla^2(1/r) = 0$ everywhere in the volume, because P is extracellular, so r never approaches zero within the cell. Thus

$$\Phi_e(P) = \frac{1}{4\pi\sigma_e}\int_v \left\{\nabla(\frac{1}{r})\cdot\nabla(\sigma_e\Phi_e - \sigma_i\Phi_i)\right\} du \tag{8.80}$$

In (8.80), suppose the variation in the term $(\sigma_e\Phi_e - \sigma_i\Phi_i)$ is along only one axis, the x axis. Then the gradient of the term can be found as its partial derivative of this term with respect to x. Then

$$\Phi_e(P) = \frac{1}{4\pi\sigma_e}\int_U \left[\frac{\partial}{\partial x}(\sigma_i\Phi_i - \sigma_e\Phi_e)\right]\frac{\vec{a}_r\cdot\vec{a}_x}{r^2}\,du \tag{8.81}$$

Equation (8.81) is an expression for the potential at point P that depends on a volume integral. Equation (8.81) identifies an equivalent dipole moment volume density source, $\overline{\tau}$, which fills the fiber. It is given by

$$\overline{\tau} = \frac{\partial}{\partial z}(\sigma_i\Phi_i - \sigma_e\Phi_e)\overline{a}_z \tag{8.82}$$

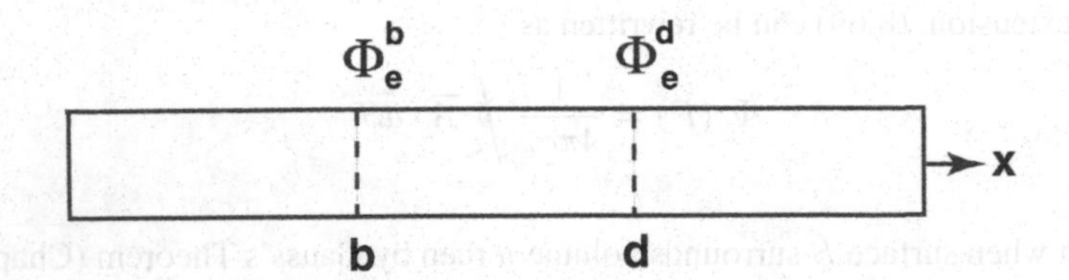

Figure 8.15. Cylindrical Fiber with Cross-Sectional Lines. The axial coordinate is x, and cross-sectional lines are indicated at $x = b$ and $x = d$. Potentials Φ_i vary with x only. Within the cell Φ_e is defined to have the value of Φ_e just outside the membrane at the same x position.

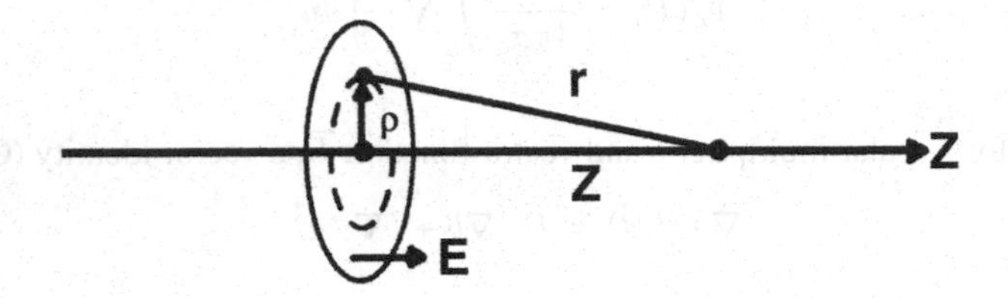

Figure 8.16. The field from a double layer planar circular disc is evaluated along the axis. The Figure shows the geometry where the disc axis is assigned the z direction, while the radial direction on the disc is ρ, so that an annular source area is $2\pi\rho\, d\rho$.

Equation (8.82) is a rigorous expression from which previous expressions may be derived by introducing one or more approximations. In vitro nerve experiments provide an opportunity to examine the model and its approximation [2].

8.3.3. Field Evaluation

The above section shows that the field from a cell that is excited along one axis can be understood with a source term given as a dipole volume density. In the following we examine more carefully the nature of this kind of source term. It is especially useful to examine the field along the axis of a uniform double-layer disc that is planar, and circular with radius a (see Figure 8.16).

Based on the results for analyzing this field given in Chapter 2, the field can be evaluated as:

$$\Phi(z) = \frac{1}{4\pi\sigma_e}\int_{\text{disc}} \frac{\tau\bar{a}_r \cdot d\bar{S}}{r^2} = \frac{\tau}{4\pi\sigma_e}\int_0^a \frac{z 2\pi\rho\, d\rho}{(\rho^2 + z^2)^{3/2}} \tag{8.83}$$

where the dipole strength per unit area (assumed uniform) is $\bar{\tau} = \tau\bar{a}_z$, $d\bar{S}$ is in the z direction, and the dot product introduces z/r.

The integral is elementary, and the result is

$$\Phi(z) = \frac{\tau}{2\sigma_e}\left[-\frac{z}{\sqrt{\rho^2 + z^2}}\right]_0^a = -\frac{\tau}{2\sigma_e}\left(\frac{z}{\sqrt{a^2 + z^2}} - \frac{z}{|z|}\right). \tag{8.84}$$

The first term within the parentheses of (8.84) is a uniform linear function of z, which for small z passes through zero at $z = 0$. The second term equals -1 for negative z and $+1$ for positive z.

Consequently, the change in field strength in crossing the disk from $z = -\epsilon$ to $z = +\epsilon$ is

$$\Phi(+\epsilon) - \Phi(-\epsilon) \equiv \Delta\Phi = \frac{\tau}{\sigma_e} \tag{8.85}$$

because the first term of (8.84) contributes nothing and the second a discontinuity of τ/σ_e.

It is of interest to examine the behavior of $\partial\Phi/\partial z$ in crossing the membrane. We note that in taking the derivative of the expression within the parentheses of (8.84), the second term drops out while the first has a constant value of $1/a$ for $-\epsilon < z < +\epsilon$.

Consequently,

$$\frac{\partial\Phi(+\epsilon)}{\partial z} - \frac{\partial\Phi(-\epsilon)}{\partial z} = 0 \tag{8.86}$$

Thus the field at P from the double-layer disc of strength τ may be written, using (8.83) through (8.86), as

$$\Phi_P = \frac{1}{4\pi\sigma_e}\int_{\text{disc}} \frac{\tau\overline{a}_r \cdot d\overline{S}}{r^2} = \frac{1}{4\pi}\int_{\text{disc}} \frac{\Delta\Phi\overline{a}_r \cdot d\overline{S}}{r^2} \tag{8.87}$$

where $\Delta\Phi$ is given by (8.85). Note that the final field expression for the double-layer source does not depend on the conductivity when expressed in terms of the potential discontinuity as a source expression.

For a double-layer surface of arbitrary shape and varying strength, the behavior of its field in crossing the double layer can be argued as follows.

- At the point of crossing the surface, a very small (double-layer) disc is cut out.
- If small enough, the double-layer strength is constant over its extent.
- If small enough, the shape can be approximated as planar.
- If the contribution of the remainder of the double layer is examined along the aforementioned path, the fields and derivatives are everywhere continuous and well-behaved (since $r \neq 0$).
- So, discontinuities are those introduced by the disc alone.
- Thus (8.85) is satisfied at every surface point, where $\tau(S)$ is evaluated at that point.
- In addition, (8.86) will also be satisfied (i.e., continuity of the normal derivative).
- Thus, if the surface is closed, the field produced by the surface can be written as (8.87), except that the integral is over the entire closed, finite, surface rather than a small disc.

That is,

$$\Phi_p = \frac{1}{4\pi} \oint \frac{\Delta\Phi \, \overline{a}_r \cdot d\overline{S}}{r^2} \tag{8.88}$$

Although derived under these specific circumstances, in fact (this can be confirmed directly) this expression describes the scalar function Φ to be simply a solution of Laplace's equation with the discontinuity $\Delta\Phi$ on S while satisfying (8.86) on S.

For the active cell described in Figure 8.11 it should be clear that both intracellular and extracellular spaces are passive (contain no sources). The active membrane supplies currents into those spaces, and the source of this field can only come from the membrane.

Suppose the membrane source was a double layer lying in the membrane. Then the generated field would necessarily satisfy (8.85) and (8.86).

If we compare (8.85) with (8.60), we see that we could find τ to satisfy (8.60). However, (8.86) does not satisfy (8.58) except in the special case that $\sigma_i = \sigma_e$ (which is unusual physiologically).

We can circumvent this difficulty by defining a related scalar function for which a double-layer source, alone, is appropriate. This function is

$$\psi = \sigma\Phi \tag{8.89}$$

where Φ is the scalar potential generated by the active cell and σ is the intracellular or extracellular conductivity. Since σ is piecewise constant and, necessarily, $\nabla^2\Phi = 0$, then ψ also satisfies Laplace's equation. In crossing the membrane of the active cell ψ must satisfy

$$\psi_i - \psi_e = \sigma_i\Phi_i - \sigma_e\Phi_e = \Delta\psi \neq 0 \tag{8.90}$$

This inequality follows, since both Φ and σ are discontinuous across the membrane of an active cell.

The condition on the normal derivative of ψ can easily be found from

$$\frac{\partial\psi_i}{\partial n} - \frac{\partial\psi_e}{\partial n} = \sigma_i\frac{\partial\Phi_i}{\partial n} - \sigma_e\frac{\partial\Phi_e}{\partial n} = 0 \tag{8.91}$$

which follows from continuity of current [as utilized in writing (8.58)]. In view of (8.90) and (8.91), a double-layer source will generate the ψ field.

This field is described by (8.88), which satisfies the boundary conditions (8.85) and (8.86) corresponding to (8.90) and (8.91). Thus

$$\psi_P = \frac{1}{4\pi} \oint \frac{\Delta\psi \, \overline{a}_r \cdot d\overline{S}}{r^2} \tag{8.92}$$

Substituting for $\Delta\psi$ from (8.90) and then replacing ψ from (8.89), we now have

$$\Phi_P = \frac{1}{4\pi\sigma} \oint \frac{(\sigma_i\Phi_i - \sigma_e\Phi_e)\overline{a}_r \cdot d\overline{S}}{r^2} \tag{8.93}$$

where the unsubscripted σ is the conductivity at the field point, P (namely, either σ_i for intracellular or σ_e for extracellular).

The result, utilizing the divergence theorem, is

$$\Phi_p = \frac{1}{4\pi\sigma_e}\int_v \frac{\partial}{\partial z}(\sigma_i\Phi_i - \sigma_e\Phi_e)\frac{\overline{a}_r \cdot \overline{a}_z}{r^2}dv \quad (8.94)$$

where the field point lies extracellularly and hence σ_e must be used in the coefficient of (8.94).

8.3.4. Inhomogeneities—Secondary Sources

Biological tissues give rise to volume conductors that are inhomogeneous (and possibly anisotropic). The torso, being made up of several organs, is a good example. In a first-order treatment one approximates each region (organ) to be uniform, so that the volume conductor is composed of two or more composite regions each of which is homogeneous.

At the interface between regions of different conductivity a boundary condition must be satisfied, namely, that the potential and normal component of current be continuous. If the two adjoining regions are designated by a prime and a double prime (i.e., their conductivities are σ' and σ''), then

$$\Phi' = \Phi' \quad \text{and} \quad \sigma'\frac{\partial\Phi'}{\partial n} = \sigma''\frac{\partial\Phi''}{\partial n} \quad (8.95)$$

These boundary conditions may also be expressed relative to the scalar function ψ, introduced in (8.89). This gives

$$\psi' - \psi'' = \Phi(\sigma' - \sigma'') \neq 0 \quad (8.96)$$

and

$$\sigma'\frac{\partial\Phi'}{\partial n} - \sigma''\frac{\partial\Phi''}{\partial n} = \frac{\partial\psi'}{\partial n} - \frac{\partial\psi''}{\partial n} = 0 \quad (8.97)$$

By applying the results in the previous section, it is seen that the boundary conditions, involving a discontinuity in potential but continuity of the normal derivative, correspond to an equivalent double-layer source at the interface (but now considering all of space to otherwise be homogeneous). The source strength is given by (8.96) and it generates a field

$$\psi_P = \frac{1}{4\pi}\oint \Phi(\sigma' - \sigma'')\frac{\overline{a}_r\, d\overline{S}}{r^2} \quad (8.98)$$

the double-layer source being

$$\Phi(\sigma' - \sigma'')\overline{a}_n \quad (8.99)$$

where $\overline{a}_n$ is a unit vector normal to the interface.

The above equivalent source is considered a *secondary source*, since it arises only when a *primary source* has established a field and current flows across the interface separating the regions of different conductivity.

This view provides a conceptual (and possibly a computational) approach to considering the effect of inhomogeneities. In this approach, one finds the primary source field as if the volume conductor were uniform and infinite and then adds the fields generated by the secondary sources.

Carrying out this concept gives the following contribution from secondary sources:

$$\Phi_P = \frac{1}{4\pi\sigma_P} \sum_i \oint_{S_i} \Phi_i(\sigma_i' - \sigma_i'') \frac{\overline{a}_r \cdot d\overline{S}}{r^2} \tag{8.100}$$

where i denotes the ith surface on which a discontinuity in conductivity occurs. Note that in obtaining (8.100) the conductivity at the field point, *P*, enters the coefficient of the expression.

One should add the homogeneous medium primary source contribution to the field evaluated in (8.100). Φ_i arises from *both* primary and secondary sources.

8.4. NOTES

1. The sources of the EEG are associated with postsynaptic potentials rather than the action potential, but the fundamental source–field relations are the same.
2. In principle "redrawing" the time waveforms in the form of a spatial distribution requires a time waveform from every spatial position. Usually fewer time waveforms are available than that, so the spatial distribution is constructed from a few temporal, together with estimates of the velocity. In Figure 8.2A the waveforms arise from a mathematical description of V_m, so the number of temporal waveforms is not an issue.
3. Actually it is more common to know $V_m(t)$ for a few points, together with some velocity estimates, from which $V_m(x)$ is inferred.
4. Note that $i_m(x)$ does not have to be found repeatedly for multiple field points as long as Φ_i (or V_m) remain the same.
5. The fact that Φ_e is not always near zero is one of the motivations for the more general mathematical approach given in the second section of this chapter.
6. Another advantage of using this terminology is that it links to the mathematical and numerical literature that evaluates properties of equations of this kind.
7. Thus lumped sources come to us as a kind of dinosaur bones from an earlier age, and a computer-oriented reader may wish to skip over them.
8. These values will be different since the potential at the peak is often given relative to a zero reference other than the baseline value.
9. We may loosely refer to $-\partial V_m/\partial x$ as a measure of the axial dipole density, but the coefficient in (8.41) should be included in any quantitative discussion.
10. Note that when I_i is positive the corresponding dipole elements are oriented in the positive *x* direction, even though $\partial V_m/\partial x$ is negative. See also Plonsey R. 1977. Action potential sources and their volume conductor fields. *Proc IEEE* **65**:601–611.
11. Of course, the time required for the gating variables to return to baseline will be longer, perhaps 12 msec, and in that sense the spatial extent of an action potential is greater.

8.5. REFERENCES

1. Deupree DL, Jewett DJ. 1988. Far field potentials due to action potentials traversing curved nerves, reaching cut ends, crossing boundaries between cylindrical volumes. *Electroencephalogr Clin Neurophysiol* **70**:355–362.
2. Plonsey R. 1974. The active fiber in a volume conductor. *IEEE Trans Biomed Eng* **21**:371–381.

3. Plonsey R. 1977. Action potential sources and their volume conductor fields. *Proc IEEE* **65**:601–611.

4. Spach MS, Barr RC, Serwer GA, Kootsey JM, Johnson EA. 1972. Extracellular potentials related to intracellular action potentials in the dog Purkinje system. *Circ Res* **30:**505–519.

5. Spach MS, Barr RC, Serwer GA, Johnson EA, Kootsey JM. 1971. Collision of excitation waves in the dog Purkinje system. *Circ Res* **29:**499-511.

9

Cardiac Electrophysiology

The application of quantitative and engineering methods in electrophysiology is extensive; its early history developed with studies on the heart. As an electrical generator the heart forces electrical currents throughout the whole body volume. Its signal strength is considerably larger than that from other bioelectric sources; body surface signals of a millivolt or more are typical.

This chapter is designed to introduce the reader to topics in cardiac bioelectricity that fall within three categories, as beginning points within what is now a huge field of studies. These categories are:

1. Intercellular communication, by which an organ with a great number of cells produces synchronized heartbeats over a lifetime, including propagation of activation from its site of origin throughout the ventricles.
2. Cardiac cellular models, by which the sequential action of cardiac membranes and groups of cells is determined. In a fundamental way, these models provide the basis from which the properties of the organ as a whole are created, and their response to stimulation is understood.
3. Electrocardiography, the measurement of cardiac events from the body surface, a subject of historical significance and broad usefulness in present application.

We have selected these categories because of their significance, and because the engineering of instruments, methods of analysis, and quantitative models has played an important role in the development of each one.[1]

The topics in this chapter are presented more or less in the order of increasing anatomical size scale. They were not, however, developed in this order historically. Thus they have many points of separation of terminology and style, and gaps in knowledge of how one ties to another.[2]

The relative ease of measurement of cardiac electrical waveforms, and the recognized importance of heart disease as one of the major causes of human morbidity and mortality in the world,

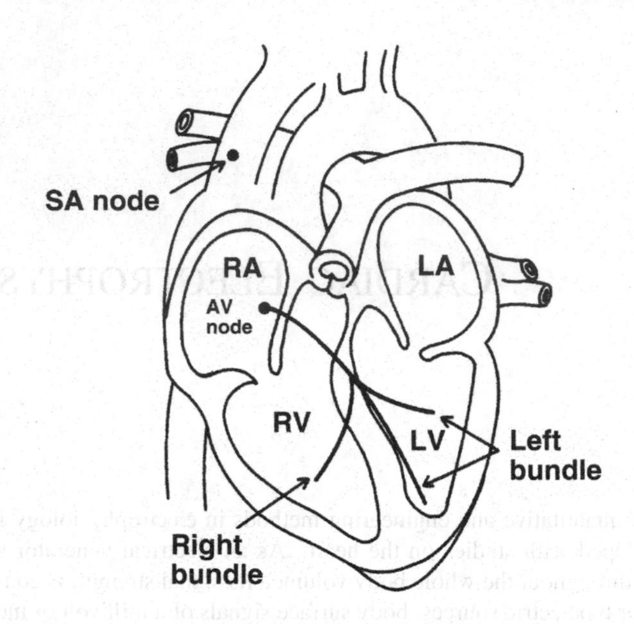

Figure 9.1. Pacemaker (SA node) and Specialized Conducting Regions (AV node, right and left bundles) of the Mammalian Heart.

has attracted the early and sustained attention of physicians, but also of quantitative physiologists, engineers, and mathematicians. The consequences have been increasingly comprehensive and detailed electrophysiological models, as well as analytical tools and medical devices, used for evaluating cardiac function and treating some cardiac conditions. At the same time, much remains unknown, so the medical and engineering understanding and treatment of cardiac electrical rhythms remains a major research enterprise.

9.1. INTERCELLULAR COMMUNICATION

How do millions of cells, each an independent physiological unit, link together to produce a coordinated, durable, and adaptable functioning whole? To address this question requires an overview of the major functional components of the heart and then a more detailed discussion of how cells communicate electrically with their neighbors.

The major cellular components of the heart are the working muscle (cells) of the atria and ventricles, the specialized conduction cells, and the pacemaker cells. The general anatomical structure is shown in Figure 9.1. An analysis of cardiac anatomy from an electrophysiology viewpoint was provided by Spach and Barr [35].

The pacemaker cells, found at the SA node, are characterized by being self-excitatory. That is, in pacemaker cells the transmembrane potential spontaneously increases until threshold is reached and an action potential takes place, as shown in Figure 9.2b. As a consequence, a regular succession of action potentials originates from the SA node. These action potentials lead to a regular series of heart beats.

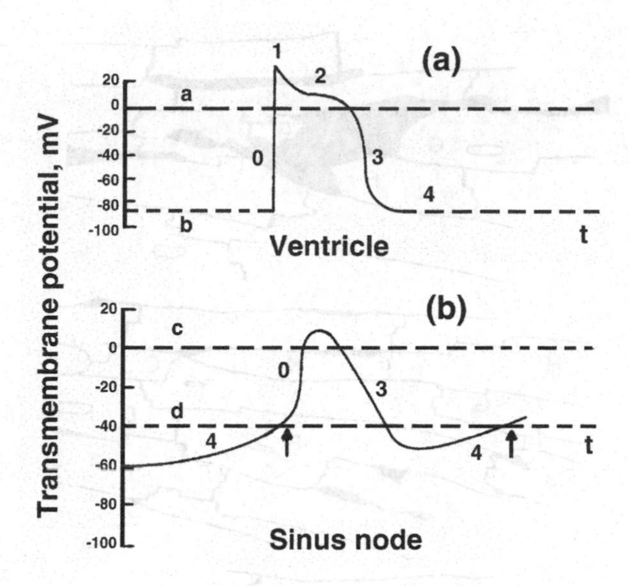

Figure 9.2. (a) Ventricular action potential; (b) pacemaker action potential. Phase 0 corresponds to activation, 1 to rapid recovery, 2 to the *plateau*, 3 recovery, 4 rest or slow depolarization (for pacemaker or *automatic* cells). From Hoffman BF, Cranefield P. 1960. *Electrophysiology of the heart*. New York: McGraw-Hill. Copyright ©1960, McGraw-Hill.

Action potentials initiated by the pacemaker cells excite neighboring cells, and these excite their neighbors, so that a spread of excitation on a cell-to-cell basis takes place in the atria. (A compendium of the action potentials recorded from different structures within the heart was provided by Hoffman and Cranefield [1960].) When this excitation reaches the AV node, specialized conduction cells carry the impulse into the ventricles. Because nonconducting fibrous tissue separates the atria from the ventricles, activity reaches the ventricles only through the AV junction. The specialized tissue of the AV junction conducts very slowly, hence introducing a latency between atrial and ventricular excitation that is useful for the pumping action of the heart.

In the ventricular region the *Purkinje* tissue conducts the impulse rapidly to numerous sites in the right and left ventricle from which further conduction takes place on a cell-to-cell basis within the *working ventricular muscle*. This latter phase of ventricular excitation is characterized, grossly, by a spreading from endocardium to epicardium and from apex to base. The transmembrane action potential of the pacemaker and the ordinary ventricular cell are shown in Figure 9.2. The Purkinje cell action potential is similar to the ventricular action potential except for a sharper initial peak. While the Purkinje tissue behaves in many ways like a nerve axon, it is simply a variety of cardiac muscle as is true of the other tissue discussed here.

That excitation in the working myocardium spreads contiguously from initiating sites to all other points has been known for a long time. In this sense the heart behaves like a single large cell having a complex shape. Because of the observed uniform contiguous activation, the heart is said to behave as a functional *syncytium*. This syncytial behavior raised questions regarding the mechanism of the observed cell-to-cell spread, because of the recognition that each cell of the

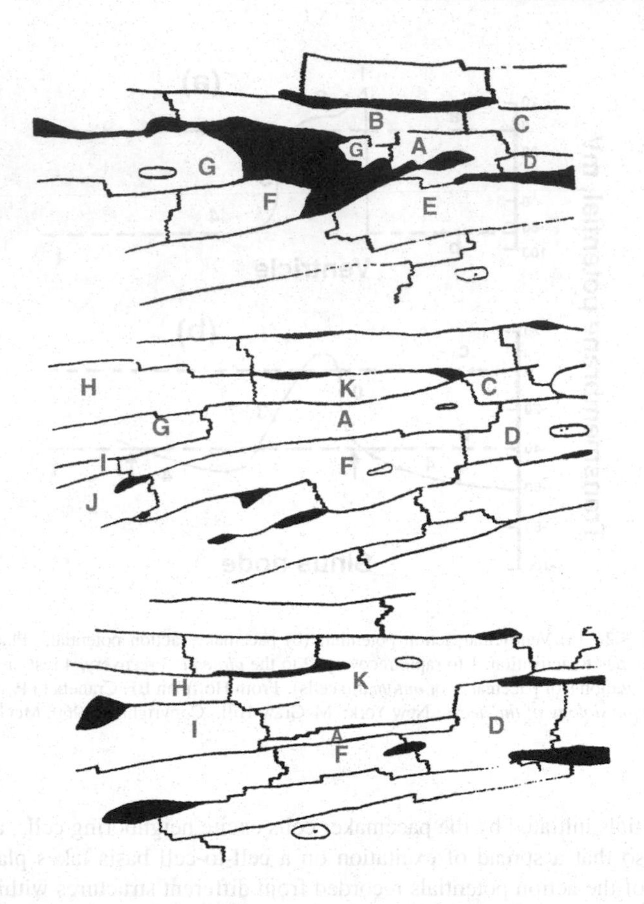

Figure 9.3. Structure of Cardiac Muscle. The Figure shows three camera lucida drawings from a series of 42 consecutive 2-μm-thick plastic sections showing multiplicity of interconnections of the myocytes at intercalated disks. Shaded areas denote prominent interstitial vessels and septae. From top to bottom are shown sections 12, 6, and 22. Myocyte A is followed in its entirety and makes contact at intercalated discs with cells B–K. Hoyt RH, Cohen ML, Saffitz JE. 1989. Distribution and three-dimensional structure of the intercellular junctions in canine myocardium. *Circ Res* **64**:563–574.

heart is surrounded by a plasma membrane. The presence of that membrane, a very good electrical insulator, would seem to preclude electrical current from moving from the interior of one cell directly into the interior of an adjacent one. It is now known that such intracellular connection is accomplished by specialized intercellular junctions. These junctions, which include cell-to-cell electrical pathways, have been the subject of much study.

9.1.1. Gap-Junctional Structure

It is known from electron microscopic studies that cardiac cells are arranged in a brick-like structure. Each cell is somewhat cylindrical in shape and roughly 100 μm long and 15 μm in diameter. Every cell is in contact with several neighboring cells. A two-dimensional projection of cardiac cells (myocytes) that illustrates their interdigitating structure is given in Figure 9.3, and this also identifies the location of intercellular junctions. The opposing cell membranes form

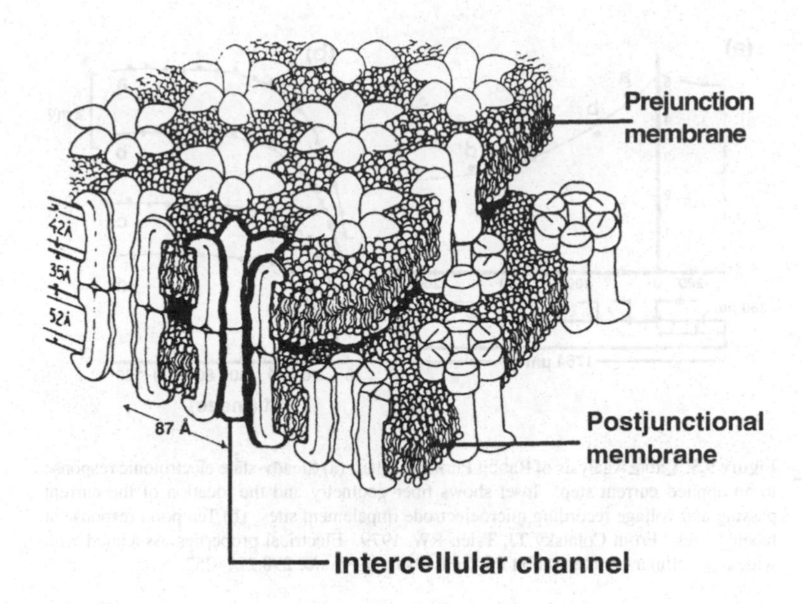

Figure 9.4. Details of the Communicating-Type Intercellular Cardiac Junction (connexon array) are shown. Each unit (connexon) is a protein channel running transverse to the opposing membranes. Connexons from abutting cells align themselves to form structural continuity. The structural detail shown is based on morphometry obtained from x-ray diffraction, electron microscopy, and chemical studies. The gap spacing is given as 35 Å. Plonsey R. 1989. The use of a bidomain model for the study of excitable media. *Lect Math Life Sci* **21**:123–149. From Makowski L, Caspar DLD, Phillips WC, Goodenough DA. 1977. Gap-junctional structures, II: analysis of x-ray diffraction. *J Cell Biol* **74**:629–645. Reproduced from the *Journal of Cell Biology* (1977) **74**:629–645, by copyright permission of the Rockefeller University Press.

what is known as an intercalated disk structure.[3] It is the specialized *gap junction* which provides for cell-to-cell transfer of ions and hence electric current. These currents behave as *local circuit currents* as they flow from an active cell into an adjoining resting cell, via gap junctions, and cause the depolarization of the adjoining resting cells.

The gap junction is characterized by the narrowing of the intercellular space from around 200 to 30 Å. EM sections transverse to the long axis reveal a hexagonal array of cylindrical elements that fill the intercellular space. Each element is called a *connexon* and has a structure somewhat similar to that for a membrane ion channel (as described in Figure 9.4), including a gate. The connexon protein is a hexamer whose six polypeptide subunits surround a core channel. The connexons from the two abutting cells align themselves to be physically continuous: their central channels form a uniform link between the cells (insulated from the extracellular medium). This (aqueous) channel runs from the intracellular space of one cell through the intercellular gap into the intracellular space of the following cell, making for actual cytoplasmic continuity. A drawing of the gap-junctional region which illustrates the hexagonal array of connexons and their morphometry is shown in Figure 9.4.

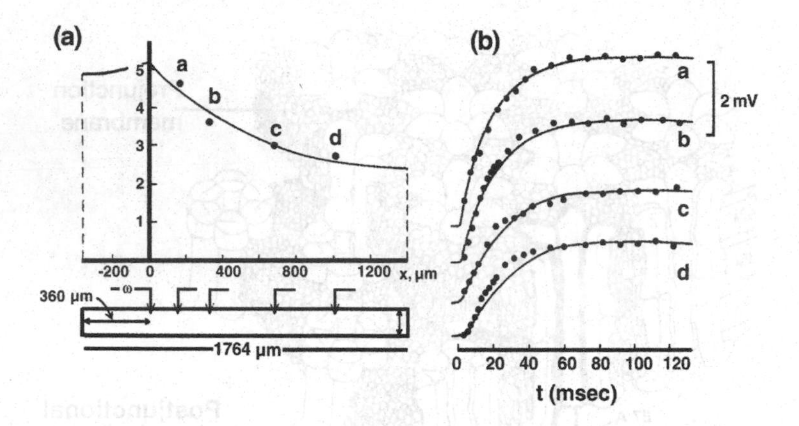

Figure 9.5. Cable Analysis of Rabbit Purkinje Fiber. (a) Steady-state electrotonic response to an applied current step. Inset shows fiber geometry and the location of the current passing and voltage recording microelectrode impalement sites. (b) Temporal response at labeled sites. From Colatsky TJ, Tsien RW. 1979. Electrical properties associated with wide intercellular clefts in rabbit Purkinje fibers. *J Physiol* **290**:227–252.

The intercellular channels are around 20 Å in diameter and have been shown to permit molecules up to 1000 Dalton to pass from cell to cell (fluorescent tracers such as Procion yellow are used to explore this pathway). These same channels explain how current introduced into one cell readily affects the voltage in an adjoining cell. The channels are characterized as aqueous, since they introduce relatively little selectivity as long as the molecule is sufficiently small so as not to be excluded sterically. The junctions are characterized as "low resistance" because the effective resistance is several orders of magnitude less than what would result simply from two plasma membranes butted together. The intercellular resistance is affected by the calcium ion concentration, the pH, longitudinal electric fields, and possibly sodium concentration.

9.1.2. Functional Continuity

Experimental evidence for the electrical continuity between the intracellular space of neighboring cardiac cells is typified by the electrotonic experiment illustrated in Figure 9.5. Here, Colatsky and Tsien [9] injected intracellular current at one end of a rabbit Purkinje fiber and measured intracellular potentials at increasingly distal points. The steady-state potential has an experimentally demonstrated exponential character even though the fiber contains between 2 and 20 separate cells in any cross-section and where individual cells (around 100 μm long) are a small fraction of the length constant of around 1 mm.

The interconnections are essential to the measured continuous character. The behavior is functionally that of a single uniform cell (cable). In fact, the spatial and temporal dependence as shown in Figure 9.5 corresponds closely (solid lines) with that seen in Figure 7.4, which is for a uniform linear cable. So in spite of the discrete multicellular structure of this heart muscle preparation, it behaves, functionally, as a single uniform cell. Such a behavior would not result if the junctional resistance were high since, for increasingly high junction resistance, only the cell into which current is injected would show significant transmembrane current.

From the data in Figure 9.5 we obtain an effective space constant, λ, of approximately 1 mm. The temporal potential curves at several axial positions (*a, b, c, d*) are well represented by a time constant $\tau = 18$ msec. From the dimensions, the authors were able to show, also, that the total effective intracellular resistivity is $R_i = 350\ \Omega$cm.

Chapman and Fry [8] investigated ventricular trabeculae[4] from frog ventricle and also observed linear cable behavior. For their preparation they found $\lambda = 0.328$ mm from measurements similar to those described in Figure 9.5. They also determined a value of τ to fit their step responses measured at five different axial positions and obtained $\tau = 4.15$ msec.

A total average (effective) value of R_i obtained was evaluated as 588 Ωcm. Chapman and Fry independently measured the cytoplasmic resistivity to be 282 Ωcm and consequently inferred that the remaining 306 Ωcm was a contribution to the total intracellular resistivity by the junctional resistance. We note that the junctional resistance contribution is roughly equal to the cytoplasmic one and is consequently half the total. The lumped junctional resistance per cell can be calculated from these data and is

$$R_d = \frac{306 \times 131 \times 10^{-4}}{\pi (7.5 \times 10^{-4})^2} = 2.27\,\mathrm{M}\Omega \tag{9.1}$$

for the estimated cell length of 131 μm and radius of 7.5 μm.

The specific junctional resistance r_d (Ωcm^2) equals the above R_d times the cellular cross-sectional area, so that

$$\begin{aligned} r_d &= R_d \times \pi \times (7.5 \times 10^{-4})^2 \\ &= 306 \times 131 \times 10^{-4} \\ &= 4\,\Omega\mathrm{cm}^2 \end{aligned} \tag{9.2}$$

The result for r_d, expressed as a specific resistance in (9.2), is independent of the actual cross-sectional area. Its size compared to the resting membrane resistance of a cardiac cell (of around 20,000 Ωcm^2) is the basis for describing it as a "low-resistance" junction, though it roughly doubles the effective intracellular resistivity. Consequently, it is not negligible but considerably less than what would be the case with the two aforementioned plasma membranes ($2 \times 20,000 = 40,000\Omega\,\mathrm{cm}^2$).

9.1.3. Gap Junction Resistance

Junctional resistance is never measured directly. Rather, values for junctional resistance are only inferred through an interpretation based on assumed structure. Invariably, this inference involves a number of approximations and uncertainties regarding cardiac morphometry. For example, one makes estimates of the number and extent of intercellular connections, the number of cells, and membrane folding. An approach which has fewer uncertainties utilizes a preparation consisting of a pair of attached myocytes and the use of two patch electrodes. This approach is described in the following paragraphs.

In Chapter 5 we noted the difficulties in the early investigation of cardiac membrane electrophysiology owing to the small size of a cardiac cell (myocyte). Experiments were performed on cell aggregates, in view of their larger size, but such multicellular preparations cannot be readily

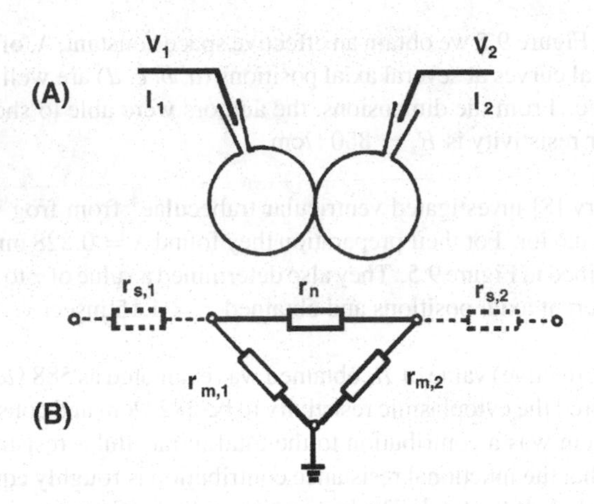

Figure 9.6. Diagram of Experimental Arrangement. Each cell of a cell pair is connected to a voltage-clamp circuit via a patch electrode in the whole-cell configuration. Separate voltages V_1, V_2 can be applied to each cell and the resulting currents, I_1, I_2, measured. (Subscripts 1 and 2 refer to cells 1 and 2, as described in A.) The equivalent circuit is shown in B. The sarcolemmal resistance is denoted by r_{m1}, r_{m2}, the junctional (nexal) resistance by r_n, and the access (pipette) resistance, shown dotted, by r_{s1}, r_{s2}. From Weingart R. 1986. Electrical properties of the nexal membrane studied in rat ventricular pairs. *J Physiol* **370**:267–284.

voltage clamped. A breakthrough came with the availability of patch-clamp electrodes, since this could be applied to individual myocytes.

A patch clamp was applied to a preparation consisting of two attached myocytes. The preparation was obtained through an enzymatic technique for isolating cardiac cells, a technique that yields cell pairs as well as single cells. In the study by Weingart [40] to be described, adult rat hearts were used. Figure 9.6A describes the experimental arrangement. Patch electrodes are shown attached to each cell in the whole-cell configuration.

Each cell is separately voltage clamped. As noted in describing spherical cells (see "Single Spherical Cell" in Chapter 7), each cytoplasmic region is approximately isopotential, so that the sarcolemmal membrane has a uniform transmembrane potential; under a time-invariant voltage clamp this entire membrane can be modeled as a lumped resistance, r_m.

The junctional membrane resistance is designated r_n, while the pipette resistance plus that introduced by the presence of membrane elements lying in the pipette tip (resulting from the whole-cell procedure) are shown as an access resistance, r_s. The resulting circuit is described in Figure 9.6B. Neglecting the access resistances and applying Ohm's law evaluates the two pipette currents I_1 and I_2 as

$$I_1 = \frac{V_1}{r_{m1}} + \frac{V_1 - V_2}{r_n} \tag{9.3}$$

and

$$I_2 = \frac{V_2}{r_{m2}} + \frac{V_2 - V_1}{r_n} \tag{9.4}$$

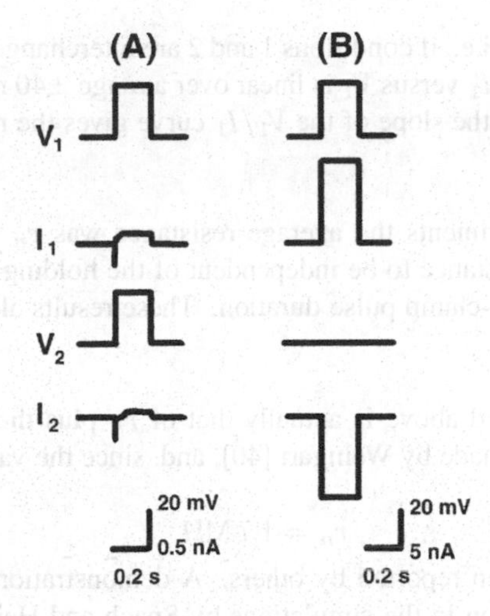

Figure 9.7. Current Flow in a Cardiac Cell Pair under Voltage-Clamp Conditions. (A) Symmetrical pulse application ($V_1 = V_2 = 27$ mV applied for 200 ms). The associated current signals I_1 and I_2 show a time-dependent inward component; this reflects a sarcolemmal current only. (B) asymmetrical pulse application [$V_1 = 27$ mV for 200 ms; as in (A), $V_2 = 0$]. The signals I_1 and I_2 now show large amplitudes and no time dependency. [Note the tenfold increase in scale compared to (A).] The holding potential $V_H = -42$ mV in (A) and (B). From Weingart R. 1986. Electrical properties of the nexal membrane studied in rat ventricular pairs. *J Physiol* **370**:267–284.

where V_1 and V_2 are the voltage-clamped potentials relative to a prior V_H holding potential. Thus currents evaluated in (9.3) and (9.4) are relative to that under the holding condition.

If V_2 is kept at the holding potential ($V_2 = 0$) while V_1 is pulsed, then

$$I_1 = \frac{V_1}{r_{m1}} + \frac{V_1}{r_n} \text{ and } I_2 = -\frac{V_1}{r_n} \tag{9.5}$$

Equation (9.5) evaluates the pipette current flowing into the pulsed cell 1 as the sum of two components (junctional and sarcolemmal), while the current of cell 2 is junctional only.

A pair of experiments is described in Figure 9.7, where A shows equal voltage-clamped pulses of 27 mV (from a −42 mV holding potential).[5] Each pipette current is similar and describes a sarcolemma component only. The holding potential of −42 mV suppresses the fast Na^+ inward current.

What is seen is the inward Ca^{++} followed by an outward K^+ current. Figure 9.7B describes the experiment giving Eq. (9.5), where I_2 is a junctional (nexal) current only. The junctional current is much greater than the sarcolemmal current, so the scale in B differs from A by a factor of ten.

Although I_1 also contains the sarcoplasmic current, its relatively small magnitude is not evident in B where the junctional current component predominates. Further experiments show each cell to behave similarly (i.e., if conditions 1 and 2 are interchanged). If in (9.5) V_1 is varied, data are found that show that I_1 versus V_1 is linear over a range ±40 mV. The linearity identifies r_n as resistive. Additionally, the slope of the V_1/I_1 curve gives the nexal membrane resistance value.

Over a number of experiments the average resistance was $r_n = 3.25$ MΩ. Additional experiments showed this resistance to be independent of the holding potential and independent of the duration of the voltage-clamp pulse duration. These results also held if cell 1 and cell 2 were interchanged.

The value of r_n evaluated above is actually that of r_n plus the access resistance r_{s1}. A correction for the latter was made by Weingart [40], and, since the value $r_{s1} \approx r_n$, a final value of r_n was

$$r_n = 1.7\,\mathrm{M\Omega} \tag{9.6}$$

Values of this order have been reported by others. A demonstration of the importance of the gap-junctional structure is seen in the simulations by Spach and Heidlage [36, 37] based on a realistic 2D cellular structure.

If one assumes the nexal resistance to represent the total end-to-end junctional resistance between two abutting cells and that the cross-sectional area is 7.5 μm × 7.5 μm ≈ 56 μm^2, [6] then a specific resistance of 1.7 × 106 × 56 ×10^{-8} = 0.95 Ωcm^2 is obtained that is somewhat lower than that found in the experiments of Chapman and Fry [8]. The uncertainty in these experimental measurements concerns the relationship between the number of open gap-junctional channels in an in-vivo preparation and the number of (open) channels in the in-vitro preparation achieved through enzymatic dissociation of cardiac cells.

A similar experiment on chick embryo cell pairs, performed by Veenstra and de Haan [39], yielded the junctional current shown in Figure 9.8. It is assumed that only a few junctions contributed to this current, and the opening and closing of a *single channel* is responsible for the observed current pulse of 5 pA magnitude. This result is obtained for $V_1 = -40$ mV and $V_2 = -80$ mV, so that the junctional voltage $V_j = -40$ mV and the single-channel conductance is evaluated as

$$g_j = \frac{5 \times 10^{-12}}{40 \times 10^{-13}} = 125\,\mathrm{pS} \tag{9.7}$$

The conductance obtained in this way is an approximate value. The value corresponds, roughly, to the ohmic conductance of an aqueous pore of length 15 nm, diameter 1 nm, and resistivity $\rho = 100\Omega$cm. If the access resistance is approximated as $2\rho/\pi d$, then with the aforementioned values a channel conductance of 94 pS is found.

If the gap-junctional region of abutting cells contains 62,000 open channels, as estimated by Weingart [40], then the value of junctional resistance using $g_j = 125$ pS is

$$r_j = \frac{1}{62,000 \times 125 \times 10^{-12}} = 0.13\,\mathrm{M\Omega} \tag{9.8}$$

an order of magnitude less than the direct measurement by Weingart. These data have been included here to demonstrate the difficulty in finding consistent values of this extremely important parameter in cardiac electrophysiological studies.

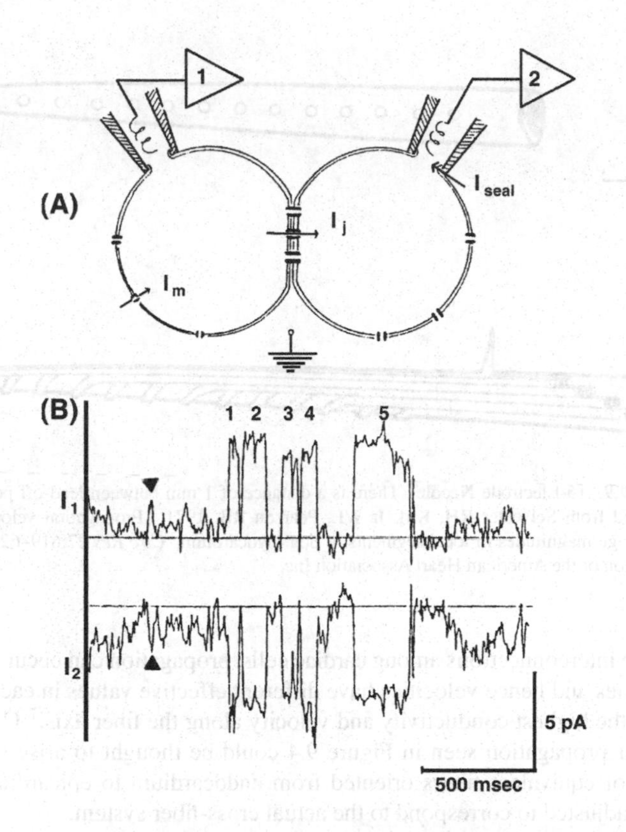

Figure 9.8. Two-Cell Preparation with $V_1 = -40$ mV, $V_2 = -80$ mV. The sarcoplasmic currents were measured separately by applying equal values of V_1 and V_2 and subtracted from the total current with the above values of V_1 and V_2, leaving only the junctional current $I_j = \pm(V_2 - V_1)/r_n$. The five distinct quantal events (numbered) are assumed to result from single channel openings. From Veenstra RD, De Haan RL. 1986. Measurements of single channel currents from cardiac gap junctions. *Science* **233**:972–974. Copyright ©1986, American Association for the Advancement of Science.

9.1.4. Propagation in the Ventricles

The cardiac fibers, grossly, spiral around the heart. Consequently, the fibers are oriented parallel to the endocardium and epicardium, the inner and outer surfaces of the heart wall, though the angle made with a fixed reference changes continuously from the inner to the outer surface. Contraction of fibers so organized achieves a wringing action thought to efficiently squeeze out (pump) blood.

While electrical activation along the fiber direction yields the highest velocity of propagation, the gross activation *wavefront* can be expected to progress in a direction orthogonal to the fiber axis (the direction of difficult or slow conduction). For example, a point stimulus initiates an elliptical wavefront with its long axis along the fiber direction so the broad front of the ellipse moves in the cross-fiber direction. In fact, we have seen that propagation does occur normal to the epicardium, hence transverse to fiber orientation.

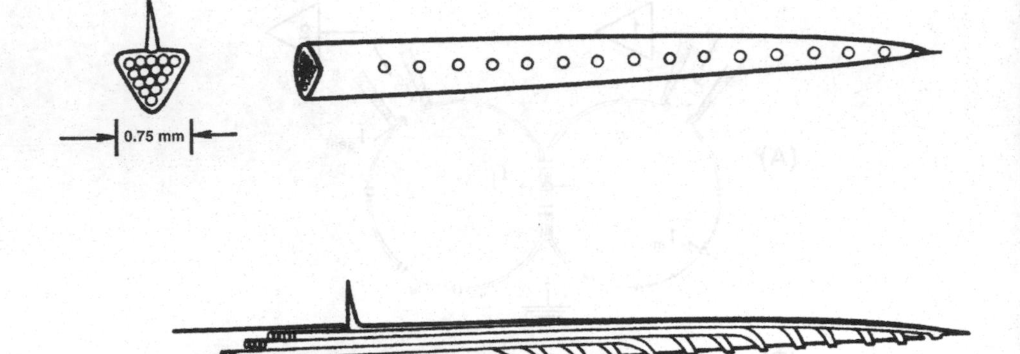

Figure 9.9. 15-Electrode Needle. There is a distance of 1 mm between lead-off points. Reprinted from Selvester RH, Kirk Jr WL, Pearson RB. 1971. Propagation velocities and voltage magnitudes in local segments of dog myocardium. *Circ Res* **27**:619–629, by permission of the American Heart Association Inc.

Because of the interconnections among cardiac cells, propagation can occur in any direction, but the conductivities and hence velocities have different effective values in each direction (i.e., anisotropic), with the highest conductivity and velocity along the fiber axis.[7] Consequently, the uniform cross-fiber propagation seen in Figure 9.4 could be thought to arise from conduction through a system of equivalent fibers oriented from endocardium to epicardium, provided its conductivities are adjusted to correspond to the actual cross-fiber system.

Intramural electrodes

Cardiac intramural electrodes are extracellular macroelectrodes. The electrode size is about 1 mm, much larger than the 1- to 10-μm diameter of a cardiac microelectrode. The larger size is needed to provide the stiffness required for insertion, as well as allow for multiple small wires. Nevertheless, they must be as small as possible, mainly to minimize tissue damage. Normally, the signals derived from these electrodes are interpreted in terms of underlying source behavior rather than as a sample of a field. Both unipolar and bipolar arrangements are used.

Cardiac needle electrodes are used to map the activation isochrones. Each needle (possibly an actual hypodermic needle) is inserted normal to the epicardium and passes completely through the wall (and possibly the septum as well). The needle interior contains a bundle of wires each of which is exteriorized through a small hole in the needle wall. The tip is uninsulated. In this way a linear array of electrodes is formed along the surface of the needle (each electrode being insulated from the needle shank). A needle electrode with a similar structure, but without use of a hypodermic needle, is shown in Figure 9.9 (the Figure was chosen because it illustrates the basic configuration).

When a needle electrode is inserted into the free wall of the left ventricle, roughly ten point electrodes spaced 1 mm apart will lie within the 1 cm of wall thickness. Assuming that an activation source is a uniform double layer, one can estimate the expected waveform along the needle and from these determine a relationship between source and signal. We assume the double layer itself to have an overall thickness of $\approx 1.0\,\mathrm{mm}$.

A calculation of the field arising from such double-layer sources can be facilitated by dividing the (thick) dipole layer into a number of component lamina layers (where the double-layer strength of each lamina can be assumed constant in the lateral direction). For each lamina the potential field can then be found from the solid angle expression of (9.25). This idea is discussed in detail below.

Reference Electrodes: For intramural measurements a second electrode is, of course, required. This reference electrode often is chosen at a location physically removed (but electrically connected) from the sources responsible for the field under examination, such as the left leg. Such a reference is a *remote reference.*

Unipolar and bipolar waveforms

In Figure 9.10 we show a needle electrode and unipolar and bipolar potentials measured with them by van Oosterom and van Dam [38]. These signals are recorded from the free ventricular wall. The bipolar waveforms were obtained from each successive pair of lead-off points and generally show the monophasic waveforms suggested in Figure 9.11.

One also notes the increasing latency in moving from endocardium to epicardium. Since the electrodes are spaced 1 mm apart, it is possible to estimate the outward (phase) velocity from these data. The (spatial) potential profiles are determined at successive instants of time. One notes the expected (though approximate) rapid potential change across the double-layer source.

Ventricular isochrones

If one uses the timing of the bipolar peaks to give a time of excitation to each site where there is an electrode on one of a number of intramural electrodes, and if one then draws lines connecting the sites excited at the same time, then a set of *excitation isochrones* are determined. Such a set is shown in Figure 9.12, as reported by Durrer et al. [12]. These isochrones shown that activation, once started at a point in the tissue, continues contiguously to all surrounding tissue.

The points of earliest activity seen in this Figure depend on the location of terminal endings of the conduction system. These are, typically, in the left-ventricular septum about a third the distance from apex to base. One should also note the regular propagation in the outer wall in which the wavefronts are more or less planar and conduction takes place from endocardium to epicardium.

This work, and that of others, shows how well cardiac intracellular space is electrically interconnected.

9.1.5. Waveforms Arising from Free Wall Activation

Measurement of the electrical waveforms arising from the activation of the free walls of the cardiac ventricles has played a critical role in understanding the way that ventricular muscle functions. Such an understanding remains fundamental to recognizing and interpreting the waveforms observed in present-day clinical procedures, where ventricular waveforms are observed

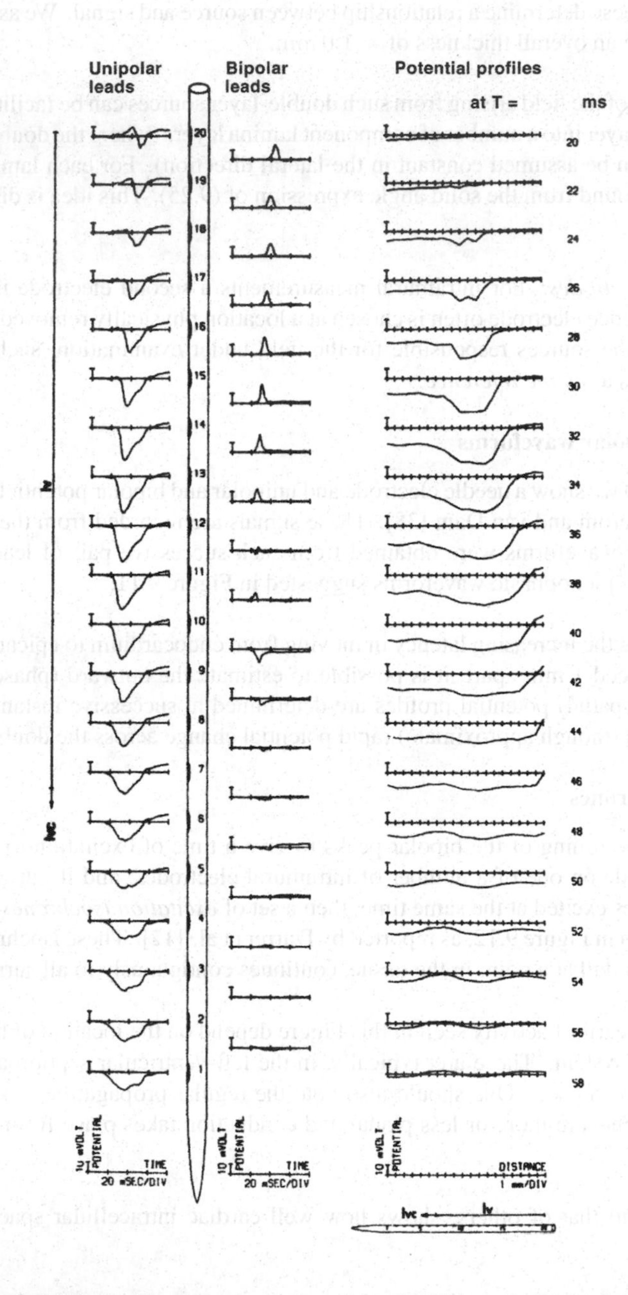

Figure 9.10. Needle Electrode (plunge electrode) with 20 Electrode Points. Associated unipolar and bipolar (adjacent pairs) signals also shown. The profile is the spatial distribution at the time shown at the right. Electrodes 7–20 lie within the ventricular wall. From van Oosterom A, van Dam R. 1976. Potential distribution in the left ventricular wall during depolarization. *Adv Cardiol* **16**:27–31, by permission from S. Karger AG, Basel.

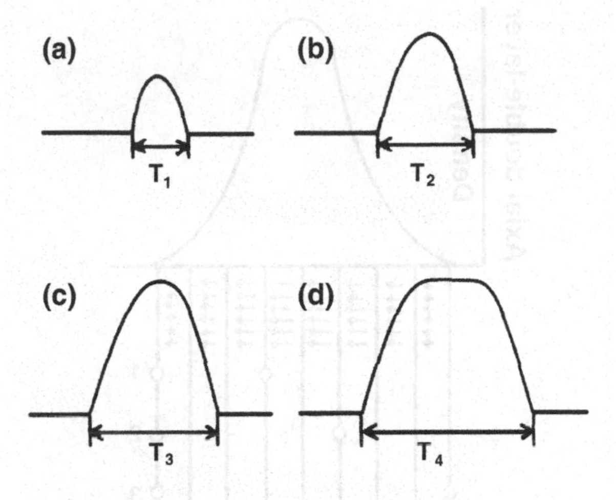

Figure 9.11. Waveforms Measured by Bipolar Needle Electrodes (similar to those described in Figure 9.9) of an activation wave in the left ventricular free wall. In (a) electrode separation corresponds to 1 of Figure 9.13, (b) corresponds to electrode separation 2, (c) corresponds to electrode separation 3, and (d) for condition 4 in Figure 9.13.

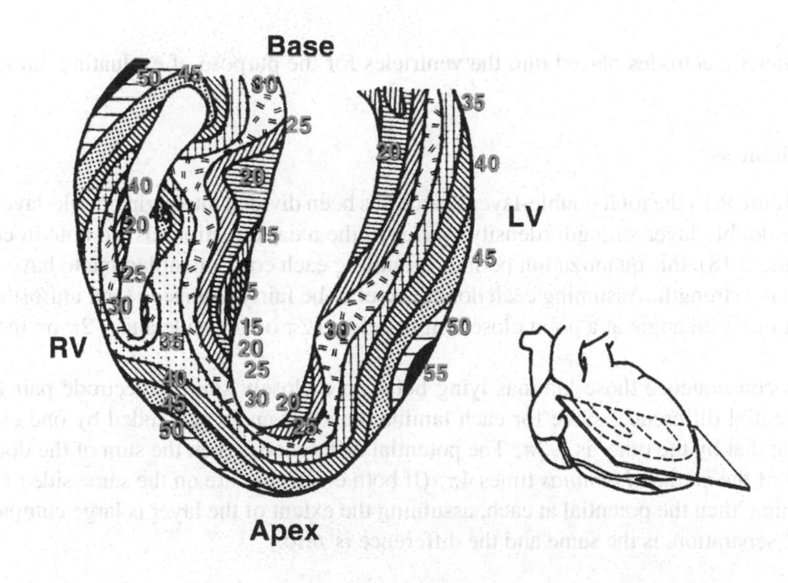

Figure 9.12. Isochronous Lines of Activation of the Human Heart (RV = right ventricle, LV = left ventricle). Values in msec. From Durrer D, et al. 1970. Total excitation of the isolated human heart. *Circulation* **41**:899–912, by permission of the American Heart Association. Redrawn in Liebman J, Plonsey R, Gillette P. eds. 1982. *Pediatric electrocardiography*. Baltimore, MD: Williams and Wilkins.

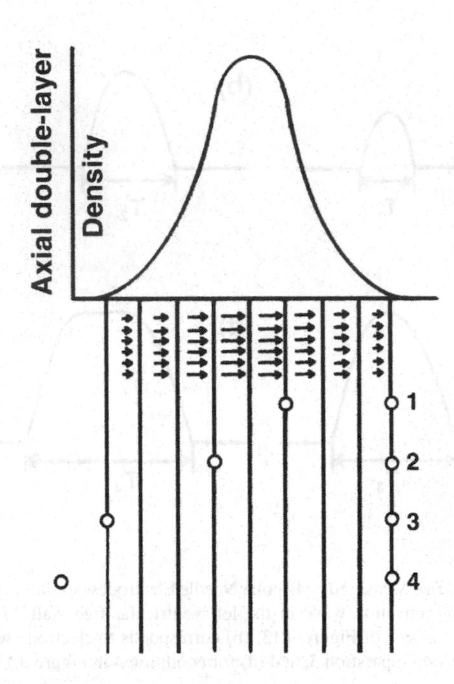

Figure 9.13. Total Activation Wave, subdivided into eight layers (lamina).

from catheter electrodes placed into the ventricles for the purpose of evaluating human cardiac function.

Wave thickness

In Figure 9.13 the total double-layer source has been divided into eight double-layer laminas. Since the double-layer strength (density) varies in the axial direction (as we note in connection with Figure 9.18), this quantization permits assuming each component lamina to have a uniform and fixed net strength. Assuming each double layer to be fairly extensive (and uniform) laterally results in the solid angle at a point close to it to being $+2\pi$ on one side and -2π on the other.

As a consequence those laminas lying between a closely spaced electrode pair contribute a net potential difference, since for each lamina the solid angle subtended by one electrode is $+2\pi$ while that by the other is -2π. The potential difference is then the sum of the double-layer strengths of the *included laminas* times 4π. (If both electrodes are on the same side of a double-layer lamina, then the potential at each, assuming the extent of the layer is large compared to the electrode separation, is the same and the difference is zero.)

For unipolar recordings the passage of the double layer across the electrode is recorded by a change in potential from $+2\pi$ times the net double-layer strength to -2π times the net double-layer strength. The rapid transition results from the electrode first being on one side and then the other side of the wave.

This transition occurs in an elapsed time equal to the wave thickness divided by the velocity. Assuming a velocity of 50 cm/sec and a thickness of 1 mm, the time is evaluated to be 1 mm/500 mm/sec = 2 msec, during which the waveform will depend on the geometry and double-layer density function in the axial direction.

If we consider configuration 1 (Figure 9.13), we expect the waveform illustrated in Figure 9.11a. The initial increase in potential comes about as the wave (double layer) enters the space between the electrodes until the space is filled. The further increase arises because successive laminas have increasing strengths.

The peak results when the center of the excitation wave is midway between the electrodes. The total signal duration, T, may be evaluated from

$$T = \frac{\text{wave thickness} + \text{electrode separation}}{\text{wave velocity}} \tag{9.9}$$

In the following, a subscript is added corresponding to each electrode separation shown in Figure 9.13. For electrode spacing 2 (Figure 9.13) we obtain the waveform shown in Figure 9.11b.

Because the electrode spacing is greater than in 1 a greater number of laminas can lie between the electrodes, so the amplitude will be greater. In addition, T determined from (9.9) will be larger on account of increased electrode spacing, so $T_2 > T_1$.

For electrode separation 3, the amplitude of the recorded signal will reach the maximum value possible. This condition arises when the total wave lies between the electrodes. It occurs for a moment only. The waveform for this case is shown in Figure 9.11c. From Eq. (9.9) we determine that $T_3 > T_2 > T_1$.

In condition 4 of Figure 9.13, a period of time arises when the wave lies entirely between the electrodes. During this interval, at least according to our idealized model, no change in potential should occur and a flat-topped signal is expected.

A flat-topped waveform is illustrated in Figure 9.11d. The extent of the flat top is dependent on the electrode spacing. From (9.9), we expect the duration to be the longest of the four examples ($T_4 > T_3 > T_2 > T_1$). Actual recordings tend to confirm the basic expectations noted here, as illustrated, in Figures 9.23 and 9.24. Thus, in Figure 9.14 we display the potentials between electrodes spaced 2 mm apart and lying on a needle placed normal to endocardium and epicardium. The electrodes are numbered consecutively.

For the recording pair 9–11 (4-mm separation) no increase in amplitude but an increase in width occurs, as expected, since the wave thickness is less than 2 mm. An even greater width is seen in recordings 7–11, as expected, since the dimensions correspond to condition 4 of Figure 9.11.

The bipolar recording resulting from two closely spaced electrodes can be expressed as a spatial derivative of the unipolar recording. Thus a determination of the time of passage of a wavefront can be linked to the appearance of the peak in the bipolar monophasic signal or, correspondingly, the maximum derivative of the unipolar signal.

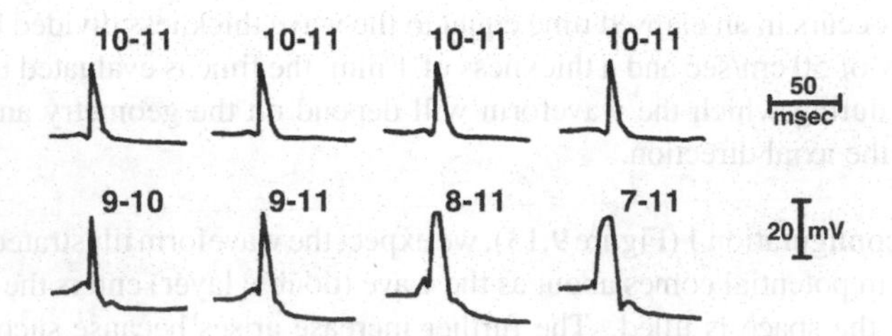

Figure 9.14. An 11-Electrode Needle with a distance of 2 mm between the lead points was used. Electrodes are numbered consecutively, and the bipolar signal measured between numbered electrodes is shown. From Durrer D, van der Tweel LH, Blickman JR. 1954. Spread of activation in the left ventricular wall of the dog, III. *Am Heart J* **48**:13–35.

The use of the bipolar signal is sometimes favored, because signals from distant sources tend to have similar values at both electrodes and subtract to zero so that only local sources are recorded. This discrimination is also evident in the dipole versus monopole lead fields. The unipolar electrode responds better to distant as well as local fields, since its lead field has the lowest order in inverse distance. If the goal is to examine timing, then the first derivative unipolar signal also is useful.

Assuming electrode separation equal to or greater than the wave width, then at the peak of the signal (when the double-layer sources are entirely between electrodes) and assuming each electrode "sees" a solid angle of $\pm 2\pi$, applying (9.25) gives the peak wave magnitude $\Delta\Phi_{pk}$ (which we also have referred to as V_{wave}) as

$$|\Delta\Phi_{pk}| = \frac{4\pi\delta}{4\pi\sigma} = \frac{\delta}{\sigma} \tag{9.10}$$

Consequently, the double-layer strength δ is given by

$$\delta = \sigma\Delta\Phi_{pk} \tag{9.11}$$

Equation (9.11) describes a method of evaluating the double-layer strength based on measured signal strengths. For measurements described in Figures 9.25 and 9.26, we note that $\Delta\Phi_{pk} \approx 40$ mV. These correspond to the estimates described in Eq. (9.13).

An equivalent transmural fiber

In Figure 9.15 we illustrate a hypothetical fiber structure that is oriented normal to the free wall. This arrangement was chosen because it utilizes the single-fiber activation studied earlier and because the resulting wave of activation (isochrones) has a shape that is essentially the one observed experimentally.

The syncytial structure of actual tissue permits this model to be advanced as "equivalent" electrophysiologically, in that it will have similar electrical characteristics to the real structure. Thereby the equivalent fiber is a useful theoretical entity, even though histologically it does not

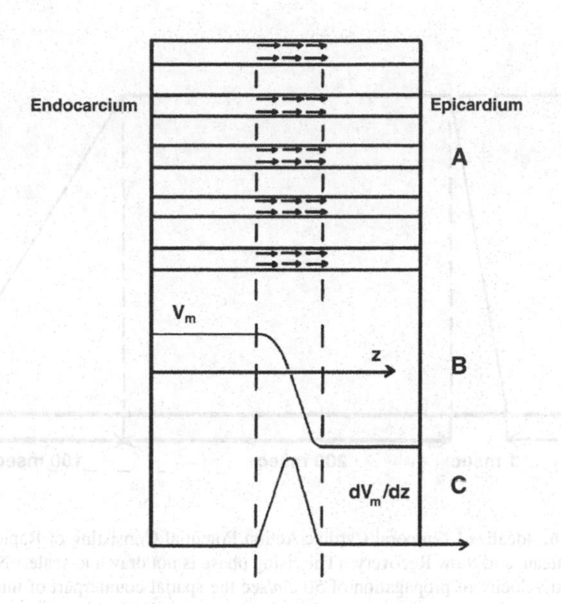

Figure 9.15. Hypothetical Fiber Orientation and Corresponding Activation Wave. A rise time of approximately 1 msec and a velocity of around 50 cm/sec means that the source region would be roughly 0.5 mm thick. From Liebman J, Plonsey R, Gillette PC. eds. 1982. *Pediatric electrocardiography*. Baltimore, MD: Williams and Wilkins.

describe the actual cell structure. We also assume an idealized temporal action potential. This action potential is in the form of a trapezoid and is illustrated in Figure 9.16. The rising phase is linear and has a duration of 1 msec, and there is a flat plateau of around 200 msec in duration. Action potential recovery is linear and is approximately 100 msec in duration.

Assuming a uniform velocity of 50 cm/sec, the spatial distribution can be determined from the temporal waveform, because a propagating wave must be of the form $f(t - z/\theta)$, where θ is the velocity. Fields arising in the structure in Figure 9.15 are well described by the linear core-conductor model, where we have shown in (6.22) that

$$\phi_e = \frac{r_e}{r_i + r_e} v_m \tag{9.12}$$

Data from Roberts and Scher [33] give the following effective resistivity parameters for propagation in the cross-fiber direction (that which normally prevails): $r_i = 1680$ Ωcm and $r_e = 1250$ Ωcm. These values consider that each region fills the entire tissue space, a *bidomain* viewpoint (to be described in a later section).

Consequently, if $v_m = 100$ mV we should measure an extracellular (interstitial) potential signal with a magnitude given by

$$[\phi_e]_{\text{peak}} = \frac{1250}{2930} \times 100 = 43\,\text{mV} \tag{9.13}$$

This value is the order of magnitude of experimentally determined values.[8]

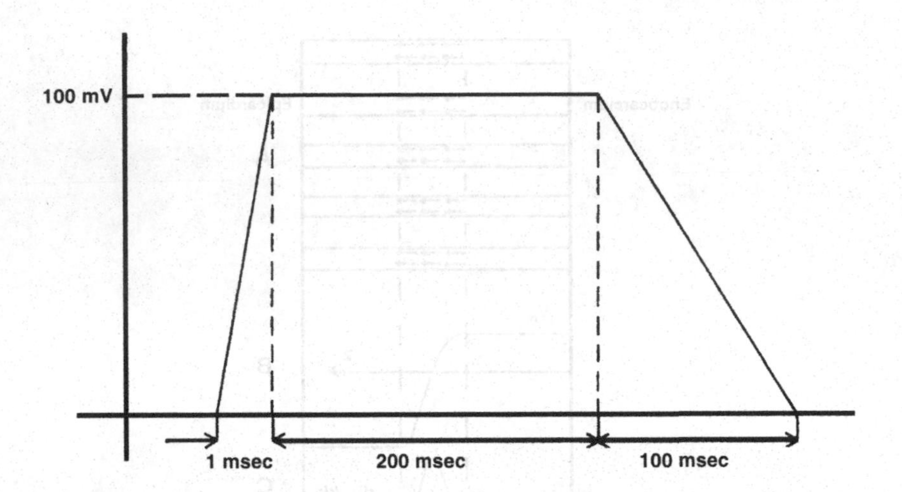

Figure 9.16. Idealized Temporal Cardiac Action Potential Consisting of Rapid Depolarization, Plateau, and Slow Recovery. (The rising phase is not drawn to scale.) Note that at an expected velocity of propagation of 50 cm/sec the spatial counterpart of this temporal action potential would require a tissue of at least 15 cm in thickness.

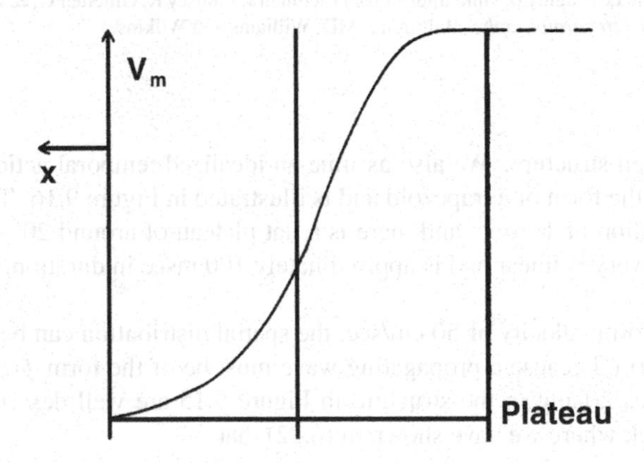

Figure 9.17. Spatial Variation of Rising Phase of Action Potential $v_m(z)$. The variation of $\phi_e(z)$ is similar, except for a change in sign and multiplication by scale factors of 0.43 [see Eqs. (9.12) and (9.13).

Based on the above we can estimate the spatial variation of v_m and ϕ_e as shown in Figure 9.17. The rising phase has been rounded from the idealized waveform in Figure 9.16; it corresponds to the experimental observation of $v_m(t)$ and the assumption of uniform conduction velocity of the activation wave.

If the temporal waveform of Figure 9.16 is converted into a spatial waveform based on a uniform velocity, the result cannot be completely correct because the wall thickness of the heart would have to be at least 15 cm to encompass the wave, while it is actually only around 1 cm.

Even more important is that, while activation is propagated, the action potential duration is determined in part by intrinsic tissue properties, so the recovery sequence may not be the same as the activation sequence. Consequently the *spatial* variation of potential during recovery cannot be found utilizing propagated behavior. The presence of the plateau shown in Figure 9.16 is correct, so long as one recognizes the limited size of the heart and that the duration of the plateau varies somewhat with position. (In fact, the cardiac action potential morphology is, itself, a function of position).

Double-layer sources

The structure of Figure 9.15 and the $v_m(z)$ described in Figure 9.17 permits applying (8.70) to determine the axial dipole source density associated with each fiber. We have (8.82), namely,

$$\overline{\tau}(z) = \frac{\partial}{\partial z}(\sigma_i \phi_i - \sigma_e \phi_e)\overline{a}_z \tag{9.14}$$

where $\overline{\tau}(z)$ is a current dipole moment per unit volume which fills each fiber. Since ϕ_i and ϕ_e have been defined to be uniform in any fiber cross-section, $\overline{\tau}(z)$ is also uniform (and hence only a function of z) in any fiber.

While in (9.14) a possible interaction between fibers appears to have been ignored, such modifications would be introduced through the behavior of ϕ_i and ϕ_e over the *tissue* cross-section. The conductivities σ_i and σ_e are the microscopic values for intracellular and interstitial space, respectively. Except near the periphery of the tissue, the linear core-conductor model should apply, assuming the tissue itself to be reasonably uniform.

Repeating (6.21) and (6.22) we have

$$\phi_i = \frac{r_i}{(r_i + r_e)} v_m \tag{9.15}$$

and

$$\phi_e = -\frac{r_e}{(r_i + r_e)} v_m \tag{9.16}$$

where r_i and r_e are the intracellular and interstitial axial resistances per unit length.

When Eqs. (9.15) and (9.16) are applicable, the source is seen to depend solely on the spatial derivative of the transmembrane action potential. In the case of uniform propagation, this can be determined from knowledge of the temporal action potential and the local conduction velocity. The extracellular field can be found from the integral of the dipole density source elements as described in (9.14), and this gives

$$(\phi_e)_P = \frac{1}{4\pi\sigma_e} \int_{\text{cells}} \frac{\partial}{\partial z}(\sigma_i \phi_i - \sigma_e \phi_e)\overline{a}_z \cdot \nabla(1/r) dv \tag{9.17}$$

where the subscript n denotes the direction of propagation. The integration is taken over the volume occupied by all active cells, of course.

If the relative volume occupied by the *cells* in Figure 9.15 (essentially, the intracellular space) is denoted F_i and the remaining *interstitial* volume designated F_e, where $(F_i + F_e = 1)$, then

$$r_i = \frac{1}{\sigma_i F_i} \text{ and } r_e = \frac{1}{\sigma_e F_e} \tag{9.18}$$

where r_i and r_e are the resistances per unit length *per unit cross-section* of tissue.

Note that this definition of the resistances *is different* from the earlier definition in which the resistances were evaluated per fiber; however, (9.15) and (9.16) continue to be correct with this new understanding since the ratio r_i/r_e is unchanged.

Substituting (9.15), (9.16), and (9.18) into (9.17) results in

$$(\phi_e)_P = \frac{1}{4\pi F_i}\int_{\text{cells}} \frac{r_e}{(r_i+r_e)}\frac{\partial v_m}{\partial z}\bar{a}_z \cdot \nabla(1/r)dv \tag{9.19}$$

Because cardiac cells occupy approximately 80% of the total tissue space (i.e., $F_i \approx 0.8$), it is convenient to regard the sources as if they occupy the entire volume (with respect to distant extracellular fields). When integrating a source assumed to fill the entire tissue, the results require multiplication by F_i to maintain the correct total source.

The result found from (9.19) is then

$$(\phi_e)_P = \frac{1}{4\pi}\int_{\text{tissue}} \frac{r_e}{(r_i+r_e)}\frac{\partial v_m}{\partial z}\bar{a}_z \cdot \nabla(1/r)dv \tag{9.20}$$

and the integration is throughout the entire tissue volume. Assuming reasonably uniform tissue, then $[r_e/(r_i+r_e)](\partial v_m/\partial z)$ should be defined and uniform over any cross-section that is normal to the direction of propagation.

The above equations, (9.14) to (9.20), are based on the equivalent tissue model in Figure 9.15, whose geometry must be evaluated from that of the actual tissue. Assuming uniformity, both the actual and equivalent tissue will permit uniform propagation of a plane wave in the endocardium to epicardium direction, and either defines the same equivalent intracellular, interstitial resistances per unit length, r_i and r_e.

So these values, which enter (9.20), can be simply considered to be evaluated from a plane wave experiment in which v_m, ϕ_i, and ϕ_e are measured and (9.15) and (9.16) used to evaluate the aforementioned resistances.

For field points within the tissue (i.e., the interstitial space), (9.19) simplifies into. (9.16). In particular, we learn from (9.16) that the difference of interstitial (i.e., measured) potential on opposite sides of the activation wave, V_{wave}, is the change in transmembrane potential v_m from rest to peak (plateau) multiplied by the extracellular resistance divider ratio.

That is,

$$V_{\text{wave}} = \Delta v_m \frac{r_e}{r_i+r_e} = (v_{\text{peak}} - v_{\text{rest}})\frac{r_e}{r_i+r_e} \tag{9.21}$$

This axial source density occupies a region of width ≈ 0.5 mm, and $\tau(z)$ as determined from an axial derivative of v_m using Figure 9.17 is error function (like) in shape. The total source is that of a thick (~ 0.5 mm) double layer whose axial density is proportional to $-\partial v_m/\partial z$.

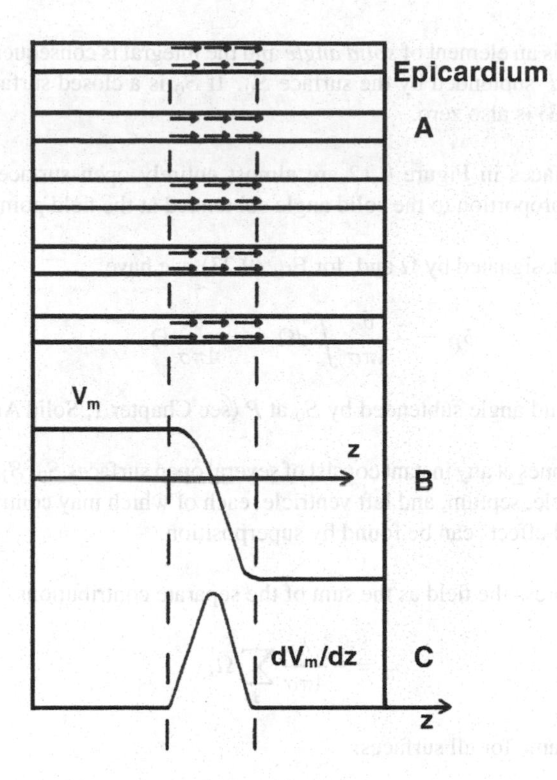

Figure 9.18. Effective Source Distribution associated with activation of the free wall of the heart. From Liebman J, Plonsey R, Gillette PC. eds. 1982. *Pediatric electrocardiography*. Baltimore, MD: Williams and Wilkins.

A description of this source is given in Figure 9.18. For field points outside the heart and at a distance large compared to 0.5 mm, perhaps at the torso surface, then the axial extent of the source is unimportant and can be considered to be a (mathematical) double-layer surface (i.e., zero axial thickness) and uniform in the lateral direction.

A consequence of the above is that the activation wavefront (isochrones) can also be interpreted as the site of a uniform double-layer source distribution with strength given by (9.21). Using the dipole expressions of Chapter 2, such as (2.29), one such surface can be designated S_0, at which the double-layer strength (dipole moment per unit area) is $\overline{\delta}$ (constant).

Then the generated field at the field point P is ϕ_P, given by

$$\phi_P = \frac{1}{4\pi\sigma} \int_{s0} \overline{\delta} \cdot \nabla\left(\frac{1}{r}\right) dS \tag{9.22}$$

When (9.21) is considered valid, $\overline{\delta} = V_{\text{wave}} \overline{a}_z$. Since $d\overline{S}$ is in the direction of the outward surface normal to the activation wavefront, we can also write (9.22) as

$$\phi_P = \frac{\delta}{4\pi\sigma} \int_{s0} \nabla\left(\frac{1}{r}\right) \cdot d\overline{S} \tag{9.23}$$

The integrand of (9.23) is an element of *solid angle* and the integral is consequently the total solid angle at the field point P subtended by the surface S_0. If S_0 is a closed surface, then since the solid angle is zero, (9.23) is also zero.

The activation surfaces in Figure 9.12 are almost entirely open surfaces and contribute, according to (9.23), in proportion to the solid angle subtended at the field point.

The solid angle is designated by Ω and, for Eq. (9.23), we have

$$\phi_P = -\frac{\delta}{4\pi\sigma}\int d\Omega = -\frac{\delta}{4\pi\sigma}\Omega_0 \tag{9.24}$$

where Ω_0 is the total solid angle subtended by S_0 at P (see Chapter 1, Solid Angles).

If activation isochrones at any instant consist of several open surfaces $S_0, S_1, S_2, \ldots$ (possibly lying in the right ventricle, septum, and left ventricle, each of which may contribute one or more surfaces), their summed effects can be found by superposition.

That is, we can express the field as the sum of the separate contributions. We have

$$\phi_p = -\frac{\delta}{4\pi\sigma}\sum_i \Omega_i \tag{9.25}$$

assuming that δ is the same for all surfaces.

Interestingly, though many arguable approximations underlie Eq. (9.25), it appears to give reasonable results in simulations of electrocardiographic fields.

9.2. CARDIAC CELLULAR MODELS

Within cardiac cells the smallest units of propagation are clusters of 50 to 100 cells. The fundamental driving force of excitation arises from action potentials across cell membranes, but the fundamental properties of propagation arise from how groups of cells are connected to one another, and the collective properties that result. The following sections give first a recent model of cardiac membrane, and follow with models of propagation based on the cardiac multicellular structure.

9.2.1. Luo-Rudy Membrane Model

Over the years there has been considerable interest in the formulation of a Hodgkin–Huxley type model for cardiac membrane. This interest arises from the attention given to heart disease, which is one of the main causes of death in humans. There have additionally been many studies of the electrocardiogram as a tool in cardiac diagnosis and direct measurements of electrical waveforms from the hearts of animals and people. These findings have, in turn, sparked interest in theoretical studies to strengthen the clinical interpretation of electrocardiograms, motivating clinical investigations of cardiac arrhythmias and other abnormal conduction processes.

Analytical studies ultimately rest on appropriate membrane models. While the Hodgkin–Huxley equations have been found useful in a variety of vertebrate excitable cells, cardiac action

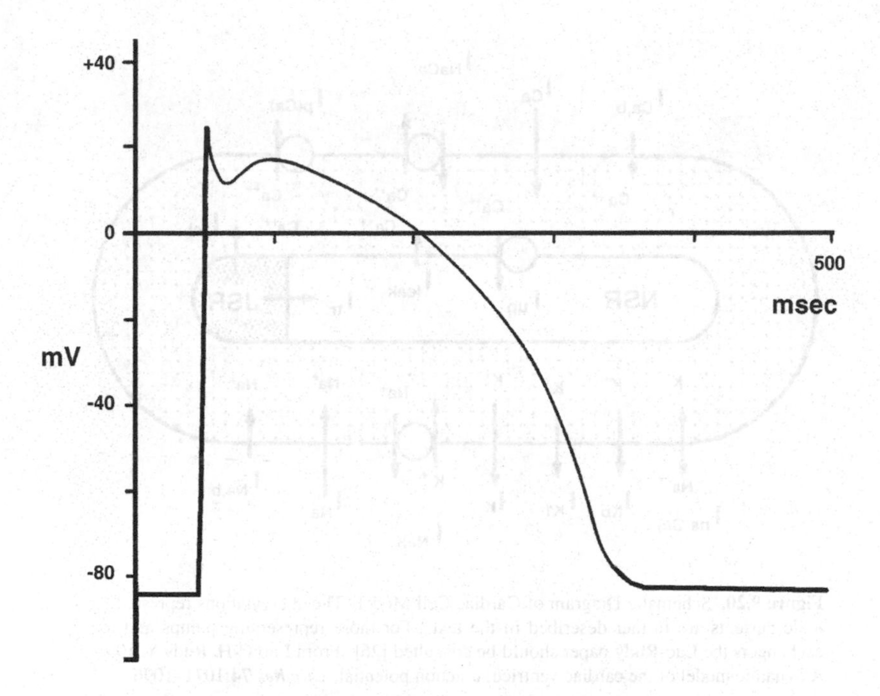

Figure 9.19. Typical Ventricular Cardiac Action Potential (Membrane Action Potential). This is obtained using the Beeler–Reuter model. From Beeler GW, Reuter H. 1977. Reconstruction of the action potential of ventricular myocardial fibers. *J Physiol* **268**:177–210.

potentials are more complex in the origin. An immediate and striking difference is that the cardiac action potential is two orders of magnitude longer in duration (see Figure 9.19) than in nerve and striated muscle.

The first of the modern cardiac models was that of DiFrancesco and Noble [13], as it included all the membrane currents for which there was experimental evidence, and it included the internal calcium currents closely associated with muscle contraction.

One of the more recent of the specialized cardiac muscle models is that of Luo and Rudy [26]. We include here only a brief description. The Luo–Rudy (L-R) model is a contemporary cardiac ventricular membrane model that extends previous ventricular models (notably those of Ebihara and Johnson [14], Beeler and Reuter [5], and DiFrancesco and Noble [13], utilizing the more recent single-channel and single-cell patch-clamp measurements.

In fundamental respects the approach of Hodgkin and Huxley, originated more than fifty years ago, remains consistent with a greatly enlarged accumulation of data, as well as some entirely new aspects. At the same time, the recognition of the greater complexity of the cardiac membrane leads to a much more extensive model, with many new features.

Earlier efforts to develop cardiac models were hampered by the difficulty in clamping cardiac cells which are individually too small (around 100 μm long and 15 μm in diameter) or pose other

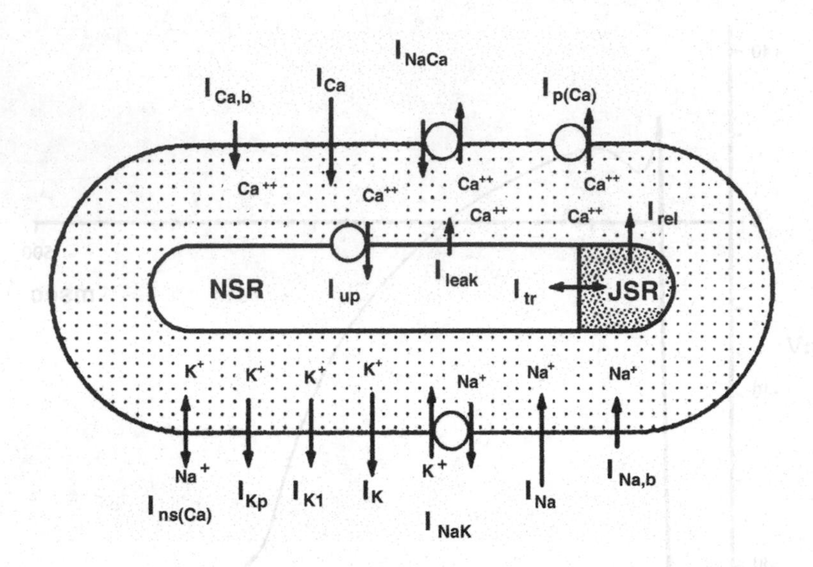

Figure 9.20. Schematic Diagram of Cardiac Cell Model. The abbreviations representing ionic currents are further described in the text. For those representing pumps and ion exchangers the Luo-Rudy paper should be consulted [26]. From Luo C-H, Rudy Y. 1994. A dynamic model of the cardiac ventricular action potential. *Circ Res* **74**:1071–1096.

difficulties (arising from cellular interconnections), such as in the use of cell clusters (discussed in Ebihara et al. [15]).

A schematic diagram of the single-cell structure and membrane paths for the L-R phase II model is given in Figure 9.20. The model contains several compartments, including the network sarcoplasm reticulum (NSR) and the junctional sarcoplasm reticulum (JSR). These compartments allow for changes in ion concentration resulting from active and passive ion movement as introduced earlier in the DiFrancesco–Noble Purkinje model [13]. Provision for both active and passive ion movement is included.

While the model reflects many interesting biophysical processes, we limit attention here to the passive K, Na, and Ca flux for comparison with the Hodgkin–Huxley model to identify where the original model is still electrophysiologically sound. Each current component is described separately below.

I_{Na}, the fast Na^+ current

The fast sodium current expression evaluates the sodium current through the fast sodium channel and accounts, mainly, for the rapid upstroke in transmembrane potential. In addition to sodium inactivation discussed for the squid axon, there is also included a slow inactivation parameter, *j*, to reflect observations of this type in cardiac cells.

The fast sodium current expression is

$$I_{\mathrm{Na}} = \bar{G}_{\mathrm{Na}} m^3 h j (V_m - E_{\mathrm{Na}}) \tag{9.26}$$

where the remaining parameters have the same meaning as in HH. All gating variables (*m, h, j*) satisfy the differential equation described earlier, namely,

$$\frac{dy}{dt} = \alpha_y(1-y) - \beta_y y \tag{9.27}$$

where y is m, h, j. Equation (9.27) is precisely the first-order rate process that we have assumed earlier to describe channel gating and the HH parameter behavior.

Rate constants for *m, h, j* are given in Luo–Rudy; we show here only that for *m*, as an illustration. This equation for m is

$$\begin{aligned} \alpha_m &= 0.32(V+47013)/\{1-\exp[-0.1(V+47.13)]\} \\ \beta_m &= 0.08\exp(-V/11) \end{aligned} \tag{9.28}$$

A comparison of (9.28) with the Hodgkin–Huxley definitions of α_m and β_m (Chapter 5) demonstrates a similar mathematical structure but quantitative differences. In both cases the expressions are, of course, empirical and reflect a process of curve-fitting to experimental data.

I_K, the time-dependent potassium current

The activation parameter is designated X and the conductance depends on X^2. An inactivation parameter, X_i, is defined and this also provides for an inward rectification for the $I_K - V$ curve. The expression for I_K is

$$I_K = \bar{G}_K X_i X^2 (V_m - E_k) \tag{9.29}$$

In comparison with the comparable expression for potassium current in the HH squid axon, n^4 is replaced by X^2, X_i is added, and also, here, $\bar{G}_\mathrm{K}$ is not assumed to be constant but its dependence on $[\mathrm{K}]_o$ is recognized through

$$\bar{G}_\mathrm{K} = 0.282\sqrt{[K]_o/5.4} \quad \mathrm{mS}/\mu\mathrm{F} \tag{9.30}$$

Also, E_K is given by

$$E_\mathrm{K} = (RT/F)\ln\left[([\mathrm{K}]_o + P_{\mathrm{Na,\,K}}[\mathrm{Na}]_o)\,/\,([\mathrm{K}]_i + P_{\mathrm{Na,\,K}}[\mathrm{Na}]_i)\right] \tag{9.31}$$

This expression differs from the potassium Nernst potential in that it allows for a small sodium current through the potassium channels. The expression is obtained from the GHK equation in evaluating the potassium reversal potential with $P_\mathrm{Na}/P_\mathrm{K} = P_\mathrm{Na,K}$.

I_{K1}, the time-independent potassium current

This potassium channel has a single-channel conductance that varies with the square root of $[\mathrm{K}]_o$ and also has a high selectivity for potassium. (The reversal potential is, accordingly, the potassium Nernst potential.)

This channel is closed for elevated transmembrane potentials. Also, since its time constant is very short ($\tau_{\mathrm{K1}} \approx 0.7$ msec), its inactivation gate, K1, can be assigned its steady-state value, $\mathrm{K1}_\infty$. Accordingly

$$I_{\mathrm{K1}} = \bar{G}_{\mathrm{K1}}\mathrm{K1}_\infty(V_m - E_{\mathrm{K1}}) \tag{9.32}$$

where

$$E_{\mathrm{K1}} = (RT/F)\ln([\mathrm{K}]_o/[\mathrm{K}]_i) \tag{9.33}$$

and

$$\bar{G}_{\mathrm{K1}} = 0.6047\sqrt{[\mathrm{K}]_o/5.4} \tag{9.34}$$

From the Hodgkin–Huxley analysis of chapter 5, we know that because an equation of the form of (9.27) describes change over time, then $\mathrm{K1}_\infty = \alpha_{\mathrm{K1}}/\alpha_{\mathrm{K1}} + \beta_{\mathrm{K1}})$. The rate constants are

$$\begin{aligned} \alpha_{\mathrm{K1}} = &\ 1.02/\{1 - \exp[0.2385(V_m - E_{\mathrm{K1}} - 59015)]\} \\ \beta_{\mathrm{K1}} = &\ \{0.49124 + \exp[0.08032(V_m - E_{\mathrm{K1}} + 5.476)]\} \\ &+ \exp[0.06175(V_m - E_{\mathrm{K1}} + 594.31)]\} \\ &\{1 + \exp[-0.5143(V_m - E_{\mathrm{K1}} + 4.753)]\} \end{aligned} \tag{9.35}$$

$I_{K\rho}$, the potassium plateau current

This channel contributes a potassium plateau current that is time independent but also insensitive to $[K]_o$. We have

$$I_{K\rho} = \bar{G}_{K\rho}K_\rho(V_m - E_{K\rho}) \tag{9.36}$$

where $E_{K\rho} = E_{\mathrm{K1}}$ and $K_\rho = 1/\{1 + \exp[(7.488 - V_m)/5.98]\}$.

I_{Ca}, the calcium current

For the cardiac cell the calcium current plays an important role in initiating ventricular muscle contraction. It contributes, electrophysiologically, to the long plateau seen in intracellular ventricular action potentials (as seen in Figure 9.19).

For reasons given above, the calcium current formulation is best described by the GHK equation. The L–R model expresses the calcium current I_{Ca} as

$$I_{\mathrm{Ca}} = df f_{\mathrm{Ca}}\bar{I}_{\mathrm{Ca}} \tag{9.37}$$

where d is an activation gate and f an inactivation gate.

Then letting the parameter s stand for Ca, K, or Na (but all flowing through the calcium channel) and using the general expression for current from the Goldman–Hodgkin–Katz equations, we get

$$\bar{I}_s = P_s z_s^2 \frac{V_m F^2}{RT} \frac{\gamma_{si}[S]_i \exp(z_s V_m F/RT) - \gamma_{so}[S]_o}{\exp(z_s V_m F/RT) - 1} \tag{9.38}$$

The ionic activity coefficient, γ, used in (9.38) has values

$$\gamma_{\mathrm{Ca}i} = 1, \gamma_{\mathrm{Ca}o} = 0.341, \gamma_{\mathrm{Na}i} = \gamma_{\mathrm{Na}o} = \gamma_{\mathrm{K}i} = \gamma_{\mathrm{K}ao} = 0.75 \tag{9.39}$$

Equation (9.38) serves to describe current flow through L-type calcium channels by Ca^{2+}, Na^+, and K^+. The calcium channel currents due to sodium and potassium are included since, while their permeability is small, their relatively high concentration gradients result in a flow that must be included (as discussed earlier in connection with Figure 9.20). These potassium and sodium currents are given by

$$I_{\mathrm{Ca,K}} = d f f_{\mathrm{Ca}} \bar{I}_{\mathrm{Ca,K}} \quad \text{and} \quad I_{\mathrm{Ca,Na}} = d f f_{\mathrm{Ca}} \bar{I}_{\mathrm{Ca,Na}} \tag{9.40}$$

In (9.37) and (9.40), as noted above, d is an activation gate and f an inactivation gate. Because of short time constants, these are given by their steady-state values, which are

$$d_\infty = 1/\{1 + \exp[-(V + 10)/6.24]\} \tag{9.41}$$

and

$$f_\infty = \{1 + \exp[(V_m + 35.06)/8.6]\}^{-1} + 0.6\{1 + \exp[(50 - V_m)/20]\}^{-1} \tag{9.42}$$

We also require f_{Ca}, which is given by

$$f_{\mathrm{Ca}} = \{1 + ([\mathrm{Ca}^{2+}]_i / K_{m,\mathrm{C_a}})^2\}^{-1}, K_{m,\mathrm{C_a}} = 0.6 \mu\mathrm{mol}/\ell \tag{9.43}$$

The total potassium current is given by four components described above. Consequently, no comparison to the HH potassium current (simulated with a single term) is possible. In this case the extended duration of the cardiac action potential is an underlying reason for the more complex formulation of potassium current. The model accounts for additional currents arising from the active processes. While our treatment here of the L–R model is incomplete, it is clear that the fundamental approach of Hodgkin and Huxley has been retained.

9.2.2. Two-Dimensional Cell Model

A two-dimensional cell model based on realistic cell size and shape was introduced by Spach and Heidlage [36]. The cell arrangement of 33 cells that formed the basic unit of the model is shown in Figure 9.21. The individual cell shapes are irregular, following the shapes of isolated cardiac myocytes, fit together as if they were pieces of a puzzle. Thus the pattern of cell positions is similar to patterns seen on histological preparations. The cell-to-cell connections likewise are located with a realistic frequency and pattern.

Simulations based on this cell pattern made use of a resistive grid formed by a large number of repeating grid elements, as shown in Figure 9.22. The Figure is drawn for a central node k. As the intracellular arrangement is two dimensional, there are connections to adjacent elements x^n and x^p along the x axis, and adjacent elements y^n and y^p along the y axis. Additionally, there is a transmembrane connection of each intracellular node to a corresponding extracellular node. (In this sense the model is what is sometimes called 2.5 dimensional.) The membrane properties are indicated in the Figure by the RC combination, where C_m is the membrane capacitance for the element and R_m represents the pathway for all ionic membrane currents of a cardiac membrane model.

One can analyze a two-dimensional grid by beginning (as was done in Chapter 6) with the membrane current equation (9.44), namely

$$I_m(x,t) = I_{ion}(x,t) + I_C(x,t) \tag{9.44}$$

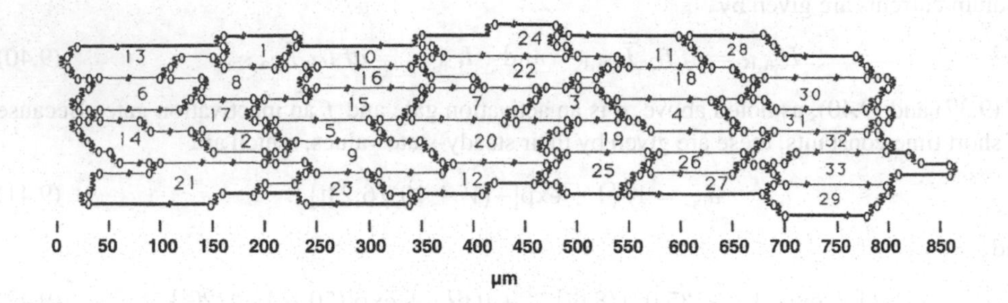

Figure 9.21. Cell Arrangement. Arrangement of 33 cells that formed the basic unit of the 2D model. The Figure shows the location of each cell (the dotted areas represent the regions of the interplicate disks), and the distribution of the connections (symbols) between adjacent cells within the 33-cell unit. Figure by MS Spach and F Heidlage.

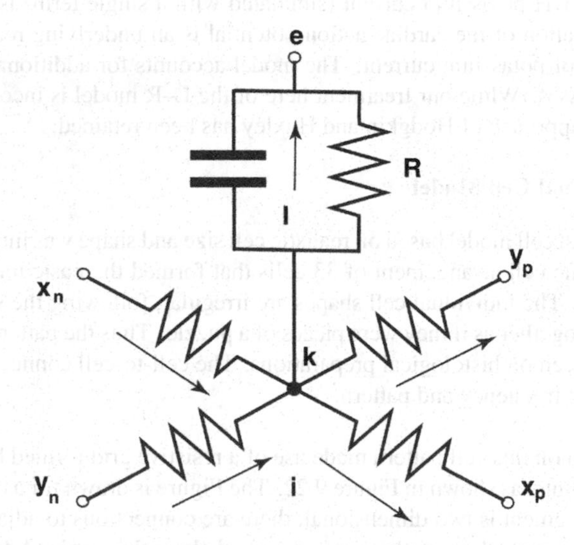

Figure 9.22. 2D Grid. Four intracellular current pathways in the x–y plane (the cell plane) connect to a central intracellular node k. The four adjacent intracellular nodes are along x in the positive direction (node at x_P) and negative direction (note x_n). Similar nodes lie along y. The resistance between node i and node x_P is R^{xp}. Analogous notation is used for all four intracellular paths. Transmembrane current I_m between the intracellular node k and adjacent extracellular node has capacitive and ionic components I_C and I_{ion}, where the latter flows (symbolically) through resistor R_m. Indices i and e signify intracellular and extracellular, respectively, while superscripts indicate position, e.g., Φ_i^k is the intracellular potential at node k. In computations by Spach and Heidlage, the extracellular potential always is assigned a potential of zero. The arrows indicate directions of positive current flow along x and y, or transmembrane.

As in Chapter 6, one may substitute for I_C, and obtain (similarly to (9.45))

$$\frac{\partial V_m^k(x,t)}{\partial t} = \frac{1}{C_m}[I_m^k(x,t) - I_{ion}^k(x,t)] \tag{9.45}$$

where the superscript k indicates that the calculation is for the kth node.

Total membrane current I_m also is equal to the net current to k from adjacent intracellular nodes. Because the cells form a two-dimensional grid, there are neighboring nodes along x and along y, as shown diagramatically in Figure 9.22. In equation form, one has

$$I_m^i = (I_i^{xn} - I_i^{xp}) + (I_i^{yn} - I_i^{yp}) \tag{9.46}$$

where, for example, I_i^{xn} is the intracellular current from node xn to node i, and I_i^{xp} is the intracellular current from node i to node xp.

Thus, where the R values are the resistances to the adjacent node, and Φ are the potentials at the adjacent segments,

$$I_m^i = \frac{\Phi_i^{xn} - \Phi_i^k}{R^{xn}} + \frac{\Phi_i^{xp} - \Phi_i^k}{R^{xp}} + \frac{\Phi_i^{yn} - \Phi_i^k}{R^{yn}} + \frac{\Phi_i^{yp} - \Phi_i^k}{R^{yp}} \tag{9.47}$$

Spach and Heidlage proceeded by assuming that $\Phi_e \approx 0$ at all extracellular nodes. That allowed the conversion of every intracellular potential in (9.47) to the corresponding transmembrane potential, so that, with V_m as the transmembrane potential,

$$I_m^k = \frac{V_m^{xn} - V_m^k}{R^{xn}} + \frac{V_m^{xp} - V_m^k}{R^{xp}} + \frac{V_m^{yn} - V_m^k}{R^{yn}} + \frac{V_m^{yp} - V_m^k}{R^{yp}} \tag{9.48}$$

The modifications in moving from the 1D model of chapter 6 to the 2D model summarized above are straightforward but may seem simpler than is actually the case. Among the challenges are the following:

- With the small spatial discretization used to represent the tissue, e.g., on a 10-micrometer grid, only a 2-mm square segment results in 40,000 elements, much larger than the several hundred often used in 1D.
- A central feature of the model is the detailed specification of the shapes and junctions of the elements. A corollary of the opportunity for detail is that the resistances connecting the elements along x and y vary depending on segment location, e.g., within the cell versus adjacent to a boundary, or a gap junction. Thus specification of the geometry (and verification that it has been specified correctly) is a major task.
- Because of the small spatial size of an element, the time step required for stability also is small, as one notes from the mesh ratio equation (6.58). Thus the simulation of a propagation sequence can become a large-scale computing task, even though the overall tissue size is small and the cell structure is only 2D.

- In view of issues of stability and computation time, careful attention must be paid to the numerical methods, e.g., with these factors in mind Spach and Heidlage used an Gauss–Seidel implicit method, rather than the explicit method of Chapter 6, to find transmembrane voltages at a following time step.

Such 2.5D models allow results that give considerable insight into the characteristics of real cardiac tissue. Spach, Heidlage, and coworkers have used this kind of model to test how the microscopic structural properties of cardiac muscle, including its anisotropic nature and detailed arrangement of gap junctions, influence the propagation of depolarization, as reported in [36] the paper cited and more recent ones, including the effects of changes seen with aging.

The model was used to demonstrate that cardiac muscle is discontinuous in nature because of recurrent discontinuities of axial resistance (i.e., does not behave as does a continuous cable), due to the cellular interconnections. More generally, it was used to demonstrate that it is not accurate to think of 2D propagation in cardiac tissue as quantitatively equivalent to continuous cables. A modified version of the model also showed that propagation is markedly influenced by connective tissue separating excitable regions, even when such connective tissue is present on a microscale.

The study of cardiac tissue electrophysiology often requires realistic cellular geometry and, in particular, model details about cellular interconnections and fiber structure, including the effects of connective tissue. With models that incorporate increasingly detailed cardiac cell structure and interconnection, the simulation of activation becomes an alternative to the measurement of isochrones and other experimental studies.

Simulation accuracy depends on an ability to include structural details, ionic behavior, etc. Simulation has an advantage in that it describes the link between sources and the cellular structure and electrophysiology and can be examined at multiple size scales, from submicron to centimeter, with no change in prescribed conditions. Furthermore, the computer programs and data can be utilized repeatedly for the study of arrhythmias, particularly fibrillation (arising from reentry), and for determining the conditions under which these are triggered.[9]

9.2.3. Bidomain Model

The *bidomain* model provides a strategy for understanding larger-scale attributes of the cardiac structure, with its huge number of individual cells, without having to describe that structure in cellular detail. The bidomain model provides a way to incorporate average properties of the intracellular and interstitial space along with the membrane, with its ionic properties, that separates these spaces. As averaged fields are uniform and continuous, they can be described using partial differential equations. Solutions to the partial differential equations can be found analytically for particular cases, though others require numerical methods).[10]

The intracellular space of cardiac tissue, including the many intercellular junctions, forms a singly connected, though complex, space. If one can forego an examination of field variations on a cellular or subcellular scale, then such a space can be approximated by a uniform and continuous region with averaged conductivity and smoothed fields. Anisotropy can be included by specifying a conductivity tensor as a function of position. Similar comments apply to the interstitial space.

It is convenient to assume that both the intracellular and interstitial regions occupy the entire heart volume. Consequently, at any point within the model heart there will be both a corresponding intracellular and interstitial bidomain potential.[11]

Bidomain mathematical description

A description of the bidomain model may best be provided by the following equations, which give the intracellular current density, $\overline{J}_i$, and the interstitial current, $\overline{J}_e$, namely,

$$\overline{J}_i = g_{ix}\frac{\partial \Phi_i}{\partial x}\overline{a}_x + g_{iy}\frac{\partial \Phi_i}{\partial y}\overline{a}_y + g_{iz}\frac{\partial \Phi_i}{\partial z}\overline{a}_z \tag{9.49}$$

and

$$\overline{J}_e = g_{ex}\frac{\partial \Phi_e}{\partial x}\overline{a}_x + g_{ey}\frac{\partial \Phi_e}{\partial y}\overline{a}_y + g_{ez}\frac{\partial \Phi_i}{\partial z}\overline{a}_z \tag{9.50}$$

In (9.49) and (9.50) we have assumed that the intracellular and interstitial spaces are anisotropic and that a total of six parameters are required to describe it. Usually, the principal axes in both spaces correspond, since they both depend on fiber orientation.

It is also frequently assumed that the conductivities transverse to the fiber direction are uniform (i.e., if the fiber lies along z, then $g_{ix} = g_{ey}$ and $g_{ex} = g_{ey}$), which reduces the number of conductivity variables to four, if the intracellular and interstitial anisotropy ratios are the same (i.e., $g_{ix}/g_{ex} = g_{iy}/g_{ey} = g_{iz}/g_{ez}$ = constant), the equations simplify to those for a single domain (monodomain). This approximation is sometimes made for the insights that may be gained while simplifying the problem considerably.

Dependable sets of cardiac conductivity values in vivo are not yet available, and the equal anisotropy assumption cannot be fully critiqued. However, it seems likely that equal anisotropy is at best a gross approximation, since intracellular and interstitial structural factors are only weakly correlated. Moreover, the junctional resistance affects only the intracellular domain.

For propagating activity, an important constraint linking (9.49) with (9.50) is that the transmembrane current leaving one space must enter the other space to satisfy the conservation of current. Thus we require

$$-\nabla \cdot \overline{J}_i = \nabla \cdot \overline{J}_e = I_v \tag{9.51}$$

where I_v is the transmembrane current per unit volume.

The conductivities in (9.49) and (9.50), designated $g_{ix}, \ldots$, are "bidomain conductivities," that is, they are coefficients appropriate to the continuous tissue space. Suppose the actual tissue were a uniform array of fibers oriented in the z direction. Suppose also that the microscopic intracellular and interstitial conductivities were σ_i, σ_e, and the fibers occupy a fraction, F, of the total volume ($0 < F < 1$). Then the bidomain conductivities g_{iz} and g_{ez} are

$$g_{iz} = \sigma_i F \qquad \text{and} \qquad g_{ez} = \sigma_e(1 - F) \tag{9.52}$$

In this example, the bidomain conductivities take into account that the intracellular and interstitial domains occupy the total tissue volume while they actually are associated with smaller spaces, i.e., F or $(1 - F)$.

For tissues with a more complex structure, a determination of the bidomain conductivity will be more difficult. Such a determination may require an investigation of the expected average current–voltage values and then the bidomain conductivities necessary to realize them. These considerations are the basis for Eq. (9.18).

Point stimulation of an isotropic bidomain

To provide some insight into the properties of the bidomain model we consider an important example, the response of an isotropic bidomain to an interstitial point source. We also assume a subthreshold stimulus and achievement of steady state. Under isotropic conditions, the current and electric field are related by a scalar constant so that (9.49) and (9.50) become

$$\overline{J}_i = -g_i \nabla \Phi_i \tag{9.53}$$

and

$$\overline{J}_e = -g_e \nabla \Phi_e \tag{9.54}$$

where g_i and g_e are the intracellular and interstitial bidomain conductivities.

The introduction of an interstitial point current source I_0 will require a modification of (9.51). Using (9.53) and (9.54), gives the following:

$$\nabla \cdot \overline{J}_e = -g_e \nabla^2 \Phi_e = I_v + I_0 \delta(r) \tag{9.55}$$

while

$$-\nabla \cdot \overline{J}_i = -g_i \nabla^2 \Phi_e = I_v \tag{9.56}$$

In (9.55) the point source location has been chosen at the coordinate origin. Since $V_m = \Phi_i - \Phi_e$, multiplying (9.55) by ρ_e and (9.56) by ρ_i and adding gives

$$\nabla^2 V_m = (\rho_i + \rho_e) I_v + \rho_e I_0 \delta(r) \tag{9.57}$$

where $\rho_i = 1/g_i$ and $\rho_e = 1/g_e$.

If R_m is the membrane resistance for a unit area of membrane, then at steady state $i_m = V_m/R_m$, where i_m is the transmembrane current per unit area. Let β be the ratio of the total tissue surface to the total volume occupied. Then, in the bidomain sense, the transmembrane current per unit volume is $I_v = i_m \beta$, or

$$I_v = \beta \frac{V_m}{R_m} \tag{9.58}$$

Setting

$$\rho_m = \frac{R_m}{\beta} \tag{9.59}$$

allows (9.57) to be written as

$$\nabla^2 V_m = \frac{V_m}{\lambda^2} + \rho_e I_0 \delta(r) \tag{9.60}$$

where $\lambda = \sqrt{\rho_m/(\rho_i + \rho_e)}$ is a space constant (for a three-dimensional isotropic bidomain).

If the Laplacian is written in spherical coordinates using (2.7) and the radial symmetry is recognized, than V_m varies with r only. Then (9.60) becomes

$$\frac{1}{r^2}\frac{\partial}{\partial r}\left(r^2\frac{\partial V_m}{\partial r}\right) = \frac{V_m}{\lambda^2} + \rho_e I_0 \delta(r) \tag{9.61}$$

We may solve (9.61) for $r > 0$ and introduce the source term through a boundary condition at $r = 0$. Multiplying (9.61) by r allows one to write the homogeneous equation as

$$\frac{d^2}{dr^2}(V_m r) = \frac{(V_m r)}{\lambda^2} \tag{9.62}$$

The solution to (9.62) is straightforward, namely,

$$(V_m r) = Ae^{-r/\lambda} + Be^{r/\lambda} \tag{9.63}$$

Because we cannot have a solution in which the potential increases indefinitely with r, (9.63) reduces to

$$V_m = A\frac{e^{-r/\lambda}}{r} \tag{9.64}$$

The boundary condition at $r \to 0$ is that the current introduced into the interstitial space is I_0, because there is no radial current or electric field in the intracellular space at the origin, the current being entirely interstitial at this point of application. Thus at $r = 0$,

$$\frac{\partial V_m}{\partial r} = \frac{\partial \Phi_i}{\partial r} - \frac{\partial \Phi_e}{\partial r} = -\frac{\partial \Phi_e}{\partial r} \tag{9.65}$$

Consequently,

$$-4\pi r^2 g_e \left.\frac{\partial \Phi_e}{\partial r}\right|_{r\to 0} = 4\pi r^2 g_e \left.\frac{\partial V_m}{\partial r}\right|_{r\to 0} = I_0 \tag{9.66}$$

Substituting (9.64) into (9.66) and solving for A gives $A = -\rho_e I_0/4\pi$, and (9.64) becomes

$$V_m = -\frac{\rho_e I_0}{4\pi}\frac{e^{-r/\lambda}}{r} \tag{9.67}$$

Starting with the differential equation for Φ_i in (9.56) and using (9.58), one can solve for Φ_i following steps similar to that given above for V_m but now using (9.67). One obtains

$$\frac{d^2}{dr^2}(\Phi_i r) = -\frac{\rho_i \rho_e}{\rho_m}\frac{I_0}{4\pi}e^{-r/\lambda} \tag{9.68}$$

The boundary condition at $r = 0$ is

$$\partial \Phi_i / \partial r = 0 \tag{9.69}$$

and at $r \to \infty$ is

$$\Phi_i \to 0 \tag{9.70}$$

Thus the solution, with integration constants evaluated from the boundary condition, is

$$\Phi_i(r) = -\frac{\rho_t I_0}{4\pi}\frac{e^{-r/\lambda}}{r} + \frac{\rho_t I_0}{4\pi r} \tag{9.71}$$

where

$$\rho_t = \frac{\rho_i \rho_e}{\rho_i + \rho_e} \tag{9.72}$$

and ρ_t is the total resistivity of the medium in the absence of the membranes. It describes the tissue, without membrane, homogenized into a monodomain.

Since $\Phi_e = \Phi_i - V_m$, then from (9.67) and (9.71) we have

$$\Phi_e(r) = \frac{\rho_t \rho_e}{\rho_i}\frac{I_0}{4\pi}\frac{e^{-r/\lambda}}{r} + \frac{\rho_t I_0}{4\pi r} \tag{9.73}$$

An examination of (9.71) and (9.73) shows a number of informative points:

- The *second* term of (9.73) is similar to the term that is present in most earlier equations for potentials, e.g., (2.21). Thus it is noteworthy that (9.73) contains an additional term, the first term.

- Both (9.71) and (9.73) contain a *differential* and a *common* component. The former [the first terms in (9.71) and (9.73)] is identical in form to the solutions of the linear core-conductor model. In this regard, compare these equations to (7.34). The close relationship of this model's result to that of the core-conductor model causes one to realize that the core-conductor model can be considered to be a one-dimensional bidomain.

- The common terms, the second terms in (9.71) and (9.73), describe the potential present in a monodomain, that is, in a monodomain $V_m = 0$, and there is no distinction between intracellular and interstitial potentials.

- In this example, when r exceeds, say, 10λ the bidomain terms can be ignored and the fields are described correctly by a monodomain model. Space constant λ is a measure of the extent of the region over which current redistribution takes place and where $V_m \neq 0$.

- The $1/r$ factor in the above equations arises from the three-dimensional nature of the problem, where current densities and other fields diminish at this rate as a result of a uniform radial divergence.

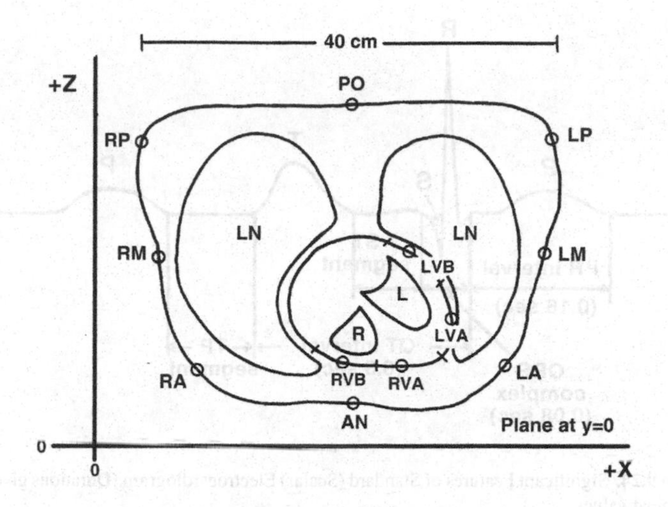

Figure 9.23. Cross-Section of the Human Torso. LN, lung; R, right heart cavity; L, left heart cavity; AN, anterior of torso; PO, posterior of torso. Other points are labels of reference sites. This Figure is a tracing from Section 26 in Eycleshymer AC, Shoemaker DM. 1970. *A cross-sectional anatomy*. New York: Appleton-Century-Crofts, Prentice-Hall.

9.3. ELECTROCARDIOGRAPHY

Electrocardiography involves understanding cardiac electrical events by means of measurements on the body surface. The goal of electrocardiography is to utilize body surface potentials, available noninvasively, to evaluate the status of the heart, especially in response to clinical questions.

Electrocardiography is made possible by the location of the human heart within the chest. An example of the location of the heart and lungs inside the torso is shown in Figure 9.23. Equivalent views are readily available from MRI. One sees on the cross-section that the heart is a large organ, with dimensions greater than the distance from its surface to the body surface. It also is located asymmetrically, both with respect to the heart, and with respect to the chest.

The implications of positioning are that electrodes on the chest are much closer to the heart, and, conversely, the RV portions of the heart are closer to the body surface than are LV portions. It is thus to be expected that electrical activity in the positions where the distances are relatively small will generate larger signals on the body surface.

9.3.1. Standard Leads

The standard or limb leads were introduced originally by Willem Einthoven ("father of electrocardiography"[12]).

One of the goals of the standard leads was to allow different investigators to compare results. Originally, electrodes were placed at the extremities (wrists and ankles). Placement was not critical since the extremities are, roughly, isopotential, and the available surface area was large.

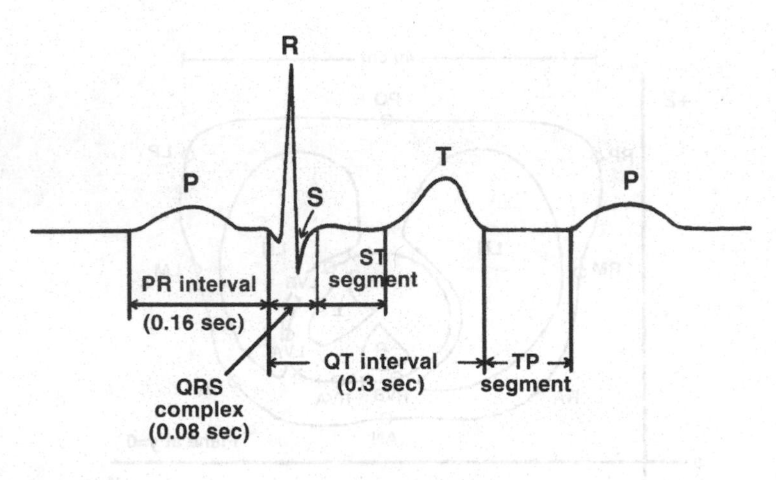

Figure 9.24. Significant Features of Standard (Scalar) Electrocardiogram. Durations given are typical values.

The right leg was normally grounded to help reduce noise. The remaining extremity potentials were paired to give the following three (lead) voltages:

$$V_{\mathrm{I}} = \Phi_{\mathrm{LA}} - \Phi_{\mathrm{RA}} \tag{9.74}$$

$$V_{\mathrm{II}} = \Phi_{\mathrm{LL}} - \Phi_{\mathrm{RA}} \tag{9.75}$$

$$V_{\mathrm{III}} = \Phi_{\mathrm{LA}} - \Phi_{\mathrm{LA}} \tag{9.76}$$

where RA is the right arm, LA the left arm, and LL the left leg.

Voltages such as V_{I} are called lead voltages as distinct from electrode voltages, since each lead voltage arises from a defined combination of electrodes in defined positions. $V_{\mathrm{I}} = \Phi_{\mathrm{LA}} - \Phi_{\mathrm{RA}}$ represents, for example, the potential of the right arm relative to that of the left arm, with the resulting voltage designated the lead V_{I}. A nominal lead voltage as a function of time is illustrated in Figure 9.24. The initial deflection is designated the P-wave, and it arises from activation of the atria. It initiates and consequently precedes contraction of the atria muscle.

The activation of the ventricles gives rise, normally, to the wave of largest magnitude and is designated the QRS. Many clinical diagnoses are based on its morphology and beat-to-beat timing. Recovery of the cardiac cells of the ventricles combines to produce the T wave. Atrial repolarization is normally masked by the QRS.

The PR interval, the time from the beginning of the P wave to the onset of the QRS complex, is a measure of AV conduction time and is useful clinically for this reason. The baseline is established by the TP segment of the wave. The TP segment is the time from the end of T to the beginning of the following P.

Normal ST segments, from the end of S to the beginning of T, are at baseline, while deviations may be useful in clinical diagnosis. The QT interval gives the total duration of the ventricular systole, while the heart rate is given by the reciprocal of the R–R interval.

Since by Kirchhoff's law the net potential drop around a closed loop is necessarily zero, then

$$(\phi_{LA} - \phi_{RA}) + (\phi_{RA} - \phi_{LL}) + (\phi_{LL} - \phi_{LA}) = 0 \tag{9.77}$$

Equation (9.77) is clearly correct in its own right, since we assume each potential to be single valued. Using (9.74)–(9.76), Eq. (9.77) can be written as

$$V_{I} + V_{III} = V_{II} \tag{9.78}$$

so that only two of the limb lead measurements are independent.

9.3.2. Precordial Leads

Additional electrocardiographic data are obtained from leads placed on the chest (precordium). Such leads were introduced to sample the electrocardiographic field close to the heart.[13] (In fact, five out of the six standard locations are on the left upper thorax.)

Each precordial lead is measured against the *Wilson's central terminal* (WCT) as a reference. WCT is formed at the junction of three 5K resistors,[14] the other end of each being connected to one of the limb leads as illustrated in Figure 9.25.

Assuming the use of a potential measuring system with a very high input impedance, then little net current is drawn from the CT and it then follows that the sum of the currents into the CT from each limb must equal zero (Kirchhoff's current law). Thus, if Φ_{CT} denotes the central terminal potential, then

$$\frac{\Phi_{CT} - \Phi_{RA}}{5000} + \frac{\Phi_{CT} - \Phi_{LA}}{5000} + \frac{\Phi_{CT} - \Phi_{LL}}{5000} = 0 \tag{9.79}$$

Solving for ϕ_{CT} gives

$$\phi_{CT} = \frac{\phi_{RA} + \phi_{LA} + \phi_{LL}}{3} \tag{9.80}$$

Because this potential is the mean of the extremity potentials, it was felt that it was a good "zero reference." It is not clear what people meant by this statement historically. Perhaps it meant that this potential, as an average of those measured in three directions from the heart, was thus not solely dependent on any one direction.

Wilson's central terminal addresses the fact that in humans no remote reference site is available, since electrical conduction is limited by the body surface's interface with air. Consequently some other electrical reference must be used. Wilson's central terminal is chosen as an practical alternative.

The most desirable condition is where the lead field of the active electrode(s) in the source region is independent of the reference electrode shape or location, a condition that is satisfied by a remote reference. Then the lead voltage is solely determined by the lead field configuration of the active electrode(s). A less satisfying reference is one whose influence on the lead field in the source region is small and where small perturbations in reference electrode location have a very small and tolerable effect on the lead field.

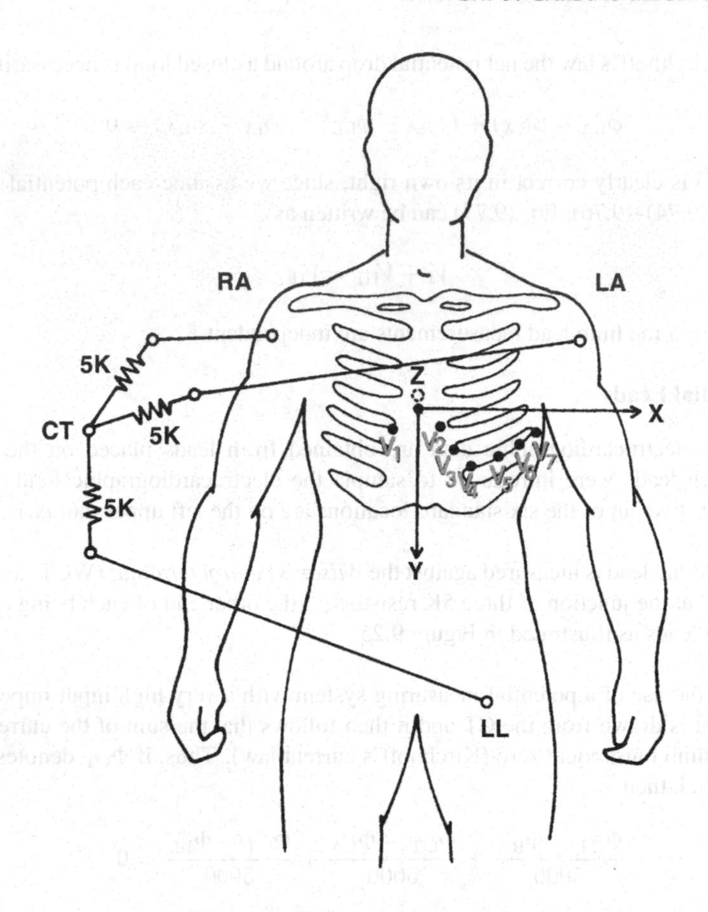

Figure 9.25. The Wilson Central Terminal (CT) is the common point connected through 5K resistors to RA, LA, and LL. It is the reference electrode for the precordial leads (V_i, $i = 1, 2, 3, 4, 5, 6$). Note that since the limbs provide little opportunity for current flow (being long and thin), the potential is approximately constant, so connection to extremities may be made near the torso without affecting the measurements much.

The Wilson central terminal fulfills none of these conditions, but since it involves electrodes at the extremities, whose location is definite, its contribution to a lead field and lead voltage is definite and reproducible. The contribution should be dependable for a given subject but will be less so in comparing subjects with different body shapes.

Three additional leads can be obtained by comparing each limb lead potential with the central terminal voltage. For example, from (9.80) we have, for the right arm,

$$v_{\mathrm{R}} = \phi_{\mathrm{RA}} - \frac{\phi_{\mathrm{RA}} + \phi_{\mathrm{LA}} + \phi_{\mathrm{LL}}}{3} = \frac{2}{3}\phi_{\mathrm{RA}} - \frac{\phi_{\mathrm{LA}} + \phi_{\mathrm{LL}}}{3} \tag{9.81}$$

If, in creating the CT voltage, the connection to RA is dropped, then in place of (9.81) we have

$$aV_{\mathrm{R}} = \phi_{\mathrm{RA}} \frac{\phi_{\mathrm{LA}} + \phi_{\mathrm{LL}}}{2} = \frac{3}{2} v_{\mathrm{R}} \tag{9.82}$$

where the final expression is found by comparison with (9.80). Here, aV_R is referred to as an *augmented* potential and is preferred over v_R since it gives a 50% stronger signal (while having precisely the same wave shape). The remaining two augmented leads are formed in the same way (by comparing a limb lead ith voltage with that of the CT formed by dropping reference to the ith limb lead). The three augmented plus six precordial plus three limb leads constitute the standard *twelve-lead system*, the backbone of current clinical electrocardiography.

9.3.3. Body Surface Potential Maps

With the string galvanometer Einthoven could record only one lead at a time. So long as successive heartbeats were similar, acquisition of additional leads could be obtained sequentially leaving, however, some ambiguity on the relative timing of each lead. With the advent of electronic amplifiers, early instruments continued the practice of containing a single amplifier, and the technician would switch to the successive leads in the 12 standard leads recorded clinically.

Present-day devices utilize small and low-noise solid-state chips that enable simultaneous recording of all leads. These same advances have also made possible simultaneous recording from multiple electrodes covering the entire torso. An example is described in Figure 9.26, which exhibits 140 leads. Assuming these signals are sampled 2000 times per second, the required data storage rate is 0.28 Mbytes/sec.

The potentials measured at many points over the torso can be considered as samples of a continuous surface distribution. The electrode spacing chosen in many systems is not uniform but closer on the anterior surface, since spatial potential gradients are greatest here. The surface potential distribution is usually displayed in the form of isopotential maps. Such maps may be examined sequentially, searching for abnormal timing or morphology.

Because the potential at the torso is derived from sources throughout the heart, there is not a one-to-one correspondence between patterns at the body surface, patterns at the epicardium, and intramural source activity. Nevertheless, some regional information can be estimated based on model studies and expected electrophysiological behavior. The possibility of obtaining regionally specific information is one reason why such maps have attracted attention.

Body surface potential data may be used to reconstruct the signals that would result with any lead system. But since the electrode positions probably do not coincide with the location of, say, the standard 12 lead positions, interpolation will be necessary; for an adequately dense electrode system this should introduce little error.

The body surface potential map consequently fulfills any and all lead systems in that it obviously contains all the data that are available at the torso. In a sense it obviates the need for further lead system development[15] and shifts attention to possibilities in further processing and analyzing these collected data to extract the maximum information about the heart.

At present, body surface potential data may be processed utilizing a number of proposed algorithms to achieve such goals as the estimation of maps on the heart (where regional information is much more accurately displayed), the susceptibility to arrhythmia, and the isochronal distribution on the heart surface.

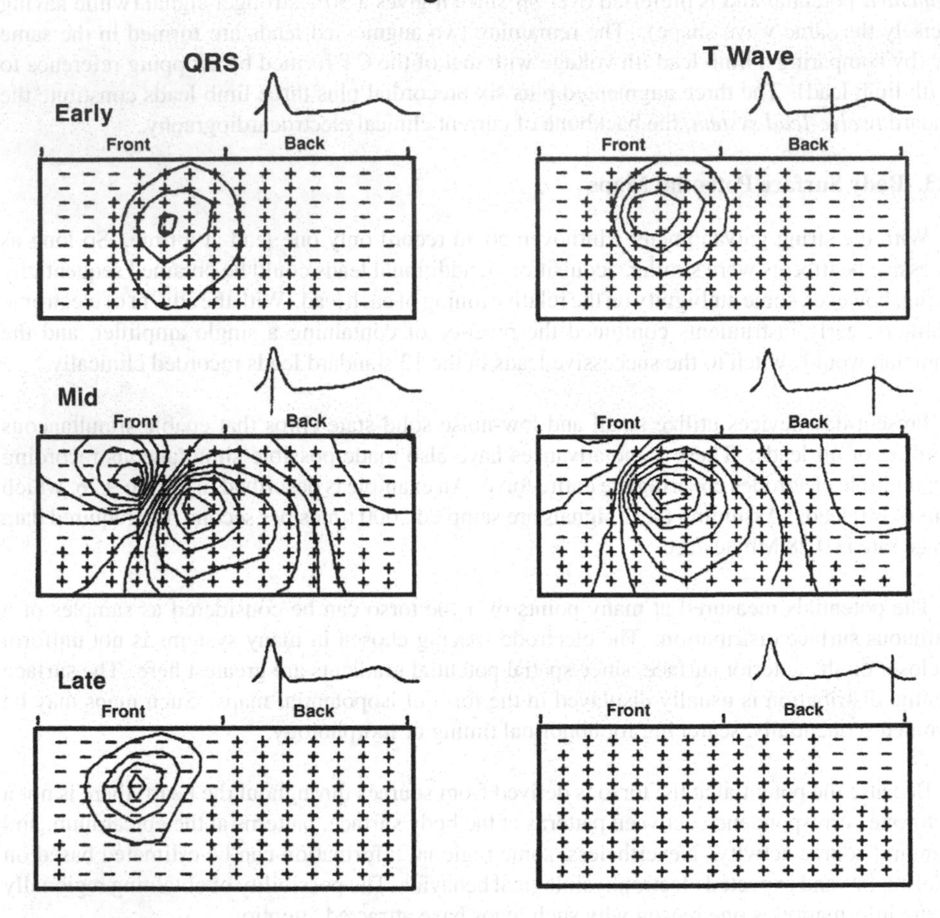

Figure 9.26. Body Surface Electrode Array and Potential Maps of QRS and T at Six Selected Time Instants. The electrodes and potentials on the anterior torso are displayed in the map from the left edge to the center; the potentials over the back (from left to right) continue in the map from the center to the right-hand edge. The map corresponds to the torso cut along the right mid-axillary line and unrolled. The top of the grid is at the level of the clavicles and the bottom at the level of the umbilicus. Shown is an average normal map compiled from subjects in the age group 30–39. Isopotential lines are 0.1 mV apart during QRS and 0.05 mV apart during T. From Green LS, Lux RL, Haws CW, Burgess MJ, Abildskov JA. 1986. Features of body surface potential maps from a large normal population. In *Electrocardiographic body surface mapping*, Ed RTh van Dam, A van Oosterom. Dordrecht: Nijhoff, with the kind permission of Kluwer Academic Publishers.

9.3.4. Interpretation by Statistical Classification

The predominant mode of clinical interpretation of electrocardiograms is by measuring specific characteristics of a new recording (features such as rate of occurrence or absolute or relative size of different deflections) and comparing these measurements to reference values stored in a database. The database gives the characteristics of the electrocardiograms of individuals whose clinical status was determined by independent measurements or clinical evaluation. The new recording can thereby be identified as closest to one of the previously identified groups.

ECG interpretation of this kind has been successful as related to gross classification (a new set of recordings placed in one of ten or fewer cardiac disease categories), but this statistical classification has not been as successful as a tool for precise evaluation. Limitations include the natural variability of humans in both cardiac and torso function, but also the expense and complexity of deriving well-documented subject groups across age, sex, and cardiac condition of the subjects used in the reference data bases.

Classification by statistical grouping dates from the time of Einthoven or before, but has become much more precise and much faster with the advent of computer systems. Reference databases, how they are constructed, and the statistical measures used to compare new recordings to characteristics used for reference are beyond the scope of this text. Pertinent information is, however, readily available in ECG teaching material as well as from more specialized references.

9.3.5. Physical and Physiological Interpretation

Although comparison with previously recorded traces has been the primary mode of ECG interpretation, investigators have always been interested in the actual physical or physiological electrical sources within the heart that gave rise to a waveform recorded on the body surface. Indeed, the ultimate achievement of electrocardiography would be a determination of the spatially distributed electrical sources in the heart from measurements on the body surface. However, electrical activity within the heart is not uniquely specified by the body surface potentials since one can always add sources that generate no field, or a field below the noise level.

One can also describe an equivalent source lying at the surface of the heart that generates potentials equivalent to those of the actual sources on the body surface. Such a source can be uniquely specified by body surface potentials. This heart source generates external fields which are identical to those generated by the actual primary sources.

While the inverse solution in the form of moving from potentials at the body surface to those on heart surface would be unique if measurements were made in a noise-free manner, in fact the body surface measurements do contain noise. The consequence is an *ill-posed* problem because small errors at the torso surface tend to be magnified at the heart (because potential magnitudes increase as the distance from the heart decreases).

Various approaches have been investigated to stabilize the inverse process, all of which involve the use of a priori limitations on allowed cardiac outcomes. If done well, such constraints reflect the expected behavior of the heart potential distribution [19].

An essential principle in the physical and physiological interpretation of electrocardiograms is the conceptual separation of what the heart is doing electrically ("the heart vector") from the selection and positioning of electrodes on the body surface ("the lead vector"). Both of these concepts are, at various times, infused with figurative meaning, but the core fact is that the signals observed on the body surface depend somewhat separably on their electrical sources within the heart versus the influence of the intervening volume conductor between the heart and the body surface, thus allowing these aspects to be examined independently.

A fundamental question is whether a physical inverse can be used to pinpoint the site of origin of heartbeats that arise from sites in the ventricles, which are called *ectopic sites.* The inverse determination of ectopic sites has been achieved successfully. Studies using epicardial potentials showed that they can be determined to within 1 cm [29]. More recent model work using dipole analysis to compare signals from a catheter with those arising from the ectopic site show accuracies within 1 mm [2]. This level of precision would, if achieved in humans, allow this kind of inverse solution to be useful in clinical procedures.

The heart vector

Since at any instant of time during activation of the heart the source is a distribution of double-layer surfaces, a first approximation to the source could be found from a simple vector sum of all elements. Such a process avoids recognizing that at most times the double-layer elements are widely distributed. Rather, for simplicity, it treats all such elements as if they were at the same location. The resultant is a single dipole, the *heart dipole or heart vector* [6].

The underlying double layers are created and undergo changes in an orderly, progressive way, so the heart dipole is also expected to vary (both in magnitude and direction) in a smooth manner. The idea that the heart behaves electrically as a dipole generator is central to clinical models in electrocardiography. In fact, *vectorcardiography* is based entirely on an evaluation of the locus of the tip of the heart dipole during the heart cycle. This space curve, called a *vector loop*, is, indeed, smooth. Vectorcardiographic devices output projections of the vector loop on the principal coordinate planes.

At any moment during cardiac activation, one or more open activation surfaces (isochrones) can be identified and each considered (at least approximately) the site of a uniform double-layer source. The sources associated with each such surface are dipoles oriented normal and outward to the surface and lying in a relatively narrow radial region constituting the wave thickness, as described in Figure 9.18. This thickness is normally ignored, as we have mentioned above. One can characterize these source dipoles that are distributed in this way throughout the heart by a density function $\overline{J}_i$, the dipole moment per unit volume as a function of position within the heart.

The heart vector (or heart dipole), $\overline{H}$, defined in the previous paragraph is related to $\overline{J}_i$ simply by

$$\overline{H} = \int \overline{J}_i \, dV \tag{9.83}$$

or, component-wise,

$$H_x = \int J_{ix} \, dV, \quad H_y = \int J_{iy} \, dV, \quad H_z = \int J_{iz} \, dV \tag{9.84}$$

where $\overline{H}$ is the heart vector and $\overline{J}_i = J_{ix}\overline{a}_x + J_{iy}\overline{a}_y + J_{iz}\overline{a}_z$. Usually, $\overline{H}$ is considered fixed in position and so is only a function of time; $\overline{J}_i$ is a function of both position and time.

The "dipole hypothesis" refers simply to the idea that $\overline{H}(t)$ is a good representation of the net source activity. The approximation clearly destroys spatial information, since the net source evaluated in (9.83) is found from component elements as if they were all at the same point. The validity of such an approximation depends on the ratio of the extent of the source distribution relative to the source field distance. Since the heart lies just beneath the anterior torso, one would conclude that the approximation should be a poor one, yet surprisingly good results are usually obtained.

A strong formal basis for understanding the heart vector has been provided by Geselowitz [22, 23]. The theoretical structure can be used to find the components of the heart vector independent of its position within the torso, and to find the location within the torso where a single vector provides the best fit to data observed on the body surface.

The lead vector

The voltage measured between two body surface electrodes is known as a *lead voltage*. For a dipole heart source, it depends on the lead location, heart location, heart vector, and torso volume inhomogeneities. Because the system (in spite of this complexity) is linear, one can split the aforementioned influences into two parts, namely, the heart vector and everything else. The "everything else" reflects the effects of geometry and inhomogeneities in conductivity.

A formal relationship can be developed in the following way. For a particular subject, the heart vector location (usually the center of the heart) and lead position are chosen. We assume first that the heart vector (dipole) is of unit magnitude in the x direction. Let the resultant lead voltage be designated ℓ_x. In a similar way, a unit dipole oriented along y produces ℓ_y, and a unit z dipole ℓ_z.

Based on linearity, it follows that for a heart vector described by

$$\overline{H} = h_x \overline{a}_x + h_y \overline{a}_y + h_z \overline{a}_z \tag{9.85}$$

the lead voltage V_ℓ is given by (superposition)

$$V_\ell = h_x \ell_x + h_y \ell_y + h_z \ell_z \tag{9.86}$$

Equation (9.86) can be interpreted as the dot product between $\overline{H}$ and a vector $\overline{\ell}$ whose rectangular components are ℓ_x, ℓ_y, ℓ_z, that is,

$$V_\ell = \overline{H} \cdot \overline{\ell} \tag{9.87}$$

Thus we may consider the lead voltage to result from the projection of $\overline{H}$ on $\overline{\ell}$ times the magnitude of $\overline{\ell}$. This expression is in the promised form and demonstrates the lead voltage dependence on the heart vector and a second vector $(\overline{\ell})$ that reflects the geometry and conductivities. The vector $\overline{\ell}$ is known as the *lead vector*. For a given lead location, if the heart vector *position* is varied, then $\overline{\ell}$ for each such position will be different. By associating the vector $\overline{\ell}$ with each point in space, a vector field (vector function of position) is generated.

While the above account depends on the approximate heart dipole model, it can be strengthened by considering $\overline{H}$ to represent a dipole source *element*, in which case V_ℓ can be found from

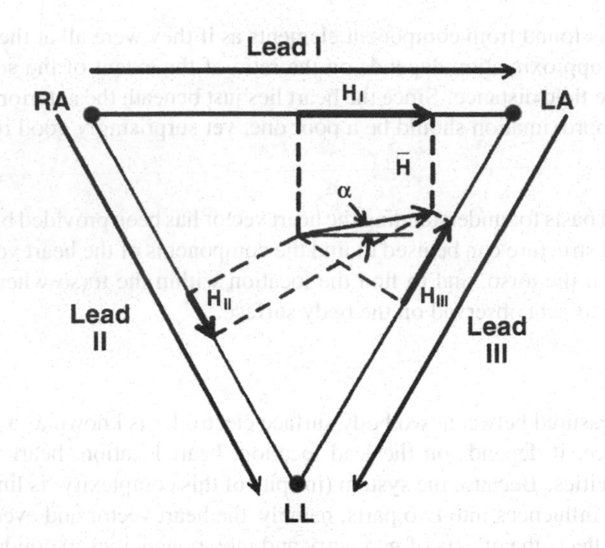

Figure 9.27. The Einthoven Triangle. The sides of the equilateral triangle describe the lead vectors for the limb leads, as shown. For the heart vector, $\overline{H}$, its projections on the triangle sides are labeled H_{I}, H_{II}, and H_{III} (choosing the subscript according to the respective lead); the sign of the lead voltages is found from the dot product of these projections and the corresponding lead vector (e.g., $V_{\mathrm{I}} = \overline{H}_{\mathrm{I}} \cdot \overline{L}_{\mathrm{I}}$). Note that in the illustrated case V_{I} is positive, V_{III} negative, and V_{II} positive. For simplicity in drawing, the lead vectors were given unit magnitudes.

(9.87) by summing over all such source elements. In this case ℓ is treated as a vector function of position, as noted above. This view will be further elaborated in a subsequent section.

Interpretation of standard leads

Einthoven hypothesized that the lead vectors for the standard leads would form an equilateral triangle. This is suggested by the physical position of each lead. The Einthoven triangle is illustrated in Figure 9.27, where an arbitrary heart vector is also depicted at the center of the triangle. (A rational basis for this arrangement will be given later in the chapter.)

According to (9.87), the lead voltages (V_{I}, V_{II}, V_{III}) are found by projecting the heart vector on the respective lead vector (and multiplying by the lead vector magnitude). In the example illustrated in Figure 9.27, since the heart vector points toward the left side of the body, the potential of LA is positive, while RA is negative, so that $V_{\mathrm{I}} = \phi_{\mathrm{LA}} - \phi_{\mathrm{RA}}$ must be positive. This is correctly evaluated by the dot product of $\overline{H}$ and the lead vector for lead I.

The geometrical relationship in Figure 9.27 leads to the expressions

$$V_{\mathrm{I}} = H \cos\alpha \tag{9.88}$$

$$V_{\mathrm{II}} = H\cos(60+\alpha) = \left(\frac{H}{2}\right)\cos\alpha - \left(\frac{\sqrt{3}}{2}\right)H\sin\alpha \tag{9.89}$$

$$V_{\mathrm{III}} = -H\sin(30+\alpha) = \left(-\frac{H}{2}\right)\cos\alpha - \left(\frac{\sqrt{3}}{2}\right)H\sin\alpha \tag{9.90}$$

where $\overline{H}$ is the heart vector magnitude, while the angle it makes with the horizontal direction is α.

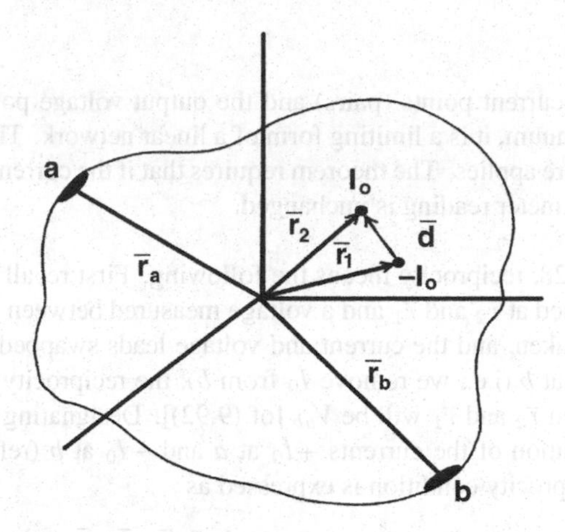

Figure 9.28. Dipole Source $I_0\overline{d}$ Giving Rise to Surface Potential Lead Voltage V_{ab}. This source–field relationship is examined by the application of reciprocity.

Note that these expressions satisfy the constraining relation (9.78), as indeed would any lead vectors forming a closed triangle. The angle α is called the *instantaneous electrical axis.*

9.3.6. Reciprocity and the Lead Field

A valuable interpretation of the lead vector was developed by McFee and Johnston [27] based on the application of reciprocity. We develop this result under somewhat more general conditions. Though the concepts of the lead field were developed first for electrocardiography, they are much more general concepts that can be a applied to the measurement of electrical signals of many other kinds.

Suppose an arbitrarily shaped volume conductor contains a point source and point sink, as described in Figure 9.28. The source is of magnitude I_0 and located at $\overline{r}_2$. The sink $(-I_0)$ is at $\overline{r}_1$. Their displacement $\overline{d} = \overline{r}_2 - \overline{r}_1$ is assumed to be very small, so that the composite source approximates a dipole. In the electrocardiographic sense this source could represent, say, a heart vector given by

$$\overline{H} = I_0\overline{d} \tag{9.91}$$

The dipole generates a field that we designate as Φ_1. The voltage that would be measured by lead *ab* (that is, with the surface electrodes *a* and *b*), as shown in Figure 9.28, is then

$$V_{ab} = \Phi_1(\overline{r}_a) - \Phi_1(\overline{r}_b) \tag{9.92}$$

Using the lead vector concept we have

$$V_{ab} = \overline{H} \cdot \overline{L}_{ab} \tag{9.93}$$

where $\overline{L}_{ab}$ is the lead vector for lead *ab*.

Reciprocity condition

Now between input current points (pairs) and the output voltage points (pairs) we have a linear system. As a continuum, it is a limiting form of a linear network. The reciprocity theorem of network theory therefore applies. The theorem requires that if the current source and voltmeter are interchanged, the voltmeter reading is unchanged.

Applied to Figure 9.28, reciprocity means the following. First recall that in the paragraphs above a current was applied at $\overline{r}_2$ and $\overline{r}_1$ and a voltage measured between a and b. Suppose now the *reciprocal* action is taken, and the current and voltage leads swapped. Thus a current I_0 is introduced at a and $-I_0$ at b (i.e., we remove I_0 from b), the reciprocity theorem says that the voltage measured between $\overline{r}_2$ and $\overline{r}_1$ will be V_{ab} [of (9.92)]. Designating Φ_2 to be the potential field arising from application of the currents $+I_0$ at a and $-I_0$ at b (referred to as *reciprocal energizing*), then the reciprocity condition is expressed as

$$V_{ab} = \Phi_2(\overline{r}_2) - \Phi_2(\overline{r}_1) \tag{9.94}$$

Finding the lead vector

Since $\overline{d}$ is assumed small, the following Taylor series expansion can be terminated at the linear term:

$$\Phi_2(\overline{r}_2) = \Phi_2(\overline{r}_1) + \left.\frac{\partial \Phi_2}{\partial d}\right|_{\overline{r}_1} d \tag{9.95}$$

or, based on the properties of the directional derivative given in the section entitled "Gradient" in Chapter 1, we have

$$\Phi_2(\overline{r}_2) = \Phi_2(\overline{r}_1) + \nabla\Phi_2 \cdot \overline{d} \tag{9.96}$$

Putting (9.96) into (9.94) gives

$$V_{ab} = \nabla\Phi_2 \cdot \overline{d} \tag{9.97}$$

Now Φ_2 arises from application of I_0 at a and $-I_0$ at b so, in view of linearity, a unit current (instead of I_0) would generate the field Φ_2^0 where

$$\Phi_2^0 = \frac{\Phi_2}{I_0} \tag{9.98}$$

Then, in place of (9.97) we have

$$V_{ab} = \nabla\Phi_2^0 \cdot I_0\overline{d} \tag{9.99}$$

or, from (9.91),

$$V_{ab} = \nabla\Phi_2^0 \cdot \overline{H} \tag{9.100}$$

A comparison of (9.100) with (9.93) shows that the lead vector associated with lead $a - b$ is the gradient of the scalar potential set up by reciprocally energizing $a - b$ using a unit current; it is evaluated at the dipole source location. The field Φ_2^0 is the *lead field* (of lead $a - b$), and because of the way it is set up it is often possible to guess its structure, at least approximately. The lead vector field, $\nabla\Phi_2^0$, is the current flow field associated with the scalar leads a to b (except for a change in sign and scale, since $\overline{J} = \sigma\overline{E} = -\sigma\nabla\Phi_2$).

Lead field extended to multiple dipoles

If the actual source is not a single dipole but rather a realistic dipole distribution defined by $\overline{J}_i\,(x, y, z)$, then (9.100) can be applied to each dipole element $\overline{J}_i dV$. Summing (superposition) gives the total lead voltage

$$V_{ab} = \int \overline{J}_i \cdot \nabla \Phi_2^0 \, dV \tag{9.101}$$

In (9.101), $\nabla \Phi_2^0$ is evaluated at each point for which $\overline{J}_i$ is nonzero, so $\nabla \Phi_2^0$ is regarded as a vector function of position or vector field. Sometimes it is referred to as a *lead vector field*, as noted above.

Lead field theory applied to lead I

We have noted that $\nabla \Phi_2^0$ is the lead field of the potential-detecting electrode(s). The application of lead field theory to electrocardiography is illustrated in Figure 9.29, where the reciprocally energized leads are $V_{\rm I} = \Phi_{\rm LA} - \Phi_{\rm RA}$. To obtain the lead field it is necessary to apply a unit current to LA and remove a unit of current from RA. The reciprocal current density within the torso is from LA to RA, but some bowing of flow lines is inevitable. The current, $\overline{J}$, is described by the expression

$$\overline{J} = -\sigma \nabla \Phi_2^0 \tag{9.102}$$

The lead vector [in (9.101)] is $\nabla \Phi_2^0$, found from (9.102) as

$$\nabla \Phi_2^0 = -\frac{\overline{J}}{\sigma} \tag{9.103}$$

So in the sketch in Figure 9.29 it is $-\overline{J}$ (which is proportional to $\nabla \Phi_2^0$) that is shown. This drawing can be interpreted as a description of the lead vector field. We note that at the heart center the direction is more or less horizontal (as approximated by Einthoven). But there is some curvature, so the results in (9.101) and (9.103) are not surprising.

Lead system design

A valuable concept of lead field theory is its characterization of the lead voltage as a weighted sum of the contributing sources. From (9.101), we note that the weighting function is $\nabla \Phi_2^0$.

For example, the reciprocal field from a point electrode varies as $1/r^2$. This field will weight sources close to the electrode more heavily by a factor that is proportional to the inverse square of the separation distance. Thus the field constitutes a quantitative description of the "sensitivity" of the unipolar electrode.

For a closely spaced electrode pair, the field $\nabla \phi_2^0$ will be a dipole field that varies as $1/r^3$. In this case the weighting is even stronger in favor of closer source elements.

One can use (9.101) not only to assist in the interpretation of what is measured as V_{ab}, but one can also specify a desired and physically realizable $\nabla \Phi_2^0$ to obtain a particular property. Then an electrode configuration that generates or approximates $\nabla \Phi_2^0$ can be sought, thus representing a rational approach to electrode system design.

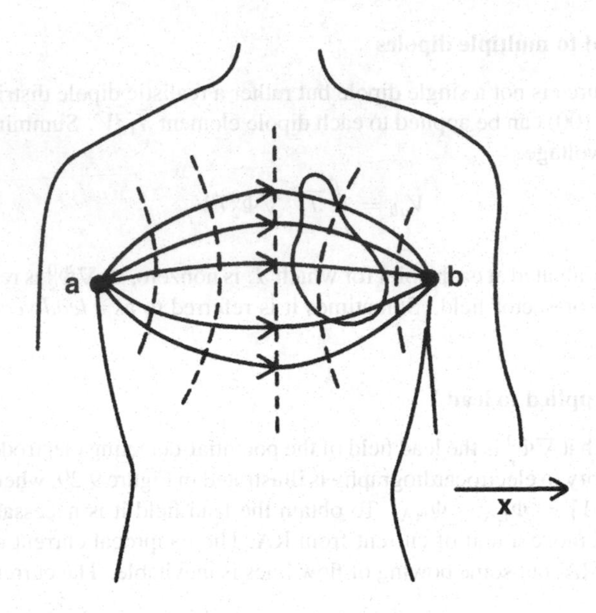

Figure 9.29. Lead Field associated with lead I is proportional to the negative of the current flow field arising with a unit current introduced at b and removed at a.

9.3.7. Isochronal Inverse Solution

Inverse solutions are estimates of objective information about the heart from measurements on the body surface [34]. An illustration is the model introduced by van Oosterom and colleagues. For simplicity, we assume the heart to be homogeneous and isotropic and to contain a uniform double-layer source that initially is closed and expanding outward.

The potential at the body surface point $\overline{y}$, assuming a double-layer magnitude of $\overline{\tau}$, is [from (9.25)]

$$\Phi(\overline{y}) = \frac{\tau}{4\pi\sigma}\,\Omega \tag{9.104}$$

So long as the activation wave is closed (i.e., $\Omega = 0$) no potential is present, if all parts of the closed surface have equal strength.

When breakthrough occurs because the excitation wave reaches the ventricular surface, the solid angle departs from zero (possibly rapidly) and a potential is seen on and outside the heart. The actual potential includes that computed from (9.104). Additional source terms arise from secondary sources generated at the various boundaries between regions of different conductivity. The heart–lung interface and the torso–air interfaces are known to be significant.

Because the system is linear, when the activation wave loses an area element ΔS in its termination at the surface, the loss is the same as the introduction of a negative dipole element.

Thus a potential over the heart surface can be written as [10, 11]

$$\Phi(\overline{y}, t) = \int_{\text{heart}} A(\overline{y}, \overline{x}) \Delta S(\overline{x}) \tag{9.105}$$

where $\overline{x}$ locates the dipole (on the heart surface). $A(\overline{y}, \overline{x})$ is a linear function that determines the potential at a body surface point $\overline{y}$ from the dipole element at $\overline{x}$ that accounts for the actual geometry and conductivities.

If S_i is the total portion of the closed surface intercepted by the epicardium and endocardium (where breakthrough also occurs), then a summation of elements as identified in (9.105) gives

$$\Phi(\overline{y}, t) = \int_{S_i} A(\overline{y}, \overline{x}) d(\overline{x}) \tag{9.106}$$

This integral can be taken over the entire heart by including the Heaviside step function $H(t)$, which is zero until arrival of the activation wave when it equals one (i.e., when the surface elemental dipole turns on). Thus

$$\Phi(\overline{y}, t) = \int_{\text{heart}} A(\overline{y}, \overline{x}) H(t - \tau(x)) d(\overline{x}) \tag{9.107}$$

where $\tau(x)$ is the arrival time of the wave at the heart surface point x. Isochrone intersections with the heart surface are given by $\tau(x) = \text{constant}$. Integration of (9.107) over the total activation time (from 0 to 1) requires an integration of $H(t - \tau)$ only, and the result is

$$w(\overline{y}) = \int_{\text{heart}} A(\overline{y}, \overline{x})(T - \tau(x)) d(\overline{x}) \tag{9.108}$$

If (9.108) is separated into two integrals, that involving the constant T is equivalent to finding the field from a uniform closed double layer and goes to zero. The result is

$$w(\overline{y}) = -\int_{\text{heart}} A(\overline{y}, \overline{x}) \tau(x) dx \tag{9.109}$$

Equation (9.109) describes the time integral of the potential at body surface points as arising from a spatial integral over the heart of activation time $\tau(x)$. The equivalent source in this model is a dipole in the heart surface whose field is $A(\overline{y}, \overline{x})$. This contribution is turned on by the expanding breakthrough periphery, and a contribution to $W(\overline{y})$ is given as the breakthrough periphery passes across x. The rime integral not surprisingly depends simply on the breakthrough time $\tau(x)$.

The function $w(\overline{y})$ can be evaluated from the body surface potential data at all N electrode sites. From (9.109), one could write N equations in N unknown sample values of $\tau(\overline{x}_i)$, $i = 1, \ldots, N$, and $i(x_i)$ determined in principle.

In this way, sample values of the continuous function $\tau(x)$ are found; with enough values the continuous function can be evaluated. The procedure requires good knowledge of the volume conductor geometrical and conductivity values from which the $A(\overline{x},\overline{y})$ are evaluated in a separate calculation. Unfortunately, the matrix formed from $A(\overline{y},\overline{x})$ is ill conditioned and regularization procedures must be applied. However, as shown in Figure 9.30, good results have been achieved.

9.3.8. Realistic Heart-Torso Models

Although it is clear from the above that the Einthoven triangle model is rather crude, it played a major role historically in unifying the concepts of electrocardiography, and it continues to have value today in understanding electrocardiograms at the top level.

Much more realistic models are available today, including many of the heart structure itself. Additionally, realistic torso models are currently available for computer simulation utilizing the finite element and (to some extent) the boundary element methods. With these, the major organs can be included as homogeneous regions of appropriate conductivity, and cardiac performance can be evaluated in a much more detailed way. There has been a focus on inverse solutions that can be verified, at least in principle, from cardiac measurements [4], rather than equivalent sources. A further description of such models can be found in Gulrajani [16] and elsewhere [3, 17, 18, 20].

More realistic models also have exploited the convergence of advances in computing and computer algorithms (e.g., Pollard, Hooke, and Henriquez [30, 7]) that allow elements with thousands of elements to be represented. Also greatly improved is knowledge of the cardiac 3D anatomy, made possible through imaging modalities such as MRI. Moreover, there is the opportunity to compare model results to experimental data acquired at high data rates on many data channels simultaneously. Such models are used not only for electrocardiography but for active modes of intervention, such as the response to electric shock as studied with a bidomain model [1] and defibrillation [24].

Accuracy of forward simulations

Simulation of the cardiac body surface potentials may begin with measured isochrones obtained from, say, a canine preparation using plunge electrodes. Such modeling experiments have as their goal an examination and elucidation of the various factors which contribute to the resulting surface potentials. Unfortunately, there is no experimental measurement of both plunge electrode and body surface potentials recorded simultaneously, and consequently no way of fully evaluating an aforementioned modeling effort.

One can readily list a number of factors that could be expected to affect the accuracy of the above forward simulation. First, the plunge electrodes sample at only a relatively few points throughout the heart, and hence isochrones and associated double-layer sources depend on interpolation and may contain significant error. The plunge electrode itself causes tissue damage and additional error as a consequence of the presence of its conducting (or nonconducting) surface, hence introducing false secondary sources. One can improve the sampling error with additional electrodes, but this increases the fraction of damaged tissue.

A second factor affecting simulation accuracy concerns the description of the geometry and conductivities of the constituents of the volume conductor, including intracavitary blood, torso shape, thoracic organs, and spine [18, 25]. A number of imaging modalities are available that can provide good geometrical values. Reported measurements of tissue conductivities vary

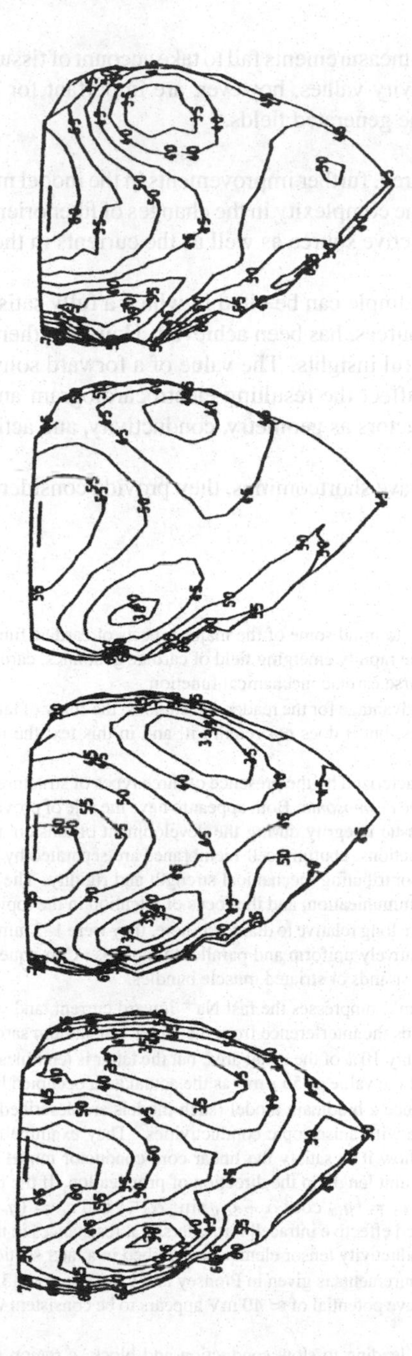

Figure 9.30. Inverse Isochrones. Left and right ventricles: (1) simulated activation sequence, (2) activation sequence computed classically for unperturbed transfer matrix A, (3) activation sequence computed classically for perturbed matrix A (A' includes additive noise), (4) inverse A' using extra constraints. Step size is 5 msec. From Huiskamp G, van Oosterom A. 1995. Four surfaces in electrocardiography. In *Proceedings of the 22nd international congress on electrocardiology*. Ed A van Oosterom, TJ Oostendorp, GJH Uijen. Nijmegen: University Press.

considerably. In many cases measurements fail to take account of tissue inhomogeneity, anisotropy, etc. [31]. Correct conductivity values, however, are important for a determination of both the source strength as well as the generated fields.

Although typically ignored, further improvements in the model may be expected by including heart tissue anisotropy and the complexity in the changes of fiber orientation. These will influence both the description of the active source as well as the currents in the volume conducting tissue.

At present, no single example can be cited in which a fully satisfactory forward simulation, based on intramural heart sources, has been achieved. However, there are some that give reasonable results and provide useful insights. The value of a forward solution is its ability to identify the significant factors that affect the resulting electrocardiogram and an ability to examine the effects of changes in such factors as geometry, conductivity, and activation pattern.

While present models have shortcomings, they provide considerable insight into these influences [16].

9.4. NOTES

1. It is thought provoking to bring to mind some of the major aspects of cardiac function that are mentioned here only minimally if at all, including the rapidly emerging field of cardiac genomics, cardiac neural interactions and control, cardiac metabolism, and of course cardiac mechanical function.
2. This independence implies an advantage for the reader and student: the study of later topics in the chapter may benefit from knowledge of earlier ones, but it does not require it, and in this text the reader may wish to focus on some sections but not others.
3. The intercellular region is characterized by the presence of three types of structures. The most numerous is the *intermediate junction* followed by the *desmosome*. Both appear to have the role of providing mechanical coupling between adjoining cells, maintaining tissue integrity during the development of tension along the fiber axis during ventricular contraction. Except at junctions, abutting cell membranes are separated by around 200 Å; however, filaments bridge the intercellular space contributing mechanical strength and rigidity. The third structure is the *gap junction*, responsible for intercellular communication, and the focus of attention in the topic of cardiac electrophysiology.
4. Trabecula tissue preparations are long relative to diameter (here, they were 1–3 mm long and 0.2–0.5 mm in diameter), while constituent fibers are relatively uniform and parallel to the axis. Consequently, they constitute a multifibered preparation similar to Purkinje strands or striated muscle bundles.
5. The holding potential of -42 mV suppresses the fast Na^+ inward current, and what is seen is a Ca^{++} inward and K^+ outward current. This avoids the interference from the larger fast sodium sarcolemmal current.
6. The gap-junctional area is roughly 10% of the nexal area, but the latter is ten times greater than the smooth transverse area due to folding. So we take the value of 56 μm^2 as the actual area occupied by the gap junctions.
7. Muller and Markin [28] introduce a bidomain model (such models are described later in this chapter) of a uniform *two-dimensional* cardiac tissue with anisotropic conductivities. They examine a uniform plane wave propagation in an arbitrary direction and show it to satisfy the linear core-conductor model with an effective intracellular and interstitial resistance value per unit length in the direction of propagation. If the fiber direction is x and propagation makes an angle α with x, then $r_i = (g_{ix} \cos^2 \alpha + g_{iy} \sin^2 \alpha)^{-1}$ and $r_e = (g_{ex} \cos^2 \alpha + g_{ey} \sin^2 \alpha)^{-1}$, where r_i and r_e are the aforementioned effective intracellular, interstitial resistances in the direction of propagation and the g's are the tissue bidomain conductivity tensor elements described in a later section.
8. A summary of cross-fiber measurements is given in Plonsey and van Oosterom [32]; none is completely satisfactory. However, the expected trans-wave potential of ≈ 40 mV appears to be consistent with a number of direct and indirect measurements.
9. Under pathological conditions leading to slow conduction and block, a region may allow activation to loop back to a previously depolarized site at which refractoriness has ended, permitting a second activation leading to an uncontrolled loop. The ultimate consequence is a number of reentry loops no longer controlled from the SA node. This may terminate spontaneously, repeat continuously (tachycardia), or go on to the lethal arrhythmia of fibrillation.

10. Major contributions to the development of bidomain models by Tung (1978), Schmitt (1969), and Miller and Geselowitz (1978) are cited by Gulrajani [16]. Spatial details can also be incorporated into bidomain models. The point here is that the model can provide useful information about electrical events over a larger spatial scale on the basis of average properties rather than *requiring* detailed specification.
11. A random intracardiac point must lie in either the intracellular or interstitial space, but such a point returns both a bidomain intracellular and interstitial potential. This seeming paradox occurs because the bidomain describes fields that are averaged over several cells. In the bidomain model at the aforementioned point, the bidomain potentials are the averaged intracellular or averaged interstitial potentials over the region surrounding the point. Whether the actual point is inside or outside a cell is thus inconsequential.
12. Among the first to measure the electrocardiogram was Augustus Waller using a Lippman electrometer. He published his findings in 1887. Waller was a London physiologist and was interested in cardiac electrophysiology; he used his dog, Jimmie, as a subject. (The dog became famous, but animal rights activists were critical.) On a visit to Einthoven at the turn of the century, he was surprised to learn of the ECG's clinical value, never having contemplated this in his own work. Einthoven's critical contribution was in the development of an improved string galvanometer, which made conventional recording of the ECG in doctor's offices possible. In addition, Einthoven developed a lead system and additional framework for evaluating an electrocardiogram. Einthoven won the Nobel Prize in medicine in 1924. An excellent history of electrocardiography has been provided by Johnson and Flowers [21]
13. V_1 is at the fourth right intercostal space at the sternal edge. V_2 is at the fourth left intercostal space at the sternal edge. V_4 is at the fifth left intercostal space at the mid-clavicular line. V_3 is midway between V_2 and V_4. V_5 is at the same level as V4 but at the anterior-axillary line. V_6 is at the same level as V_4 but at the mid-axillary line.
14. The 5 K value was chosen as a compromise. It was intended to be high in relation to skin-electrode resistances and low in relation to the amplifier impedances of the day. Modern systems more frequently use higher resistor values, perhaps 50K or even greater.
15. At least it does so in principle, though in practice the discussion shifts to how many leads are required to map the torso.

9.5. REFERENCES

1. Ashihara T, Trayanova NA. 2004. Asymmetry in membrane responses to electric shocks: insights from bidomain simulations. *Biophys J* **87**:2271–2282.
2. Barley ME, Cohen RJ. 2006. High precision guidance of ablation catheters to arrhythmic sites. In *Proceedings of the 28th Annual International Conference of the IEEE on Engineering in Medicine and Biology*, pp. 6297–6300. Washington, DC: IEEE.
3. Barnard ACL, Duck IM, Lynn MS. 1967. The application of electromagnetic theory to electrocardiology, II: numerical solution of the integral equation. *Biophys J* **7**:463–491.
4. Barr RC, Spach M. 1978. Inverse calculation of QRS-T epicardial potentials from body surface potential distribution for normal and ectopic beats in the intact dog. *Circ Res* **42**:661–675.
5. Beeler GW, Reuter H. 1977. Reconstruction of the action potential of ventricular myocardial fibers. *J Physiol* **268**:177–210.
6. Burger H, van Milaan JB. 1947. Heart vector and leads: II and III. *Brit Heart J* **9**:154–160; **10**:229–233 (1948).
7. Cates AW, Pollard AE. 1998. A model study of intramural dispersion of action potential duration in the canine pulmonary conus. *Ann Biomed Eng* **26**:567–576.
8. Chapman RA, Fry CH. 1978. An analysis of the cable properties of frog ventricular myocardium. *J Physiol* **283**:263–282.
9. Colatsky TJ, Tsien RW. 1979. Electrical properties associated with wide intercellular clefts in Purkinje fibers. *J Physiol* **290**:227–252.
10. Cuppen JJM. 1985. Calculating the isochrones of ventricular depolarization. *SIAM J Sci Stat Comput* **5**:105–120.
11. Cuppen JJM, van Oosterom A. 1984. Model studies with the inversely calculated isochrones of ventricular depolarization. *IEEE Trans Biomed Eng* **31**:652–659.
12. Durrer D. van Dam RT, Freud GE. Janse MJ, Meijier FL, Arzbaecher RC. 1970. Total excitation of the isolated human heart. *Circulation* **41**:899–912.

13. DiFrancesco D, Noble D. 1985. A model of cardiac activity incorporating ionic pumps and concentration charges. *Philos Trans Roy Soc London* **307**:353–398.

14. Ebihara L, Johnson EA. 1980. Fast sodium current in cardiac muscle. *Biophys J* **32**:779–790.

15. Ebihara L, Norikazu S, Lieberman M, Johnson EA. 1980. The initial inward current in spherical clusters of chick embryonic heart cells. *J Gen Physiol* **75**:437–456.

16. Gulrajani RM. 1998. *Bioelectricity and biomagnetism*. New York: John Wiley & Sons.

17. Gulrajani R, Mailloux GE. 1983. A simulation study of the effects of tissue inhomogeneities on electrocardiographic potentials using realistic heart and torso models. *Circ Res* **52**:43–56.

18. Horacek BM. 1971. The effect of electrocardiographic lead vectors of conductivity inhomogeneities in the human torso. PhD dissertation. Dalhousie University. Halifax, Canada. See also Lead theory. In *Comprehensive electrocardiography*. Ed PW Macfarlane, TD Veitch Lawrie. New York: Pergamon Press (1989).

19. Huiskamp GJM, van Oosterom A. 1988. The depolarization sequence of the human heart surface computed from body surface potentials. *IEEE Trans Biomed Eng* **35**:1047–1058.

20. Hyttinen J, Kauppinen P, Köbbi T, Malmivuo J. 1997. Importance of the tissue conductivity values in modeling the thorax as a volume conductor. In *Proceedings of the 19th International Conference of the IEEE on Engineering in Medicine and Biology*, pp. 2082–2085. New York: IEEE.

21. Johnson JC, Flowers NC. 1976. History of electrocardiography and vectorcardiography. In *The theoretical basis of electrocardiology*, pp. 381–396. Ed CV Nelson, DB Geselowitz. Oxford: Clarendon Press.

22. Gabor D, Nelson CV. 1954. Determination of the resultant dipole of the heart from measurements on the body surface. *J Appl Phys* **25**:413–416.

23. Geselowitz DB. 1965. Two theorems concerning the quadrupole applicable to electrocardiography. *IEEE Trans Biomed Eng* **12**:164-168.

24. Newton JC, Knisley SB, Zhou X, Pollard AE, Ideker RE. 1999. Review of mechanisms by which electrical stimulation alters the transmembrane potential. *J Cardiovasc Electrophysiol* **10**:234–243.

25. Klepfer RN, Johnson CR, MacLeod RS. 1997. The effects of inhomogeneities and anisotropies on electrocardiographic fields: a 3D finite element study. *IEEE Trans Biomed Eng* **44**:706–719.

26. Luo C-H, Rudy Y. 1994. A dynamic model of the cardiac ventricular action potential. *Circ Res* **74**:1071–1096.

27. McFee R, Johnston FD. 1953. Electrocardiographic leads, I, II, III. *Circulation* **8**:554–567; **9**:255–266 (1954); **9**:868–880 (1954).

28. Muller AL, Markin VS. 1978. Electrical properties of anisotropic nerve-muscle syncytia, II: spread of flat front of excitation. *Biophysics* **22**:536–541.

29. Oster H, Taccardi B, Lux RL, Ershler PR, Rudy Y. 1997. Noninvasivc electrocardiographic imaging: Reconstruction of epicardial potentials, electrograms, and isochrones and localization of single and multiple cardiac events. *Circulation* **46**:1012–1024.

30. Pollard AE, Hooke NF, Henriquez CS. 1992. Cardiac propagation simulation. *Crit Rev Biomed Eng* **20**:319–358.

31. Plonsey R, Barr RC. 1986. A critique of impedance measurements in cardiac tissue. *Ann Biomed Eng* **14**:308–322.

32. Plonsey R, van Oosterom A. 1991. Implications of macroscopic source strength on cardiac cellular activation models. *J Electrocardiol* **24**:99–112.

33. Roberts DE, Scher A. 1982. Effect of tissue anisotropy on extracellular potential fields in the canine myocardium in situ. *Circ Res* **50**:342–351.

34. Rudy Y, Oster H. 1992. The electrocardiographic inverse problem. *Crit Rev Biomed Eng* **20**:25–45.

35. Spach MS, Barr RC. 1976. Cardiac anatomy from an electrophysiologic viewpoint. In *The theoretical basis of electrocardiology*, pp. 3–20. Ed CV Nelson, DB Geselowitz. Oxford: Clarendon Press.

36. Spach MS, JF Heidlage. 1982. A multidimensional model of cellular effects on the spread of electrotonic currents and on propagating action potentials. *Crit Rev Biomed Eng* **20**:141–169.

37. Spach MS, Heidlage JF. 1993. A multidimensional model of cellular effects on the spread of electrotonic currents and on propagating action potentials. In *High-performance computing in biomedical research*, pp. 289–318. Ed TC Pilkington et. al. Boca Raton, FL: CRC Press.

38. van Oosterom A, van Dam R. 1976. Potential distribution in the left ventricular wall during depolarization. *Adv Cardiol* **16**:27–31.

39. Veenstra RD, DeHaan EL. 1986. Measurement of single channel currents from cardiac gap junctions. *Science* **233**:972–974.

40. Weingart R. 1986. Electrical properties of the nexal membrane studied in rat ventricular cell pairs. *J Physiol* **370**:267–284.

Additional References

Frank E. 1954. General theory of heart-vector projection. *Circ Res* **2**:258–270.

Frank E. 1957. Volume conductors of finite extent. *Ann NY Acad Sci* **65**:980–1002.

Gulrajani R, Savard P, Roberge FA. 1988. The inverse problem in electrocardiography: solutions in terms of equivalent sources. *CRC Crit Rev Biomed Eng* **16**:171–214.

Hoffman BF, Cranefield PF. 1960. *Electrophysiology of the heart.* New York: McGraw-Hill.

Johnson CR, MacLeod RS, Ershler PR. 1992. A computer model for the study of electrical current flow in the human thorax. *Comput Biol Med* **22**:305–323.

Liebman J, Plonsey R, Gillette P. 1982. *Pediatric electrocardiography.* Baltimore: Williams and Wilkins, Baltimore.

Nelson CV, Geselowitz DB, eds. 1976. *The theoretical basis of electgrocardiology.* Oxford: Clarendon Press.

Plonsey R. 1969. *Bioelectric phenomena.* New York: McGraw-Hill.

Plonsey R. 1965. Dependence of scalar potential measurements on electrode geometry. *Rev Sci Instrum* **36**:1034–1036.

Rosen MR, Janse MJ, Wit AL. 1990. *Cardiac electrophysiology: a textbook.* New York: Futura Publishing.

Selvester RH, Solomon JC, Gillespie TL. 1968. Digital computer model of a total body electrocardiographic surface map. *Circulation* **38**:684–690.

10

THE NEUROMUSCULAR JUNCTION

10.1. INTRODUCTION

It is a nerve bundle or nerve trunk that conveys excitatory impulses from the central nervous system to a target muscle. Each nerve fiber is normally myelinated, but at its terminal end it becomes unmyelinated. Also at the end the nerve fiber branches and each branch contacts a single (specific) muscle fiber within the whole muscle.

This anatomical arrangement is illustrated in Figure 10.1, where a total of three muscle fibers are shown being activated by a single nerve. The set of muscle fibers activated by one motor neuron, as illustrated, constitutes a *motor unit*. The individual fibers of one motor unit are found interdigitated with the fibers of other motor units.

The number of fibers in a motor unit may be small (around 5) for muscles with finely graded performance. Conversely, the number may be large (around 150) for muscles requiring less precision.

As shown in Figure 10.1, the muscle fibers are excited near their center. Thereafter action potentials propagate in both directions from this site toward the respective ends.

Each unmyelinated terminal branch of a motor (nerve) fiber will have a diameter on the order of 1.5 μm. The action potential propagating along this fiber does so because of the presence of local circuit currents, as described for fiber propagation in Chapter 6.

If such a nerve fiber were to be directly joined to a muscle fiber of, say, 50 μm in diameter, then as the propagating action potential reached the nerve–muscle junction, the local circuit current density set up in the muscle would be considerably reduced from that in the nerve. This reduction would occur because the nerve and muscle must have the same total current, a consequence of current conservation. However, since the muscle fiber/nerve fiber effective membrane area is at least the ratio of their respective radii, namely, (50/1.5) = 33.33, the ratio of their current densities must be in the reciprocal, that is, 0.03.

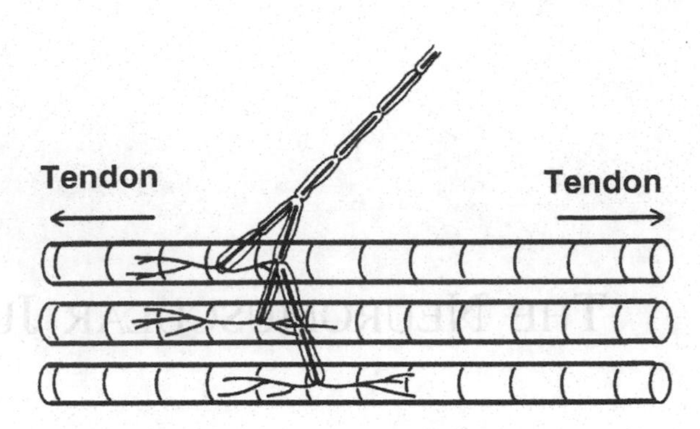

Figure 10.1. Motor Nerve Showing Branching to Activate Three Muscle Fibers. The nerve endings are unmyelinated, as shown. From Keynes RD, Aidley DJ. 1981. *Nerve and muscle*. Cambridge: Cambridge UP. Reprinted with the permission of Cambridge University Press.

Because λ for the muscle is greater than for the nerve, the effective muscle/nerve area ratio will be even larger. Thus the local circuit current generated by the nerve will result in a hundredfold or so lower current density in the muscle (assuming the nerve to muscle ratio of the product of diameter times length constant to be around 100). Consequently, the membrane depolarization produced in the muscle by the nerve action potential would be reduced by 1/100 times the depolarization the nerve induces in itself.

Although a safety factor exists in both the nerve and muscle, it is typically less than 10. Consequently, direct transfer of electrical activity from nerve to muscle cannot take place.

It is therefore no surprise to learn that the mechanism whereby the motor neuron stimulates its target muscle is not electrical but depends, instead, on a chemically mediated action. This chapter is devoted to the electrophysiology of the *neuromuscular junction*. In an engineering sense one can think of the neuromuscular junction in the role of an impedance transformer (from high-impedance nerve to low-impedance muscle).

The cell-to-cell coupling of the neuromuscular junction is an example of *synaptic transmission* of excitation from one cell to an adjoining cell. Such coupling arises not only between nerve and muscle, but also muscle to muscle and nerve to nerve. In most cases transmission is based on a chemical mediator, but there are examples where the coupling is electrical (as we found in cardiac muscle where the *synapse* is the gap junction).

We have limited our consideration in this chapter to the neuromuscular junction because that topic provides the needed background for the next two chapters. Actually, its general considerations would also apply at neural–neural synapses as well. In fact, much of the accumulated research on synaptic behavior has been derived from the study of the neuromuscular junction.

10.2. NEUROMUSCULAR JUNCTION

A sketch of a neuromuscular junction is shown in Figure 10.2 and a cross-sectional view of a single nerve terminal is shown in Figure 10.3. Illustrated in both figures is the very ending of

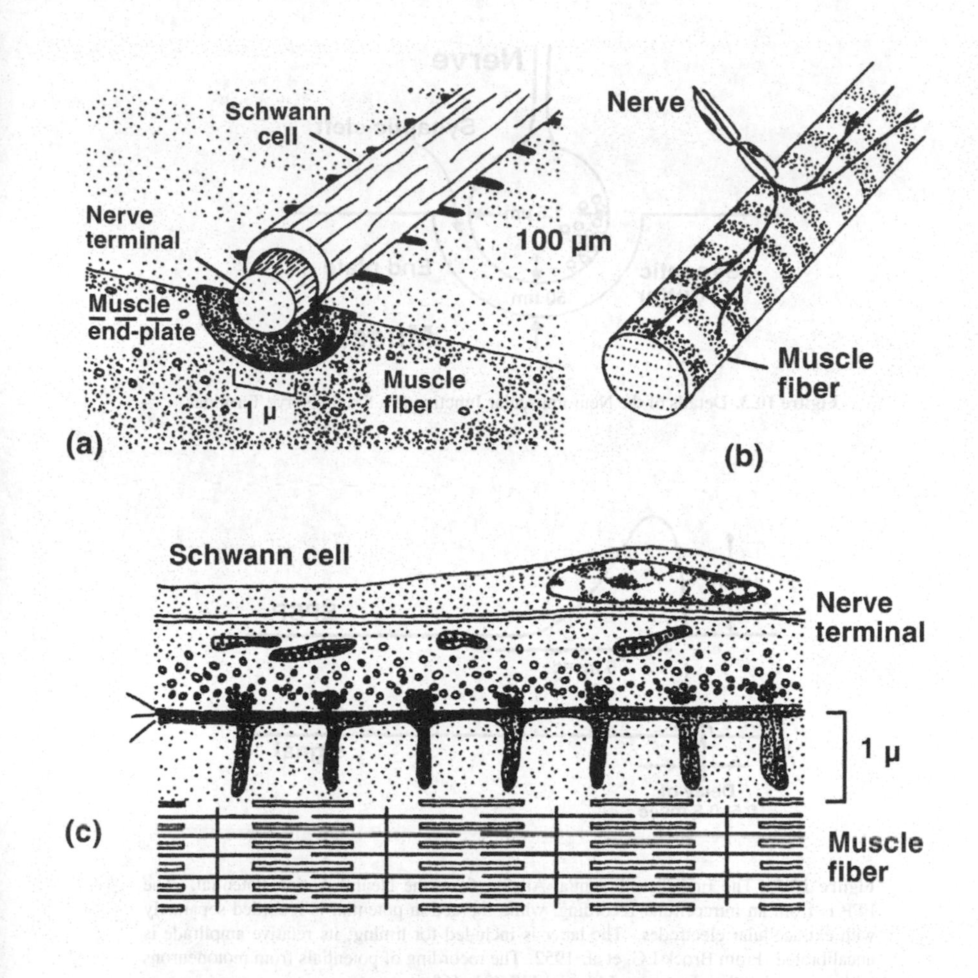

Figure 10.2. Neuromuscular Junction of Frog. (a) One portion of the junction. (b) General position of endings of motor axon on muscle fiber, showing portion (a) as a small rectangle. (c) Schematic drawing from electron micrographs of a longitudinal section through the muscle fiber. 1, terminal axon membrane; 2, "basement membrane" partitioning the gap between nerve and muscle fiber; 3, folded postsynaptic membrane of muscle fiber. From Katz B. 1966. *Nerve, muscle, and synapse*. New York: McGraw-Hill. Copyright ©1966, McGraw-Hill, with permission of the McGraw-Hill companies.

a single nerve branch at the muscle fiber it will activate. This region of neuromuscular contact is known as an *end plate*. The nerve cell is actually separated from the muscle cell by a gap of around 50 nm between their respective membranes. This neuromuscular structure exemplifies the broader class of *synaptic* junction, which characterizes transfer of excitation from one cell to another cell. The nerve lies in "synaptic gutters," which are infoldings in the postsynaptic (muscle) membrane. The length of a typical "gutter" is 100 μm, while its diameter is around 2 μm.

Within the nerve ending one sees the presence of spherical vesicles (around 50 nm in diameter) that contain a chemical transmitter. When a nerve impulse arrives at the nerve ending, some of

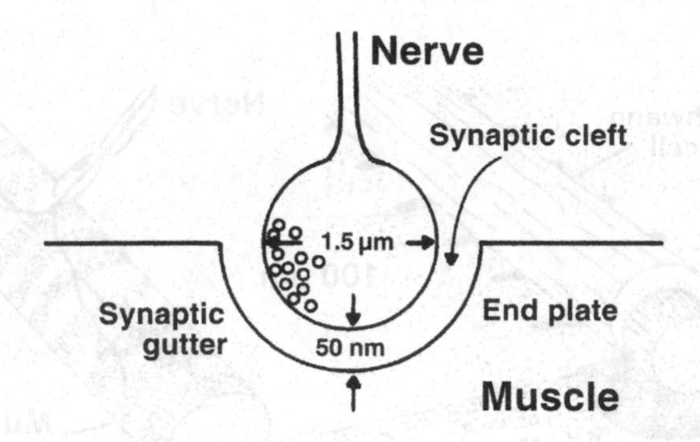

Figure 10.3. Details of the Neuromuscular Junction at a Single Nerve Terminal.

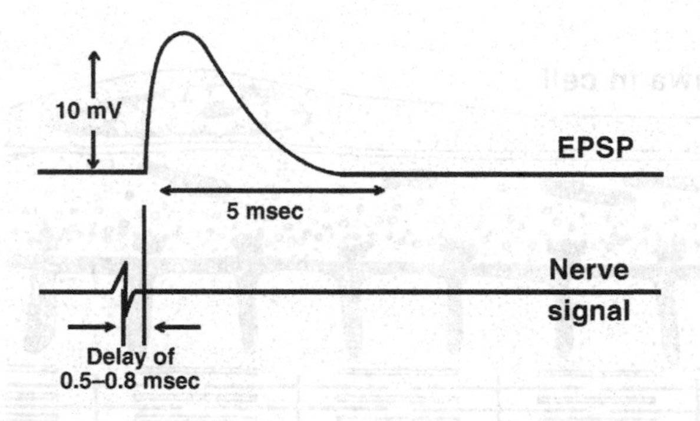

Figure 10.4. The End-Plate Potential Arising from the Neural Action Potential. The EPP is from an intracellular recording, while the action potential is recorded separately with extracellular electrodes. The latter is included for timing; its relative amplitude is uncalibrated. From Brock LG, et al. 1952. The recording of potentials from motoneurons with an intracellular electrode. *J Physiol* **117**:431–460.

these vesicles will be caused to fuse with the end-plate membrane and then empty their contents into the synaptic cleft (a process called *exocytosis*). The transmitter diffuses to the post-junctional membrane, where it complexes with receptors resulting in permeability changes, giving rise to the depolarization of the post-junctional membrane. This depolarization initiates the activation of the entire post-junctional cell. A microelectrode placed in the post-junctional cell can measure the ultimate effect, as shown in Figure 10.4.

In the general case of a synapse, an *excitatory* or *inhibitory* electrical response is elicited, depending on the nature of the transmitter; the neuromuscular junction is always excitatory. A typical EPSP, or *excitatory postsynaptic potential*, is illustrated in Figure 10.4; it follows the (illustrated) nerve signal that gives rise to it. The neuromuscular (excitatory) transmitter, as mentioned above, is acetylcholine (ACh).

The ACh channel is one of the earlier membrane channels to be studied with the patch electrode to reveal single-channel behavior. In addition, its complete amino acid and gene sequences have also been completely determined. In fact, a functioning ACh channel has been obtained by reinserting a purified macromolecule into a lipid membrane [6]. A model of this receptor is shown in Figure 4.1.

This channel differs from ion channels that are voltage gated (discussed in Chapter 4) in that gate opening and closing depends on a *ligand* (i.e., the transmitter ACh) rather than a voltage gradient. Other ligands include glutamate, glycine, and γ-aminobutyric acid, as well as acetylcholine.

Figure 10.5 gives additional details on the acetylcholine receptor. In panel A, a three-dimensional reconstruction of the channels is shown, adapted from Toyoshima C, Unwin N. 1982. Ion channel of acetylcholine receptor reconstructed from images of postsynaptic membranes. *Nature* **336**:247–250. Panel B presents an AChR model consisting of the 5 subunits. The internal pore admits Na^+ and K^+ ions, following binding of ACh to each of the α subunits. Panel C gives the molecular structure of one of the two α subunits, which has four membrane spanning regions, showing its five subunits consisting of two α plus β, γ and δ polypeptides. The channel protein has a molecular weight of 280,000 Da. Channel opening and closing requires the binding of both α units with an ACh molecule (i.e., two ACh required for each channel opening).

10.3. QUANTAL TRANSMITTER RELEASE

Even in the absence of stimulation of the presynaptic nerve, *miniature end-plate potentials* (MEPP) of 0.5 mV per MEPP, appearing every second on the average, are seen. These potentials are of interest, since they cast light on the mechanism of transmitter release and its take-up by receptors. These potentials are blocked by curare, which is known to interfere with the ability of postsynaptic receptors to respond to ACh. The MEPP simply arises from the random subthreshold release of ACh.

In fact, the aforementioned release is *quantized*, since each MEPP is of similar amplitude, except for a rare occasion when an MEPP of double value arises. This effect is easy to understand, since each vesicle holds about the same quantity of transmitter and therefore elicits a similar MEPP (except when two vesicles are simultaneously released and a double amplitude MEPP is seen).

The less likely possibility that the receptor response is quantized was discarded when it was observed that iontophoretic application of different quantities of ACh directly to the receptor resulted in EPPs whose magnitudes are proportional to ACh size, as shown, for example, by Kuffler and Yoshikami [3].

A postsynaptic potential waveform similar to the MEPP arises as a response to stimulation when the end-plate region lies in a reduced Ca^{++} (and/or increased Mg^{++}) medium, since these ionic conditions greatly attenuate release of ACh, thereby permitting an examination of the response from smaller numbers of ACh packets. In response to a nerve stimulus one may observe no EPP (i.e., no end-plate potential, hence zero quantal release), or a response similar to an MEPP (i.e., a single quantum of ACh), and responses that are small integral multiples of an MEPP.

The normal EPP is thus simply the sum of a large number of MEPPs, where the single (isolated) MEPP arises from the release of a single quantum. (For human muscle, the average number of ACh quanta is around 30.)

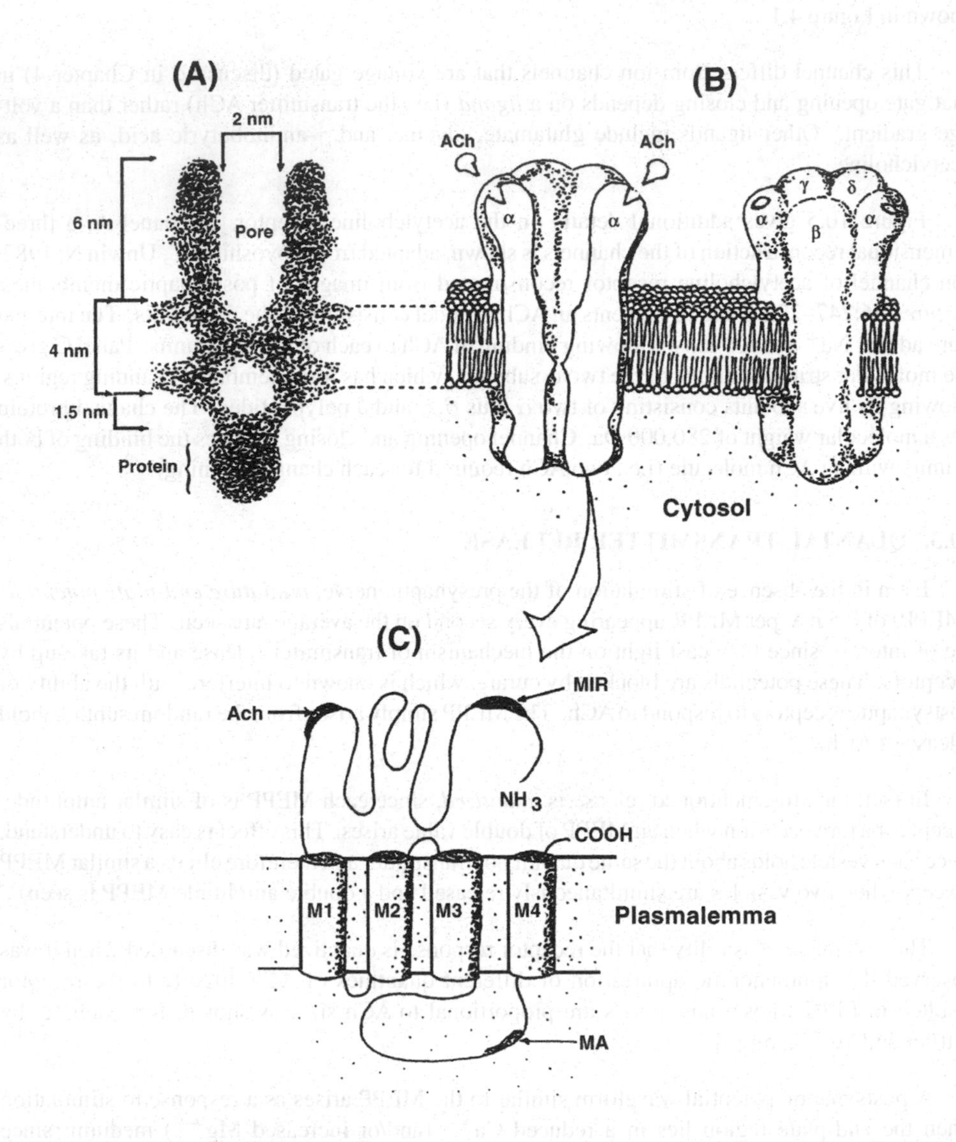

Figure 10.5. Details of the ACh Receptor. ACh binding to both α units is necessary for gate opening. Also shown is the amino acid sequence of the α subunit and their membrane crossings. Adapted from Beeson D, Barnard E. 1970. Acetylcholine receptors at neuromuscular junctions. In *Neuromuscular transmission: basic and applied*. Ed A Vincent and D Wray. Manchester: Manchester UP. Reprinted with permission from McComas AJ. 1996. *Skeletal muscle: form and function*. Champaign, IL: Human Kinetics.

10.4. TRANSMITTER RELEASE, POISSON STATISTICS

The picture of transmitter release arising from nerve stimulation suggests that a quantitative description can only be given probabilistically. Since there are a large number of vesicles and the probability of release of any one vesicle is small, the overall response to any stimulus can be described by a Poisson distribution. The details are established as follows.

Suppose there are *n* release sites and that the probability of release at any specific site is *p*, where *p* assumes the presence of a nerve stimulus and depends also on the concentration of Ca^{++}, Mg^{++}, and other ions.

The probability of *exactly x* release sites (of *n*) is given by the binomial distribution

$$f(x) = \frac{n!}{x!(n-x)!} p^x q^{n-x} \tag{10.1}$$

where $q = 1 - p$ and *n* is the total number of sites. This expression is derived by first noting that the number of different ways *x* sites can be release sites and $(n - x)$ non-release sites is

$$\binom{n}{x} = \frac{n!}{x!(n-x!} \tag{10.2}$$

Furthermore, the probability of realizing any *one* of the aforementioned is $P^x q^{n-x}$. Since $\binom{n}{x}$ is the number of equally likely ways of realizing exactly *x* of *n* release sites, and each configuration is statistically independent, then the probability of exactly *x* release sites is the product of $\binom{n}{x}$ with $P^x q^{n-x}$, and this is given in (10.1).

If we consider $n \to \infty$, $p << 1$, $np \to m$, $x << n$ (*x* remains small and independent of *n*), then we have, approximately,

$$n(n-1)\cdots(n-x+1) \simeq n \cdot n \cdots n = n^x \tag{10.3}$$

Also, for $p << 1$, we have

$$q = (1-p) \simeq e^{-p}, \quad q^{(n-x)} \simeq e^{(n-x)p} \simeq e^{-np} \tag{10.4}$$

The second expression in (10.4) follows, since we assume that $n >> x$. Consequently,

$$\frac{n!}{x!(n-x)!} = \frac{n(n-1)\cdots(n-x+1)}{x!} = \frac{n^x}{x!} \tag{10.5}$$

and

$$f(x) = \frac{n^x p^x e^{-np}}{x!} = \frac{e^{-m} m^x}{x!} \tag{10.6}$$

where $m = np$. The quantity *m* is the average quantal release per trial (under normal conditions). The final result in Eq. (10.6) is a *Poisson distribution*.

One can estimate m from experimental measurements by comparing the mean EPP amplitude (proportional to m quanta) with the mean MEPP amplitude, where the latter arises from a single quanta. Thus,

$$m = \frac{\text{mean amplitude of EPP}}{\text{mean amplitude of MEPP}} \tag{10.7}$$

This result assumes that EPP is linearly proportional to quantal release. This assumption is satisfactory for small amounts. For large release, a saturation effect is noted in the resulting EPP. In this case a better estimate is given by

$$m = \frac{\overline{\upsilon}}{\overline{\upsilon}_1}\left(1 - \frac{\overline{\upsilon}}{V_0}\right)^{-1} \tag{10.8}$$

where $\overline{\upsilon}$ is the average EPP, $\overline{\upsilon}_1$ is the average MEPP, and V_0 is the maximum possible EPP. This expression is described by Martin [5].

An alternate evaluation of m is to count the number of failures, n_0, to elicit any response in N trials, where N is reasonably large. Then, using (10.6), we have

$$f(0) \approx \frac{n_0}{N} = \left(\frac{e^{-m}m^x}{x!}\right)_{x=0} = e^{-m} \tag{10.9}$$

Taking the log of (10.9) and solving for m yields

$$m = \ln\left(\frac{N}{n_0}\right) \tag{10.10}$$

Because values of m found from (10.7) and (10.10) agree in experiments that have been conducted, the process appears to be, indeed, Poisson. The factor m is referred to as the *quantal content.*

The effect of elevating Mg^{++} or depressing Ca^{++} is to reduce p (increase q). The accuracy of the Poisson model is diminished by temporal variations in n and spatial variations in p. The above statistical behavior is usually studied in reduced Ca^{++} or elevated Mg^{++}, to attenuate release greatly so that observations are simplified.

The assumption that each quantum is associated with the release of the contents of a single vesicle appears to be confirmed by direct experiments. If the presynaptic membrane is removed and ACh applied directly to the postsynaptic terminal, one can calibrate the amount of ACh needed for an MEPP and thereby evaluate the amount of one quantum.

In the experiments of Kuffler and Yoshikami [3] it was determined that one quantum represented around 10,000 ACh molecules. This agrees, roughly, with the 6000 molecules of ACh contained in a vesicle that is 50 nm in diameter. Since approximately 1700 ACh molecules actually bind to receptor sites, the process may be said to be 10–20% efficient.

While the transmitter content of a single vesicle and the MEPP it produces have been regarded as fixed quanta in the foregoing discussion, in fact there are small variations in these quantities. The result is that distributions in the EPP amplitudes in a large number of trials comprise a smooth curve (with many peaks) rather than a discontinuous one. This fact does not alter the basic ideas presented above but adds an additional complication in evaluating end-plate response.

10.5. TRANSMITTER RELEASE, Ca^{++} AND Mg^{++}

The required presence of extracellular calcium for release of transmitter is demonstrated by the absence of release when Ca^{++} is not present in the perfusate (even though postganglionic cells could still be directly depolarized by ACh).

More recent experiments also demonstrate that intracellular calcium must bind to intracellular membrane proteins (release sites) for release to take place. The pre-junctional terminal contains large numbers of Ca^{++} channels, and these facilitate Ca^{++} entry near release sites.

Dodge and Rahamimoff [2] proposed a model in which it is assumed that Ca^{++} and Mg^{++} bind to a presynaptic structure, X, so that

$$\mathrm{Ca} + X \leftrightarrows \mathrm{Ca}X(K_1) \tag{10.11}$$

and

$$\mathrm{Mg} + X \leftrightarrows \mathrm{Mg}X(K_2) \tag{10.12}$$

Where K_1 and K_2 are dissociation constants, and CaX facilitates while MgX inhibits quantal release.

If X_0 is the total amount of X available both in its unbound and bound form, then by definition (with [ ] denoting concentration),

$$[\mathrm{X}_0] = [X] + [\mathrm{Ca}X] + [\mathrm{Mg}X] \tag{10.13}$$

or, rearranging, we have

$$[X] = [X_0] - [\mathrm{Ca}X] - [\mathrm{Mg}X] \tag{10.14}$$

From the law of mass action applied to (10.11) and (10.12),

$$[\mathrm{Ca}][X_0] - [\mathrm{Ca}X] - [\mathrm{Mg}X] = K_1[\mathrm{Ca}X] \tag{10.15}$$

and

$$[\mathrm{Mg}][X_0] - [\mathrm{Ca}X] - [\mathrm{Mg}X] = K_2[\mathrm{Ma}X] \tag{10.16}$$

Solving for [MgX] from (10.15) gives

$$[\mathrm{Mg}X] = -\frac{K_1[\mathrm{Ca}X]}{[\mathrm{Ca}]} + [X_0] - [\mathrm{Ca}X] \tag{10.17}$$

Substituting (10.17) into (10.16) results in

$$K_1[\mathrm{Mg}][\mathrm{Ca}X] = K_2[X_0][\mathrm{Ca}] - K_2[\mathrm{Ca}X][\mathrm{Ca}] - K_1K_2[\mathrm{Ca}X] \tag{10.18}$$

Rearranging (10.18) leads to

$$[\mathrm{Ca}X]\{K_1[\mathrm{Mg}] + K_2[\mathrm{Ca}] + K_1K_2\} = K_2[X_0][\mathrm{Ca}] \tag{10.19}$$

Solving for $[\mathrm{Ca}X]$, and then dividing numerator and denominator on the right-hand side by K_1K_2, yields

$$[\mathrm{Ca}X] = \left(\frac{[X_0]}{K_1}\right)\frac{[\mathrm{Ca}]}{\left(1 + \frac{[\mathrm{Mg}]}{K_2} + \frac{[\mathrm{Ca}]}{K_1}\right)} \tag{10.20}$$

The concentration of $[\mathrm{Ca}X]$ is a measure of the facilitation of transmitter release. From (10.20), we note that it is reduced by [Mg], increased by $[X_0]$, and depends on the dissociation constants K_1 and K_2. It also increases with an increase in $[\mathrm{Ca}^{++}]$ by an amount n that depends on Mg^{++} as well as the dissociation constants. This equation is in a form that is familiar in enzyme kinetics.

Experimental evidence from Dodge and Rahamimoff [2], who developed the above expressions, suggests that the dependence of EPP amplitude on $[\mathrm{Ca}^{++}]$ can be given by

$$\text{EPP amplitude} = k[\mathrm{Ca}X]^m \tag{10.21}$$

where k and n are constant. Curve fitting leads to $n = 4$. Using (10.20) and (10.21) leads to

$$\text{EPP amplitude} = k\left(\frac{W[\mathrm{Ca}]}{1 + \frac{[\mathrm{Ca}]}{K_1} + \frac{[\mathrm{Mg}]}{K_2}}\right)^4 \tag{10.22}$$

where

$$W = \frac{[X_0]}{K_1} \tag{10.23}$$

Taking the fourth root of (10.22) and inverting gives, finally,

$$\frac{1}{\text{EPP amplitude})^{1/4}} = \frac{1}{K_1k^{1/4}W} + \frac{1}{K^{1/4}W}\left(\frac{[\mathrm{Mg}]}{K_2} + 1\right)\frac{1}{[\mathrm{Ca}]} \tag{10.24}$$

so that $(\text{EPP amplitude})^{-1/4}$ is linear with $[\mathrm{Ca}]^{-1}$.

Equation (10.24) fits the experimental data of Dodge and Rahamimoff [2] quite well (with $n = 3.8$). These results support the notion that elevation of extracellular calcium concentrations increases EPP amplitude, with magnesium playing an inhibitory role. That n has the value of ~ 4 was interpreted by Dodge and Rahamimoff as requiring the simultaneous presence of four calcium ions at each X receptor site to facilitate ACh release.

While the presence of extracellular calcium is required for ACh release, it is the intracellular calcium that actually triggers it. A direct confirmation of the facilitative role of calcium can be obtained by the use of a calcium-sensitive photoprotein *aequorin*. Measurement of the light signal from the aequorin gives a quantitative measure of calcium entry into the presynaptic terminal of a cell upon stimulation.

If Ca entry is blocked by depolarizing to a more positive potential than the calcium Nernst potential E_{Ca} then the EPSP should be abolished. Depolarizations to greater than +130 mV do, in fact, accomplish this, thereby confirming that it is the calcium entry resulting from the arrival of the action potential at the end plate that is responsible for transmitter release.

10.6. POST-JUNCTIONAL RESPONSE TO TRANSMITTER

The postsynaptic membrane contains a localized region in which there are specialized ACh receptors and associated channels (as described in Figure 10.5). The receptor density is estimated at $104/\mu m^2$, which is an order of magnitude greater than that of sodium and potassium channels of squid axon. ACh molecules released at the pre-junctional membrane diffuse across the synaptic cleft and form a transmitter–receptor complex at the receptor sites that opens the associated channels (specifically, the binding of two ACh molecules being required to open the ACh channel). When open, ionic flow can take place.

The behavior of the synaptic membrane is shown symbolically in Figure 10.6. A portion of the postsynaptic membrane contains receptor channels. This portion is available for current flow only when the channels are open (hence the inclusion of a switch that closes only during ACh release).

The associated ion currents are described by the parallel-conductance model introduced in Chapter 3. The current flow of the xth ion is $I_x = g_x(V_m - E_x)$, where V_m is the transmembrane potential and E_x is the Nernst potential. The remaining postsynaptic membrane is described by a normal subthreshold passive muscle membrane with a resting conductance and resting potential characterized by (g_r, E_r).

The conductance values for the receptor channels in Figure 10.6 are not to be confused with those in excitable membranes but are unique to synaptic membranes, where they depend on the properties of the ACh channel. One important difference is that the temporal behavior of the synaptic conductance is a simple exponential decay. Magleby and Stevens [4] show that $G_s = ke^{-\alpha t}$, where $\alpha = Be_m^{AV}$. So the transmembrane potential affects only the decay rate.

As we know, the ACh channel is ligand (rather than voltage) sensitive. The binding of ACh molecules to receptors occurs very rapidly and the conductance reaches its maximum value essentially instantaneously.

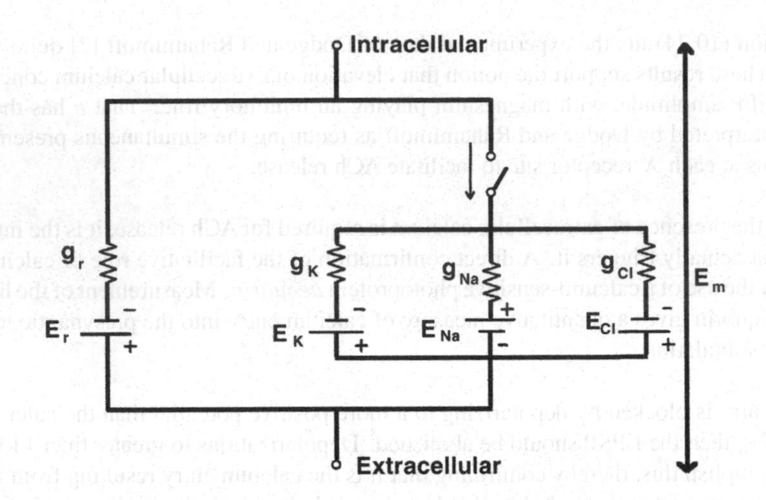

Figure 10.6. Parallel-Conductance Model of Postsynaptic Membrane that is influenced by transmitter, and remaining cell membrane (in resting state) [6]. The switch closes at the point of arrival of ACh when the circuit describes the instant of maximum conductance. From Junge D. 1981. *Nerve and muscle excitation*. Sunderland, MA: Sinauer Associates. Based on Takeuchi A, Takeuchi J. 1960. On the permeability of end-plate membrane during the action of transmitter. *J Physiol* **154**:52–67.

In addition to the localized synaptic membrane, Figure 10.6 shows the remaining membrane of the post-junctional cell described by the lumped elements E_r and g_r, which represent the resting transmembrane potential and resting membrane conductance. This membrane is more completely characterized by normally excitable properties, such as described in Chapter 5. The model in Figure 10.6 requires inclusion of active elements when the transmembrane potential approaches threshold.

Upon transmitter release, and hence the closing of the switch in Figure 10.6, the current, I_s, in the synaptic channels can be expressed as

$$I_s = g_K(V_m - E_K) + g_{\mathrm{Na}}(V_m - E_{\mathrm{Na}}) + g_{\mathrm{Cl}}(V_m - E_{\mathrm{Cl}}) \tag{10.25}$$

where E_K, E_{Na}, and E_{Cl} are the Nernst potentials of potassium, sodium, and chloride ions.

The EPP reversal potential (the voltage clamp value at which no current flows when the transmitter is applied) is found by setting $I_s = 0$ in (10.25) to yield

$$V_m^{\mathrm{rev}} = \frac{g_{\mathrm{K}}E_{\mathrm{K}} + g_{\mathrm{Na}}E_{\mathrm{Na}} + g_{\mathrm{Cl}}E_{\mathrm{Cl}}}{g_{\mathrm{K}} + g_{\mathrm{Na}} + g_{\mathrm{Cl}}} \tag{10.26}$$

Experiments performed with different extracellular ion concentrations (i.e., with variations in $[\mathrm{K}]_e$, $[\mathrm{Cl}]_e$, and $[\mathrm{Na}]_e$) permit g_{K}, g_{Na}, and g_{Cl} in (10.26) to be estimated.

Such experiments were performed by Takeuchi and Takeuchi [7] and show that for the frog neuromuscular junction the reversal potential is independent of the chloride Nernst potential, E_{Cl}.

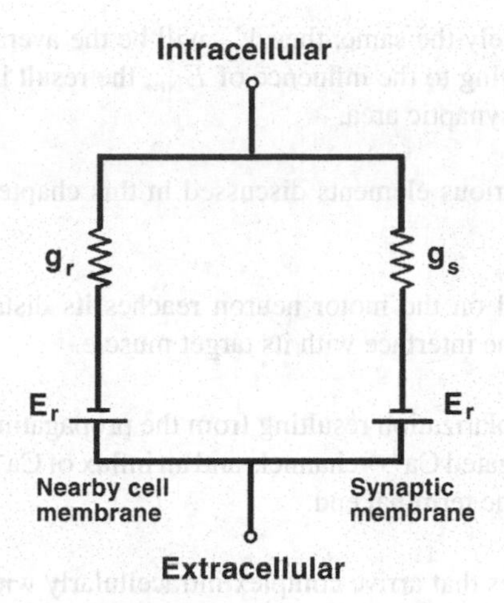

Figure 10.7. Simplified Electrical Model of Postsynaptic Junction and Adjoining Cell Membrane Following Release of Transmitter and Activation of Synaptic Channels. From Junge D. 1981. *Nerve and muscle excitation*. Sunderland, MA: Sinauer Associates. Used with permission.

It was assumed, consequently, that $g_{\mathrm{CL}} \approx 0$. Changes in $[\mathrm{Na}]_e$ and $[\mathrm{K}]_e$ followed the predictions of (10.26) provided $g_{\mathrm{Na}}/g_{\mathrm{K}} \simeq 1.29$.

Both sodium and potassium permeabilities are elevated almost equally by ACh when acting at the post-junctional membrane. The effect is a true *depolarization* of this membrane. The specific change in ionic conductance depends on the species and tissue being studied.

Substances such as α-bungarotoxin and curare also combine with the postsynaptic receptors, but do so without opening the channels. They consequently compete with the ACh for binding sites and hence interfere with the normal functioning of the neuromuscular junction. This effect is reflected quantitatively by reductions in the potassium and sodium conductances of the synaptic region for the same amount of transmitter release.

If all elements in the network of Figure 10.6 are taken into account, then (since the total current into any node must be zero to satisfy conservation of current)

$$g_r(V_m - E_r) + g_{\mathrm{K}}(V_m - E_{\mathrm{K}}) + g_{\mathrm{Na}}(V_m - E_{\mathrm{Na}}) = 0 \tag{10.27}$$

and consequently

$$V_m = \frac{g_r E_r + g_{\mathrm{K}} E_{\mathrm{K}} + g_{\mathrm{Na}} E_{\mathrm{Na}}}{g_r + g_{\mathrm{K}} + g_{\mathrm{Na}}} \tag{10.28}$$

where V_m is the postsynaptic potential.

With the simplifying assumption that $g_{\mathrm{K}} = g_{\mathrm{Na}} = g_s/2$ and $(E_{\mathrm{K}} + E_{\mathrm{Na}})/2 = E_s$ [where, based on (10.25), E_s is the reversal potential], we get an alternate expression for the postsynaptic potential, reflecting the simplified circuit in Figure 10.7, namely,

$$V_m = \frac{E_r g_r + E_s g_s}{g_r + g_s} \tag{10.29}$$

If g_r and g_s are approximately the same, then V_m will be the average of E_r and E_s. Since E_s is more positive than E_r owing to the influence of E_{Na}, the result is a depolarization of the cell membrane that adjoins the synaptic area.

Putting together the various elements discussed in this chapter we have the following sequence:

1. The action potential on the motor neuron reaches its distal ending at the presynaptic terminal, which is the interface with its target muscle.

2. The membrane depolarization resulting from the propagating action potential causes the opening of voltage-gated Ca^{++} channels and an influx of Ca^{++}, the Ca^{++} channels being relatively dense at the terminal end.

3. The Ca^{++} molecules that arrive complex intracellularly with the release sites and result in the fusing of synaptic vesicles with the terminal membrane.

4. Through the process of exocytosis, ACh from synaptic vesicles is released into the synaptic cleft.

5. ACh diffuses across the synaptic gap to the postsynaptic ACh channels lying in the muscle membrane. When a pair of ACh molecules bind with the channel protein, the channel opens.

6. A consequence of ACh channel opening is that sodium and potassium are both free to flow based on high and nearly equal conductances, thereby depolarizing the post-junctional membrane and initiating an action potential in the muscle fiber.

10.7. REFERENCES

1. Boyd IA, Martin AR. 1956. The end-plate potential in mammalian muscle. *J Physiol* **132**:74–91.

2. Dodge FA, Rahamimoff R. 1967. Co-operative action of calcium ions in transmitter release at the neuromuscular junction. *J Physiol* **193**:419–432.

3. Kuffler SN, Yoshikami D. 1975. The number of transmitter molecules in a quantum: an estimate from iontophoretic application of acetylcholine at the neuromuscular synapse. *J Physiol* **251**:465–482.

4. Magleby KL, Stevens CF. 1972. The effect of voltage on the time course of end-plate currents. *J Physiol* **223**:151–171.

5. Martin AR. 1955. A further study of the statistical composition of the end-plate potential. *J Physiol* **130**:114–122.

6. Montal M, Anholt R, Labana P. 1986. The reconstituted acetylcholine receptor. In *Ion channel reconstitution*, pp. 157–204. Ed C Miller. New York: Plenum Press.

7. Takeuchi A, Takeuchi N. 1960. On the permeability of end-plate membrane during the action of transmitter. *J Physiol* **154**:52–67.

Additional References

Aidley DJ. 1978. *The physiology of excitable cells*. Cambridge: Cambridge UP.

Junge D. 1981. *Nerve and muscle excitation*. Sunderland, MA: Sinauer Associates.

Katz B. 1966. *Nerve, muscle and synapse*. New York: McGraw-Hill.

Kenes RD, J. 1981. Aidley D. *Nerve and muscle*. Cambridge: Cambridge UP.

Stein RB. 1980. *Nerve and muscle*. New York: Plenum Press.

11

SKELETAL MUSCLE

The goal of the material in this chapter is to provide a very brief introduction to *skeletal muscle*, its structure, and its electrophysiological and contractile properties.[1]

11.1. MUSCLE STRUCTURE

A whole muscle can be divided into separate bundles. Each bundle contains many individual fibers. The fiber is the basic (smallest) functional unit (it constitutes a single cell). It is bounded by a plasma membrane and a thin sheet of connective tissue, the *endomysium*. The bundles are also surrounded by a connective tissue sheet, the *perimysium*, which delineates specific *fascicles*. The whole muscle is encased in its connective tissue sheet, namely, the *epimysium*.

Most skeletal muscles begin and end in tendons. Muscle fibers lie parallel to each other, so the force of contraction contributed by each is additive. The general features noted above are illustrated in Figure 11.1. In this chapter attention will be primarily directed to the electromechanical properties of the single muscle fiber.

Each muscle fiber is made up of many fibrils, each of which, in turn, is divisible into individual filaments. The filaments are composed of contractile proteins, essentially myosin, actin, tropomyosin, and troponin.

Mature fibers may be as long as the muscle of which they are a part (tens of centimeters); they vary in diameter from 10 to 100 μm. As noted above, each fiber contains myofibrils, which are proteins and which lie in the cytoplasm. The cytoplasm also contains mitochondria, the SR and T systems, plus glycogen granules. When examined under light microscopy (LM), the myofilaments show characteristic cross-striations (banding), which are in register in all myofilaments (see Figure 11.1.) It is the latter property from which skeletal muscle derives the alternate name of *striated* (muscle).

The overall physical features of a muscle fiber are shown in Figure 11.2. This shows the dense packing of myofibrils, the *transverse tubular system* (TTS), and the *sarcoplasm reticulum*

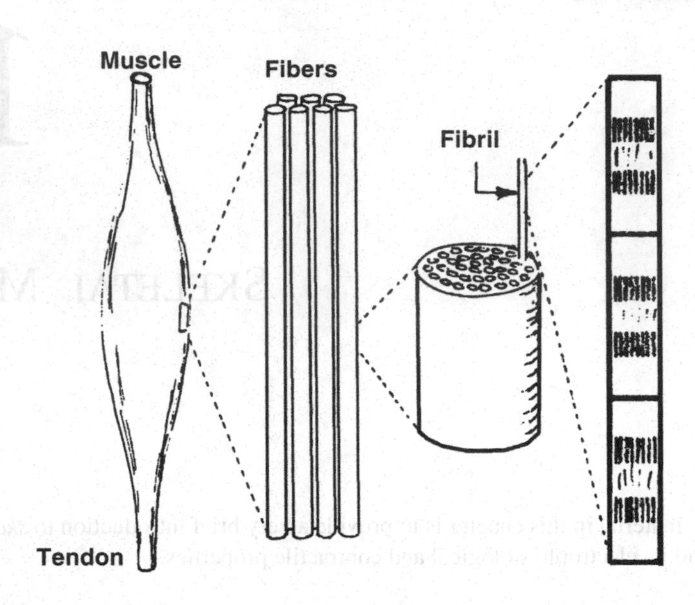

Figure 11.1. Structure of a Whole Muscle and Its Components. The cross-striations are visible under light microscopy. From Keynes RD, Aidley DL. 1981. *Nerve and muscle*. Cambridge: Cambridge UP. Based on Schmidt-Nielsen K. 1979. *Animal physiology*. Cambridge: Cambridge UP. Reprinted with the permission of Cambridge University Press.

(SR). Both the bounding membrane and the TTS membrane are excitable and play an important part in the process whereby contraction is initiated.

11.2. MUSCLE CONTRACTION

Each mammalian muscle fiber is contacted by a single nerve terminal. The muscle fiber is known as a *twitch* fiber, since the response to a single nerve stimulus is a twitch. The time to reach the peak of a typical twitch contraction is around 200 msec, while recovery requires an additional 600 msec.

In normal activity a muscle will shorten as it develops force (tension). However, experiments are often carried out under conditions of constant muscle length (isometric) as well as under conditions of constant muscle load (isotonic). To study behavior under isometric conditions, a transducer is needed that converts force into an electrical signal while itself undergoing very little deflection.

If a second stimulus is applied before the effect of the previous twitch has ended, then the second (twitch) response will build on the residual of the first and *summation* results. Corresponding to a long inter-stimulus interval, a "bumpy" response is seen. For increasing stimulus frequency, a value will be reached where the bumps disappear and a smooth buildup to a maximum steady level results, as illustrated in Figure 11.3. The frequency is known as the *fusion* frequency, and the muscle is said to be in *tetanus*. The peak twitch tension to the maximum tetanus tension is the twitch/tetanus ratio, which is about 0.2 for mammalian muscle.

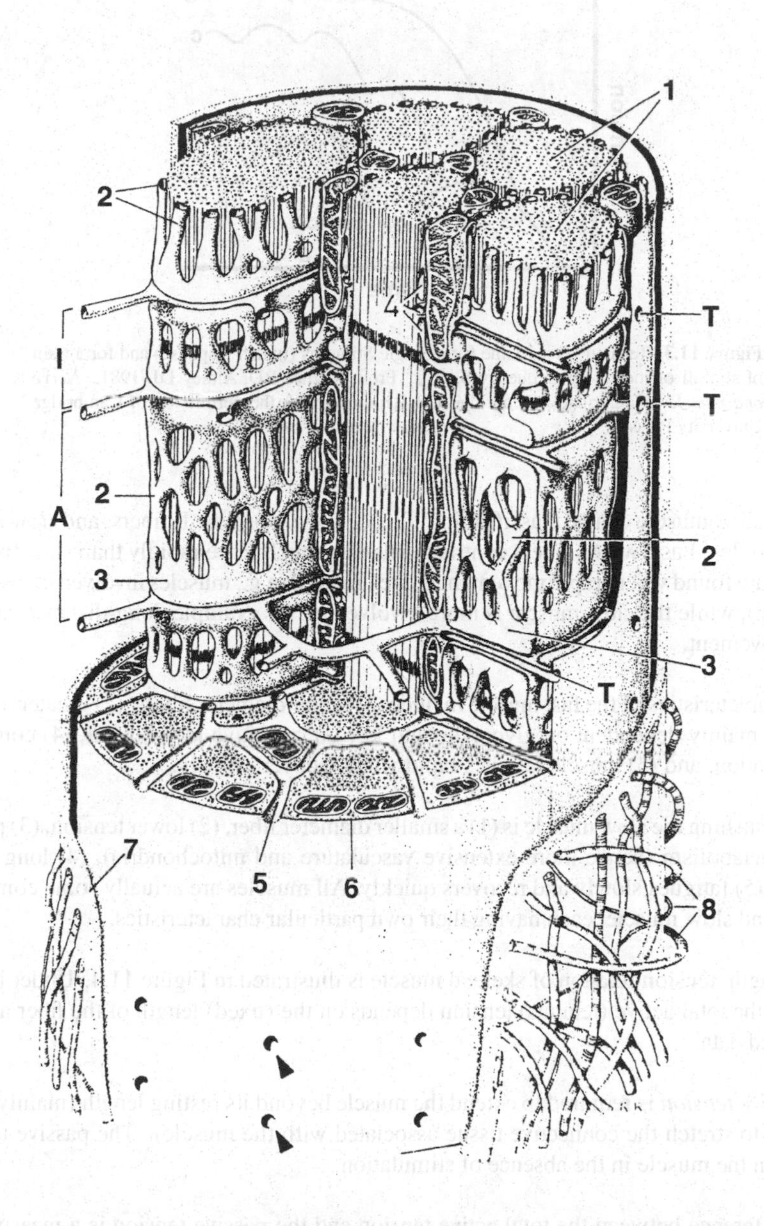

Figure 11.2. Magnified View of the Structure of a Single Muscle Fiber, with a cutaway view of the myofibrillar structure. Each fibril is surrounded by a sarcoplasmic reticulum (SR) and by the transverse tubules system (TTS), which opens to the exterior of the fiber. From Krstic RV. 1970. *Ultrastructure of the mammalian cell.* Berlin: Springer-Verlag, with permission.

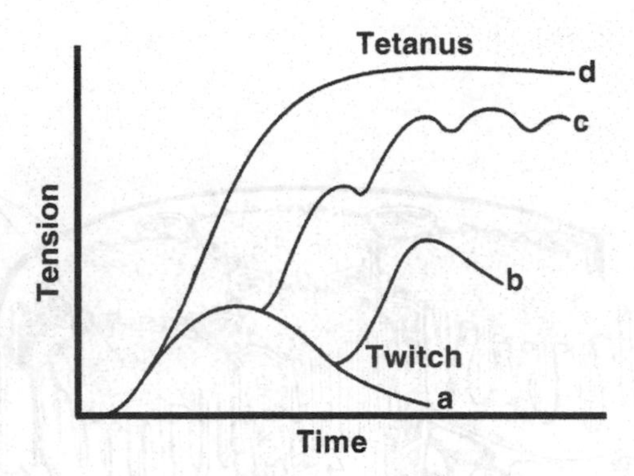

Figure 11.3. Tension versus Time for a Single Stimulus (twitch response) and for a train of stimuli of increasing frequency b, c, d. From Keynes RD, Aidley DJ. 1981. *Nerve and muscle*. Cambridge: Cambridge UP. Reprinted with the permission of Cambridge University Press.

Mammalian muscle can be classified into *fast glycolytic* or type II fibers, and *slow oxidative* or type I muscle.[2] Fast (white) fibers contract and relax much more rapidly than slow (red) ones. The former are found where rapid movement is encountered (e.g., muscles involved in fast running and jumping), while the slow muscle is more involved in, for example, long-distance running or postural movement.

The characteristics of the fast muscle include (1) larger diameter fibers, (2) greater developed tension, (3) mainly dependent on glycolytic and less on oxidative metabolism, (4) contractions of short duration, and (5) muscle fatigues rapidly and recovers slowly.

Distinguishing the slow muscle is (1) a smaller diameter fiber, (2) lower tension, (3) primarily oxidative metabolism (hence more extensive vasculature and mitochondria), (4) long-duration twitch, and (5) fatigues slowly and recovers quickly. All muscles are actually some combination of the fast and slow muscle, each having their own particular characteristics.

The length–tension relation of skeletal muscle is illustrated in Figure 11.4. Under isometric conditions, the total active (tetanus) tension depends on the (fixed) length of the fiber according to the plotted data.

A *passive tension* is required to extend the muscle beyond its resting length (mainly because of the need to stretch the connective tissue associated with the muscle). The passive tension is measured on the muscle in the absence of stimulation.

The difference between the total active tension and the passive tension is a measure of the contractile force derived from stimulation and is called the *active increment*. The latter quantity reaches a maximum at the resting length and is lower for either greater or lesser lengths. An explanation of this behavior is given in a subsequent section.

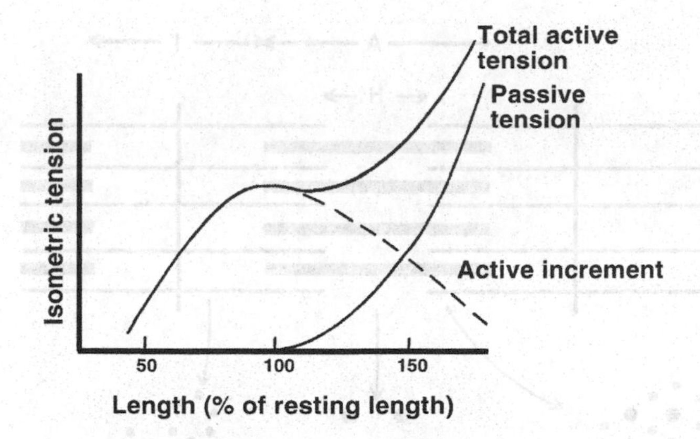

Figure 11.4. Length–Tension Relationship for a Skeletal Muscle under Isometric Conditions. From Keynes RD, Aidley DJ. 1981. *Nerve and muscle*. Cambridge: Cambridge UP. Reprinted with the permission of Cambridge University Press.

11.2.1. Structure of the Myofibril

Each fiber contains a large number of cylindrical (protein) constituents called *myofibrils*. The banded structure seen for the fiber as a whole is, in fact, a consequence of the Saffie banding and alignment of the individual fibrils. The banding corresponds to the structure of the protein components of the myofibril, namely, the *thick* and *thin* filaments.

The thick filaments are around 11 nanometers in diameter, while the thin filaments are around 5 nm in diameter.

The arrangement of these filaments is shown in Figure 11.5a, where it is seen that in the cross-section they are interdigitated in a hexagonal array, while along the axis they lie in a recurring pattern of overlapping and non overlapping regions. When viewed lengthwise, the banding effect arises from the relative amounts of transmitted light permitted by the thick and thin filaments.

In Figure 11.6 we show both the structural organization of the thin and thick filaments and the associated banding that would be observed in the LM. The two main bands are the dark A band and the lighter I band. The bands alternate regularly along the myofibril. In the middle of the I band is the *Z line* (dark line), while the middle of the A band has a lighter region, the *H zone*. The H zone is bisected by a darker *M line* surrounded by a lighter region, the L zone (not always seen). The repeating unit (Z–Z distance) is the *sarcomere*.

These characteristic bands of different light intensity derive from the underlying thin and thick filament structure, the major elements of which can be recognized in Figure 11.6.

- The dark A band arises from the overlapping thin and thick filaments, while the lighter H zone reflects the presence of thick filaments alone.

- The M line and L zone derive from the structural details of the thick filament at its center, the M line from crosslinks at the center.

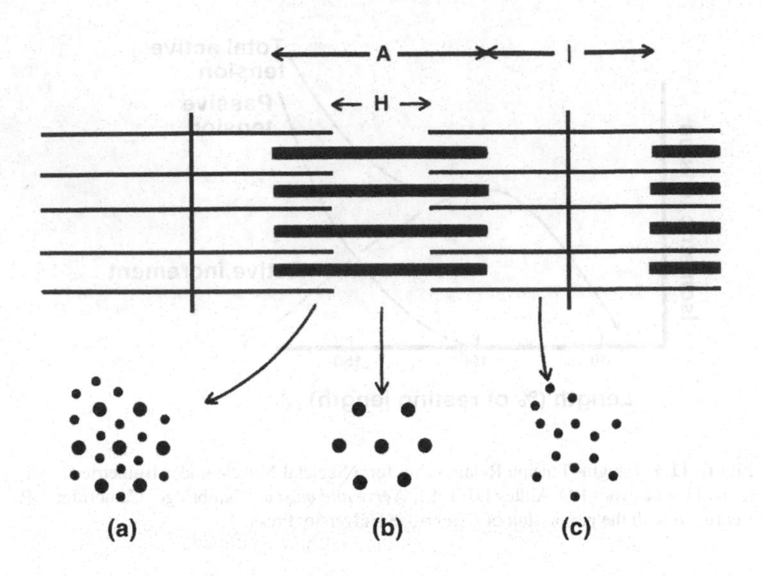

Figure 11.5. Axial and Cross-Sectional View of a Portion of the Array of Thin and Thick Filaments that constitutes a single fibril. The cross-section at (a) registers the presence of both thin and thick filaments, while that at (b) thick filaments only, and at (c) thin filaments only. From Aidley DJ. 1978. *The physiology of excitable cells*. Cambridge: Cambridge UP. Reprinted with the permission of Cambridge University Press.

- The L zone is due to the absence of projections on either side of the thick filament (to be described later); the L zone is around 0.15 μm in width.
- The Z line reflects the interconnection of the I filaments from the region to its left and its right.
- The above letters are derived from the German and reflect certain properties of their designated regions. They are A = anisotropic (polarizes light), I = isotropic, Z = zwischenscheibe, H = Henrens disc, and M = mittlemembrane

The thick filament is made up of *myosin*, a complex protein. Trypsin splits it into *light meromyosin* (LMM) and *heavy meromyosin* (HMM). The latter has a short tail and two globular heads; it has an ATPase behavior (i.e., it hydrolyzes ATP into ADP + P with the release of large amounts of energy). The light meromyosin is rod-like and does not split ATP.

The thin filament is *actin*, which is also a protein. There are two forms, but neither has ATPase behavior. (The important ATPase activity is actually confined only to the globular sub fragments.) The LMM and the tail of the HMM are composed of two α-helices that coil around each other. When combined in a solution, the actin and myosin form a complex called *actomyosin* (a quite viscous material). A description of the myosin structure is given in Figure 11.7.

Glycerol-extracted fibers are prepared by soaking muscle fibers in 50% glycerol for several weeks, a process that removes most sarcoplasmic material except for the contractile elements. The fibers are found to be in *rigor* (they are stiff and resist contraction, a result of the formation

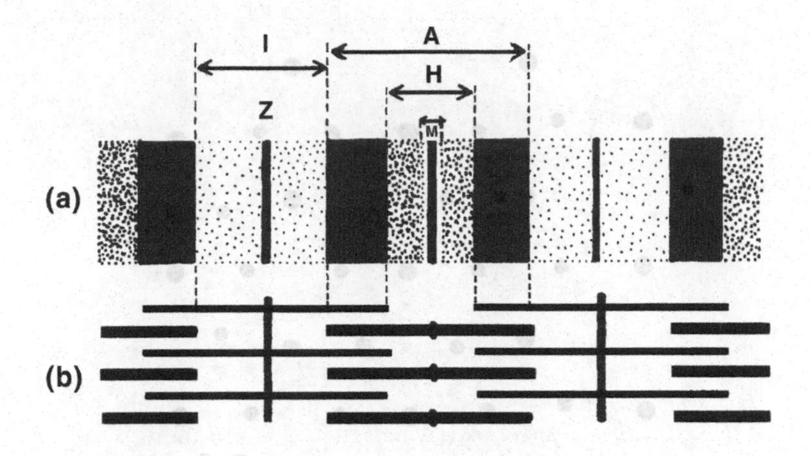

Figure 11.6. Myofibrillar Structure and Associated Pattern Seen in a Light Microscope. The banding nomenclature is given. The observed pattern of light intensity in (a) can be explained by the underlying structure shown in (b). From Keynes RD, Aidley J. 1981. *Nerve and muscle*. Cambridge University Press. (a) is based on a photograph by Dr. HE Huxley. Reprinted with the permission of Cambridge University Press.

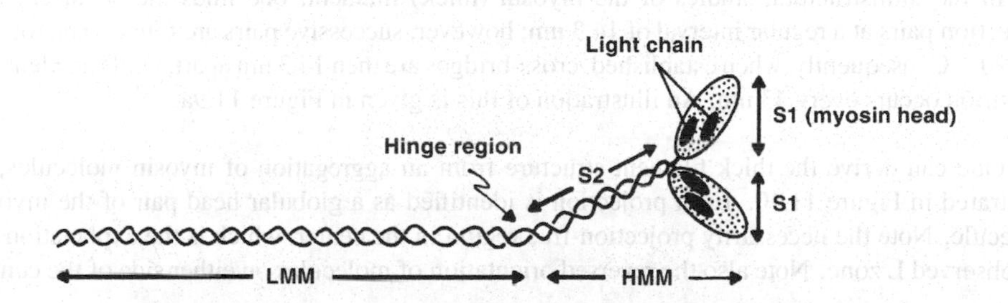

Figure 11.7. Different Components of the Myosin Molecule. Proteolytic enzymes cleave the molecule into heavy meromyosin (HMM) and light meromyosin (LMM). The HMM comprises a short segment of the α-helical rod (S2) and the two globular heads (S1), to which the light chains are attached. The globular heads form the cross-bridges. Reprinted by permission from McComas AJ. 1996. *Skeletal muscle*, Champaign, IL: Human Kinetics. Based on Vibert P, Cohen C. 1988. Domains, motions, and regulation in the myosin head. *J Muscle Res Cell Motility* **9**:296–305, and Rayment I, et al. 1993. Structure of the actin–myosin complex and its implications for muscle contraction. *Science* **261**:58–65.

of cross-bridges between the actin and myosin). If ATP and magnesium are added, the fibers become readily extensible due to the breakage of crosslinks by the ATP. If Ca^{++} is also added, then contraction takes place.

In the transverse plane, the relative positions of the thin and thick filaments in a region of overlap is as illustrated in Figure 11.8. One notes that each thick filament is surrounded by six thin filaments, while each thin filament is surrounded by three thick ones. Hence, there are twice as many thin as thick filaments.

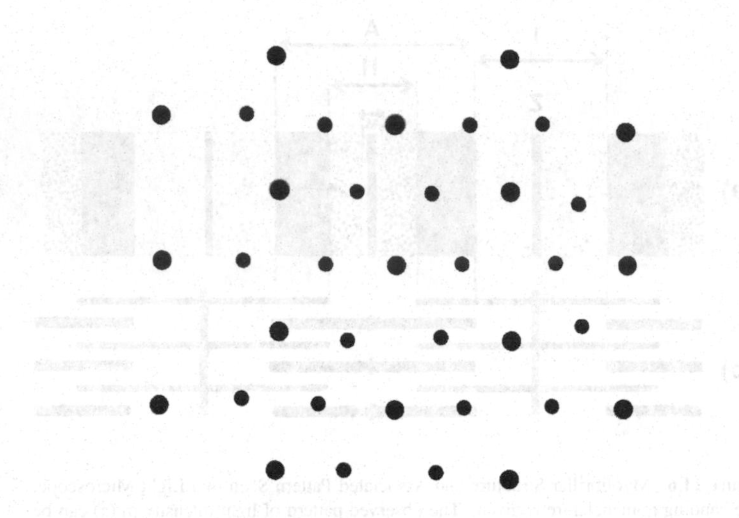

Figure 11.8. Transverse Plane View of the Thin and Thick Filament Structure in an axial plane in which they overlap (see Figure 11.6).

In the ultrastructural studies of the myosin (thick) filament, one finds the occurrence of projection pairs at a regular interval of 14.3 nm; however, successive pairs are found to be rotated by 120°. Consequently, when established, cross-bridges are then 14.3 nm apart, while an identical repetition occurs every 43 nm. An illustration of this is given in Figure 11.9a.

One can derive the thick filament structure from an aggregation of myosin molecules, as illustrated in Figure 11.10. Each projection is identified as a globular head pair of the myosin molecule. Note the necessarily projection-free region in the center, which is the explanation for the observed L zone. Note also the reversed orientation of molecules on either side of the center.

11.3. SLIDING FILAMENT THEORY

The idea that muscular contraction is a consequence of the contraction of protein units patterned after that of a helical spring had to be abandoned when measurements revealed that the A band does not change length during contraction or lengthening.

In fact, in frog muscle, as the sarcomere length is varied from 2.2 to 3.8 μm, the I filaments remain essentially at 2.05 μm in length and the A filaments at around 1.6 μm. (The Z line is $\approx$ 0.05 μm wide, and each side of the I filament has a length of 1.0 μm, to account for the total of 2.05 μm.)

As a consequence of the above, the *sliding-filament* model was advanced. According to this idea, contraction involves the relative movement of the thin and thick filaments, as illustrated in Figure 11.11, where contraction yields a reduced sarcomere length while the filaments are unchanged in length.

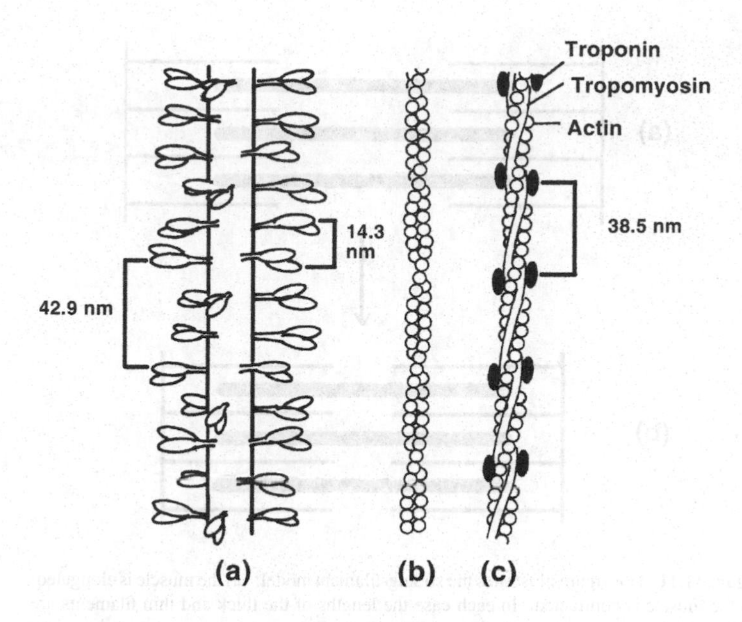

Figure 11.9. Models of the Structure of the Thick and Thin Filaments: (a) myosin; (b) F-actin; (c) thin filament. In (a) the two globular heads of myosin, which split ATP, are shown (a more detailed view is given in Figure 11.11). From Keynes RD, Aidley DJ. 1981. *Nerve and muscle*. Cambridge: Cambridge UP. Reprinted with the permission of Cambridge University Press. Based on Offer G. 1978. The molecular basis of muscular contraction. In *Companion to biochemistry*, Ed AT Bull et al. London: Longman; Huxley HE, Brown W. 1967. The low angle x-ray diagram of vertebrate striated muscle and its behavior during contraction and rigor. *J Mol Biol* **30**:383–434; and Huxley HE. 1972. Molecular basis of contraction in cross-striated muscles. In *Structure and function of muscle*, 2nd ed., pp. 301–387. Ed GH Bourne. New York: Academic Press.

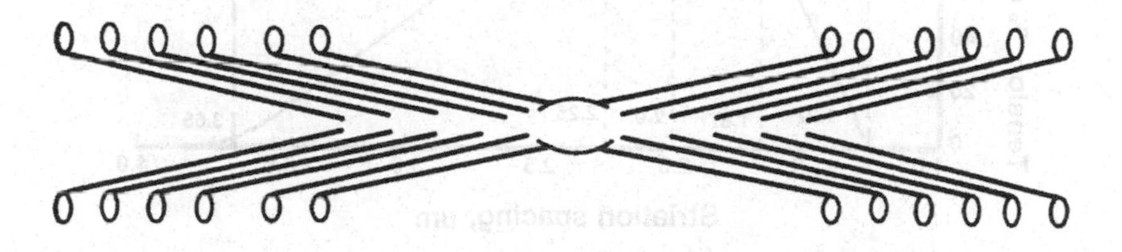

Figure 11.10. Huxley's Suggestion as to How Myosin Molecules Aggregate to Form a Thick Filament. See also Figure 11.2 for details of myosin structure. From Huxley HE. 1971. The structural basis of molecular contraction. *Proc R Soc* **178**:131–149. Redrawn in Aidley DJ. 1978. *The physiology of excitable cells*. Cambridge: Cambridge UP.

The sliding itself is thought to be produced by reactions between the projections on the myosin filaments and active sites on the thin filament. Each projection first attaches itself to the actin filament to form a *cross-bridge*, then pulls on it, causing the sliding of the actin, then releases it, and finally moves to attach to another site which is further along the thin filament.

The sliding filament theory is generally (though not universally) accepted. Accordingly, one expects isometric tension to depend on the degree of overlap in the thin and thick filaments. This

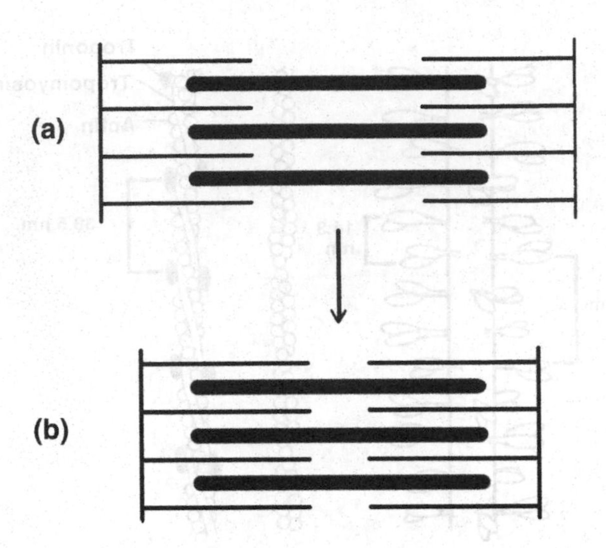

Figure 11.11. This figure illustrates the sliding-filament model: (a) the muscle is elongated; (b) the muscle is contracted. In each case the lengths of the thick and thin filaments are unchanged.

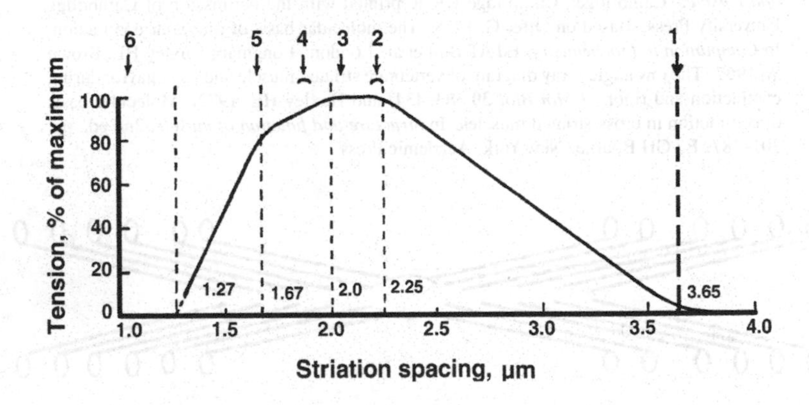

Figure 11.12. Isometric Tension of a Frog Muscle Fiber, measured as a percentage of its maximum value at different sarcomere lengths. The numbers 1–6 refer to the myofilament positions illustrated in Figure 11.13. Note that the general shape is anticipated in Figure 11.13. From Gordon AM, Huxley AF, Julian FJ. 1966. The variation in isometric tension with sarcomere length in vertebrate muscle fibers. *J Physiol* **184**:170–192. Redrawn by Aidley DJ. 1978. *The physiology of excitable cells*. Cambridge: Cambridge UP.

result is supported by the study illustrated in Figures 11.12 and 11.13 and can be understood in the following discussion.

Stage 1 (in Figures 11.12 and 11.13) refers to full extension of the myofibril. Using the dimensions given above for the thin and thick filament lengths, the sarcomere length is 2.05 +

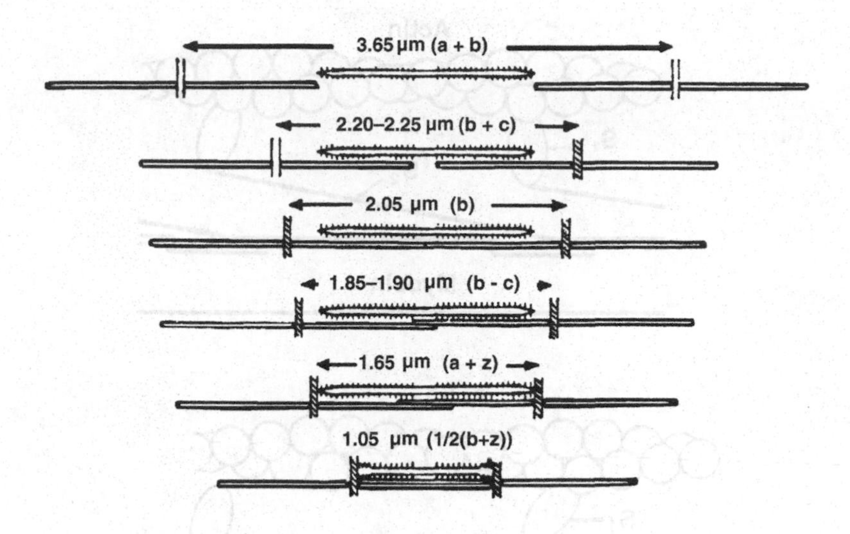

Figure 11.13. Myofilament Arrangements at Different Lengths. The numbers are the positions corresponding to the curve given in Figure 11.12. a = thick filament length (1.6 μm); b = thin filament length including z line (2.05 μm); c = thick filament region base of projections (0.15 μm); and z = z line width (0.05 μm). From Gordon AM, Huxley AF, Julian FJ. 1966. The variation in isometric tension with sarcomere length in vertebrate muscle fibers. *J Physiol* **184**:170–192. Redrawn in Aidley DJ. 1978. *The physiology of excitable cells*. Cambridge: Cambridge UP.

1.60 = 3.65 μm, which is the sum of the length of the thin plus thick filament. There can be no cross-bridges and the observed zero tension is explained on this account.

As the myofibril shortens so that the sarcomere diminishes from 3.6 to 2.2–2.25 μm (stage 2), the number of cross-bridges increases linearly with decreasing length. Therefore, the isometric tension should show a similar increase. In fact, such an increase in tension with decreased length is seen in Figure 11.12. This linear behavior ends at stage 2, when the Z–Z distance equals 2.05 μm plus the L zone width ($\approx$ 0.15 μm), or 2.20 μm.

With further shortening, the number of cross-bridges remains unchanged and a plateau in tension is both expected and observed. Stage 3 is reached when the thin filaments touch. The sarcomere equals the length of the thin filament, namely, 2.05 μm at this point.

From stage 3 to stage 4 one anticipates some internal resistance to shortening to develop, since actin filaments now overlap.

Beyond stage 4 this overlap not only constitutes a "frictional" impediment, but it also interferes with cross-bridge formation. When stage 5 (1.65 μm) is reached, the myosin filaments hit the Z line and a further increase in resistance is associated with the deformation that results beyond this point.

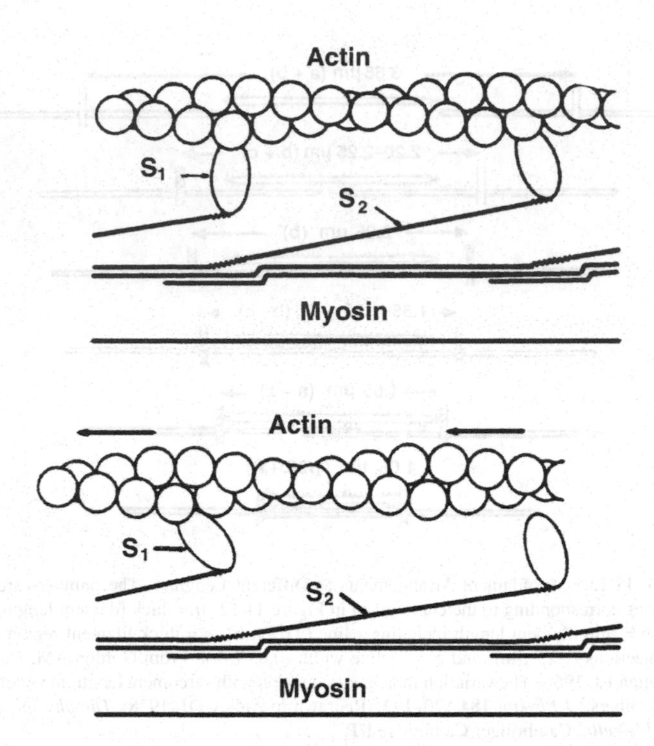

Figure 11.14. Interaction of Actin and Myosin on a Molecular Level. From Huxley HE. 1975. The structural basis for contraction and regulation in skeletal muscle. In *Molecular basis of motility*. Ed LMG Heilmeyer et al. Berlin: Springer.

The curve in Figure 11.12 shows a break point at stage 5 and a rapid decrease in tension with further shortening. Zero tension is reached at a sarcomere length of 1.3 μm, which designates stage 6.

The actin structure is described in Figure 11.9b,c and in Figure 11.14 as a double helix involving chains of monomers. The thin filament is made up of actin, troponin, and tropomyosin, as shown in Figure 11.9c. The thick filament is shown in Figure 11.14 as containing an S_2 filament subunit and the S_1 (globular head) subunits. The S_1 subunits can rotate about their point of attachment with S_2. Together, S_1 and S_2 make up the heavy meromyosin (HMM) portion of the myosin molecule; the remainder of the molecule is filamentary and constitutes the light meromyosin (LMM) (see Figure 11.7).

Sliding is accomplished by the rotation of S_1 about S_2, as noted earlier. In the upper portion of Figure 11.14, the left-hand cross-bridge has just attached while the S_1 subunit of the right-hand one has nearly completed its rotation. The lower diagram, which illustrates conditions a moment later, shows the S_1 subunit on the left-hand cross-bridge having rotated to cause the actin filament to slide leftward; the right-hand cross-bridge is now separated.

There are two S_1 cross-bridges for each myosin molecule, and each cross-bridge is relatively independent of the other, though each behaves as described here.

The biochemical events associated with these mechanical events can be described according to the following sequence:

1. Myosin is released from a cross-bridge with actin. This results from the action of ATP with which the myosin combines. That is,

$$\mathrm{AM} + \mathrm{ATP} \rightarrow \mathrm{A} + \mathrm{M} \cdot \mathrm{ATP}$$

   where A $\equiv$ actin and M $\equiv$ myosin).
2. ATP is split into ADP + P, while the myosin (S_2) repositions for reattachment with the thin filament. The products remain attached to the myosin, which now has a high affinity for actin.
3. Myosin cross-bridges attach to a new actin monomer.
4. This results in products being released and the energy so derived utilized as the power stroke (rotation of S_2 and linear movement of actin). At this point, return to step 1.

While actin will react with pure myosin so as to split ATP in the absence of calcium ions, when tropomyosin and troponin are also present, calcium ions are required. In the case of muscle, the tropomyosin and troponin are, in fact, always present and appear to exert a regulatory (control) role.

11.4. EXCITATION–CONTRACTION

The details of the process, starting with propagation of an action potential along a muscle fiber and ending with contraction of the target muscle, can now be examined. The possibility that the influx of calcium ions, associated with the membrane depolarization, is the primary initiator of the contractile mechanism has to be discarded since only about a 0.2 picomole Ca^{++}/cm^2 influx is observed (frog sartorius). This amount corresponds to an increase in internal calcium ion concentration of only 0.08 μmole (assuming a fiber diameter of 50 μm).

To better understand contemporary ideas, one must include the presence of the sarcoplasmic reticulum (SR) and the transverse tubular system (TTS). The T-system lies transverse to the fiber axis and consists of tubules that are open to the extracellular space and form a meshwork shaped somewhat as the spokes of a wheel (described in Figure 11.2).

The TTSs are located at the Z lines of frog muscle and the A–I boundary in most other striated muscle. The SR is in close proximity to the T-system but extends in the axial direction, mainly. It constitutes a network of vesicular elements surrounding the myofibrils. It is not directly connected to the TTS and is otherwise isolated from extracellular space.

Excitation propagating along the surface membrane of the muscle fiber passes the outside opening of each of the many T-tubules. It is believed that this excitation can propagate inward, that the membranes defining the T-tubules are excitable in the usual way. The inward speed of conduction has, in fact, been measured and is about 7 cm/sec (in a fiber 100 μm in diameter a latency of 0.7 msec from outside to inside would consequently be observed).

The SR, while not continuous with the TTS, in places, is in close proximity via a structure called "feet." The SR sequesters Ca^{++} (which is pumped into the SR vesicles by an ATP-driven calcium pump). This sequestration can reduce the calcium ion concentration in the muscle to a point below that necessary for contraction (i.e., it results in the relaxation of the muscle).

Then activation results from the action potential propagating throughout the TTS, which in turn results in a movement of ions to open the calcium channels in the SR membrane. This results in a release of Ca^{++} from the SR into the myoplasm. The consequent contractile process then arises as described earlier.

We assumed in the above that in the presence of tropomyosin and troponin Ca^{++} is required for ATP to be split. The tropomyosin and troponin appear, in fact, to be a structural component of the thin filament, as described in Figure 11.9c. The tropomyosin in the resting muscle is positioned to prevent the myosin heads combining with the actin monomers, but it can be moved out of the way by a conformational change in the troponin complex when calcium binds to troponin C.

11.5. NOTES

1. The interested reader can find further information in [1, 2, 3]. These works were the primary sources for the material of this chapter.
2. A continuum of fiber types actually exists and this classification describes those at each end of this "spectrum." Other classification schemes have been proposed, but this choice identifies the basic differences that are found.

11.6. REFERENCES

1. Junge D. 1981. *Nerve and muscle excitation*. Sunderland, MA: Sinauer Associates.
2. Katz B. 1966. *Nerve, muscle and synapse*. New York: McGraw-Hill.
3. Kenes RD, Aidley DJ. 1981. *Nerve and muscle*. Cambridge: Cambridge UP.

Additional References

Aidley DJ. 1978. *The physiology of excitable cells*. Cambridge: Cambridge UP.
Stein RB. 1980. *Nerve and muscle*. New York: Plenum Press.
Kenes RD, Aidley DJ. 1991. *Nerve and muscle*, 2nd ed. Cambridge: Cambridge UP.

12

FUNCTIONAL ELECTRICAL STIMULATION

12.1. INTRODUCTION

If a motor nerve is stimulated from an external electrode, the resulting action potential will propagate to the innervated muscle and a twitch will be produced. The muscle responds to the artificially initiated nerve signal just as it would a naturally occurring signal.[1]

For patients with, for example, spinal cord injury, signals originating in the brain may be unable to reach the desired motoneuron because of a transected cord. In this case, the affected muscle is paralyzed although it may otherwise be healthy and capable of excitation and contraction. In this situation an artificial signal initiated in the nerve will evoke a response. Devising strategies for the stimulation of motoneurons or the muscle itself to effect desired muscle contraction is the goal of functional neuromuscular stimulation (FNS), and the subject of this chapter.

This topic was selected for two reasons. First, it presents real human needs calling for solutions in which biomedical engineering can play an important role. Second, it represents an interesting and challenging application of much of the material presented in this text.

12.2. ELECTRODE CONSIDERATIONS

A key element in functional electrical stimulation (FES) is the initiation of an action potential on a desired nerve, while at the same time refraining from stimulating other nerves nearby. To work toward this goal requires consideration of the effect of electrode(s) size, shape, and location, and the strength and waveform of the stimulating current. Of course, one also needs to know the nerve geometry, its electrical properties, and that of the volume conductor. In addition to depolarizing currents, hyperpolarizing signals must also be considered when the goal is to block unwanted traffic or for prepolarizing purposes.

When a stimulating current is applied at or within a volume conductor, a solenoidal (closed loop) current field is established. Within the wires carrying current to the electrodes and including the electrodes themselves, current is in the form of a metal conduction current and the carrier is the

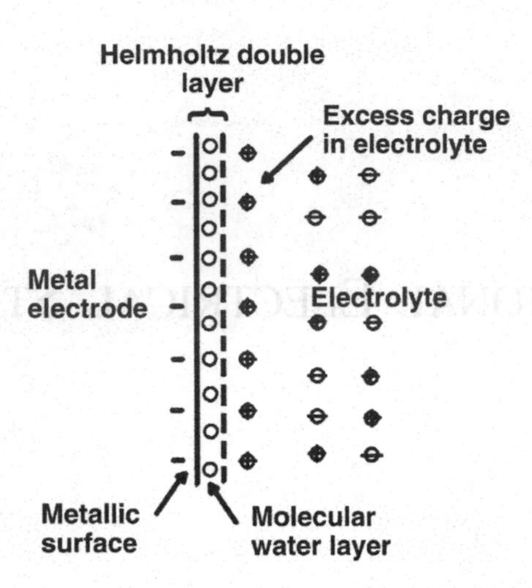

Figure 12.1. Idealized Cross-Sectional View of the Metal–Tissue Interface of an Electrode (cathode) under very low (zero) current conditions. From AM Dymond. 1976. Characteristics of the metal–tissue interface of stimulation electrodes. *IEEE Trans Biomed Eng* **23**:274–280. Reprinted with permission, copyright ©1976, IEEE.

(conduction) electron. Within the tissue, current is carried by ions, primarily sodium, potassium, and chloride.

At the electrode tissue interface an electrochemical reaction is necessary that, in part, exchanges metal electrons for ions in solution. An important consideration is that the reaction not produce products that are toxic to the tissue or have deleterious effects on the electrode itself.

12.2.1. Electrode–Tissue Interface

As illustrated in Figure 12.1, at the metal–electrolyte interface the battery (generator) is the source of a net charge in the metal phase which is positive (at the anode) and negative (at the cathode). In the electrolyte, an opposite charge develops that is separated from the electrode itself by a molecular layer of water that is adsorbed on the metal surface. The charged layer in the metal and the electrolyte constitutes a (charged) capacitance; the charges are held together by electrostatic forces.

The magnitude of the layer's capacitance depends inversely on the separation of the charged surfaces, as is known from fundamentals of electricity. Since in this case the thickness is molecular, the capacitance is remarkably high. An estimate may be found as follows.

Our model is highly idealized, but an order-of-magnitude estimate of capacitance can be obtained by regarding the separation of capacitor "plates" to be the diameter of a water molecule (0.35 nm) and assigning a relative permittivity, $\kappa = 6$, to the adsorbed water. This permits an estimate of capacitance per unit area, *C*, as described in Chapter 2, and given by

$$
\begin{aligned}
C &= \frac{\varepsilon_0 \kappa}{d} \\
&= \frac{6 \times 8.84 \times 10^{-6}}{3.5 \times 10^{-10}} \\
&= 15.2 \times 10^4 \mu\text{F/m}^2 \\
&= 15.2\, \mu\text{F/m}^2
\end{aligned} \tag{12.1}
$$

where $\varepsilon_0 = 10^8/(36\pi) = 8.84 \times 10^6$ μF/m^2. A capacitance of $C = 10$ to 20 μF/cm^2 is seen experimentally, but this close agreement with the above estimate is only fortuitous and not a confirmation of the model.

The capacitance evaluated by (12.1) appears to be constant, but it is not found to be so in practice. The reason is that κ and d depend on electrolyte concentration, electrode material, electric potential, etc. This dependence can be recognized because, while the electronic charge must reside at the electrode surface, the electrolyte charges need not (and will not) lie on a surface.

At low electrolyte concentration with a low density of charge carriers, a substantial thickness of solution may be necessary to accumulate the required charge. (The charge density will be greatest adjacent to the electrode where the electrostatic forces are greatest, while further away the relative effect of thermal forces increases.)

The consequence is to introduce additional factors affecting the effective capacitance. A simple, but not completely correct, picture is given by two capacitors in series, the Guoy–Chapman–Stem (GCS) model. One capacitor, C_H, reflects a compact charge layer close to the electrode, while the second, C_D, reflects the diffuse charge layer; C_H is relatively independent of potential, while C_D behaves in a complex way with potential [2].

The equivalent circuit describing an electrode lying in an electrolyte and a reference electrode in the same medium consists of the aforementioned capacitance, a series resistance representing the resistance of the electrolyte, and a resistance in parallel with the capacitance that reflects electrode electrolyte charge movement, beyond that associated with charge–discharge of the capacitance. This non-capacitive charge flow results from both reversible and nonreversible electrochemical *Faradic* reactions [4].

A sketch of the electrode arrangement is given in Figure 12.2a, and the equivalent circuit is shown in Figure 12.2b. In the latter, it is assumed that the reference electrode will ordinarily be physically large; since its current density will therefore be very small, its contribution to the total measured voltage (relative to the remaining factors) will be negligible and can be neglected.[2]

The membrane capacitance has, in effect, a point of voltage breakdown. Within this bound the electrode behaves capacitively, which implies linearity and reversibility. Outside this range the parallel resistance pathway of Figure 12.2b becomes effective.

The involvement of the parallel resistance signifies the occurrence of electrochemical processes. Such electrochemical processes can be a problem if irreversible. A necessary condition for reversibility is that reaction products remain at the electrode, hence available for a reverse reaction.

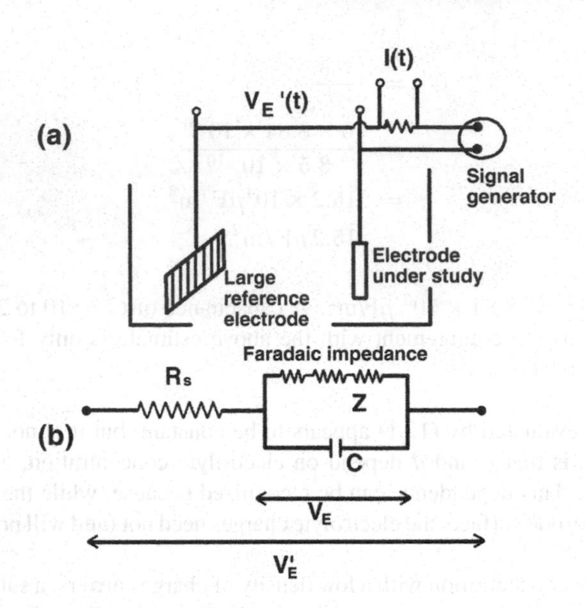

Figure 12.2. (a) Apparatus used in biomedical studies of electrode impedance where current $I(t)$ and total electrode voltage $V_E'(t)$ are monitored. (b) Equivalent circuit for the system in (a). R_s is the solution resistance, C is the double-layer capacitance, and Z is the Faradaic impedance (the latter consisting of charge-transfer resistance, diffusional impedance, and reaction impedance). From Dymond AM. 1976. Characteristics of the metal–tissue interface of stimulation electrodes. *IEEE Trans Biomed Eng* **23**:274–280, copyright ©1976, IEEE.

Irreversibility is assured if reaction products are able to diffuse away. The negative implications will be discussed presently.

When a stimulating current is introduced into a region containing nervous tissue, the passage of the current through the membranes of the nerve fibers results in a transmembrane potential (which may be depolarizing or hyperpolarizing). For subthreshold conditions, where the nerve membrane is linear, the tissue can be described with a linear electrical network.

Under these conditions the induced transmembrane potential will be proportional to the stimulating current amplitude. Usually, the applied current is maintained at a constant value, in which case the induced (transient) transmembrane potential will depend on the *RC* character of the nerve membrane.

The response to a current pulse is an *RC* transient at the electrode–tissue interface. This transient is depicted in Figure 12.3. Both the constant current pulse and the electrode voltages are shown. (Figure 12.2 gives related nomenclature.) In Figure 12.3, V_0 is the voltage across the electrolyte (IR_s), while V_E is the voltage building up across the capacitance. At the termination of the current pulse $IR_s = V_0$ is instantly ended, and the remaining voltage (V_E) leaks off slowly. The charging time constant is quite short compared to that under discharge.

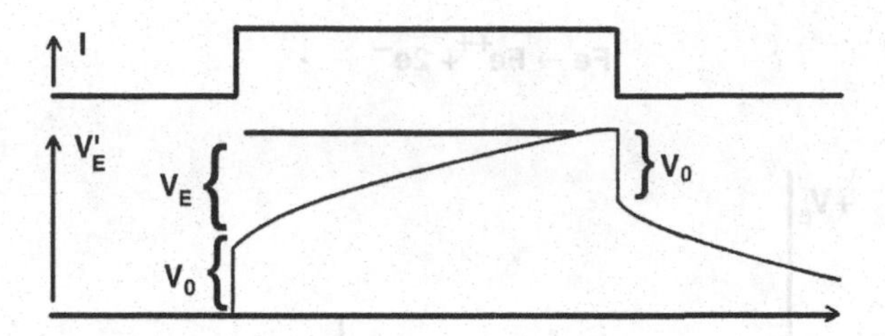

Figure 12.3. Voltage Waveform Observed between Test Electrode and Reference Electrode in response to the constant current pulse shown. V_o is the voltage across the electrolyte path (IRs), while V_E is that across the electrode–electrolyte capacitive interface. From Mortimer JT. 1981. Motor prostheses. In *Handbook of physiology*, Section I: *The nervous system*, Vol. II, *Motor control*, Part I, pp. 155–187. Bethesda, MD: American Physiological Society.

12.2.2. Electrode Operating Characteristics

Reversible–irreversible electrode analysis placed within a graphical framework was used to analyze electrode performance, in a plan developed by Mortimer [1]. A modified version is illustrated in Figure 12.4.

In Figure 12.4 the relationship between V_E and Q/A is described as linear in the central region, where the slope could be interpreted as the effective electrode capacitance (per unit area). The width of this region depends on the electrode material, its surface treatment, and on the electrolyte. In this region metal electrons pile up on the electrode considered as one plate of a capacitor.

While this capacitance is large, the maximum charge remains in the linear region and proves to be inadequate to achieve tissue activation. Exceeding this limit drives the operation beyond points I or II, or both, and hence introduces Faradaic conditions, i.e., electrochemical reactions. Increasing anodic potential drives the electrode state beyond point I in Figure 12.4 and causes electrochemical reactions of the kind that may result in electrode damage.

As an illustration, for the type of reaction (shown for a stainless steel electrode)

$$\mathrm{Fe} \rightarrow \mathrm{Fe}^{++} + 2e^- \tag{12.2}$$

the result is the dissolution of the iron, and the process is irreversible. For cathodic potentials driven beyond II (in Figure 12.4) the reactions may be of the form

$$2\mathrm{H_2O} + 2e^- \rightarrow \mathrm{H_2} \uparrow \mathrm{OH}^- \tag{12.3}$$

This reaction also is irreversible. The resulting increase in pH can result in tissue damage (though some buffering can take place).

On the other hand, for a platinum electrode, the anodic reaction may be

$$\mathrm{Pt} + \mathrm{H_2O} \rightarrow \mathrm{PtO} + 2\mathrm{H}^+ + 2e^- \tag{12.4}$$

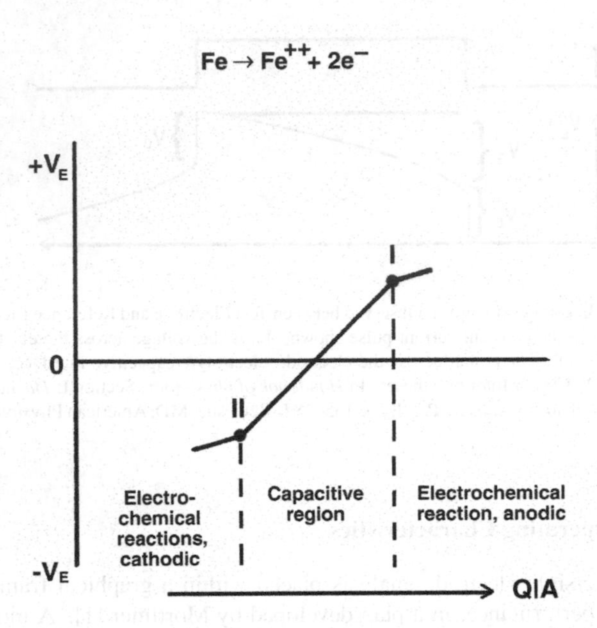

Figure 12.4. Idealized Representation of Relationship between Electrode Potential V_E and Charge Density (charge per unit of real electrode area, Q/A). Charge injection in the central region involves processes that are capacitive and therefore completely reversible. Charge injection in regions to right of point I or left of point II involve electrochemical reactions. These are reversible if, by driving current in the opposite direction, no new species are introduced. Irreversibility involves diffusion of new chemical species away from the electrode. Modified from Mortimer JT. 1981. Motor prostheses. In *Handbook of physiology*, Section I: *The nervous system*, Vol. II, *Motor control*, Part I, pp. 155–187. Bethesda, MD: American Physiological Society.

the result being the replacement of a platinum molecule by platinum oxide. No new chemical species results in the bulk medium (in this case, PtO remains bound to the electrode).

The lack of a new chemical species characterizes a reversible process. Reversibility is demonstrated by a reversal of the reaction in (12.4), which is achieved by passing current in the opposite direction. An example of a reversible cathodic reaction (with a platinum electrode) is

$$\mathrm{Pt} + \mathrm{H}^+ + e^- \rightarrow \mathrm{Pt} - \mathrm{H} \tag{12.5}$$

known as H-atom plating [25]. Here, again, no new chemical species is introduced into the bulk solution.

For monophasic stimulation, we normally have a continual buildup of charge at the electrode interface since this charge leaks away very slowly during typical inter-pulse intervals. For anodic pulses, the buildup of charge reaches point I (in Figure 12.4), from which point further pulses cause chemical reactions associated with the loss of the additional charge.

The operating point then centers at point I, so that any and all positive excursions result in chemical reactions (which may be irreversible). An equivalent consequence of monophasic

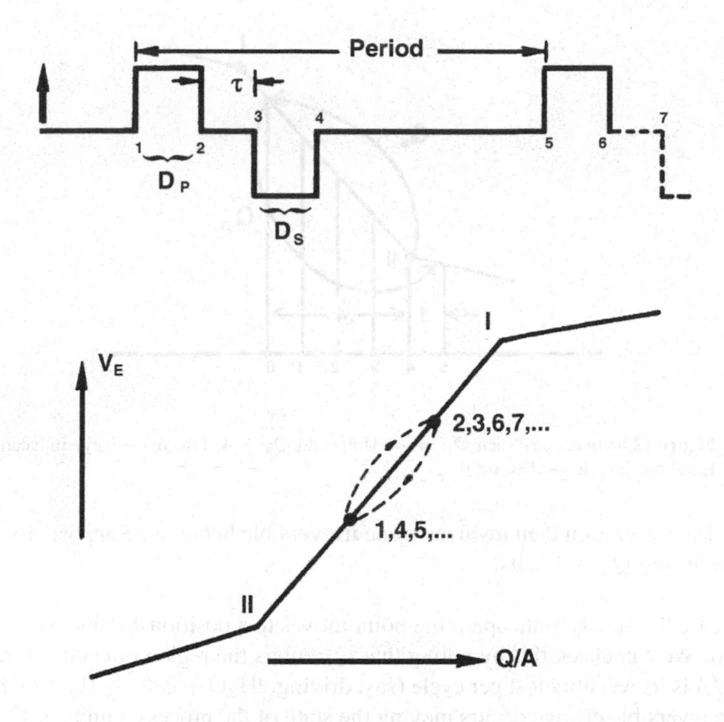

Figure 12.5. Balanced-Charge Biphasic Stimulation. (a) Stimulus waveform with zero net charge transfer per cycle ["period" >> $(D_p+\tau+D_s)$]. (b) Variation in electrode potential, for conditions where charge is accommodated entirely within capacitive region. I and D refer to current pulse amplitude and pulse duration. Subscripts P and S refer to primary and secondary stimulus pulses, respectively. Parameter τ is the time delay between the end of the primary pulse and the beginning of the secondary pulse. Balanced charge requires that $I_P D_P = I_S D_S$. Points 1–7 in (a) correspond to points in (b). From Mortimer JT. 1981. Motor prostheses. In *Handbook of physiology*, Section I: *The nervous system*, Vol. II, *Motor control*, Part I, pp. 155–187. Bethesda, MD: American Physiological Society.

cathodic pulses takes place, driving the operating point to II. This results no matter how small the injected charge is. Consequently, monophasic stimulation is rarely used.

The monophasic accumulation of charge can be avoided by using *biphasic stimulation*. In the ideal case, the charge density introduced in each phase is less than the reversible limit and the total process is then reversible and repetitive. (This condition is termed *balanced-charge biphasic*.)

The *primary pulse* is the initial one, and the effects of charge introduction are countered by the following *secondary pulse*. No imbalance can be tolerated, since it would be cumulative in time and lead to a drift toward I or II in Figure 12.4. Imbalance can be avoided by applying current to the electrodes through a series input capacitance, which ensures a balanced charge. A single balanced cycle is illustrated in Figure 12.5.

If incomplete balance does not exist, or if the primary and secondary pulses are frankly unequal, then the operating point will drift toward I or II, as noted above.

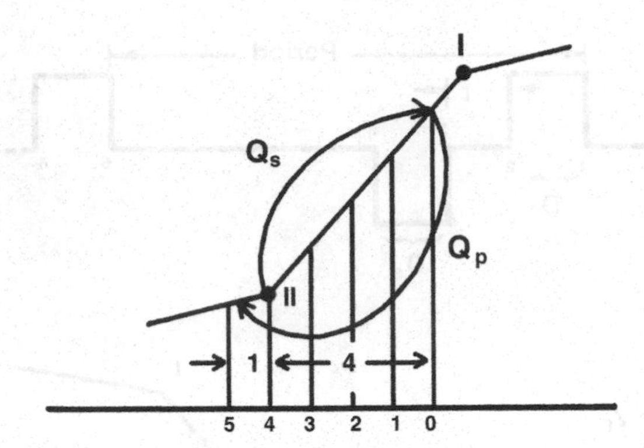

Figure 12.6. Behavior When $Q_p = -5$ Units and $Q_s = 4$. Owing to charge imbalance I, cathodic unit is lost beyond II.

Steady-state operation then involves some irreversible behavior. Suppose, for example, that $Q_p = -5$ units and $Q_s = 4$ units.

In this case the steady-state operating point moves to a position 4 units to the right of II (see Figure 12.6). We can check this by noting that Q_p enters the region of chemical reaction where 1 unit of Q/A is irreversibly lost per cycle (say, driving $2H_2O + 2e^- \rightarrow H_2 + OH^-$). When Q_s is applied, a reversible change occurs moving the state of the process 4 units of Q/A to the right of II (to the initial, operating point). The cycle repeats indefinitely.

This process may be tolerated, since buffering of OH^- by the blood is possible, within limits. The advantage is that, to this extent, one can use a larger stimulus.

12.3. OPERATING OUTSIDE REVERSIBLE REGION

Whether the goal of operating within the reversible region can be achieved depends on whether this allows a high enough current density to stimulate the desired nerve fibers. Normally, operating within this limit is not possible. Several actions that can be taken when this condition is not satisfied including the following.

12.3.1. Imbalanced Biphasic Stimulation

Some irreversible cathodic behavior can be tolerated because of buffering by the blood of the OH^- products of the electrochemical processes. Thus, some increase in the stimulus into the irreversible region may be acceptable. Note that a comparable anodic irreversibility is never tolerated, since the result is irreparable electrode damage.

12.3.2. Expanding Capacitive Region

An expansion of the capacitive region can be achieved (to some extent) by adding dielectric to the electrode or by roughening its surface. The latter effect increases its surface area, while keeping its geometric area unchanged. (The ratio of real to geometric area is the *roughness factor*.) Anodized sintered tantalum has a roughness factor of 10–100, as does tantalum pentoxide.

Current from a rough surface will not have a microscopic uniform density, as surfaces near the bottom of valleys contribute relatively little current. That is, the roughness factor may greatly exaggerate the ratio of reversible total current under rough to smooth conditions [11].

By covering the electrode with a high dielectric (insulating) film, one can ensure the absence of oxidation–reduction reactions at the interface (in effect, the capacitance may be increased by a factor of 5–10). Eventually, with increasing current, dielectric breakdown is reached and, unlike the breakdown of the aforementioned water layer, dielectric breakdown will be irreversible. Thus, operation is strictly limited by the capacitive region. A discussion of the capacitive electrode is given by Guyton and Hambrecht [10].

Animal studies show that there are additional factors that contribute to neural damage. Damage is observed even though a stimulus is capacitively coupled and charge is balanced. Such neural damage appears to arise from sustained hyperactivity of the axons. In general, these additional effects come from the passage of currents through the tissue rather than from actions at the electrode–electrolyte interface [1].

12.4. ELECTRODE MATERIALS

In choosing the material for an electrode, factors of importance include the following:

1. Passive compatibility of the material with tissue.
2. Extent of reversible behavior (capacitative region and region of reversible electrochemical reactions).
3. Mechanical compatibility with the tissue.

The materials most in use are platinum, platinum-iridium, and 316 stainless steel (SUS 316L). These materials have a history of satisfactory behavior. The charge storage capacity of platinum is stated to be 2.1 μC/mm^2 generally, but for applications in the cerebral cortex a lower limit of 0.3 μC/mm^2 is recommended. For 316 stainless steel, 0.4–0.8 μC/mm^2 (geometrical area) appears to be the limit.

Mechanical compatibility is important, and stabilization of the implant is highly desired, based on tissue growth. This can be enhanced by use of porous electrodes, where the tissue can more readily grow into it. It is desired to have good stabilization, but also not too thick an encapsulation since this increases the voltage necessary to achieve the same current. The electrode shape also is important; helical wires help convert the forces of bending into torsion, which is much more easily tolerated, and with a greatly improved life.

In the following subsections we list particular applications and the electrode type often chosen. In each case, special requirements may suggest an electrode type other than that listed. Additional discussion of electrode types is given later in the chapter.

12.4.1. Brain (Surface Electrodes)

(a) Passive Implants. Surface electrodes can be placed on the brain with a minimum resulting trauma. The end result is encapsulation with a greater thickness on the superficial side (≈ 400 μm) than on the side in contact with the brain (thickness of around 160 μm).

(b) Active Implants. Mainly platinum is used. Based on blood–brain barrier breakdown studies, only the lowest values of *balanced charge biphasic* stimulation tested (with an upper limit of 0.3 μc/mm^2) were found safe. Ta–Ta_2O_5 electrodes have a roughness factor of 100 and were found safe at 0.934 μc/mm^2 (geometric area).

If the difference in material is ignored in the above comparison, one notes that the high roughness factor of 100 stands in marked contrast to the improvement by a factor of only 3 in use of the roughened electrode. The explanation is probably, as mentioned earlier, that the (rough) surface current density is nonuniform close to the electrode surface. Since portions of the roughened area carry little current, then while they add to the true area (increasing the roughness factor) they do not significantly add to the effective capacity of the electrode [11].

12.4.2. Nerve (Cuff Electrode)

Cuff electrodes are insulating hollow cylinders with embedded (internal) circular electrodes. The cuff is placed around the nerve to be stimulated. This particular configuration essentially confines the stimulating current to the target nerve alone, and there is no extraneous excitation. Furthermore, the concentration of current flow minimizes the total current required for stimulation (hence, it reduces the electrode current density).

Transmembrane (stimulating) current is maximally enhanced in the nerve by a closely fitting cuff; however, this is poorly tolerated by the nerve trunk and mechanical trauma dictates a loose-fitting cuff.

12.4.3. Muscle (Coiled Wire Electrode)

Since the current required to stimulate the motor nerve is very much lower than that required for direct simulation of the muscle, one may choose to stimulate either the peripheral nerve or the nerve in the vicinity of the neuromuscular junction. The latter site avoids possible injury to the peripheral nerve and also permits a higher degree of selectivity, since the nerve fibers are arborized within the muscle. Because intramuscular electrodes are subject to considerable bending and flexing, they are most satisfactory when made of stainless steel helical coils. A specific practical design is one in which insulated wire (except for the tip) is wound on a ≈ 100 μm-diameter mandril and introduced into the muscle with a hypodermic needle. The electrode tip is formed into a barb, so that when the needle is removed the electrode remains in place.

(a) Passive Implants. These become encapsulated, usually with only a mild foreign-body response. For a 200-μm-diameter coil formed from 45-μm-diameter 316 stainless steel, a capsule thickness of from 50 to 300 μm develops.

(b) Active Implants. Active implants must be subdivided as follows:

(1) Monophasic: Some irreversible cathodic processes are tolerated if charge injection is sufficiently low (≤ 0.2 μC/mm^2 per pulse at a stimulation rate of 50 Hz). The afore-

mentioned implies that an average hydroxy lion generation rate of 10 μA/mm^2 could be buffered by the blood.

(2) Balanced-charge biphasic: A density of 0.1–2.0 μC/mm^2 for stimulus current should be satisfactory for minimizing electrode corrosion, under balanced charge conditions. However, beyond a primary pulse charge density ≥ 0.4 μC/mm^2 some degree of electrode corrosion may be seen.

(3) Imbalanced biphasic: Because of blood buffering one can operate with a 0.6 μC/mm^2 primary, cathodic, pulse followed by an anodic, secondary, pulse of 0.4 μC/mm^2. The imbalance permits an incremental stimulus intensity of 50%. At 50 Hz and for a typical coiled wire electrode with an area of 10 mm^2, the above conditions can be achieved with a current of 20 ma for 300 μsec (primary pulse) and 20 ma for 200 μsec (secondary pulse). A motor axon within 1–2 cm of a point-source electrode should be easily stimulated under these conditions. The net cathodic imbalance is 0.2 μC/mm^2, and for 50 Hz and an area of 10 mm^2 we evaluate

$$0.2 \times 10^{-6} \times 10 \times 50 = 100 \mu\text{C/sec} \tag{12.6}$$

or 10 μA/mm^2, and this is believed to represent the maximum that is tolerable.

12.5. NERVE EXCITATION

To evaluate the capabilities of an electrode system to stimulate a nerve, we must consider the geometry of both the electrical conductivities of the medium in which the nerve fibers and electrode(s) lie and the nonlinear behavior of the excitable membranes. Most systems have been evaluated experimentally. Only very simple configurations have been modeled and studied through simulation.

For the stimulation of a single myelinated nerve fiber by a pair of electrodes that directly contact the nerve, once can use the linear core-conductor model (Chapter 6) as the basis for a model, as shown in Figure 12.7. In the internodal region the transmembrane admittance, being very low because of the myelin sheath, can to a good approximation be set equal to zero. The model consequently admits transmembrane current only at the nodes, and this is reflected in the detailed structure of Figure 12.7.

For subthreshold conditions, Z_m (membrane impedance) is composed of a parallel R_m and C_m, but near and beyond threshold a Hodgkin–Huxley type circuit is required. For the node of Ranvier, a modification of the Hodgkin–Huxley network, due to Frankenhaeuser and Huxley (FH) [5], is frequently used. In the model of Figure 12.7, excitation occurs under the cathode if the stimulating current is large enough.

The strength–duration curve [see Eq. (7.7)] can be found experimentally or approximated from the equation for the threshold current, $I_{\text{th}}(t)$. As symbolized here, it is given by (7.8):

$$I_{\text{th}} = \frac{I_R}{(1 - e^{-Kt})} \tag{12.7}$$

where I_R is the rheobase current (that current magnitude that will cause the transmembrane potential to reach the excitation threshold at infinite time). K can be regarded as an experimentally

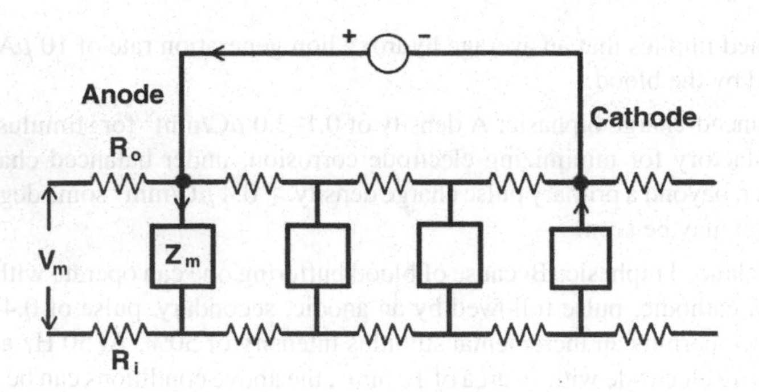

Figure 12.7. Linear Core-Conductor Model of a Myelinated Fiber. Since transmembrane current is assumed to flow only at the nodes, the axial resistances (R_O, R_i) are finite (and not infinitesimal) and represent the axial resistances in the internode. The Figure depicts the condition where subthreshold stimulating current is introduced at two separated nodes.

determined constant that depends on the membrane as well as the electrode/nerve geometry, conductivities, etc.

This relationship was discussed in Chapter 7, along with its approximate nature (it neglects accommodation) and recognition that the nodal membrane elements form a network, etc. Reilly [23] evaluated $\tau_m = 1/K = 92.3$ μsec compared to $R_m C_m = 66$ μsec. The τ_m was the least-squares error between (12.7) and the measured strength–duration curve.

One can also obtain a measured τ_m from the ratio $\mathcal{Q}_{\min}/I_{\min} = (I_R/K)/I_R \cdot T$ [an expression obtained from (12.7) giving this result for small t). Time constant τ_m was determined to be 92 μsec. The desirability of minimizing charge injection associated with a stimulating pulse has been discussed in terms of minimizing adverse electrode and/or tissue reactions.

So in the present context, Eq. (12.7) can be examined with minimum charge injection as a goal. The total charge per pulse, $\mathcal{Q}_{\text{th}}$, can be found from (12.7) as

$$\mathcal{Q}_{\text{th}} = I_{\text{th}} t = \frac{I_R t}{(1 - e^{-Kt})} \tag{12.8}$$

where t is the pulse duration. If this is very long ($t \to \infty$), then

$$\mathcal{Q}_{\text{th}} = I_R t \tag{12.9}$$

For very short pulse durations $t \to 0$, $e^{-Kt} \to 1 - Kt$, so

$$\mathcal{Q}_{\text{th}} = \frac{I_R}{K} = \mathcal{Q}_{\min} \tag{12.10}$$

Since the minimum current, $I_{\min}$, is the rheobase current I_R, then $\mathcal{Q}_{\min}/I_{\min} = 1/K = \tau_m$. The conclusion that minimum charge, $\mathcal{Q}_{\min}$, equals I_R/K can be confirmed by evaluating $d\mathcal{Q}_{\text{th}}/dt$, and showing that it goes to zero when $t \to 0$ (so $\mathcal{Q}_{\text{th}} = I_R/K$ is, in fact, the minimum charge that results in excitation).

In general, for a pulse duration t and the corresponding current amplitude that just excites, we have from (12.8) and (12.10)

$$\frac{\mathcal{Q}_{\rm th}}{\mathcal{Q}_{\rm min}} = \frac{Kt}{1 - e^{-Kt}} \tag{12.11}$$

which gives a measure of the charge injection in excess of the minimum when $t \geq 0$.

When $I_{\rm th} = I_R$, the current is rheobase. The minimum time for excitation when $I_{th} = 2I_R$ is called the *chronaxie*, t_c. From Eq. (12.7), we can evaluate t_c since

$$1 - e^{-Kt_c} = \frac{1}{2} \tag{12.12}$$

The solution of Eq. (12.12) for t_c is

$$t_c = \frac{\ln 2}{K} \tag{12.13}$$

For $t = t_c$, Eq. (12.11) gives

$$\frac{\mathcal{Q}_{\rm th}}{\mathcal{Q}_{\rm min}}\Big|_{t=t_c} = \frac{Kt_c}{1 - e^{-Kt_c}} = 2\ln 2 = 1.39 \tag{12.14}$$

or a 39% excess charge.

Substitution of (12.13) into (12.11) yields a more general result:

$$\frac{\mathcal{Q}_{\rm th}}{\mathcal{Q}_{\rm min}} = \frac{\ln(2)(t/t_c)}{1 - 2^{-(t/t_c)}} \tag{12.15}$$

A plot of Eq. (12.15) is given in Figure 12.8. Note that the percent excess charge increases rapidly beyond $t = t_c$. The use of narrow pulses appears highly desirable, based on this criterion. For the McNeal model, discussed in the next section, and based on the Reilly simulation [23], the chronaxie is $(\ln 2)\tau_m = 0.693 \times 92 = 64$ μsec.

An integrated overall system for measurement and stimulation, including both electrodes and amplifiers, is described by Pancrazio et al. [21].

12.5.1. Secondary Pulse Considerations

The injected primary current pulse is designed to achieve nerve excitation. It is followed by a secondary pulse solely to achieve reversibility. Since it is desirable to design the primary pulse so that it results in excitation with relatively little excess charge, the immediate injection of a secondary pulse, which is necessarily hyperpolarizing, can interfere with the initiation of an action potential.

The waveform of an action potential resulting from a stimulus where the strength duration is just beyond threshold has the configuration shown in Figure 12.9, and shows an initial dip. If the secondary pulse is applied during this dip, then the action potential may be extinguished. To avoid this problem, either the stimulus strength of the primary pulse must be increased or a delay must be introduced between primary and secondary pulses.

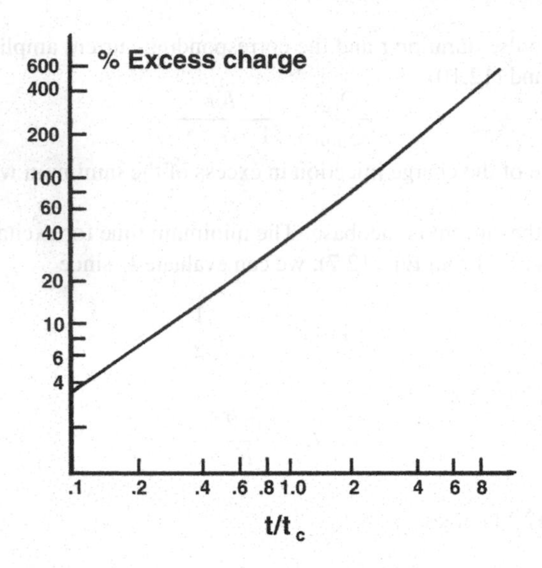

Figure 12.8. Charge Injected to Reach Membrane Threshold in Excess of Theoretical Minimum as a Function of Pulse Duration. Pulse duration (t) has been normalized to the chronaxie value (t_c). See Eq. (12.5), where percent excess charge is a function of (t/t_c) and is equal to $[0.693(t/t_c)/(1 - 2^{-t/t_c}) - 1]100\%$.

The former option is disadvantageous, since it results in an increase in the injected charge. In this example, a 100-μsec delay resulted in only a 10% reduction in the monophasic force arising from the (compound) action potential. Note that for the two shortest delays shown in Figure 12.9 the action potential is extinguished. There is no penalty for inserting a delay, since even with a value of 100 μsec the electrochemical remedy afforded by the secondary pulse (i.e., reversibility) does not significantly reduce the activation.

12.5.2. Excitation of Myelinated Nerve

The excitation of a single myelinated nerve fiber from a point current source has been examined through the use of models of a type introduced by McNeal [18]. The model assumes that transmembrane current flows between the intracellular and extracellular medium only at the nodes.[3]

Furthermore, since conditions are sought where the most proximal node will just fire, the more distal nodes will consequently be subthreshold and can be approximated by a passive *RC* network.[4]

This model is shown in Figure 12.10. The nodal model just beneath the stimulating electrode, since it will be followed to the point of threshold, is that of Frankenhaeuser and Huxley.

The applied potential field at the extracellular nodes/sites is approximated by that which exists in the absence of the fiber, based on the argument that the fiber is of small size and of relatively high resistance.[5]

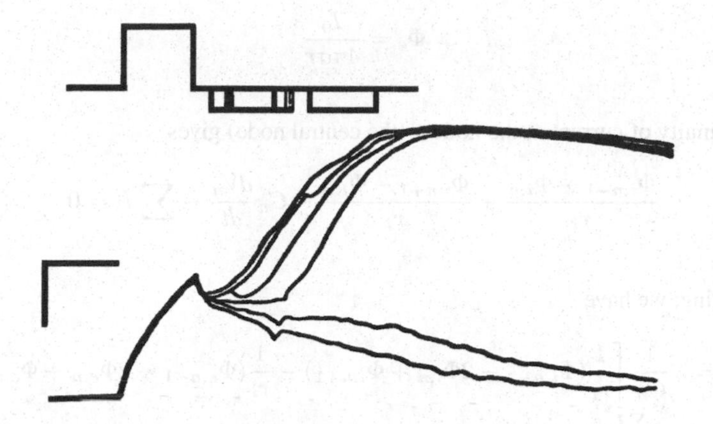

Figure 12.9. Transmembrane Voltage Response of Myelinated Nerve to Short Pulse Stimuli. Shown is the effect of an increasing delay between primary and secondary pulses. Vertical calibration bar is 20 mV, and the horizontal bar is 50 μsec. Reprinted with permission from van den Honert C, Mortimer JT. 1979. The response of the myelinated nerve fiber to short duration biphasic stimulating currents. *Ann Biomed Eng* 7:117–125, copyright ©1979, Biomedical Engineering Society.

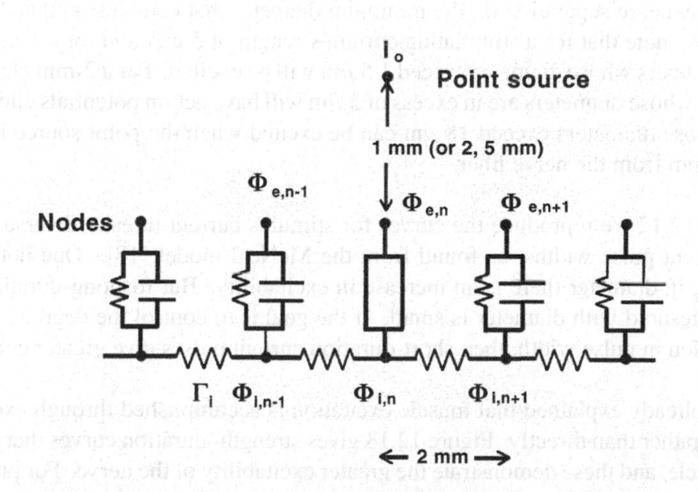

Figure 12.10. Model to Study Response of Myelinated Nerve Fiber to a Point-Source Stimulus. Source is 1, 2, or 5 mm from nerve [18]. The central node is described by Frankenhaeuser–Huxley equations, while lateral nodes are assumed to remain subthreshold and to be adequately described by *RC* elements. Based on McNeal D. 1978. Analysis of model for excitation of myelinated nerve. *IEEE Trans Biomed Eng* **BME-23**:329–377. Copyright ©1978, IEEE.

If the current at the point electrode is I_0, then the potential field, from (2.21), is simply

$$\Phi_e = \frac{I_0}{4\pi\sigma r} \tag{12.16}$$

The continuity of current at the node *n* (the central node) gives

$$\frac{\Phi_{i,n-1} - \Phi_{i,n}}{r_i} + \frac{\Phi_{i,n+1} - \Phi_{i,n}}{r_i} - C_n \frac{dV_n}{dt} - \sum I_i = 0 \tag{12.17}$$

Rearranging, we have

$$\begin{aligned} \frac{dV_n}{dt} &= \frac{1}{C_n}\left[\frac{1}{r_i}(\Phi_{i,n-1} - 2\Phi_{i,n} + \Phi_{i,n+1}) - \frac{1}{r_i}(\Phi_{e,n-1} - 2\Phi_{e,n} + \Phi_{e,n+1})\right] \\ &\quad - \frac{\Sigma I_i}{C_n} \end{aligned} \tag{12.18}$$

which is similar to (6.50). Using the model of Figure 12.10, McNeal [18] studied the effect of electrodeposition, through its effect on Eq. (12.16), and fiber diameter, through its affect on r_i in Figure 12.10. The fiber diameter also indirectly affects the structure of the myelinated fiber, since the internodal spacing in micrometers is assumed equal to 100*D*, where *D* is the diameter in micrometers.

Some results of this simulation are given here in Figures 12.11 and 12.12. In Figure 12.11 the effect of source–nerve separation on the minimum diameter that can be stimulated successfully is described. We note that for a stimulating current strength of 5 mA and for a 1-mm electrode–nerve spacing, fibers whose diameter exceed 1.5 μm will be excited. For a 2-mm electrode–nerve spacing, fibers whose diameters are in excess of 5 μm will have action potentials elicited. Finally, only fibers whose diameters exceed 18 μm can be excited when the point source is moved to a distance of 5 mm from the nerve fiber.

In Figure 12.12 we reproduce the curves for stimulus current threshold versus diameter of fiber for different pulse widths, as found from the McNeal model [18]. One notes again that for an increase in diameter there is an increase in excitability. But for long-duration pulses the variation in threshold with diameter is small. If the goal is to control the degree of recruitment through variation in pulse width, then short-duration current pulses give greater controllability.

We have already explained that muscle excitation is accomplished through excitation of its motor neuron rather than directly. Figure 12.13 gives strength–duration curves that are typical of nerve and muscle, and these demonstrate the greater excitability of the nerve. For pulse durations of 100 μsec, for example, the current required for direct muscle stimulation is around 20 times greater than that for stimulation of the associated motoneuron.

12.5.3. Nerve Trunk Anatomy

The anatomy of a peripheral nerve trunk is described in Figure 12.14. Such trunks contain both efferent fibers (signals to muscle and other organs) and afferent fibers (incoming sensory

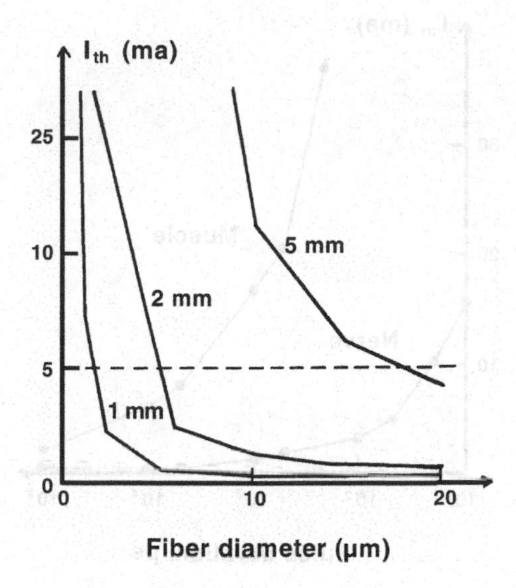

Figure 12.11. Effect of Increasing Separation between Electrode and Axon on Stimulus Threshold to Fiber Diameter Relationship. The stimulus pulse width is fixed at 100 μS. Calculated by Mortimer [19] from the model by McNeal [18]. From Mortimer JT. 1981. Motor prostheses. In *Handbook of physiology*, Section I: *The nervous system*, Vol. II, *Motor control*, Part I, pp. 155–187. Bethesda, MD: American Physiological Society.

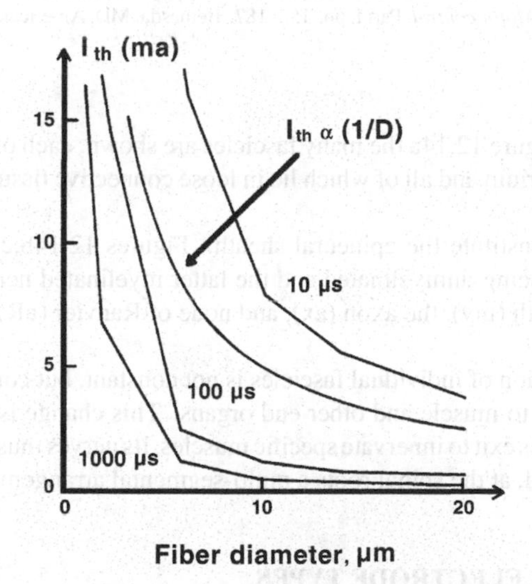

Figure 12.12. Stimulus Threshold as a Function of Nerve Diameter for Several Values of Stimulus Pulse Width. Stimulus–fiber distance is 2 mm. Calculated by Mortimer [19] using the model of McNeal [18]. From Mortimer JT. 1981. Motor prostheses. In *Handbook of physiology*, Section I: *The nervous system*, Vol. II, *Motor control*, Part I, pp. 155–187. Bethesda, MD: American Physiological Society.

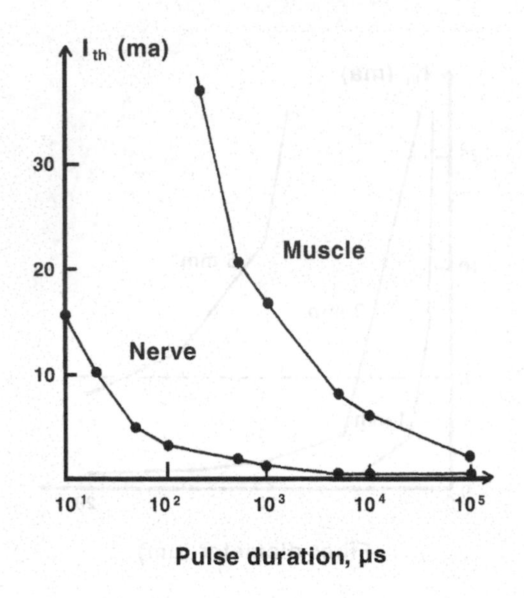

Figure 12.13. Strength–Duration Relationship for Nerve Excitation (indirect muscle excitation) and Direct Muscle Excitation. During these experiments, evoked muscle response was held constant at small fraction of total possible muscle force. The stimulus was delivered through an intramuscular electrode before and after administration of curare. (Data are representative of those collected from laboratory experiments and are, in principle, identical to the type of curves classically presented for innervated and denervated muscle.) From Mortimer JT. 1981. Motor prostheses. In *Handbook of physiology*, Section I: *The nervous system*, Vol. II, *Motor control*, Part I, pp. 155–187. Bethesda, MD: American Physiological Society.

and other signals). In Figure 12.14a the many fascicles are shown, each of which is bounded by a multi-laminated perineurium and all of which lie in loose connective tissue, the epineurium (epi).

The outer layers constitute the epineural sheath. Figures 12.14b,c show individual nerve fibers, with the former being unmyelinated and the latter myelinated nerve. The Schwann cells (Schw), the myelin sheath (my), the axon (ax), and node of Ranvier (nR) are also labeled.

The nerve composition of individual fascicles is not constant, but continually changes in the course from spinal cord to muscle and other end organs. This change is necessary since, at the periphery, where fascicles exit to innervate specific muscles, its nerves must contain all innervating fibers. On the other hand, at the spinal roots a multi-segmental arrangement of fascicles must be accommodated [16].

12.6. STIMULATING ELECTRODE TYPES

Peripheral stimulation of muscle by means of intramuscular or epimysial electrodes has the advantage of a high degree of selectivity, because the electrodes lie at or near the target muscle whose activation is desired. While described as muscle electrodes, in fact it is not the muscle but rather the peripheral nerve which is excited.

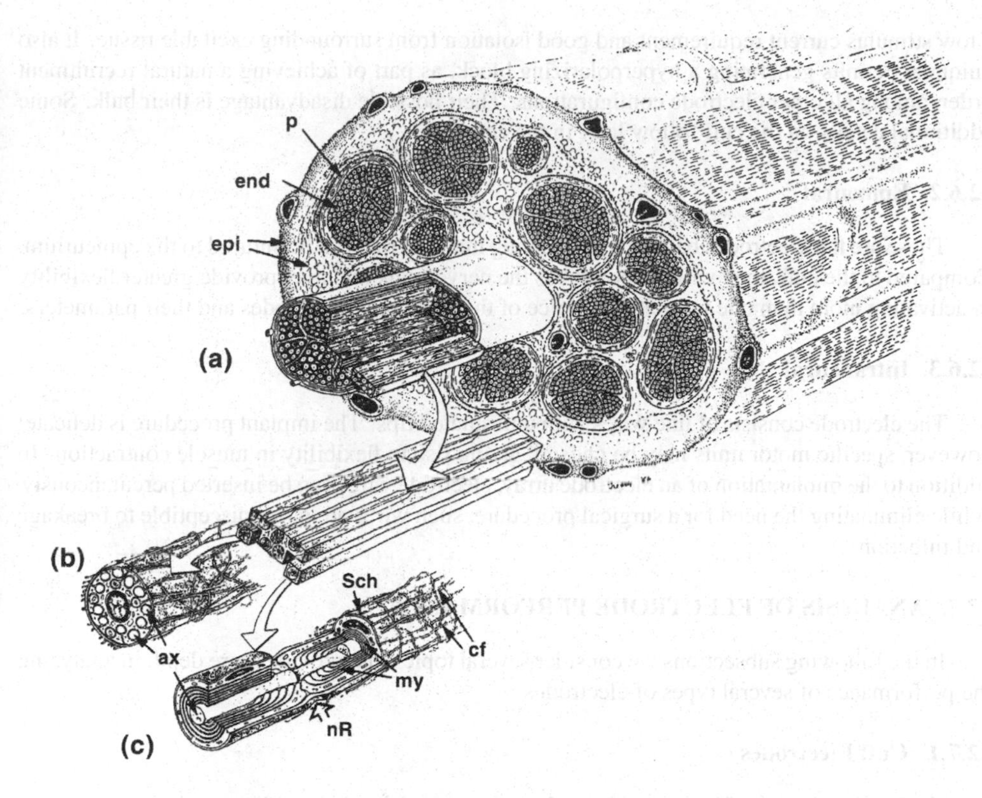

Figure 12.14. Microanatomy of Peripheral Nerve Trunk and Its Components. (a) Fascicles surrounded by a multi-laminated perineurium (p) are embedded in a loose connective tissue, the epineuium (epi). The outer layer of the epineurium is condensed into a sheath. (b) and (c) illustrate the appearance of myelinated and unmyelinated fibers, respectively. Schw, Schwann cell; my, myelin sheath; ax, axon; nR, node of Ranvier. From Lundborg G. 1988. *Nerve injury and repair*. London: Churchill-Livingston, by permission of the publisher.

Such electrodes have important disadvantages, however. The intramuscular electrodes are subject to mechanical forces that may result in electrode (wire) breakage. Both intramuscular and epimysial electrodes have varying geometrical relationships to the muscle, and the muscle's size and shape is continually changing during activity. As a result, the motor units that are stimulated will vary with time giving rise to what is called length-dependent recruitment. In view of their negative features, nerve electrodes are generally preferred. On the other hand, the electrode position makes it unlikely that they will cause neural damage.

In the following sections we consider several neural electrode types. Brief comments on their advantages and disadvantages of each type are included [20].

12.6.1. Cuff

This consists of circular wire electrodes embedded in the inner wall of an insulating hollow cylinder. The cuff electrode is placed in the tissue with the nerve trunk along the axis. They are relatively easy to implant and provide excellent confinement of the stimulating field. The result is a low stimulus current requirement and good isolation from surrounding excitable tissue. It also uniquely permits generating a hyperpolarizing block, as part of achieving a natural recruitment order, among all other electrode configurations. Their possible disadvantage is their bulk. Some additional details on the cuff follow later in the chapter.

12.6.2. Epineural

These are flat electrodes backed by insulating material that can be sutured to the epineurium. Compared to the cuff, they will not compress the nerve trunk and may provide greater flexibility in activating more than one fascicle by choice of the number of electrodes and their parameters.

12.6.3. Intraneural

The electrode consists of fine wires with uninsulated tips. The implant procedure is delicate; however, specific motor units may be chosen, giving rise to flexibility in muscle contraction. In addition to the implantation of an electrode array, electrode wires can be inserted percutaneously. While eliminating the need for a surgical procedure, such wires are more susceptible to breakage and infection.

12.7. ANALYSIS OF ELECTRODE PERFORMANCE

In the following subsections we consider several topics that provide more depth in analyzing the performance of several types of electrodes.

12.7.1. Cuff Electrodes

Cuff electrodes are efficient configurations for nerve stimulation, since excitatory current is concentrated within a confined region surrounding the nerve whose excitation is desired. The arrangement, furthermore, reduces the likelihood of unwanted excitation of other nerves in the vicinity. While near the spinal cord, nerve bundles (motoneurons) contain fibers with different targets, and stimulation results in a diffuse response.

Near the periphery, nerve bundle fascicles tend to be selective to specific muscles and parts of muscles. At this point it may be difficult to discriminate among them with surface electrodes, but with a multiple cuff, a multi-groove electrode [13], or a multiple contact electrode [9], one can obtain selective stimulation of individual fascicles. The major disadvantage in the use of the cuff or cuff-type electrode is that a surgical procedure is required for their installation.

The bipolar configuration is illustrated in Figure 12.15 and gives rise to two basic current pathways, namely, internal and external. For sites 1 and 4, these currents add to the desired hyperpolarization and depolarization achieved mainly by the internal component of current described in Figure 12.15.

The depolarization at 2 and hyperpolarization at 3 are anomalous, however. The region at 2 is described as a *virtual* cathode (i.e., depolarization occurs in the region as if an overlying cathode were present).

The region at 3 is, conversely, described as a *virtual anode*. These above-mentioned apparent cathode and anode, under the right conditions, can excite and/or block and hence become factors

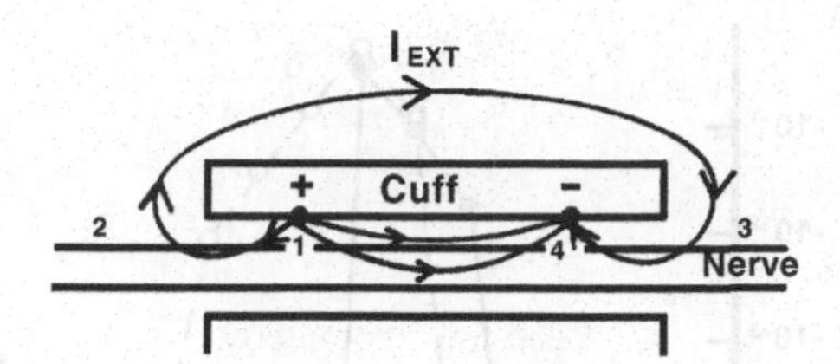

Figure 12.15. External paths 1 and 2 illustrate current that flows from anode to cathode around the outside of the cuff. Components will also enter the nerve and result in a virtual cathode and anode. External path 3 describes current from anode to cathode within the cuff but which does not enter the nerve; it will be greater for cuffs that fit loosely. The internal current illustrates the component lying within the cuff that links with the nerve; this is the desired depolarizing or hyperpolarizing pathway.

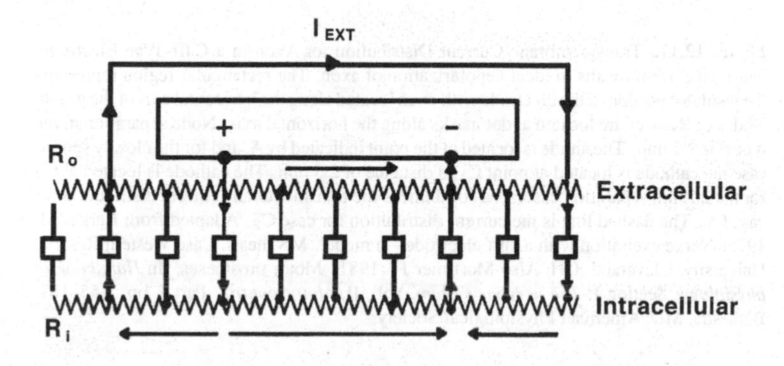

Figure 12.16. Ladder Network Model of the System, Including Current Pathways External to Electrode Cuff. A quantitative analysis of nerve response to current stimulation can be obtained from this approximating model (which lumps the internal and external paths into the two shown).

in the performance of the cuff electrode. These *virtual* sites are enhanced in importance when the bipolar electrode separation is large compared with the distances from each electrode to the nearest end of the cuff (conditions that enhance the external relative to the internal path).

The separation of the applied current into two parts is also described in Figure 12.16, which shows a path that first enters the intracellular space, flows distally, and then emerges flowing around the outside of the cuff; its path then continues as a mirror image. This single path is an illustration of the multiple current paths for the external currents in Figure 12.15.

An external path that flows within the cuff but does not link the membrane, shown in Figure 12.15, is not shown in Figure 12.16. It is noteworthy only in that it increases the total current supplied by each electrode and hence moves the operating point toward the irreversible region.

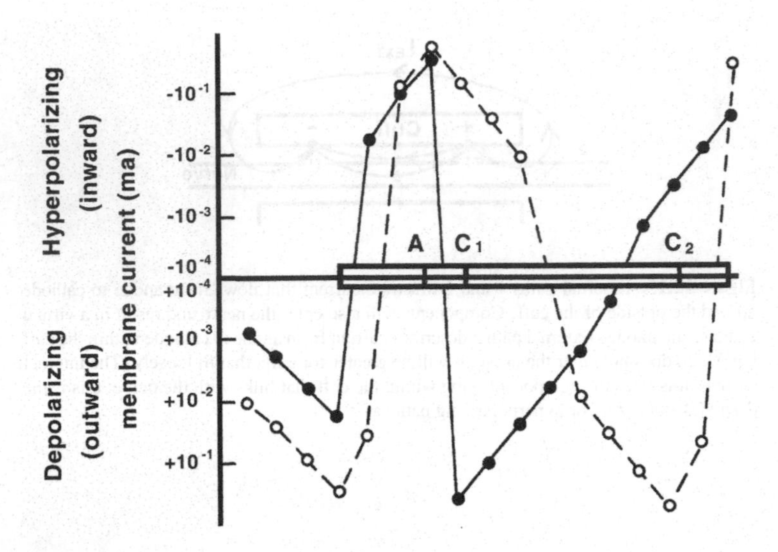

Figure 12.17. Transmembrane Current Distribution for Axon in a Cuff-Type Electrode. Outward current results in local depolarization of axon. The rectangular region represents the insulator portion of the electrode with axon located along the horizontal axis of the graph. Nodes of Ranvier are located at dot marks along the horizontal axis. Node separation in the model is 2.5 mm. The anode is located at the point indicated by A, and for the closely spaced case the cathode is located at point C_1, a distance of 2.5 mm. The cathode is located at C_2 for the 20-mm separation case. The solid line is the transmembrane current distribution for case C_1. The dashed line is the current distribution for case C_2. Adapted from Karkar M. 1975. Nerve excitation with a cuff electrode—a model. MS thesis, Case Western Reserve University, Cleveland, OH. Also Mortimer JT. 1981. Motor prostheses. In *Handbook of physiology*, Section I: *The nervous system*, Vol. II, *Motor control*, Part I, pp. 155–187. Bethesda, MD: American Physiological Society.

Figure 12.16 can be analyzed quantitatively to find regions of depolarization and hyperpolarization and their magnitudes as a function of current stimulus and cuff geometry. This analysis can be undertaken for both active and passive nodal membrane models. The aforementioned adds to the desired internal pathway, shown.

A quantitative analysis of the transmembrane current based on a steady-state linear core-conductor model was performed using the model described in Figure 12.16 for closely and widely spaced electrodes within a fixed cuff [12]. The result is illustrated in Figure 12.17, which describes the transmembrane current under large and small spacings. Near the anode, A, the current behaves as it would for a monopolar electrode, when the cathode is at C_2 (wide spacing and dotted curve). We note an influx of current near A while an outflow occurs lateral to A.

The outflow near the real cathode C_2 results in cathodal depolarization. However, note that depolarization is also exhibited to the left of A, and this marks the presence of a virtual cathode. Usually, the virtual cathode is weaker than the real cathode, as it is here (though the difference is not great). For the wide-spaced cathode C_2 the depolarization under the virtual cathode is much greater than for the narrow-spaced C_1; these considerations suggest a preference for C_1.

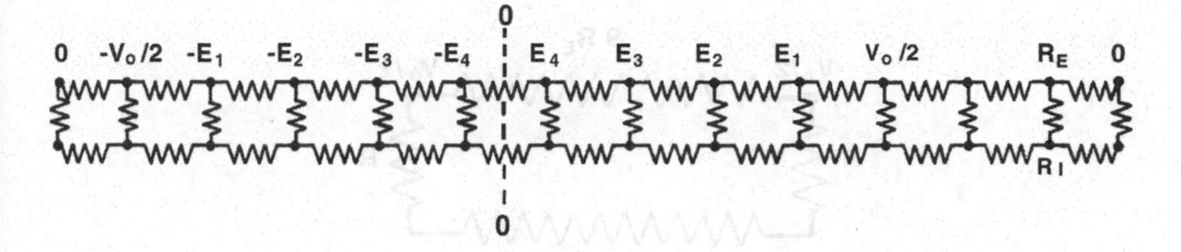

Figure 12.18. Steady-State Model of Myelinated Nerve Fiber in a Cuff Electrode. Applied potential is V_0 and 9 nodes lie between electrodes. An (anti)-symmetrical solution is assumed from the symmetrical structure and (anti)-symmetrical excitation. (Two nodes are missing on the left.)

A plan for analysis of the response of the myelinated fiber, shown in Figure 12.18, under subthreshold or even near-threshold conditions will recognize the following elements:

- The electrodes are assumed to lie at nodes that are separated here by nine internodal spaces. The network continues for three nodes beyond the electrodes, which represents the distance from each node to its respective end of the cuff.
- The external resistance (per unit length) is R_E; its value is that expected from a uniform field in the extracellular space within the cuff.
- The intracellular resistance (per unit length) is R_I, while R is the transmembrane resistance per node.
- The extracellular pathway is introduced by assuming the leftmost and rightmost extracellular node to be connected together [both sites should be set at the (same) reference, zero, potential].
- The behavior of the network in Figure 12.18 can be described by writing a series of loop and nodal equations.
- The McNeal approach, wherein a myelinated fiber is placed in an applied field evaluated in the absence of the fiber, is embodied in Figure 12.18.

Note that the elements above still neglect the membrane capacitance. Consequently, analysis based on those factors alone does not allow one to explore the transient response. Thus one cannot develop strength–duration relationships.

The network can be further simplified by the model given in Figure 12.19. This model neglects the external path and hence cannot examine the virtual anode and cathode. However, it can generate order-of-magnitude figures for the steady-state depolarization and hyperpolarization of cathode and anode, respectively.

For a surface electrode designed to excite a nerve oriented at right angles to the surface, such as illustrated in Figure 12.20, the induced transmembrane potential is as shown in Figure 12.21. This result is for a surface anode relative to a remote reference electrode. These results may

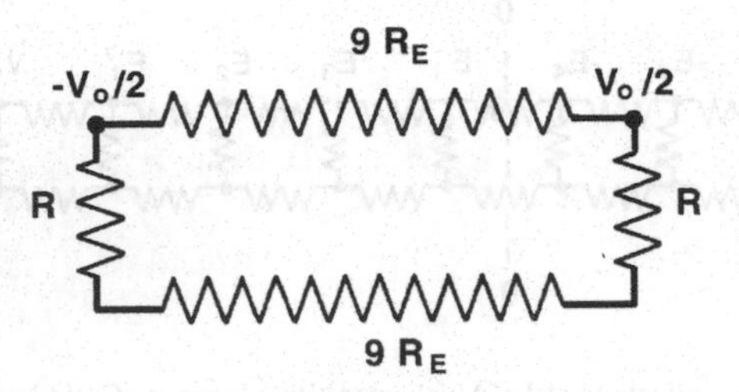

Figure 12.19. Nerve Network Model Approximation to the cuff electrode model described in Figure 12.17.

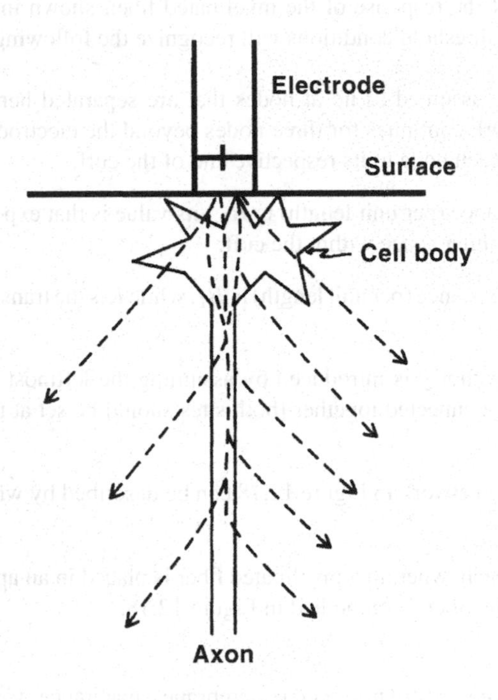

Figure 12.20. Current Path for a Surface Electrode (anode) relative to a remote reference, in the vicinity of a neuron oriented normal to the surface.

seem a bit surprising until the current pathways themselves are examined and the transmembrane potential produced by them is evaluated.

For example, the anodal extracellular current flow, as shown in Figure 12.20, is radially outward from the surface electrode. So far as the nerve cell is concerned, some of this current will enter the neuron in the region of the cell body (proximal end) and leave in the more distal region. Consequently, a hyperpolarization arises near the surface and depolarization in the distal region.

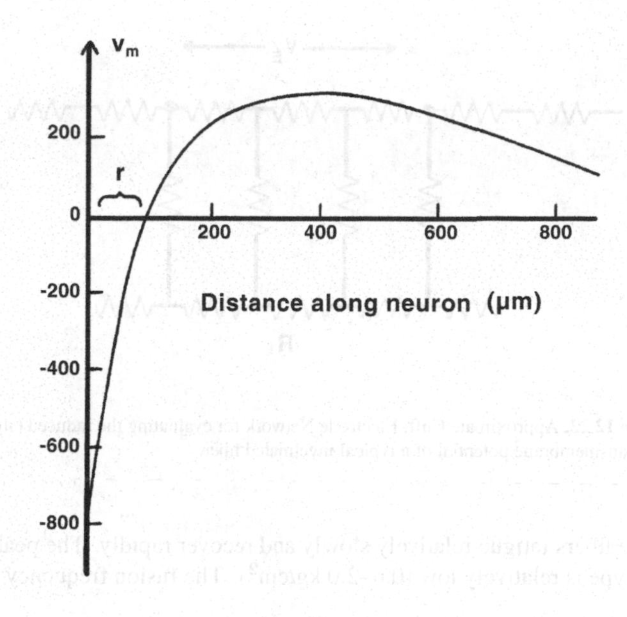

Figure 12.21. Transmembrane Potential Along Nerve Cell (with geometry as described in Figure 12.20). Stimulating electrode is located at zero, and indifferent electrode is located a great distance to the right Note that the change in transmembrane potential reverses sign at distance r to the right of the stimulating electrode. Adapted from Marks WB. 1977. Polarization changes of stimulated cortical neurons caused by electrical stimulation at the cortical surface. In *Functional electrical stimulation*. Ed JB Reswick, FT Hambrecht. New York: Marcel Dekker. Described in Mortimer JT. 1981. Motor prostheses. In *Handbook of physiology*, Section I: *The nervous system*, Vol. II, *Motor control*, Part I, pp. 155–187. Bethesda, MD: American Physiological Society.

Marks [17] shows that the changeover from hyperpolarization to depolarization occurs at a fixed distance r (see Figure 12.21). Since distal elements of a neuron are more excitable than proximal ones, *anodal* stimulation may give a lower threshold than cathodal stimulation (observe the latter by reversing the sign in Figure 12.21), a seemingly anomalous result.

12.7.2. Recruitment

We review below the two basic muscle fiber types introduced in the previous chapter.

1. *FG.* Fast-twitch glycolytic fibers are characterized by metabolism being mainly glycolytic rather than oxidative and by a very-short-duration twitch. With repeated stimulation they fatigue rapidly and then recover slowly. The force produced has a high peak value (1.5–2.0 kg/cm^2). The fusion frequency is $\approx$ 40 Hz. Innervated by large-diameter nerve fibers, the FG fiber is also of large diameter.

2. *SO.* The slow oxidative fiber has a high capacity for oxidative and low capacity for glycolytic metabolism and has a twitch contraction that is relatively long in duration. Its cross-section is the smallest, and it is innervated by the smallest-diameter nerve fibers. Upon repetitive

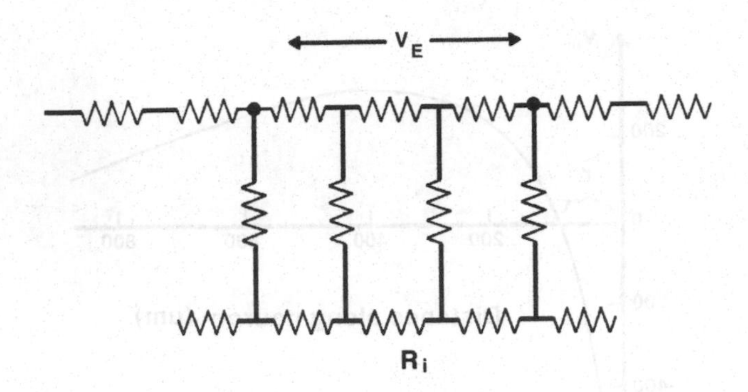

Figure 12.22. Approximate Cuff–Electrode Network for evaluating the induced (stimulating) transmembrane potential of a typical myelinated fiber.

stimulation, these fibers fatigue relatively slowly and recover rapidly. The peak force developed by fibers of this type is relatively low (0.6–2.0 kg/cm^2). The fusion frequency is $\approx$ 10 Hz.

In mixed muscle, motor units producing the greatest force are innervated by axons of large diameter and motor units producing the smallest force are innervated by axons of small diameter. Under natural conditions small motor units are recruited before the large motor units. Thus the natural recruitment order begins with the small-diameter SO units, and ends with recruitment of the large-diameter and concomitant large force provided by the FG motor units.

We consider in contrast, below, the behavior of the three main electrode configurations for functional neuromuscular stimulation.

12.7.3. Nerve Cuff Electrode

Since the field between the bipolar electrodes within the cuff is approximately uniform, we can imagine each fiber subject to the same driving potential (V_E) as illustrated in Figure 12.22. If D is the intracellular diameter and ρ_i the specific intracellular resistance, then the intracellular axial resistance per unit length, obtained from the formula for resistance of a uniform circular cylinder, is

$$R_i = \frac{4\rho_i}{\pi D^2}\Omega/\text{cm} \tag{12.19}$$

The nodal transmembrane resistance (assuming that the specific resistance is a property of the membrane and independent of fiber diameter) is

$$R_N = \frac{r_m}{\pi D l}\Omega/\text{node} \tag{12.20}$$

where r_m is the membrane leakage resistance (Ω cm^2) and l is the node width.

From the circuit in Figure 12.22 we obtain the transmembrane potential to be, roughly,

$$V_m = \left(\frac{R_N}{2R_N + LR_i}\right)V_{\text{peak}} \tag{12.21}$$

where the peak value of V_e, the applied voltage, is the designated V_{peak}, and where L is the electrode spacing.

If $LR_i >> R_N$, then

$$V_m \approx \left(\frac{V_{\text{peak}} r_m}{4 l \rho_i L} \right) D \tag{12.22}$$

For the cuff electrode, the largest-diameter fibers will experience the largest induced voltage and therefore be excited most easily. Thus, at low stimulus levels only FG units will be activated and the whole muscle behavior will be dominated by FG properties. And even at higher stimulus levels, if the FG population is relatively large, it will tend to dominate the overall performance.

12.7.4. Surface Electrode

The stimulating current density from a surface electrode decreases with increasing distance (roughly as $1/R$ due to the radial flow pattern noted earlier). Consequently, superficial fibers tend to be excited first. These turn out to be the FG fibers, and they will be excited ahead of the deeper, more fatigue-resistant fibers. An analytic consideration for fibers of varying diameters with a fixed electrode fiber distance follows.

We first note that the transmembrane potential induced in an unmyelinated fiber by an external point source consists of both depolarized and hyperpolarized regions. These regions are identified in Figure 12.23. In this Figure we note a large depolarization beneath the cathode (electrode) from current leaving the fiber's intracellular space, while laterally there is hyperpolarization resulting from the current necessarily entering the intracellular space (the anode being located remotely).

Using the McNeal model, Reilly [24] demonstrated that (peak depolarization)/(peak hyperpolarization) is 4.2–5.6. For increasing stimulus current, an excitation threshold is reached at the position of peak depolarization. But for a yet further increase in stimulus, a point is eventually reached where lateral propagation is blocked by the large hyperpolarization, and the system no longer responds functionally to the stimulating field.

This possibility of block is described in Figure 12.24, where, for example, with an electrode–fiber distance of 0.75- and 9.6-μm-diameter fibers, a current threshold of 0.65 mA is seen. However, if the current exceeds 1.4 mA, a lateral propagating action potential will be blocked and hence, functionally, no excitation will be available. The shaded region in Figure 12.24 describes the fixed (finite) region of excitation.

For a point source–fiber distance of $h = 0.75$ mm and a stimulus strength 4.0 mA $> |I_0| >$ 2.5 mA, the 9.6 μm-diameter fibers will be stimulated while the larger 38.4-μm fiber will be stimulated *and* blocked.

The larger-diameter fiber is first stimulated at $I_0 \geq 0.3$ mA, while the smaller fibers require $I_0 \geq 0.65$ mA. This comparison suggests the threshold stimulus is proportional to $1/\sqrt{D}$. In fact, the relationship can be demonstrated more generally as follows. If the more accurate expression

$$I_m = \frac{a}{2\rho_i} \frac{\partial^2 \Phi_i}{\partial x^2}$$

replaces (6.29), then (6.31) becomes, where the intracellular resistivity is here designated by ρ_i:

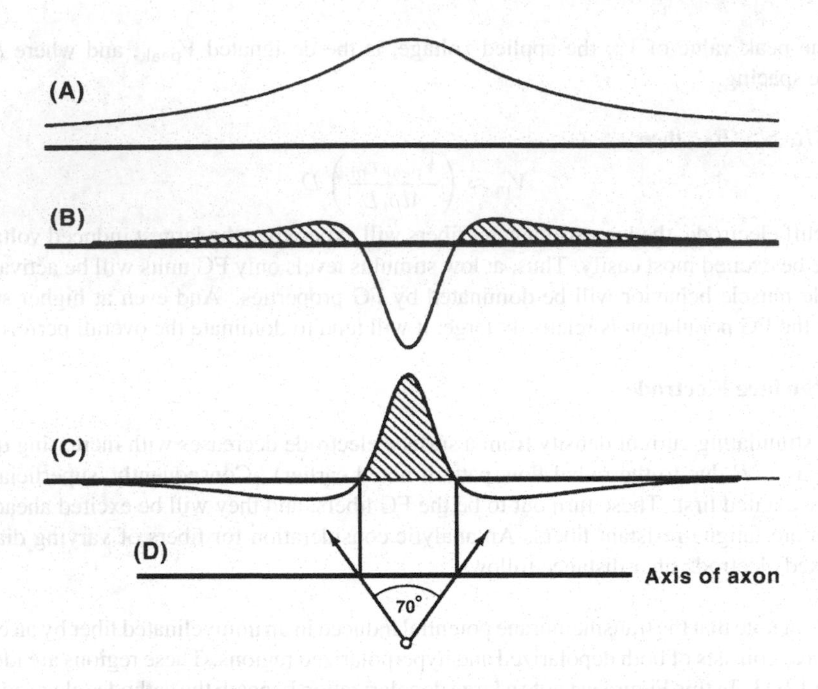

Figure 12.23. Induced Transmembrane Potential with a Point-Source Electrode for (B) anodal and (C) cathodal stimulation. The depolarized regions are shaded. (A) describes the applied field over the surface of the axon with anodal stimulation (the field would have the opposite sign for cathodal stimulation). The electrode position is shown in (D). The border between hyperpolarizing and depolarizing regions of 70.5° is independent of fiber parameters or extracellular conductivity. The reader can generate this Figure using (7.86). From Rattay F. 1987. Ways to approximate current–distance relations for electrically stimulated fibers. *J Theor Biol 125*:339–349.

$$\frac{\partial V_m}{\partial t} = \frac{1}{C_m}\left(\frac{a}{2\rho_i}\frac{\partial^2 V_m}{\partial x^2} + \frac{a}{2\rho_i}\frac{\partial^2 \Phi_e}{\partial x^2} - I_i\right) \tag{12.23}$$

A scale transformation (with new variables designated by a superscript asterisk) is

$$x = \sqrt{a}x^*, \quad h = \sqrt{a}h^*, \quad I_0 = \sqrt{a}I_0^*, \quad \Phi_e^* = \frac{I_0^*}{4\pi\sigma\sqrt{h^{*2} + x^{*2}}} \tag{12.24}$$

which converts (12.23) into

$$\frac{\partial V_m^*}{\partial t} = \frac{1}{C_m}\left(\frac{1}{2\rho_i}\frac{\partial^2 V_m^*}{\partial x^{*2}} + \frac{1}{2\rho_i}\frac{\partial^2 \Phi_e^*}{\partial x^{*2}} - I_i^*\right) \tag{12.25}$$

This demonstrates that a curve, such as that given in Figure 12.24, can be treated as a "universal curve" with regard to fiber diameter. It can be interpreted for another fiber diameter by applying the scale transformation given in (12.24), whereby spatial scaling goes as $\sqrt{a}$ and

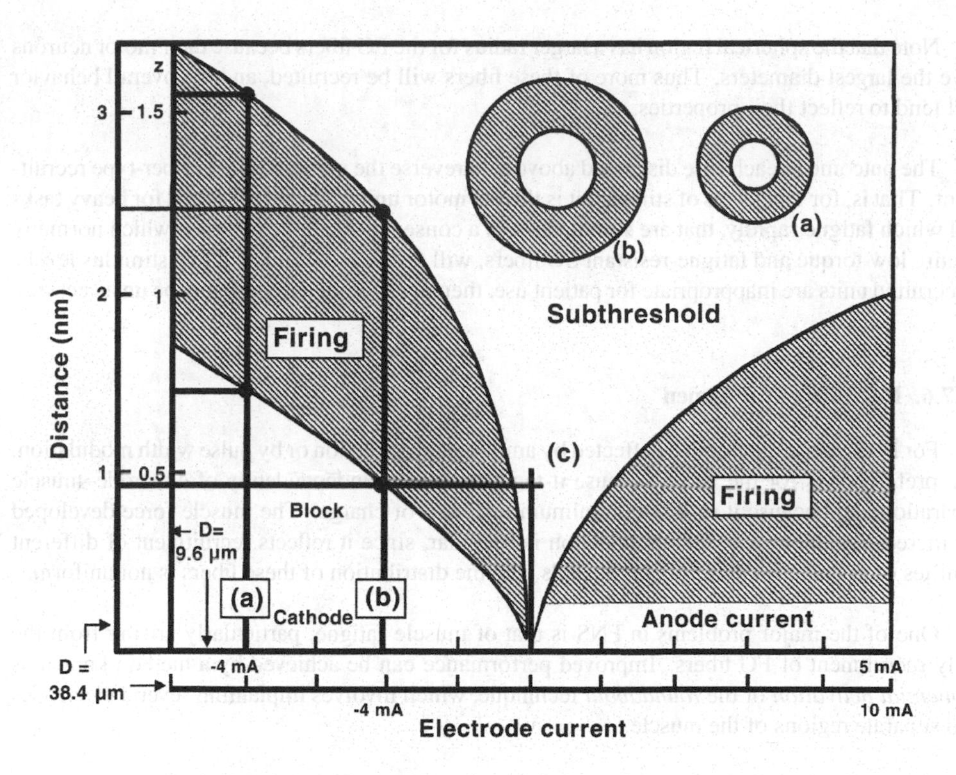

Figure 12.24. Stimulation with a Monopolar Electrode arises for points lying in the shaded areas. The inner scales are for a fiber diameter of 9.6 μm, while the outer are for a diameter of 38.4 μm. For a stimulus of –4.0 mA and $d = 9.6\ \mu$m [line (*a*)], the lower and upper limit of the electrode–fiber distance, while still achieving excitation, is roughly 0.75–1.6 mm. For the same stimulus with $d = 38.4\ \mu$m [line (b)] the interval is roughly 0.9–2.5 mm, and hence more distant fibers are reached. Line (c), for a fixed distance, shows the upper and lower stimulus current magnitude for excitation. Computation was conducted with the Hodgkin–Huxley membrane model, $T = 27°$C, $\rho_e = 300\ \Omega$ cm, and a square pulse of 100 msec duration was chosen. From Rattay F. 1987. Ways to approximate current–distance relations for electrically stimulated fibers, *J Theor Biol* **125**:339–349.

stimulus current as $\sqrt{a}$. In fact, this scaling is evident in the comparison between the 9.6- and 38.4-μm fibers, where the scaling goes as $\sqrt{(4)} = 2$. While recruitment for this geometry goes as the square root of the radius (and not the radius itself), it is still opposite to the natural order.

12.7.5. Intramuscular Electrode

The intramuscular electrode lies deep within a muscle. The current density arising from stimulation decreases with distance from the electrode (by roughly 1/distance), so that for each set of nerve fibers of given diameter a spherical region can be defined within which these fibers will be at or above threshold and outside of which excitation will not occur. Activation of all corresponding muscle fiber types (FG and SO) can be described in this way.

Note that the spherical region has a larger radius for the FG fibers because their motor neurons have the largest diameters. Thus more of these fibers will be recruited, and the overall behavior will tend to reflect their properties.

The outcome in each case discussed above is to reverse the natural order or fiber-type recruitment. That is, for low levels of stimulus it is the FG motor units, generally desired for heavy tasks and which fatigue rapidly, that are stimulated. As a consequence, delicate tasks, which normally require low-torque and fatigue-resistant SO fibers, will fail to be recruited at low stimulus levels. If recruited units are inappropriate for patient use, then use of FNS will probably be unsuccessful.

12.7.6. Recruitment Regimen

For FNS, recruitment can be effected by amplitude modulation or by pulse width modulation. The preference is for the latter, because it tends to operate independently of electrode–muscle separation and because it results in minimum injection of charge. The muscle force developed for increasing amounts of charge injection is irregular, since it reflects recruitment of different families of fibers at increasingly high levels, and the distribution of these fibers is nonuniform.

One of the major problems in FNS is that of muscle fatigue, particularly arising from the early recruitment of FG fibers. Improved performance can be achieved by a method known as *sequential activation* or the *roundabout* technique, which involves implanting several electrodes into separate regions of the muscle.

If, say, the fusion frequency is f_f, and if there are n electrodes stimulating n separate portions of the muscle, a fused contraction of the whole muscle will result if each electrode is stimulated at f_f/n and a phase shift of $360\circ/n$ is introduced between electrodes. The result is a fused force, yet each fiber group, since it is stimulated below its own fusion frequency, has a much improved fatigue resistance. Overlap of regions needs to be carefully avoided, since such a region will fatigue more quickly (being, in effect, stimulated at a higher rate).

A graded contraction can be obtained with either amplitude or frequency modulation. The range of operation with this can be from the lowest level to the point where overlap exceeds an acceptable limit.

The presence of any significant overlap means that the overall performance is limited by a fraction of the muscle being excited at a higher rate efficiency is increased by frankly exciting the entire muscle at this higher rate. One possible regimen adapted is that shown in Figure 12.25. The lowest force is brought about by pulse width modulation. The "switch point" is at the point of significant overlap, where further increase in force is obtained by an increase in stimulus frequency.

A more desirable approach to recruitment is based on mimicking the natural order. In one method, two stimulating electrodes are placed within a cuff. The proximal stimulus is very strong and excites all fibers within the nerve (or muscle). The second electrode is hyperpolarizing and hence will first block propagation on the large-diameter FG fibers, allowing propagation of the small SO fibers. For increasing hyperpolarization, propagation continues but only on decreasing-diameter fibers. In this way a normal recruitment order is achieved [14].

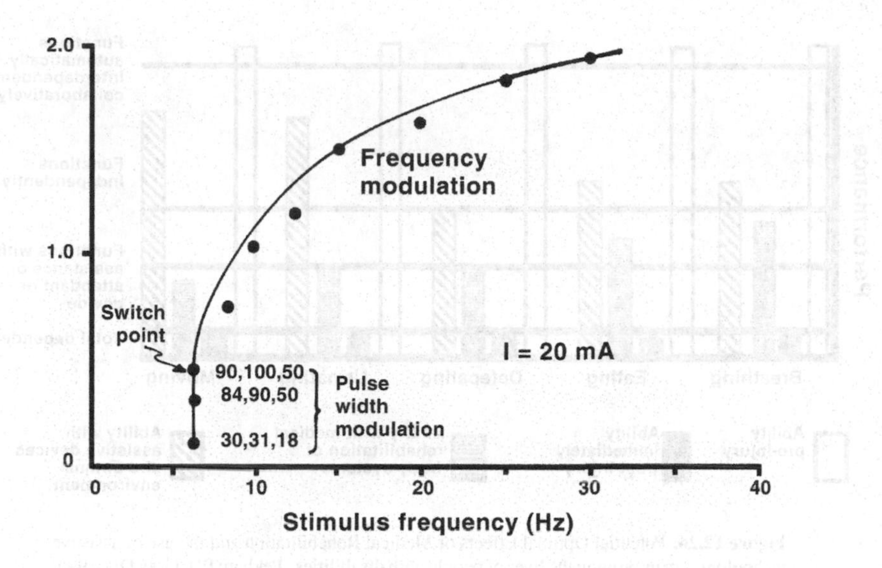

Figure 12.25. Force Characteristics of Muscle Controlled by Pulse-Width Modulation and Pulse-Rate Modulation. Numbers shown in brackets indicate pulse width, in μsec, for each of three electrodes in that particular force level [19]. Since the fusion frequency is 18 Hz, the minimum stimulus frequency for this three-electrode arrangement is one-third of 18, or 6. At the switch point, overlap has reached a level where a further increase in force is achieved by an increase in frequency.

The hyperpolarizing pulse waveform must be designed with a ramp structure at the pulse termination to avoid anode-break excitation, a so-called "quasi-trapezoidal" pulse [6]. In the stimulation of mixed nerve diameters in a nerve bundle, since small-fiber internodal spacing is less than for large-diameter fibers, the closest node to a stimulating electrode is then more likely from fibers of small diameter, where the stimulating field strength is greater. This factor works to improve the likelihood of stimulation of the SO fibers before the FG.

12.8. CLINICAL APPLICATIONS

FNS is an important tool in restoring some degree of function in cases of spinal cord injury (SCI), stroke, cerebral palsy, etc. It is a valuable technique that inserts artificial electrical stimuli where natural signals from the brain have been blocked from reaching either intact motoneurons or their target muscle. An increasing number of individuals have one or another prosthetic device contributing a functional neuromuscular stimulation (FNS), thus demonstrating their effectiveness and acceptance over ever increasing periods of time.

The value of FNS is depicted in Figure 12.26 (from David Gray). This Figure describes performance levels for five important life activities and is given for individuals undergoing a traumatic injury from the period just following the injury and the subsequent degree of recovery. In each case an assistive device is shown to restore an otherwise unreachable *independence*, thus demonstrating its importance.

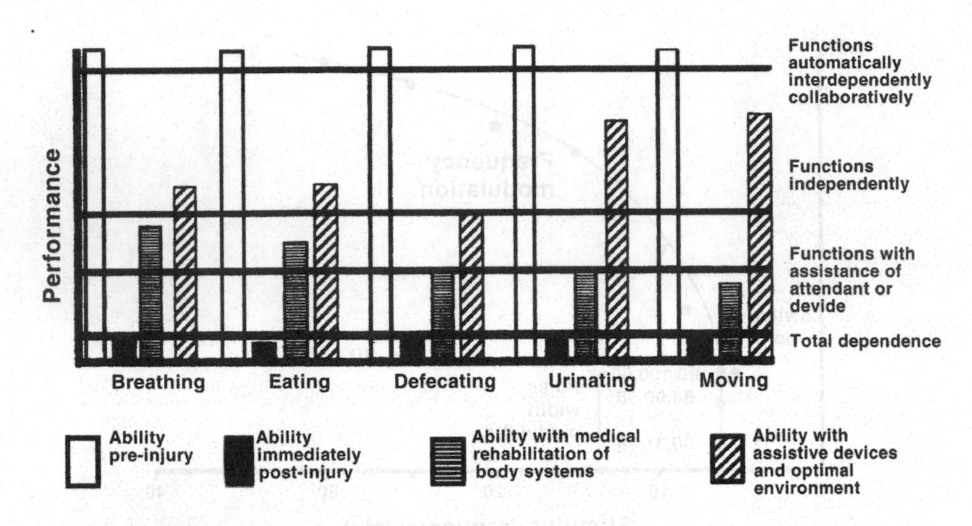

Figure 12.26. Potential Optimal Effects of Medical Rehabilitation and the use of assistive technologies in improving the lives of people with disabilities. Pekham PH, Gray DB. 1996. Functional neuromuscular stimulation (FNS). *J Rehab Res Dev* **33**(2):ix–xi.

An examination of the clinical details is beyond the scope of this book. For the reader wishing to pursue this topic, the sources in the References section should be helpful.[6]

12.9. FRANKENHAEUSER–HUXLEY MEMBRANE

In a fashion similar in form to the Hodgkin and Huxley membrane model, the Frankenhaeuser–Huxley membrane model is described by a set of mathematical equations that is partly basis on analysis of the structure, and partly a mathematical embodiment of experimental results. The Frankenhaeuser–Huxley (FH) equations [5] have been used preferentially to describe the nodal currents in myelinated nerve. The FH model makes use of GHK current expressions such as (5.91), which were presented in Chapter 5. In the F–H model the total membrane current includes potassium ($I_{\rm K}$), sodium ($I_{\rm Na}$), nonspecific delayed (I_p), and leakage current components (I_L). Thus total membrane current is given by

$$I_m = I_{\rm K} + I_{\rm Na} + I_p + I_L \tag{12.26}$$

The equations that describe the potassium behavior suffice to illustrate this model.[7] As used in the FH model, the GHK current equation (5.91)with potassium valence $z_{\rm K} = +1$ gives

$$I_{\rm K} = P_{\rm K} \frac{V_m^2 F^2}{RT} \frac{[{\rm K}]_o - [{\rm K}]_i e^{V_m F/RT}}{1 - e^{V_m F/RT}} \tag{12.27}$$

The potassium permeability is evaluated from its maximum value, $\bar{P}_{\rm K}$, and gating variable n as

$$P_{\rm K} = \bar{P}_{\rm K} n^2 \tag{12.28}$$

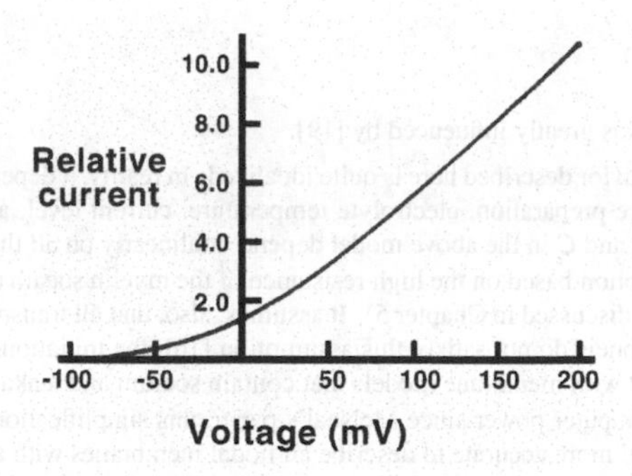

Figure 12.27. *I–V* Curve for the Potassium Channel as described by the GHK Current Equation. Parameter values are given in (12.32).

where n satisfies (5.14), namely,

$$\frac{dn}{dt} = \alpha_n(1-n) - \beta n \tag{12.29}$$

The rate constants are, in turn, evaluated (empirically) from

$$\alpha_n = 0.02(v_m - 35)\left(1 - e^{\frac{35-v_m}{10}}\right)^{-1} \tag{12.30}$$

and

$$\beta_n = 0.05(10 - v_m)\left(1 - e^{\frac{v_m-10}{10}}\right)^{-1} \tag{12.31}$$

In the above equations, as elsewhere in the text, $V_m = \Phi_i - \Phi_e$, and $v_m = V_m - V_{\text{rest}}$.

An open-channel "instantaneous" current–voltage curve from (12.27) is useful to describe its departure from linearity. We take $n = 1$ (so that the macroscopic equations describe the single open channel) and use values for a nodal membrane from Frankenhaeuser and Huxley [16] (also summarized in Schoepfle et al. [12]), namely,

$$[\mathrm{K}]_o = 2.5\,\mathrm{mM},\ [\mathrm{K}]_i = 120\,\mathrm{mM},\ T = 295.18\,\mathrm{K},$$

$$V_{\text{rest}} = -70\,\mathrm{mV},\ \bar{P}_{\mathrm{K}} = 1.2\times10^{-3}\ \mathrm{cm/sec} \tag{12.32}$$

for which Figure 12.27 results. Noteworthy is the rectification property of the GHK formulation.

12.10. FES OUTLOOK

The material in this chapter (and this text) has presented selected topics of established value from a broad field of study. Research still is moving rapidly in gaining understanding and developing bioelectric devices for recognized applications such as management of the cardiac rhythm, limb movement, and bladder control.

Present research is rapidly pushing in new directions, such as deep brain stimulation, e.g., reports by Grill and coworkers [15,8]. This text has only touched the surface of this very deep pool, partly known and with much still to be learned.

12.11. NOTES

1. The material in this chapter was greatly influenced by [19].
2. The electrode/electrolyte behavior described here is quite idealized. In reality, it depends on many variables including the electrode material, surface preparation, electrolyte temperature, current level, and frequency. Most effects are nonlinear and the values of R and C in the above model depend nonlinearly on all the aforementioned.
3. This is a conventional assumption based on the high resistance of the myelin sheath (containing, as it does, the high-resistivity lipid membrane as discussed in Chapter 5). It assumes, also, that all transmembrane ion channels lie in the nodes. While potassium channels do not satisfy this assumption [16], the initiation of the action potential (i.e., via stimulation) can be examined with membrane models that contain sodium and leakage channels only.
4. With the great increase in computer power since McNeal's paper, this simplification should be viewed historically; today, it would be simpler and more accurate to describe all nodal membranes with an active model.
5. This argument is not correct. It can be shown that the *axial* secondary sources arising from the axial transmembrane potential gradient produce little effect on the extracellular field, but this is not true of *transverse* secondary sources. For a uniform applied field, they contribute to a doubling of the field at the fiber surface (no matter how small the fiber diameter). However, what is important here is the axial field variation, which is little affected.
6. In addition, two pertinent rehabilitation journals may be helpful: the *Journal of Rehabilitation and Development* and *IEEE Transactions on Neuroscience and Rehabilitation Engineering*.
7. A complete description of the FH model is included in Johnson [3], Rattay and Aberham [22], and Frijns, Mooij, and ten Kate [7]. There one finds FH model equations and reference values, somewhat updated from the original FH publication [5], e.g., in the adjustments for temperature.

12.12. REFERENCES

1. Agnew WF, McCreery DB, eds. 1990. *Neural prostheses*. Englewood Cliffs, NJ: Prentice-Hall. Chapters 6, 9, 11.
2. Bard AJ, Faulkner LR. 1980. *Electrochemical methods: fundamentals and applications*. New York: Wiley.
3. Johnson CR. 2005. *A mathematical model of peripheral nerve stimulation in regional anesthesia*, pp. 359–367. PhD dissertation. Department of Biomedical Engineering, Duke University.
4. Dymond AM. 1976. Characteristics of the metal–tissue interface of stimulation electrodes. *IEEE Trans Biomed Eng* **23**:274–280.
5. Frankenhaeuser B, Huxley A. 1964. The action potential in the myelinated nerve fiber of *Xenopus laevis* as computed on the basis of voltage clamp data. *J Physiol* **171**:302–315.
6. Fang ZP, Mortimer JT. 1991. A method to effect physiological recruitment order in electrically activated muscle. *IEEE Trans Biomed Eng* **38**:175–179.
7. Frijns JH, Mooij J, ten Kate JH. 1994. A quantitative approach to modeling mammalian myelinated nerve fibers for electrical prosthesis design. *IEEE Trans Biomed Eng* **41**:556–566.
8. Grill WM, Snyder AN, Miocinovic S. 2004. Deep brain stimulation creates an informational lesion of the stimulated nucleus. *NeuroReport* **15**:1137-1140.
9. Grill WM, Mortimer JT. 1996. Quantification of recruitment properties of multiple contact cuff electrodes. *IEEE Trans Rehab Eng* **4**:49–62.
10. Guyton DL, Hambrecht FT. 1974. Theory and design of capacitor electrodes for chronic stimulation. *Med Biol Eng* **12**:613–619.
11. Henneberg K, Plonsey R. 1993. Boundary element analysis in bioelectricity. In *Industrial applications of the boundary element method*. Ed CA Brebbia, MH Aliabadi. Southampton: Computational Mechanics Publications.
12. Karkar M. 1975. Nerve excitation with a cuff electrode—a model. MS thesis. Case Western Reserve University, Cleveland, Ohio.
13. Koole P, Holsheimer J, Struijk JJ, Verloop AJ. 1997. Recruitment characteristics of nerve fascicles stimulated by a multigroove electrode. *IEEE Trans Rehab Eng* **5**:40–50.
14. Kuffler SW, Vaughn Williams EM. 1953. Small nerve functional potentials: the distribution of small motor nerves to frog skeletal muscle, and the membrane characteristics of the fibers they innervate. *J Physiol* **121**:289–317.

15. Kuncel AM, Grill WM. 2004. Selection of stimulus parameters for deep brain stimulation. *Clin Neurophysiol* **115**:2431–2441.

16. Lundborg G. 1988. *Nerve injury and repair*. London: Churchill-Livingston.

17. Marks WB. 1977. Polarization changes of stimulated cortical neurons caused by electrical stimulation at the cortical surface. In *Functional electrical stimulation*. Ed JB Reswick, FT Hambrecht. New York: Academic Press.

18. McNeal D. 1976. Analysis of a model for excitation of myelinated nerve. *IEEE Trans Biomed Eng* **23**:329–337.

19. Mortimer T. 1981. Motor prostheses. In *Handbook of physiology*, Section I: *The nervous system*, Volume II: *Motor control*, Part I, pp. 155–187. Bethesda, MD: American Physiological Society.

20. Naples G, Mortimer JT, Yuen TG. 1990. Overview of peripheral nerve electrode design and implantation. In *Neural prostheses*. Ed WF Agnew, DB McCreery. Englewood, Cliffs, NJ: Prentice-Hall.

21. Pancrazio JJ, Bey Jr PP, Loloee A, Manne S, Chao HC, Howard LL, Gosney WM, Borkholder DA, Kovacs GT, Manos P, Cuttino DS, A. Stenger D. 1998. Description and demonstration of a CMOS amplifier-based system with measurement and stimulation capability for bioelectrical signal transduction. *Biosens Bioelectron* **13**:971–979.

22. Rattay F, Aberham M. 1993. Modeling axon membranes for functional electrical stimulation. *IEEE Trans Biomed Eng* **40**:1201–1209.

23. Reilly JH. 1988. Electrical models for neural excitation studies. *APL Digest* **9**:44–58.

24. Reilly JP. 1998. *Applied electricity*. New York: Springer-Verlag.

25. Robblee LS, Rose TL. 1990. Electrochemical guidelines for selection of protocols and electrode materials for neural stimulation. In *Neural prostheses*. Ed WF Agnew, DB McCreery. Englewood Cliffs, NJ: Prentice-Hall.

Additional References

Grill WM, Kirsch RF. 1999. Neural prostheses. In *Encyclopedia of electrical and electronics engineering*, Vol. 14, pp. 339–350. New York: John Wiley & Sons.

Rattay F. 1990. *Electrical nerve stimulation*. Berlin: Springer-Verlag.

Stein RB, Peckham PH, Popović DP. 1992. *Neural prostheses*. Oxford: Oxford UP.

13

EXERCISES

The following sections provide exercises for some of the material in each of the chapters in the text.

13.1. EXERCISES, CHAPTER 1: VECTOR EXERCISES

1. Vector $\overline{u}$ has x, y, z components (3.9,6). Vector $\overline{v}$ has x, y, z components (5,3,2). What is the magnitude (a) of $\overline{u}$? and (b) of $\overline{v}$?

2. Vector $\overline{u}$ has x, y, z components (1,5,7). Vector $\overline{v}$ has x, y, z components (5,1,1). What is the dot product of these two vectors?

3. Vector $\overline{u}$ has x, y, z components (1,7,6). Vector $\overline{v}$ has x, y, z components (7,3,5). What is the cosine of the angle between these two vectors?

4. In the following questions, $\overline{A}$ and $\overline{B}$ are the described vectors and $\overline{a}_x$, $\overline{a}_y$, and $\overline{a}_z$ are unit vectors along the rectangular coordinate axes x, y, z.

$$\overline{A} = 1\overline{a}_x + 2\overline{a}_y + 3\overline{a}_z \quad \overline{B} = 4\overline{a}_x + 5\overline{a}_y$$

   a. What is the dot product of $\overline{A}$ and $\overline{B}$?

   b. What is the magnitude of $\overline{A}$?

c. What is the angle between $\overline{A}$ and $\overline{B}$?

d. What is the area of the triangle formed by $\overline{A}, \overline{B}$, and a line connecting their endpoints (assuming each begins at the origin)? (Use the cross product.)

5. Consider the following two vectors:

$$\overline{A} = 4\overline{a}_x + 2\overline{a}_y - \overline{2a}_z, \ \overline{B} = 2\overline{a}_x - 5\overline{a}_y - \overline{a}_z$$

a. What is the dot product of the vectors?

b. Are the two vectors are orthogonal (perpendicular)?

6. A five-sided prism has its corners at (0,0,0), (2,0,0), (0,2,0), (0,2,3), (0,0,3), and (2,0,3).

a. Make a vector for each edge of the prism.

b. Use the cross product to find the vector area of the triangle formed by the first two edges. The magnitude of the vector will be the area of that face. Set the order of multiplication so that the direction is that of the outward normal.

c. Do all five sides have the same area?

d. Is the total surface area (sum of sides) of the prism zero?

e. Is the total vector area (sum of sides) of the prism zero?

7. Throughout an unbounded region ϕ is given by $\phi = x^4$.

a. At $x = 1$, what is the x component of the gradient of ϕ?

b. the y component?

8. Throughout an unbounded region ϕ is given by $\phi = x^5$.

   a. At $x = 1$, what is the x component of the gradient of ϕ?

   b. the y component?

9. Throughout an unbounded region ϕ is given by $\phi = x^7$. At $x = 1$, what is the Laplacian of ϕ?

10. The following questions require consideration of a function, $\phi(x)$, that is defined as follows: For $x \leq -1$, ϕ is -1. For $-1 < x < 1$, ϕ equals x^3. For $x \geq 1$, ϕ is 1.

    a. Draw the graph of this function, with axes x and ϕ.

    b. On a distinct axis below the graph, indicate the regions where $\nabla\phi$, the gradient of ϕ, is zero, positive, or negative.

    c. On another axis, indicate the regions where the divergence of the gradient ($\nabla^2\phi$) is zero or nonzero.

11. In the following, consider the scalar field Ψ to be defined by $\Psi = 1/r$, where r is the distance from the origin to an arbitrary (x, y, z).

    a. Compute the gradient of Ψ. Since the gradient is a vector, it can be expressed in terms of components of unit vectors $\overline{a}_x$, $\overline{a}_y$, and $\overline{a}_z$.

    b. In a sentence, describe the direction in which the gradient ($\nabla\Psi$) points. Is it toward (x, y, z)? Perpendicular to r? Other?

12. Consider the scalar function $\Psi = x^2yz$.

    a. Find the gradient of the scalar function

    b. Find the directional derivative of Ψ in the direction given by the following unit vector: $3/\sqrt{50}\,\overline{a}_x + 4/\sqrt{50}\,\overline{a}_y + 5/\sqrt{50}\,\overline{a}_z$.

c. Evaluate the derivative at the point (2,3,1).

13. Using the dot product, square both sides of the equation $\overline{C} = \overline{A} + \overline{B}$. What is the relationship between the result and the law of cosines?

14. For $\Phi = 2x^3y^2z^4$, find $\nabla^2\Phi$ for any x, y, z.

15. Define R as $R = \sqrt{(x - x')^2 + (y - y')^2 + (z - z')^2}$.

    a. Is $\nabla(1/R) = \nabla'(1/R)$?

    b. For any $f(R)$, is $\nabla f(R) = -\nabla' f(R)$?

16. Prove that $\nabla \cdot (\Phi\nabla\Psi - \Psi\nabla\Phi) = \Phi\nabla^2\Psi - \Psi\nabla^2\Phi$.

17. Consider the vector function $\overline{A}$, where

$$\overline{A} = x^2\overline{a}_x + (xy)^2\overline{a}_y + 24x^2y^2z^3\overline{a}_z$$

    a. Find the divergence of the function.

    b. Evaluate the volume integral of $\nabla \cdot \overline{A}$ throughout the volume of a unit cube centered at the origin.

    c. Evaluate the outflow of A over the surface of the unit cube.

    d. Compare the results for the preceding parts. Does Gauss's law hold for this function?

18. Prove that

$$\int_V \Psi \nabla \cdot \overline{F}\, dV = \int_s \Psi \overline{F} \cdot \overline{dS} - \int_V \overline{F} \cdot \nabla \Psi \, dV$$

This equation is the vector equivalent of integration by parts.

19. Green's second identity relates two scalar fields, ϕ and ψ:

$$\int_V (\psi \nabla^2 \phi - \phi \nabla^2 \psi)\, dV = \oint_s (\psi \nabla \phi - \phi \nabla \psi) \cdot \overline{dS} \qquad (1)$$

a. Derive the equation that results if $\phi = 1$ (i.e., a constant) in (1).

b. What is the name of the equation that is the result of part (a)?

c. Derive the equation that results if one sets $\psi = 1$ in (1).

20. Does the following function satisfy Laplace's equation, if $h^2 = \ell^2 + k^2$?

$$\phi = \sin(kx)\sin(\ell y)\, e^{-hz}$$

21. Calculate the following surface integrals for a constant vector $\overline{K}$. Surface S is an arbitrary closed surface, and $\overline{r}$ is a vector arising from the volume within the surface.

a. $\oint \overline{K} \cdot d\overline{S}$.

b. $\oint \overline{r} \cdot \overline{K} d\overline{S}$.

22. A triangle's corners are 3 points equally spaced on a unit circle. The circle lies in the (x, y) plane at $z = 0$. The triangle's vector points in the $+z$ direction. With respect to a field point at (0,0,-0.75), what is the solid angle of this triangle?

Use the centroid method to compute the solid angle. The centroid method is the name of the approximate method that assumes that distance r is the same to all points in the triangle. It is called the centroid method because the approximate r value usually is computed as the distance from the field point to the center (centroid) of the triangle.

23. A triangle's corners are 3 points equally spaced on a unit circle. The circle lies in the (x, y) plane at $z = 0$. The triangle's vector points in the $+z$ direction. With respect to a field point at (0,0,-0.75), what is the solid angle of this triangle? Use the triangle method to compute the solid angle.

Exercises 24–26 deal with the solid angle from a square formed in the xy plane. The surface vector for the square points in the positive z direction. Corners of the square have (x, y, z) coordinates of

- $(1, 1, 0)$
- $(-1, 1, 0)$
- $(-1, -1, 0)$
- $(1, -1, 0)$

24. The observer is located at field point $(0, 0, -20)$. What is the solid angle of the square, as seen by the observer?

25. The observer is located at field point $(-20, -20, -20)$. What is the solid angle of the square, as seen by the observer?

26. The observer is located at field point $(0, 0, 4)$. What is the solid angle of the square, as seen by the observer?

13.2. EXERCISES, CHAPTER 2: SOURCES AND FIELDS

The following guidelines apply to all exercises unless otherwise specified. The conducting medium is uniform, homogeneous, and of infinite extent. Conductivity $\sigma = 0.01$ S/cm. A given potential, often designated Φ, extends throughout the medium. A set of three numbers in parentheses, such as $(3, 2, 1)$, comprise the (x, y, z) coordinates of a point of interest. For the value of the permittivity of free space, use $\epsilon_0 \approx 10^{-9}/36\pi$ Farads/meter.

1. The resistivity of a medium is 100 Ωcm.

   a. What are the corresponding units for conductivity?

   b. In these units, what is the value of the conductivity?

2. Potential $\phi = k/r$ exists throughout an infinite medium with uniform conductivity σ. In the equation for ϕ, k (= 2 mV-cm) is a constant, and r (in cm) is the distance from the origin. What is the current source density at (5,2,1)?

3. Point source $I_o = 12.57$ mA is located at coordinates $(0, 0, 0)$ cm. The surrounding medium has a conductivity of 0.1 S/cm. An imaginary sphere centered on the source point is described by the equation $x^2 + y^2 + z^2 = 1$ cm^2. What is the magnitude of the current density at any point on the sphere? Give a numerical answer, and its units.

4. Throughout a uniform medium the potential in millivolts is given by

$$\Phi = k(xy + z^2)$$

   If k is a constant with magnitude 1, and the conductivity σ in S/cm is 0.2, then

   a. Give an expression, with units, for the density of current $\overline{J}$;

   b. Determine the magnitude of $\overline{J}$ at $x = 1$, $y = 2$, and $z = 3$ cm.

5. Suppose potential ϕ in millivolts is given by the expression

$$\phi = k(x^3 - x^2)$$

where x is in cm and $k = 1$ mV-cm. Where in the medium does the net source equal zero? (Give the x value.)

6. A point current source at $(1, 0, -1)$ cm has a strength of 1 mA, with a conducting medium that has resistivity 100 Ωcm.

   a. What is the potential at point (3,2,2)? (Give units.)

   b. What is the current density at this point? (Give units.)

7. A point current source with magnitude of 4π microamperes is located at the coordinate origin. The resistivity is 50 Ωcm. What is the resulting potential at (5,5,0) cm?

8. A point current source I_o is placed at (2,7,10) cm in a uniform homogeneous medium of infinite extent. What is the relationship between the magnitudes of current density at point A and point B, if A is located at (1,5,7) cm and B at (3,9,13) cm? (That is, is the current density at A less than, equal to, or greater than that at B?)

9. Current density $\overline{J}$ μA/cm^2 (microamperes/cm^2) and two geometric vectors $\overline{U}$ and $\overline{V}$ cm are:

$$\overline{J} = 2\overline{a}_x + 3\overline{a}_y + 4\overline{a}_z$$

$$\overline{U} = 3\overline{a}_x + 4\overline{a}_y$$

$$\overline{V} = 2\overline{a}_x + 3\overline{a}_z$$

   a. What is the magnitude of the current density?

   b. What is the vector area of the triangle formed by $\overline{U}$ and $\overline{V}$ if two sides, each beginning at the origin, are formed by these vectors, and the third side by joining their end points?

   c. What is the current through the triangle?

10. Potential from E: Stimulus electrode E supplying current I_o is located at coordinates $(e, h, 0)$. Imagine the x axis to be a fiber of very small diameter. The distance from electrode E to position x along the fiber is r. What is the potential $\phi(x)$ along the surface of the fiber (i.e., along the x axis)? Answer with a mathematical expression that is a function of electrode position e, h, and axial position x.

11. Activating function from E: Stimulus electrode E supplying current I_o is located at coordinates $(e, h, 0)$. The distance from electrode E to position x along the fiber is r. A stimulus current, applied as in the preceding question, creates potential $\phi(x)$ along the x axis. What is $A(x)$, the second derivative with respect to x of the potential $\phi(x)$? The result should be an equation that is a function of e, h, and x. (Here this result is called $A(x)$, from "activating function," because of related work in a later chapter.)

12. In a homogeneous passive medium, can the conductivity be negative?

13. In the following analysis of a potential field, the potential is given by

$$\Phi = k \tanh(x)$$

where $k = 1$ millivolt and x is in mm. Unless stated otherwise, assume the region of interest is $x = -4$ to $x = 4$ cm. (The hyperbolic tangent function is used here because of the similarity of its wave shape to that of the rising phase of an action potential, studied in a later chapter.)

   a. Find an expression for $\nabla\Phi$ as a function of x.

   b. Find an expression for $\overline{J}$ as a function of x.

   c. Find an expression for $\nabla \cdot \nabla\Phi$ as a function of x.

   d. Find an expression for $\nabla^2\Phi$ as a function of x.

14. Use the results from the preceding question to make calibrated graphs. Let the x axis be about 6 inches long and the vertical axis about 6 inches high, so that the results can be easily visualized. Use multiple graphs as needed.

   a. Plot Φ as a function of x,

b. Plot the gradient of Φ as a function of x.

c. Plot the divergence of the gradient as a function of x.

15. Determine, either by reading the plots or analytically:

a. At what x value does the gradient of Φ reach its highest value?

b. At what x value is the slope of Φ at its maximum value?

c. What is the numerical value of the maximum slope?

16. On a plot of $\Phi(x)$, and at the correct x coordinates:

a. Draw arrows identified as $\overline{J}$ that show the direction of current, at the edges of the plot and at values of x that correspond to peak magnitudes of $\overline{J}$.

b. Draw an encircled plus sign to identify where the current sources are maximum.

c. Draw an encircled minus sign to identify where the current sinks have largest magnitude.

17. Ratios: Point current sources of 8 and –8 mA are located at $z = 2$ mm on the z axis and $y = -2$ mm on the y axis, respectively. At (4,0,0) potential ϕ_1 and current density $\overline{J}_1$ are computed for a medium with resistivity of 100 Ωcm. With the resistivity made twice as much (200 Ωcm), potential ϕ_2 and current density $\overline{J}_2$ are found, for the same point.

a. What is the ratio ϕ_2/ϕ_1?

b. What is the ratio $|\overline{J}_2|/|\overline{J}_1|$?

c. What is the dot product of unit vectors in the directions of $\overline{J}_1$ and $\overline{J}_2$? (i.e., are they in the same direction).

18. Answer the series of questions (13–16) if

$$\phi = k\, e^{-(x-1)^2/2}$$

19. Divergence: The current density $\overline{J}$ in μA/cm^2 is

$$\overline{J} = k(|x|\,\overline{a}_x + |y|\,\overline{a}_y + |z|\,\overline{a}_z)$$

where k is a constant with magnitude 1 and x, y, z are in cm.

a. What are the units of k?

b. What is the magnitude of the divergence of $\overline{J}$ at (1,1,1) cm? Include units.

c. Is the divergence of $\overline{J}$ at the origin less than zero, zero, or greater than zero?

20. A dipole is at the center of the coordinate system and in the z direction. Its strength is 5 mA-cm. The dipole lies in a uniform conducting medium of unlimited extent that has resistivity 100 Ωcm. Determine the voltage (the potential difference) of point A at coordinates (cm) of (10,10,10) with respect to point B at (10,10,0).

21. Voltage between two electrodes: A stimulus is given between two electrodes. One serves as a current source and the other as a sink. Each electrode is a spherical gold conductor that has a radius of 1.5 mm. The two electrodes are separated by 4 cm, center to center. The stimulus produces a rectangular current pulse of duration 2 milliseconds and current amplitude of 15 milliamperes. (The current is injected at the source electrode and withdrawn at the sink electrode.) The medium has a resistivity of 75 Ωcm. During the stimulus, what is the voltage between the electrodes? Assume that there is good contact between the electrodes and medium, with no voltage lost across a boundary layer.

22. A current source of density of $I_v = 1$ ma/cm^3 is uniformly distributed in a spherical region of radius 1.0 cm centered at the origin. The potential field arising from this source is given by

$$\Phi = (3a - br^2)/(6\sigma) \quad 0 \le r \le 1$$
$$\Phi = c/(3\sigma r) \quad r \ge 1$$

where the conductivity $\sigma = 10^{-2}$ S/cm and *r* is in cm. Constants a, b, c have magnitude one and dimensions such that Φ is in mV.

a. What are the units of constants a, b, c?

b. Evaluate $\nabla^2\Phi$ at $r = 0.5$ cm.

c. Evaluate $\nabla^2\Phi$ at $r = 1.5$ cm.

d. What is the voltage (difference in potential) between a point at $r = 0.5$ cm and one at $r = 1.5$ cm?

e. In a sentence or two, explain the different results of parts b and c.

23. Assume that the potential Φ has units of mV and is given by

$$\Phi = k(x^3)$$

where *x* is in cm, and *k* is a constant with magnitude 1 and suitable units. Consider the range $-2 < x < 2$.

a. What are suitable units for *k*?

b. What expression gives the electric field versus x?

c. What expression gives the current source density versus x?

d. Interpret the meaning of (c) in one or two sentences. Are there sources? If so, where and of what sign?

24. Point current sources of 8 mA and –8 mA are located at $z = 2$ mm on the *z* axis and $y = -2$ mm on the *y* axis, respectively. The medium has a resistivity of 100 Ωcm.

a. What is the electrical potential at point (4,0,0) cm, relative to a potential of zero at infinity, if the potential is computed as the sum from two monopoles?

b. What is the electrical potential at point (4,0,0) cm, relative to a potential of zero at infinity, if the potential is computed as the sum from two monopoles?

c. What is the difference in the result from [a] versus [b]?

d. Find the current density at the point $x = 4$, $y = z = 0$ cm.

25. For an arbitrary spherical surface located in a source- free region, confirm that the average value of potential over the spherical surface equals the value of potential at the center.

26. Two point sources have magnitudes $+I_o$ and $-I_o$. These sources are located at $(0, 0, 1)$ and $(0, 0, -1)$ mm, respectively; $I_o = 1$ nA, and the conductivity is .01 S/cm.

a. What is the potential at (5,0,0) cm, when the result is computed as the sum of the potentials from each monopole?

b. Expressed as a dipole vector, what is the source?

c. What is the vector from the center of the dipole to (5,0,0) cm?

d. What is the potential at (5,0,0) cm, computed from the dipole?

27. Two adjoined fish tanks, each a vat of saltwater, are separated by a very thin uniform insulator (of only 100 Å thickness). The insulator is flat with parallel sides. The dielectric constant of the insulator is 2. What is the capacitance across each square centimeter of the insulator?

28. For the two-box geometry of Figure 2.7, $R_i = 100$ Ωcm, $R_e = 600$ Ωcm, $a = 10$ μm, $b = 20$ μm, $L = 100$ μm, $R_m = 10000$ Ωcm^2, and $C_m = 1.2$ μF/cm^2. Consider current flow in the direction $\overline{a}_x$.

a. What is the axial resistance R^S of the small box, in Ohms?

b. What is the axial resistance of the big box, R^B, in Ohms (excluding the small box within it)?

c. What is the conductance of the two boxes, taken together as a parallel combination?

29. For the two-box geometry of Figure 2.7, $R_i = 100\ \Omega$cm, $R_e = 400\ \Omega$cm, $a = 10\ \mu$m, $b = 20\ \mu$m, $L = 100\ \mu$m, $R_m = 10000\ \Omega\text{cm}^2$, and $C_m = 1.2\ \mu\text{F/cm}^2$. Along the x axis, what is:

   a. r_i, the resistance per unit length of the small box?

   b. r_e, the resistance per unit length of the big box?

   c. r_m, the membrane resistance per unit length?

   d. c_m, the membrane capacitance per unit length?

30. For the two-box geometry, a current $I = 10$ mA is applied across the membrane separating the small box from the big box. $R_i = 100\ \Omega$cm, $R_e = 400\ \Omega$cm, $a = 10\ \mu$m, $b = 20\ \mu$m, $L = 100\ \mu$m, $R_m = 10000\ \Omega\text{cm}^2$, and $C_m = 1.2\ \mu\text{F/cm}^2$. The current is applied uniformly through all parts of the membrane. When the current starts, the voltage V across the membrane is zero.

   a. Per square centimeter, what is the applied membrane current?

   b. What is the steady-state voltage V_{SS} across the membrane.

   c. How long does it take (in seconds) for the voltage to rise from zero to half the final voltage?

31. Consider the small box, surrounded by the membrane, in isolation. A voltage of 100 mV is applied across two parallel plates, one underneath the bottom of the box and the other lying on top of the box. Box measurements are $R_i = 100\ \Omega$cm, $R_e = 400\ \Omega$cm, $a = 10\ \mu$m, $b = 20\ \mu$m, $L = 100\ \mu$m, $R_m = 10000\ \Omega\text{cm}^2$, and $C_m = 1.2\ \mu\text{F/cm}^2$. Both plates are outside the membrane but in good contact everywhere on the bottom or top, respectively. What is the steady-state current?

32. A large thin sheet of conducting material has conductivity σ Siemens/cm^2. There is a point current source at the center of the sheet of magnitude I_0 and a distal sink (also of magnitude

I_0). Derive an expression for the electrical potential variation near the source. (Assume that the current is radial and two- dimensional.)

33. Two conducting, concentric spheres are separated by a thin insulator. The inner sphere has an outer radius of 10 micrometers. The insulator is only 50 Åthick, and has a dielectric constant of 3. What is the capacitance between the spheres?

34. For a uniform conducting medium of infinite extent, the flow lines of current density are given by the differential equation

$$\frac{dx}{E_x} = \frac{dy}{E_y} = \frac{dz}{E_z}$$

For a dipole $\overline{p}$ oriented along the z axis and located at the origin, find the equation $f(y,z) = \text{constant}$ that gives the pattern of currents in the $x = 0$ plane.

35. Four point *charges* lie in the xz plane as follows:

$$+Q \text{ at } (d, 0) \text{ and at } (0, d)$$
$$-Q \text{ at } (0, 0) \text{ and at } (d, d)$$

Show that for $r >> d$ the potential Φ is

$$\Phi = -\frac{3Qd^2xz}{4\pi\varepsilon_0 r^5}$$

where

$$r = \sqrt{x^2 + y^2 + z^2}$$

13.3. EXERCISES, CHAPTER 3: BIOELECTRIC POTENTIALS

In the following questions, give a numerical answer with one percent accuracy, with units, unless otherwise instructed.

1. What is the (net) amount of positive charge in the positive ions of 90 grams of table salt? Assume the salt is fully dissociated. A mole of salt weighs 58.5 grams.

2. A voltage of 0.3 volts is uniformly applied across a width of 70 Å, within a conducting solution. The conducting ions come from 42 grams of table salt, fully dissociated. The electric field exerts a strong force on the ions. What is the force that the resulting electric field would place on the (net) positive charge in the positive ions? Compute the force in pounds, using the fact that 1 Newton is 0.2248 lbs-force.

3. The "mobility" of ions quantifies their movement when they are in an electric field. The units of mobility are (m/sec)/(V/m). (One might think of these units as velocity per unit of electric field.) Suppose a uniform field is created by placing a voltage of 1 Volt across a distance of 3 m. The temperature is 30° C, and the diffusion coefficient is 1.33E-9 m^2/sec. What is the corresponding mobility of sodium ions?

4. If the temperature is 7° C, what is the value of RT/F?. Use a precise conversion between degrees Celsius and Kelvin, and use precise values of R and F so as to find a result within 0.3%. Compare the result to the one in the table in the text, and explain the difference.

5. For a temperature of 8° C, estimate the root-mean-squared (rms) velocity of sodium ions. Begin analysis at a macro level with the energy in a mole of ions. Find the average energy of an ion. To avoid the complexity of a fluid, assume the energy of the ions is entirely kinetic energy, as in a gas. The mass of a sodium ion is approximately 3.82E-26 kg. (Velocity in m/sec can be converted to miles per hour by multiplying by 3600 sec/hour and then dividing by 1850 m/mile.)

Exercises 6–8: The Nernst–Planck Equation. Solutions to each of these questions can be found by reference to the chapter. These questions are emphasized because they are the foundation for many other and thus are worth committing to memory.

6. Begin by stating equations for (1) flow due to diffusion, and (2) flow due to an electric field.

   a. Derive the Nernst–Planck equation.

   b. Show that the units balance.

7. Write the Nernst–Planck equation. For each variable or constant, give its units.

8. In the Nernst–Planck equation, the term $(-D_i F Z_i) Z_i C_i (F/RT) \nabla \Phi$ gives the flow in response to an electric field. What is the expression for the conductivity σ? (Be sure to get the sign correct).

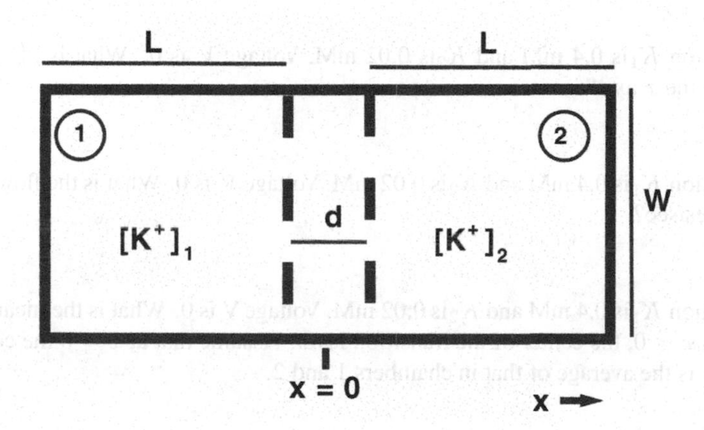

Exercises 9–15. Concentration cells.

Exercises 9 to 15: Concentrations and Flux: Two chambers numbered 1 and 2 have the following dimensions (see the figure). Dimension L is 100 microns, and W is 100 microns. Height H (out of surface) is 100 microns. The dotted lines show the edges of a boundary layer that has a thickness of 40 Å. (The dotted lines identify the position of a transition region, not a physical boundary.) The temperature is 300 degrees Kelvin, and the diffusion coefficient D_K is 1.96E-5 cm2/sec. Within either chamber the concentration and electric potential are uniform, though different between chambers. The K^+ concentrations are K_1 in chamber 1 and K_2 in chamber 2. A linear change occurs in concentration and potential across the transition region. The potential difference is the voltage V, where $V = \phi_1 - \phi_2$. Note the polarity, i.e., a positive value for V occurs when the potential in chamber 1 is the higher.

9. Concentration $K_1$is 0.22 mM and $K_2$is 0.02 mM. V is –0.04 Volts.

   a. At $x = 0$, what is I_K^D, the potassium current along x due to diffusion?

   b. At $x = 0$, what is I_K^E, the potassium current along x due to the electric field?

10. Concentration $K_1$is 0.4 mM and $K_2$is 0.02 mM. What is concentration K_1 in moles/cm^3?

11. Concentration $K_1$is 0.2 mM and $K_2$is 0.02 mM. How many K^+ ions are then in volume 1? Respond in units of moles.

12. Concentration $K_1$is 0.4 mM and $K_2$is 0.02 mM. Voltage V is 0. What is j_K^D (the particle flux) along the x axis?

13. Concentration $K_1$is 0.4 mM and $K_2$is 0.02 mM. Voltage V is 0. What is the flow of particles i_K^D, in moles/sec?

14. Concentration $K_1$is 0.4 mM and $K_2$is 0.02 mM. Voltage V is 0. What is the mean velocity of K^+ ions at $x = 0$, the center of the transition layer. Assume that at $x = 0$ the concentration of K^+ ions is the average of that in chambers 1 and 2.

15. Concentration $K_1$is 0.4 mM and $K_2$is 0.02 mM. Voltage V is 0. How long will it take for the concentration of K^+ on side 1 to decline by 10%? Assume that throughout this interval flow maintains its initial value.

Exercises 16-21: Equilibrium Voltages across Membranes. Two compartments A and B are separated by a thin membrane. Transmembrane voltages are measured as the potential in compartment A minus that in B.

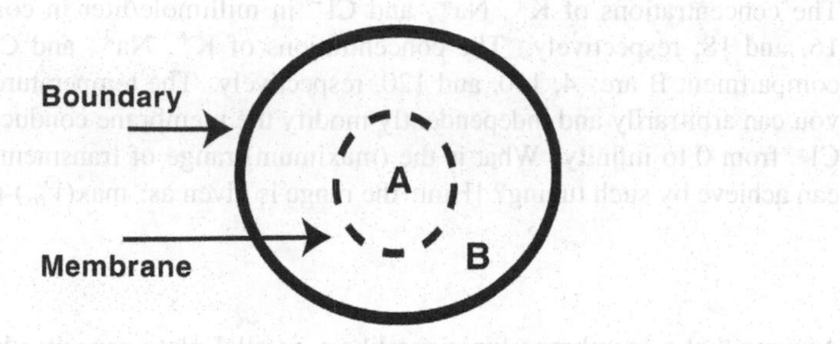

Exercises 16–20: The membrane is indicated by the dashed line. No flow occurs across the outer boundary.

Exercises 16–21: Concentrations (mM)[a]

Ion	A	B
K^+	280	10
Na^+	61	485
Cl^-	51	485

[a] mM means millimoles per liter.

16. Only sodium may cross the membrane separating the two compartments. The temperature is 24°C. What is the equilibrium potential?

17. Only potassium ions may cross the membrane. The temperature is 18°C. What is the equilibrium potential?

18. Only Cl^- ions may cross the membrane. The temperature is 36°C. What is the equilibrium potential?

19. The transmembrane potential is 1 millivolt, and the temperature is 18°C. How many ion species (Na^+, K^+, or Cl^-) are NOT in equilibrium? (Allow a small tolerance for round-off.)

20. The temperature is 36°C, and the transmembrane potential is –60 millivolts. How many ion species are NOT in equilibrium? (Allow a small tolerance for round off.)

21. The concentrations of K^+, Na^+, and Cl^- in millimole/liter in compartment A are: 120, 16, and 18, respectively. The concentrations of K^+, Na^+, and Cl^- in millimole/liter in compartment B are: 4, 140, and 120, respectively. The temperature is 26°C. Assume that you can arbitrarily and independently modify the membrane conductance for K^+, Na^+, and Cl^- from 0 to infinity. What is the (maximum) range of transmembrane voltages that you can achieve by such tuning? [Hint: the range is given as: max(V_m)-min(V_m)].

22. Assume that a membrane functions like a parallel plate capacitor having the same surface area, and assume the membrane is a nonconducting lipid with a relative permittivity of 3.

    a. The capacitance of a particular biological membrane is 1.1 microFarads per square centimeter. What is the effective membrane thickness?

    b. A membrane has an effective thickness of 23 Å. What is this membrane's capacitance per unit area, C_m?

Exercises 23–29: Cardiac Cell. A cardiac cell is satisfactorily represented by the shape of a brick with edges defined by a length, width, and depth. In these questions the membrane is considered passive in that it has unchanging values of membrane resistivity (in Ωcm^2) and capacitance (microFarads per cm^2). The cell's surface area is taken to be that of the four sides excluding the ends (width by depth). In some cases it is necessary to select from among the given parameters those needed to answer the question. Abbreviation μm is used as an abbreviation for micrometer.

23. The cell's length, width, and depth are 100, 9, and 11 μm, respectively. What is the cell's surface area? (Compute the area as the total membrane surface of all four sides, ignoring the small area at the ends of the brick.)

24. The cell has length, width, and depth of 116, 7, and 12 μm, respectively. Membrane capacitance (microFarads per cm^2) is 1.1. What is the cell's capacitance? (That is, what is the capacitance of the inside with respect to the outside, considering the surface to behave uniformly.)

25. The cell has length, width, and depth of 112, 5, and 11 μm, respectively. At rest the membrane resistivity (Ωcm^2) is 20,000. What is the cell's membrane resistance? That is, what is the resistance from inside to the outside of the cell, at rest?

26. A cardiac cell has length, width, and depth of 90, 5, and 8 μm, respectively. At rest the membrane resistivity (Ωcm^2) is 20,000 and capacitance (microFarads per cm^2) is 1.1. Suppose an amount of charge is instantly moved across the membrane and changes the transmembrane voltage by 20 millivolts. How much charge moved?

27. A cardiac cell has length, width, and depth of 106, 6, and 11 μm, respectively. At rest the membrane resistivity (Ωcm^2) is 10,000 and its capacitance is 1.1 microFarads per cm^2. What is the cell's membrane conductivity, in Siemens/cm^2?

28. A cardiac cell has length, width, and depth) of 82, 12, and 5 μm, respectively. At rest the membrane resistivity (Ωcm^2) is 20,000 and the capacitance is 1 microFarad per cm^2. If the cell membrane functions as a passive, parallel resistor–capacitor combination, what is the time constant of this cell?

29. A cardiac cell has length, width, and depth of 80, 5, and 12 μm, respectively. At rest the membrane resistivity (Ωcm^2) is 20,000, and the capacitance is 1.1 microFarads per cm^2. Suppose the membrane instantly acquires enough charge to rise from its resting potential by 20 millivolts. If the cell remains passive, i.e., continues to have the same membrane resistance and capacitance, how long will it take for the 20 milliVolt initial rise to decay to 1 mV?

30. What are suitable units for each of the following:

a. J, the current flux?

b. j, the particle flux?

c. C, the capacitance?

d. D, the diffusion coefficient?

e. ϕ, the electric potential?

f. μ, the mobility of an ion in response to an electric field?

Exercises 31–33: Segment of Nerve Fiber (Equations). A short segment of nerve fiber is cylindrical, being d cm in diameter and L cm in length. Assume there is no variation of membrane potential along the axial coordinate. The temperature is T degrees Kelvin.

- ☐ The membrane capacitance is C_m microFarads per cm^2.
- ☐ The transmembrane potential V_m is in millivolts.
- ☐ Equilibrium potentials E_K, E_{Na}, and E_L are known, with values in millivolts.
- ☐ Membrane conductances are g_K, g_{Na}, and g_L ms/cm^{2}for ions of K^+, Na^+, Cl^- respectively.
- ☐ In these exercises the conductances are constants.

Give the equation that might be used to answer the exercise as a line in a standard computer language, such as the C language. Show multiplication explicitly with an asterisk. Preserve upper and lower case letters as given in the exercise or text; that is, do not consider upper and lower case letters to be equivalent.

31. What is J_K, the current density across the membrane of K^+ ions?

32. What is J_{Na}, the current density of Na^+ ions?

33. What is J_L, the current density for Cl^- and other leakage ions?

Exercises 34–45: Short Nerve Segment. These exercises involve a short segment of an unmyelinated nerve. Each nerve segment has a circular cross-section with diameter d and length L. The temperature is 300 degrees Kelvin, and the membrane capacitance is 1.2 microFarads per cm^2. The transmembrane potential (inside potential minus outside) is clamped to –40 millivolts. There is no variation of membrane potential along the axial coordinate. Use the parallel-conductance model for analysis, as needed. Membrane conductances are: g_K 0.375, g_{Na} 0.01 mS/cm^2, g_L 0.57 mS/cm^2. The membrane conductances are assumed to be constants, in these exercises. If more values are given than are needed to find the answer, a part of the exercise is selecting the relevant information.

Concentrations (mM)

Ion	Intra	Extra
K+	280	10
Na+	50	437
Cl-	51	485

34. What is the resting potential?

35. What is E_K, the equilibrium voltage for K^+ ions?

36. What is E_{Na}, the equilibrium voltage for Na^+ ions?

37. What is E_L, the equilibrium voltage for Cl^- ions?

38. What is J_K, the current density for K^+ ions?

39. What is J_{Na}, the current density for Na^+ ions?

40. What is J_L, the current density for Cl^- ions?

41. What is dV_m/dt, the time rate of change of transmembrane potential V_m, at time t^{0+}, if t^{0+} is the time immediately following the instant when the voltage clamp is removed?

42. A segment of nerve is 7 μm in diameter and 0.7 cm in length. Membrane conductances are g_K 0.2, g_{Na} 0.25, and g_{Cl} 0.4 mS/cm^2. What is J_{ion}, the total ionic current density? Find the total as the sum of J's for K^+, Na^+, and Cl^- ions.

43. A segment of nerve is 8 μm in diameter and is 1.5 cm in length. The transmembrane potential V_m is –40 millivolts. The membrane conductances are: g_K 0.375, g_{Na} 0.02, and g_{Cl} 0.57 mS/cm^2. The resting potential is –34.7 mV. What is E_L, the equilibrium voltage for Cl^- ions?

Ex. 43 Concentrations (mM)

Ion	Intra	Extra
K+	300	35
Na+	90	450

44. A segment of nerve has a diameter of 5 and is 1.1 cm in length. The transmembrane potential is –40 millivolts. The extracellular potassium concentration has risen to 50 mM. What is the resting potential?

45. In a short segment of nerve, the transmembrane potential V_m is –56 millivolts. Values of conductances are g_K 0.375, g_{Na} 0.06, and g_L 0.8 mS/cm^2, and here membrane conductances are constant. What is dV_m/dt, the rate of change of transmembrane potential V_m?

***Examplus* giant nerve:** This entirely imaginary nerve is referenced in the three exercises that follow. In *Examplus* K_e^+ = 10 mM, Na_e^+ = 500 mM, K_i^+ = 300 mM, and Na_i^+ = 50 mM.

46. For *Examplus*, if the membrane is permeable to potassium only, what transmembrane potential, V_m, will exist at equilibrium?

47. For *Examplus*, if the membrane is permeable to sodium only, what transmembrane potential, V_m, will exist at equilibrium?

48. For *Examplus*, if the membrane is permeable both to potassium and to sodium, what information does the Nernst equation provide about the magnitude of the current flow from the intracellular to the extracellular space?

***Aplysia* cell.** *Aplysia* cells have been used in a number of electrophysiological studies.

49. In an *Aplysia* cell, values used with the parallel conductance model are:

$$E_K = -83.9 \text{ mV}, E_{Na} = 52.2 \text{ mV}, E_{Cl} = -56.7 \text{ mV}$$
$$g_K = 0.177 \ \mu S, g_{Na} = 0.116 \ \mu S, g_{Cl} = 0.707 \ \mu S$$

a. What is the resting potential?

b. If $[K]_e$ is increased tenfold, how is the result in (a) changed?

The remaining exercises touch on a variety of topics related to material in this chapter.

50. Write the Nernst–Planck equation. Which constant stands for the amount of charge in a mole of ions?

51. The diffusion coefficient is a measure of how rapidly a particle moves due to (random) thermal motion (the diffusion process). Thus the presence of the diffusion coefficient in the Nernst–Planck equation in the term of the equation dealing with flow due to diffusion is not surprising. But why does it also appear in the other (electrical) term? Explain in a sentence.

Exercises 52–55: Capillary Membrane and Donnan Equilibrium. Movement across capillary membrane is an important topic that is not the focus of this text but that does have closely related aspects, a few of which are explored in these exercises. A membrane is at a *Donnan equilibrium* when the ion flow at rest of *all* permeable ions is zero. (Before the advent of intracellular electrodes, many theories assumed the existence of a Donnan equilibrium across active and passive membranes.)

52. What must be true of the resting transmembrane potential and the Nernst potential of all ions, at a Donnan equilibrium? That is, given the ion composition of Table 3.1, why cannot nerve and muscle membranes be at a Donnan equilibrium?

53. A cell lies in an extensive, uniform, extracellular medium. The intracellular and extracellular space contain KCl, NaCl, and possibly NaX and KX (only). The membrane is permeable to Na^+, K^+, Cl^-, but not to X^-. The system is at a *Donnan* equilibrium.

    a. What does X^- represent and what is the likely reason it is impermeable?

    b. What is the basic constraint, and what is the relationship satisfied by ionic concentrations that arise in the *Donnan* equilibrium? [Hint: What must be true of permeable ion Nernst potentials?]

    c. Given that $[K^+]_e = 50$ mM, $[K^+]_i = 5$ mM, $[N^+]_e = 35$ mM, and $[Cl^-]_e = 0.85$ mM; determine the remaining ionic constituents.

    d. Determine the transmembrane potential.

54. The capillary endothelium, while restraining the movement of protein, permits the free flow of water and solute. A Donnan equilibrium of diffusible ions comes about as a result.

    a. Based on this information, complete Table 13.1.

    b. What is the transcapillary potential?

Table 13.1. Ion, Plasma, and Interstitial Fluid, for Exercise 54

Ion	Plasma Water	Interstitial fluid
Na^+	150 mM	144 mM
K^+		4.0
Cl^-		114
HCO_3^-	28	

55. The red blood cell is permeable to anions (only), since passive cation flux is balanced by active transport. The major anions are Cl^- and HCO_3^-, and these reach a Donnan equilibrium (see Exercise 31). Assume that

$$\frac{[Cl^-]_{cell}}{[Cl^-]_{plasma}} = 0.63$$

    a. What is the distribution of bicarbonate ion?

    b. What is the membrane potential (give polarity)?

13.4. EXERCISES, CHAPTER 4: CHANNELS

Exercises 1–5: Open and closed channels. Use data from Table 4.1 as needed. Abbreviation μm stands for micrometers (sometimes called microns).

1. Assume the membrane of frog skeletal muscle is at rest, and that, at rest, only potassium channels are open, with probability 0.06. What is the resistance of 0.5 cm^2 of this membrane?

2. During the upstroke of an action potential in squid giant axon, the probability of a sodium channel being open changes. At rest the probability of an open channel is 0.01, while at the peak of an action potential it is 0.2. Consider 0.6 mm^2 of cell membrane. On the average, how many channels change from closed to open as the membrane moves from rest to peak, i.e., during the upstroke of the action potential?

3. The membrane of the squid giant axon has a resting membrane resistance of 1,400 Ωcm^2. If potassium channels are the only contributor to the resting membrane resistance, what fraction of the potassium channels are open?

4. A frog skeletal muscle cell is at rest, with a rate of potassium channel opening of 2E-4 $msec^{-1}$, and a rate of potassium channel closing of 0.3 $msec^{-1}$. The cell has the shape of a brick with edges (length, width, and depth) of 2,000, 20, and 30 μm. Suppose potassium channels at the density given in Table 4.1 exist on the cell surface over all sides and ends. What is the steady state number of open potassium channels, at rest, on the average?

5. Suppose a frog skeletal muscle cell is satisfactorily represented in the shape of a brick with edges (length, width, and depth) of 2,000, 25, and 10 μm. At rest the rate of potassium channel opening is 0.005, and the rate of channel closing is 0.48 $msec^{-1}$. The resting membrane voltage is –96 mV, and the temperature is 28 degrees C. What is the (macroscopic) potassium current in at rest?

Exercises 6–13 involve the probabilities of channels being open and closed. In all questions, the number of channels is N, the probability of an open channel is p, the probability of a closed channel is q, and the channel density is D.

6. What is the formula for the expected number of open channels?

7. What is the formula for the expected number of closed channels?

8. What is the sum $p + q$?

9. What is the formula for the standard deviation of the number of open channels?

Exercises 10 to 13: Imagine that it becomes possible to design cells of varying size. A cell is to be designed that is spherical, with a membrane that has stable number of open channels. To this end, the number of channels must meet the following condition: The standard deviation of the number of open channels may be no more than one percent of the expected number of open channels.

10. In terms of p and q, give the formula for M, where M is the minimum number of channels that the cell must have.

11. Give a formula for A, where A is the minimum surface area that the cell must have to hold the required number of channels.

12. Give a formula for a, where a is the minimum radius that the cell must have to hold the required number of channels, at density D.

13. What is the minimum radius that the cell must have to hold the required number of channels? (Give the value of the radius a, in μm.)

   a. Probability p is fixed at 0.05 and D is 200 channels per μm^2.

   b. D is 200 channels per μm^2, and p varies from .05 to .95. (Radius a must cause the condition to hold for any p.)

Exercises 14–35: Model Cell in a Voltage Clamp. These exercises are framed mostly in terms of a fictitious model cell. The reason such a cell is used here is to focus on the main ideas of this chapter within a relatively simple quantitative framework. With some increase in complexity, the exercises can be brought closer to reality by substituting experimental values from Tables 3.1 and 4.1 and using actual cell dimensions.

The model cell has a surface area of 600 μm^2, a channel density $D = 40$ channels per μm^2, and a conductivity for a single open channel of $\gamma = 10$ pS. The cell surface holds distinct populations of channels for potassium and sodium ions, but both populations have the same density D and conductivity γ. The cell's resting potential V_r is –60 mV. Surrounding concentrations create a potassium equilibrium potential E_K of –80 mV and a Na^+ equilibrium potential of +60 mV.

The solutions that are provided make use of the α and β rate constants and the structure of the Hodgkin–Huxley model. Correspondingly, in these exercises, find the rate constants α_n and β_n as needed by converting absolute membrane voltages V_m to voltage offsets from the resting potential, $v_m = V_m - V_r$. Then use v_m in the reference equations for α and β.

Rate coefficients α and β in the Hodgkin– Huxley Model: The values of α_n and β_n are based on curve-fitting to experimental data for potassium channels. A similar plan is used for sodium channels. The equations for α_n and β_n and their place in the overall HH mathematical model are presented in Chapter 5. Here what is needed for these exercises are those equations in computational form. They are

```
$an=.01*(10.-$vm)/(exp((10-$vm)/10)-1);
$bn=0.125*exp(-$vm/80);
$am=0.1*(25-$vm)/(exp((25-$vm)/10)-1);
$bm=4*exp(-$vm/18);
$ah=0.07*exp(-$vm/20);
$bh=1/(exp((30-$vm)/10)+1);
```

Computational variables **an** and **bn** correspond to α_n and β_n, and similarly for m and h. Note that the transmembrane voltage, v_m, the argument to the expressions, is specified relative to the membrane's rest voltage, i.e., $v_m = V_m - V_r$. All a's and b's have units of msec^{-1}.

Evaluations of the model cell involve times t_1, steady state in phase 1, t_a, immediately after the voltage transition, t_b, which is 0.5 msec after the voltage transition, and t_2 in the phase 2 steady-state.

When an exercise asks for a potassium or sodium current, find that current by performing the following steps: Find the number of open channels, the resulting membrane conductivity, and the driving voltage (e.g., $V_m - E_K$) for the ion species.

Exercises 14 to 23: Examine the K^+ channels in the model cell. The cell is subject to a voltage clamp with V_m^1 of –40 mV and V_m^2 of 0 mV.

14. During the steady state of phase 1, what is:

    a. The probability p_1 that a K channel is open?

    b. The expected number of open K^+ channels?

    c. The fluctuation in the number of open channels, if the fluctuation is considered to be four times the standard deviation?

    d. During steady state for phase 1, what is the cell's K^+ current?

15. During steady state in phase 2, what is:

    a. The probability p_2 that a K^+ channel is open?

    b. The expected number of open K^+channels?

    c. The K^+ current for the cell?

16. As judged by the results of exercises 14 and 15, if $V_m^2 > V_m^1$ is the K^+ current in phase 2, is it also greater than that of phase 1?

17. Compare the number of open channels at phase 1 in steady state to the number of open channels in phase 2 at steady state. What is the ratio of the expected numbers of open K^+channels $N_{\text{open}}^2/N_{\text{open}}^1$?

18. Compare the cell's K^+ current (I_K^1) in phase 1 at steady state to the K^+ current (I_K^2) in phase 2 at steady state. What is the ratio I_K^2/I_K^1?

19. Consider the time $t = t_a$, which is just **after** the voltage transition but prior to any time having elapsed during phase 2. At $t = t_a$, what is the cell's K^+ current?

20. Why is the ratio found in exercise 18 different from the ratio found in exercise 19?

Exercises 21–26: Again examine the potassium channels and currents in the model cell, this time with V_m^1 of –55 mV and V_m^2 of 55 mV.

21. At steady state in phase 1, what is:

a. The probability p_1 that a K^+ channel is open?

b. The expected number of open K^+ channels?

c. The fluctuation in number of open channels, if the fluctuation is considered to be four times the standard deviation?

d. During steady state for phase 1, what is the cell's K^+ current?

22. During steady state in phase 2 what is:

a. The probability p_2 that a K^+ channel is open?

b. The expected number of open K^+ channels?

c. The cell's K^+ current?

23. As judged by the results of exercises 21 and 22, if V_m^2 is greater than V_m^1, are more K^+ channels open?

24. Quantitatively compare the number of open channels during the steady state of phase 1 to the number open in the steady state of phase 2. What is the ratio of the expected numbers of open K^+ channels $N_{\text{open}}^2/N_{\text{open}}^1$?

25. Compare the K^+ current at steady state in phase 1 (I_K^1) to that at steady state in phase 2 (I_K^2). What is I_K^2/I_K^1?

26. Compare the I_K current at time $t = t_a$, I_K^a, evaluated immediately after the transition of V_m, to I_K^1, the current just before. What is the ratio I_K^a / I_K^1?

27. In a few sentences, explain why there is a difference between the answers to Exercises 25 and 26.

Exercises 28–30 inquire about currents for sodium ions (Na^+) as well as potassium ions. Sodium ions cross the membrane in different channels than those of potassium. For Na^+ the probability that a channel is open is $m^3 h$. Gating variables m and h have values determined by equations of the same form as used for gating variable n for potassium; however, their α and β values are different. Time t_1 is a time during the steady state of phase 1, and time t_2 is during the steady state of phase 2.

28. In the model cell, examine Na^+ and K^+ currents with V_m^1 of –57 mV and V_m^2 of –20 mV.

    a. At t_1, what is I_{K}^1?

    b. At t_1, what is I_{Na}^1?

    c. At t_2, what is I_{K}^2?

    d. At t_2, what is I_{Na}^2?

29. In the model cell, examine Na^+ and K^+ currents with V_m^1 of –58 mV, and V_m^2 of –30 mV.

    a. At t_1, what is I_{K}^1?

    b. At t_1, what is I_{Na}^1?

    c. At t_2, what is I_{K}^2?

    d. At t_2, what is I_{Na}^2?

30. As judged by the results of Exercises 28 and 29, is $|I_{\mathrm{K}}| > |I_{\mathrm{Na}}|$ in all cases?

Exercises 31–35 include K^+ and Na^+ currents as before, and add the evaluation of both at a time t_b, less than a millisecond after the voltage transition. The addition is of interest in two ways, first, in how the calculation is done, and second, in the nature of the outcome, especially as compared to the preceding exercises.

31. The number of channels that are open, N_o, is given in equation (4.11) as a function of time t after a voltage transition at $t = 0$. That equation includes undetermined constant A, to be found from the boundary condition. [Equation (4.12) shows the particular form of (4.11) that results for the boundary condition of all channels closed at $t = 0$.] Define N_1 as a constant equal to the number of open channels at the end of phase 1.

    a. Give an equation for N_1 in terms of quantities known in phase 1.

    b. Derive an equation for $N_o(t)$ in phase 2 that depends on N_1 and that contains no undetermined parameters. The goal is to have an equation that can be used to find the number of open channels in phase 2 at times before phase 2's steady state.

32. In the model cell, evaluate the requested currents if V_m^1 is –55.5 mV, and V_m^2 is –24.5 mV.

    a. At t_1, what is I_K^1?

    b. At t_1, what is I_{Na}^1?

    c. At time t_b, what is I_K^b?

    d. At time t_b, what is I_{Na}^b?

    e. At t_2, what is I_K^2?

    f. At t_2, what is I_{Na}^2?

33. The model cell is clamped with V_m^1 of –59 mV and V_m^2 of –15 mV. Evaluate the requested currents.

    a. At t_1 and for K^+, what is I_K^1?

b. At t_1 and for Na^+, what is I^1_{Na}?

c. At time t_b and for K^+, what is I^b_{K}?

d. At time t_b and for Na^+, what is I^b_{Na}?

e. At t_2 and for K^+, what is I^2_{K}?

f. At t_2 and for Na^+, what is I^2_{Na}?

34. As judged by the results of exercises 32 and 33, does I_{Na} ever have the same sign as I_{K}?

35. As judged by the results of Exercises 32 and 33, is it always true that $|I_{\mathrm{K}}| > |I_{\mathrm{Na}}|$?

Exercises 36–40 explore quantitative aspects of the extraordinary experimental techniques used for the study of channels.

36. Based on the γ_{K} value found from Figure 4.9 (265 pS), evaluate the minimum gigaseal resistance that gives $\mathrm{S/N} \geq 30$. Assume that $\Delta f = 1$ kHz, $T = 293$ K, and for the size signal take $V_m = 50$ mV.

37. From Eq. (4.5) and assuming $Q_g = 10 \times \varepsilon$ (electronic charge), plot the fraction of open channels as a function of V_m (–100 mV < V_m < 50 mV). Choose $T = 293$ K.

38. When a microelectrode is inserted into the intracellular space of a cell, current from the electrode to that space may pass through the electrode resistance, as desired, or through the capacitance across the electrode wall. Figure 13.1 describes the physical arrangement (left) and the equivalent circuit (right). Although the capacitance is distributed along the length of the electrode that penetrates the cell, a simple lumped capacitance is frequently assumed, as shown in the figure (right). An estimate of the capacitance per unit length can be obtained by

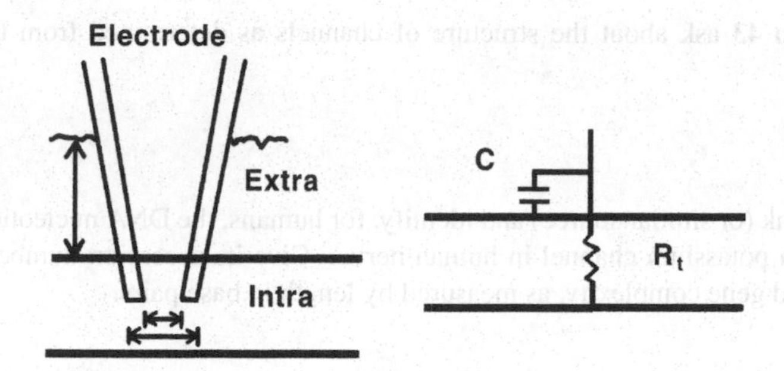

Figure 13.1. Microelectrode–Tissue Capacitance.

assuming the tip to be cylindrical, in which case the "well-known" formula is

$$\frac{C}{\ell} = \frac{2\pi\varepsilon_r\varepsilon_0}{\ln(D/d)} \tag{13.2}$$

where ε_r is the relative permittivity of the glass wall, ε_0 the permittivity of free space, D the outer diameter, and d the inner diameter of the micropipette tip. If $D = 1.3$ μm, $d = 1.0$ μm, $\varepsilon_r = 1.8$, $\sigma = 0.5$ S/cm, and $\ell = 2$ mm, determine C. Estimate the resistance of the microelectrode assuming the internal wire to end 2.1 mm from the tip. What is the electrode time constant? Comment on its significance in the measurement of time-varying voltage and current.

39. Consider a channel of 5 Å radius and 100 Å long. Assuming ohmic behavior, evaluate the channel conductance (assume the medium is Ringer solution with a resistivity of $\rho = 60$ Ωcm). Now, in addition to the channel's resistance, an additional resistance arises due to the "access resistance" of the surrounding medium. Assume that the current approaches and leaves the channel via radial paths and with a uniform current density on concentric hemispherical surfaces with origin at the center of the terminal aqueous channel surface. Find the value of the resistance from the channel to a reference electrode at infinity. (In finding the result, it is useful to picture the current flow as passing through a series of hemispherical surfaces, with the smallest having the same radius as the channel, and successive hemispheres growing larger to infinity.)

40. In the use of fluctuation analysis of single-channel measurements, the presence of thermal noise sets a limit on the resolution of the rate constants associated with open and closing of channel gates. This is illustrated in the following example: Consider recording from a 1-μm^2 patch with resting membrane resistance of 1000 Ωcm^2 and bandwidth of 2000 Hz. Evaluate σ^2 from (4.2) and compare σ with typical single-channel currents (where σ^2 is the mean-square noise power).

Exercises 41 to 43 ask about the structure of channels as determined from their nucleotide sequence.

41. Use GenBank (or similar source) and identify, for humans, the DNA nucleotide sequence for a gene for a potassium channel in human nerve. Give its accession number, chromosome location, and gene complexity, as measured by length in base pairs.

42. Repeat exercise 41 for a sodium channel.

43. Based solely on the gene complexity as indicated by length, which gene has the most complex structure, that for the potassium channel or that for the sodium channel?

13.5. EXERCISES, CHAPTER 5: ACTION POTENTIALS

Exercises 1–25 deal with the Hodgkin–Huxley membrane model. The earlier exercises deal with individual elements of the model, while later exercises deal with larger-scale composite events. Exercises requiring computer code for their solution are included thereafter. Many of these exercises refer to an HH membrane defined by the set of parameters given in Table 13.2.

Table 13.2. HH Membrane and Environmental Parameters

$\bar{g}_K$	36	mS/cm^2	maximum K^+conductivity
$\bar{g}_{Na}$	120	mS/cm^2	maximum Na^+ conductivity
g_L	0.3	mS/cm^2	leakage conductivity
C_m	1.0	μF/cm^2	membrane capacitance
E_K	-72.1	mV	K^+ Nernst potential
E_{Na}	52.4	mV	Na^+Nernst potential
E_L	-49.2	mV	leakage Nernst potential
V_r	-60	mV	resting potential
I_s	0	μA/cm^2	stimulus current
I_m	0	μA/cm^2	total membrane current for patch if no stimulus

Table 13.3. HH State Variables, Set A

V_m	-11.5	mV	transmembrane potential
n	0.378	—	gating probability n
m	0.417	—	gating probability m
h	0.477	—	gating probability h

1. Conductivity and current: Write a formula that can be used to determine each of the quantities listed. The result should be a function of the parameters given in Table 13.2 and state variables of Table 13.3. It may also be a function of one of the preceding items, e.g., the formula for I_K may include g_K.

   a. g_K, the K^+ conductivity.

   b. g_{Na}, the Na^+ conductivity.

   c. I_K, the K^+ current.

d. I_{Na}, the Na^+ current.

e. I_L, the leakage current.

f. I_{ion}, the total ionic current.

g. $V_{dot} \equiv \dot{V}_m \equiv dV_m/dt$, the time derivative (recall that Newton's calculus used a dot over a variable to signify a time derivative) of V_m at $t = t_0$. Include stimulus current I_s in the expression.

2. Channel probabilities n, m, and h: Write a formula that can be used to determine the value of each one of the quantities listed. The result should depend on the parameters given in Table 13.2, the state variables of Table 13.3, and the results of any preceding items on the list.

a. v_m, the transmembrane potential relative to the resting value.

b. α_n, the rate constant for n for opening channels.

c. β_n, the rate constant for n for closing channels.

d. α_m, the rate constant for m for opening channels.

e. β_m, the rate constant for m for closing channels.

f. α_h, the rate constant for h for opening channels.

g. β_h, the rate constant for h for closing channels.

h. $\dot{n} \equiv dn/dt$ at $t = t_0$, the time derivative of n evaluated at $t = t_0$.

i. $\dot{m} \equiv dm/dt$ at $t = t_0$, the time derivative of n at $t = t_0$.

j. $\dot{h} \equiv dh/dt$ at $t = t_0$, the time derivative of n at $t = t_0$.

3. Time shift Δt: Write formulas for the changes in V_m and probabilities n, m, and h during an interval Δt beginning at time $t = t_0$. Also give formulas for the new values. Assume that Δt is small enough that derivatives retain their initial values throughout the interval.

   a. ΔV_m, the change of V_m from $t = t_o$ to $t = t_0 + \Delta t$.

   b. Δn, the change of probability n from $t = t_o$ to $t = t_0 + \Delta t$.

   c. Δm, the change of probability m from $t = t_o$ to $t = t_0 + \Delta t$.

   d. Δh, the change of probability h from $t = t_o$ to $t = t_0 + \Delta t$.

   e. $V_m(t_0 + \Delta t)$, the value of V_m at $(t_0 + \Delta t)$.

   f. $n(t_0 + \Delta t)$, the value of n at $(t_0 + \Delta t)$.

   g. $m(t_0 + \Delta t)$, the value of m at $(t_0 + \Delta t)$.

   h. $h(t_0 + \Delta t)$, the value of h at $(t_0 + \Delta t)$.

4. Applied membrane current. In Table 13.2 a value of stimulus current $I_S = 0$ was given. That choice resulted in a changing value of V_m, as found in the preceding exercise. Suppose that one wishes to apply a stimulus to maintain V_m at a constant value, i.e., make $\dot{V}_m = 0$. Write a suitable formula for the stimulus current I_S that must be applied across the membrane. (Such a current is applied, for example, in a voltage clamp.)

Exercises 5–8 ask that numerical values be found for the quantities defined by the formulas found in the preceding exercises, 1 to 4.

5. Conductivity and current: Find the numerical value (with units) of each of the following, using the parameter values given in Table 13.2 and the values for the state variables of set A, which are given in Table 13.3. For this illustrative example, use Δt equal to 50 μsec. A Δt of 50 μsec works well for hand calculations and for initial computer runs done for testing

algorithms. When accurate results become the goal, a smaller Δt of 1 to 10 μsec is more frequently chosen, because the time integration is then done more precisely. In these initial exercises the emphasis is on instruction, so a larger Δt is used, thus making the nature of the step by step evolution more obvious.

a. g_K, the K^+ conductivity.

b. g_{Na}, the Na^+ conductivity.

c. V_K, the transmembrane voltage relative to E_K.

d. V_{Na}, the transmembrane voltage relative to E_{Na}.

c. I_K, the K^+ current.

d. I_{Na}, the Na^+ current.

e. I_L, the leakage current.

f. I_{ion}, the total ionic current.

g. $V_{dot} \equiv \dot{V}_m$, the time derivative of V_m, evaluated at t_0.

6. Channel probabilities n, m, and h: Find values for each of the quantities listed. The result should depend on the parameters given in Table 13.2, the state variables of Table 13.3, and, as needed, the results of any preceding items on the list.

a. v_m, the transmembrane potential relative to the resting value.

b. α_n, the rate constant for n for opening channels.

c. β_n, the rate constant for n for closing channels.

d. α_m, the rate constant for m for opening channels.

e. β_m, the rate constant for m for closing channels.

f. α_h, the rate constant for h for opening channels.

g. β_h, the rate constant for h for closing channels.

h. $\dot{n} \equiv dn/dt$ at $t = t_0$, the time derivative of n at $t = t_0$.

i. $\dot{m} \equiv dm/dt$ at $t = t_0$, the time derivative of n at $t = t_0$.

j. $\dot{h} \equiv dh/dt$ at $t = t_0$, the time derivative of n at $t = t_0$.

7. Time shift Δt: Use the results of Exs. 5 and 6 to find values for each of the following for the interval Δt beginning at time $t = t_0$. Assume that Δt is small enough that derivatives retain their initial values throughout the interval.

a. ΔV_m, the change of V_m from $t = t_o$ to $t = t_0 + \Delta t$.

b. Δn, the change of probability n from $t = t_o$ to $t = t_0 + \Delta t$.

c. Δm, the change of probability m from $t = t_o$ to $t = t_0 + \Delta t$.

d. Δh, the change of probability h from $t = t_o$ to $t = t_0 + \Delta t$.

e. $V_m(t_0 + \Delta t)$, the value of V_m at $(t_0 + \Delta t)$.

f. $n(t_0 + \Delta t)$, the value of n at $(t_0 + \Delta t)$.

g. $m(t_0 + \Delta t)$, the value of m at $(t_0 + \Delta t)$.

h. $h(t_0 + \Delta t)$, the value of h at $(t_0 + \Delta t)$.

8. What stimulus current must be applied across the membrane to keep the transmembrane voltage constant, if V_m and all other parameters have the values given in the preceding exercise? That is, what stimulus current is necessary to prevent the change in V_m that was found in the previous exercise? Give the value of the needed stimulus current, in microAmperes per cm^2.

Table 13.4. HH State Variables, Set B

V_m	-11.5	mV	Transmembrane potential
n	0.759	—	gating probability n
m	0.955	—	gating probability m
h	0.104	—	gating probability h
I_m	0	mA	total membrane current if no stimulus

Exercises 9–12 use set B of the state variables, as given in Table 13.4. The questions are otherwise the same as in Exs. 5–8. These set B exercises are included because the numerical results are materially different from those of set A. They allow an interesting qualitative as well as quantitative comparison.

9. Conductivity and current: Find the numerical value (with units) of each of the following, using the parameter values given in Table 13.2 and the values for the state variables given in Table 13.4. For this illustrative example, use Δt equal to 50 μsec.

   a. g_K, the K^+ conductivity.

   b. g_{Na}, the Na^+ conductivity.

   c. V_K, the transmembrane voltage relative to E_K.

   d. V_{Na}, the transmembrane voltage relative to E_{Na}.

   c. I_K, the K^+ current.

   d. I_{Na}, the Na^+ current.

   e. I_L, the leakage current.

f. I_{ion}, the total ionic current.

g. $V_{dot} \equiv \dot{V}_m$, the time derivative of V_m, evaluated at t_0.

10. Channel probabilities n, m, and h: Find values for each of the quantities listed. The result should depend on the parameters given in Table 13.2, the state variables of Table 13.4 and, as needed, the results of any preceding items on the list.

a. v_m, the transmembrane potential relative to the resting value.

b. α_n, the rate constant for n for opening channels.

c. β_n, the rate constant for n for closing channels.

d. α_m, the rate constant for m for opening channels.

e. β_m, the rate constant for m for closing channels.

f. α_h, the rate constant for h for opening channels.

g. β_h, the rate constant for h for closing channels.

h. $\dot{n} \equiv dn/dt$ at $t = t_0$, the time derivative of n evaluated at $t = t_0$.

i. $\dot{m} \equiv dm/dt$ at $t = t_0$, the time derivative of n at $t = t_0$.

j. $\dot{h} \equiv dh/dt$ at $t = t_0$, the time derivative of n at $t = t_0$.

11. Time shift Δt: Use the results of Exs. 9 and 10 to find values for each of the following for the interval Δt beginning at time $t = t_0$. Assume that Δt is small enough that derivatives retain their initial values throughout the interval.

a. ΔV_m, the change of V_m from $t = t_o$ to $t = t_0 + \Delta t$.

b. Δn, the change of probability n from $t = t_o$ to $t = t_0 + \Delta t$.

c. Δm, the change of probability m from $t = t_o$ to $t = t_0 + \Delta t$.

d. Δh, the change of probability h from $t = t_o$ to $t = t_0 + \Delta t$.

e. $V_m(t_0 + \Delta t)$, the value of V_m at $(t_0 + \Delta t)$.

f. $n(t_0 + \Delta t)$, the value of n at $(t_0 + \Delta t)$.

g. $m(t_0 + \Delta t)$, the value of m at $(t_0 + \Delta t)$.

h. $h(t_0 + \Delta t)$, the value of h at $(t_0 + \Delta t)$.

12. What stimulus current must be applied across the membrane to keep the transmembrane voltage constant? That is, what stimulus current is necessary to prevent the change in V_m that was found in the previous exercise? Give the value of the needed stimulus current in milliamperes per cm^2.

13. The above results show that ΔV_m for the state variables of set A (Ex. 8) has the opposite sign from the ΔV_m as found for the state variables of set B (Ex. 12), that is, in one case V_m is going up, while in the other it is going down. How does this happen? V_m itself is the same in both cases, so V_m's value cannot be the factor that makes the difference. Answer this question in a sentence or two, making specific reference to all three of the individual ionic currents, and the total ionic current, comparatively.

Exercise 14 deals with special cases of α, a particularly pernicious source of errors in automated computation.

14. Special cases for α: Two of the equations for α have special cases for particular values of v_m. Each equation has a denominator that goes to zero at a particular v_m. Examine each equation listed near the value given. (It may be helpful to use a power-series expansion of e^u for u near to zero.) Find a simplified form of the equation for each α that applies for v_m near the given value of v_m. The simplified form should make clear the value of α at and near the special v_m value.

a. n: find an equation for α_n near $v_m = 10$.

b. n: determine α_n at $v_m = 10$.

c. m: find an equation for α_m near $v_m = 25$.

d. m: determine α_m at $v_m = 25$.

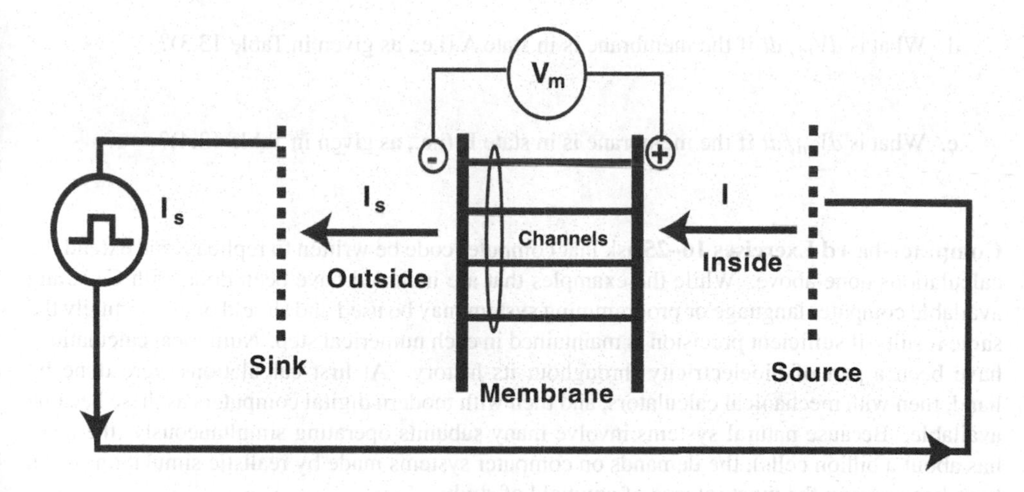

Figure 13.2. Schematic of Membrane Stimulation. The figure provides a schematic drawing of the stimulator (circle) and its connection to source and sink electrodes (dotted lines). Current I_s flows from these electrodes from the source to the sink. At the membrane, charges move through the several kinds of ion channels, symbolized by the thin lines, or accumulate on the membrane surfaces, thus increasing the transmembrane voltage. Note that V_m has a positive value when the membrane is more positive on the inside than on the outside, and the current has a positive value when it flows from inside to outside. This schematic figure may be compared to the diagram of the experimental setup in Figure 5.7.

Exercise 15 deals with the initial depolarizing current, a stimulus current. Stimuli are analyzed at length in a later chapter. Here the topic is examined only as needed for some relatively simple exercises.

15. Suppose one considers again Eq. (5.39) from the text, which is essentially

$$C_m \frac{dV_m}{dt} + I_{ion} = I_m$$

A stimulus is applied to a membrane having the characteristics of Table 13.2. Thus $I_m = I_s$ (and thereby has a nonzero amplitude), and during the period the stimulus is applied. By the convention used here, the stimulus starts at $t = t_0$. Unless specified otherwise, $I_s = 50$ μA/cm^2, and the membrane is at rest. Examine the instant after the stimulus begins.

a. What is dV_m/dt?

b. What is dV_m/dt if $I_s = -50\ \mu\text{A/cm}^2$?

c. What is ΔV_m for the first 100 msec after the stimulus starts, if dV_m/dt retains its initial value throughout that interval?

d. What is dV_m/dt if the membrane is in state A (i.e., as given in Table 13.3)?

e. What is dV_m/dt if the membrane is in state B (i.e., as given in Table 13.4)?

Computer-based Exercises 16–25 ask that computer code be written to replicate and extend the calculations done above. While the examples that are included have been done with C++, any available computer language or programming system may be used and should obtain virtually the same results, if sufficient precision is maintained in each numerical step. Numerical calculations have been a part of bioelectricity throughout its history. At first calculations were done by hand, then with mechanical calculators, and then with modern digital computers as these became available. Because natural systems involve many subunits operating simultaneously (the heart has about a billion cells), the demands on computer systems made by realistic simulations often have been among the most intense of any field of study.

Exercises 16–20 focus on developing a computer program for performing Hodgkin–Huxley calculations for periods of time that include many Δt cycles.

16. Examine the results of exercise 14. Write a computer procedure that, given v_m, computes α and β for n, m, and h.

    a. Give a listing of the code needed. It should handle satisfactorily the special cases of v_m.

    b. Use the code to make an 8-column table, where the columns are (1) the line number, (2) the v_m value, (3)–(8) α and β for n, m, and h. Let the rows extend from v_m of 0 up to v_m of 50, in 5 mV steps.

17. Write a computer program that begins with the data given in Table 13.2 and the state variables of set A (Table 13.3). Write code that finds and displays the results of all parts of Exs. 4–8. After completing the program, use it again with the state variables changed to those of set B (Table 13.4), but no other changes. The computer program then should find the results of Exs. 9–12.

18. Write a computer program that begins with the following input values.

```
Membrane v17.1 March 28, 2005 RB

deltaTime: 50 microseconds
StimAmplitude: 200
StimDuration: 150

HH: EK -72.100 ENa 52.400 EL -49.187
HH: gbarK 36.0 gbarNa 120.0 gL 0.3
HH: n 0.31768 m 0.05293 h 0.59612
HH: Cm 1.00 Temper 6.30
HH: gK 0.367 gNa 0.011
HH: Vm -60.000 Vr -60.000
```

Write code so that the program operates cyclically through successive Δt intervals. Print an output table corresponding to the output table below. Compare carefully the results for all lines, but compare results especially carefully for lines 5 and 8.

The output table is constructed in the following way: The value of loopcnt (column 1) is the number of computation loops completed by the program. The corresponding time, given in integer microseconds, is shown in column 2. Note that the time origin is selected so that the first stimulus occurs at a time of zero; however, some calculation prior to that time is useful to establish the baseline.

The stimulus current at each moment is tabulated in column 3. A nonzero value of stimulus current is present in the output table when the stimulus will be nonzero during the **subsequent** Δt. Thereby, the total current is listed as nonzero at time zero and zero at 150 microseconds, when the stimulus ends.

The v_m value in the 4th column produces the n, m, h values in columns 6, 7, and 8. The v_m time derivative v_{dot} (5th column) is determined in the program once v_m, n, m, and h are known.

```
Output Table, deltaTime=50 usec
loopcnt time Is Vm vdot n m h
0 -200 0.0    -60.000 -0.0   0.31768 0.05293 0.59612
1 -150 0.0    -60.000 -0.0   0.31768 0.05293 0.59612
2 -100 0.0    -60.000 -0.0   0.31768 0.05293 0.59612
3 -50  0.0    -60.000 -0.0   0.31768 0.05293 0.59612
4 0    200.0 -60.000 200.0 0.31768 0.05293 0.59612
5 50   200.0 -50.000 193.2 0.31768 0.05293 0.59612
6 100  200.0 -40.339 187.5 0.31934 0.06726 0.59343
7 150  0.0    -30.965 -16.1 0.32308 0.09804 0.58617
8 200  0.0    -31.769 -3.2   0.32925 0.14894 0.57257
```

19. Continuing from Ex. 18, extend the duration of the time period that is simulated.

   a. After 200 μsec, does V_m continue to become more negative, as v_m does between 150 and 200 microseconds?

b. What is V_m at 1000 μsec?

20. Reduce Δt. Begin with these membrane parameters:

```
Membrane v17.1 March 28, 2005 RB

deltaTime: 50 microseconds
StimAmplitude: 200
StimDuration: 150

HH: EK -72.100 ENa 52.400 EL -49.187
HH: gbarK 36.0 gbarNa 120.0 gL 0.3
HH: n 0.31768 m 0.05293 h 0.59612
HH: Cm 1.00 Temper 6.30
HH: Vm -60.000 Vr -60.000
```

21. Compare program results at 700 μsec from the start of the stimulus for two choices of dt, to see how much the results change. Compute the difference as the value when $\Delta t = 2$ minus that when $\Delta t = 50$.

    a. V_m.

    b. n.

    c. m.

    d. h.

Use these values to start the simulation:

```
deltaTime: 2 or 50 microseconds
StimAmplitude: 200
StimDuration: 150

EK -72.100 ENa 52.400 EL -49.187 mV
gbarK 36.0 gbarNa 120.0 gL 0.300 mS/cm2
n 0.31768 m 0.05293 h 0.59612
Vm -60.00 Vr -60.00 MV
Cm 1.00 uF/cm2
T 6.30 degrees C
```

Exercises 22–27 focus on using a computer program to understand what happens in a membrane that is characterized by the Hodgkin–Huxley equations.

22. Is there a linear relationship between stimulus and response: Rerun the computer code from Ex. 20, for 2000 μsec. Then make two additional simulation runs, one with a stimulus amplitude of half as much and the other for a stimulus amplitude of twice as much. Compare the three outputs (original stimulus amplitude, half, twice). Determine if the output (v_m) is proportional to the input (to the stimulus amplitude) for the time periods identified. Linearity means that when the input has k times the amplitude, then the output wave shape is unchanged and has k times the amplitude. (The response can be linear even though the output wave shape is different from that of the input.) Here there are several outputs, but in this exercise examine v_m. Consider the response to be linear if v_m is scaled (within a tolerance of 10%) by the same factor as the stimulus, for the time period identified.

    a. For $t \leq 100$ μsec.

    b. For $t > 200$ μsec.

23. Threshold: Use repeated computer runs to find the threshold amplitude that produces an action potential. Limit the range of evaluation to integer values of the stimulus amplitude, in units of μA/cm^2. Determine the "just-above-threshold" stimulus amplitude for a stimulus of 150-μsec duration that is minimally sufficient to produce an action potential. That is, when this stimulus amplitude is reduced by one μA/cm^2, no action potential follows.

    a. What is the just-above-threshold stimulus amplitude, in μA/cm^2?

    b. What is the just-above-threshold membrane voltage, i.e., v_m when the stimulus (that is just-above-threshold) ends?

Exercises 24 through 27 focus on the time elapsed between the start of the stimulus and subsequent events of membrane excitation and recovery. To complete each exercise, perform a simulation with the same starting conditions as in exercise 20, except as otherwise given. Answer the question by inspecting results at each 50 μsec multiple.

24. Time to peak: How long is the time interval from the start of the stimulus to the peak of the subsequent action potential? For simplicity, judge the peak to occur when v_m reaches its largest positive value when tabulated every 50 μsec. The stimulus amplitude, in μA/cm^2, is

    a. 50

    b. 200

    c. 500

25. Time to return to initial conditions: How much time elapses from the start of the stimulus to the time when the membrane is stable within the initial-condition envelope? Here "stable within" means values within the envelope that do not spontaneously move outside the envelope thereafter. Assume that the membrane is within the initial-condition envelope when the values of v_m are within 0.1 mV of the initial value, **and** n, m, and h are within .01 of their initial values.

26. Leakage g_L: Leakage conductivity g_L seems uninteresting because "leakage" is not gated and not specific to movement of a named ion species. Nonetheless, the value of g_L has ramifications. In this exercise, some are explored. Reduce the value of g_L to 0.01 mS/cm^2, keeping the other membrane parameters the same as in Table 13.2. With the new, lower value for g_L, determine the following:

    a. According to the *parallel-conductance* equation, a constraint among I_{Cl}, I_{K}, and I_{Na} exists at rest. Making use of this constraint, find the new value for E_L that is required to maintain the same resting potential, with the new g_L.

    b. With the new values of g_L and E_L used in place of those of Table 13.2, how long now is required, from the start of the stimulus, for the membrane to return to stability within the initial-condition envelope? (See Ex. 24.) Stimulus: $I_s = 200\ \mu\text{A/cm}^2$ for 150 μsec.

27. AP from 2nd stimulus: How long must one wait after an initial stimulus to give a second stimulus that leads to an action potential? Evaluate with the HH model. Give the first stimulus under the same conditions as Ex. 20. Measure the stimulus interval as the time from the start of the first stimulus to the start of the second stimulus. To simplify this exercise, limit consideration to intervals that are integer multiples of 50 μsec. Set a program flag, INaflag, to 1 when $-I_{\text{Na}}$ exceeds I_{K}, and set it to zero otherwise. Judge an action potential to occur when INaflag changes from 0 to 1.[1] What is the earliest time that a 2nd stimulus can be given, and produce a 2nd action potential, if

    a. the stimulus current is $I_s = 50\mu\text{A/cm}^2$.

    b. the stimulus current is $I_s = 200\mu\text{A/cm}^2$.

    c. the stimulus current is $I_s = 500\mu\text{A/cm}^2$.

[1] This condition is an arbitrary one chosen for use in this simulation exercise. In an experimental study, a condition might be based on $\dot{v}_m$. If the preparation were a fiber, a condition might be based on whether the stimulus produced a propagated response.

28. Sketch an action potential for nerve.

a. Indicate periods of rest and action.

b. On the sketch, indicate the periods when sodium current has higher magnitude than potassium or leakage.

c. On the sketch, indicate the periods when potassium current has twice or more the magnitude of leakage.

d. On the sketch, use labeled arrows to identify a time when n has the value of n_∞ and n at rest. In two sentences, justify using n_∞ as the value of n at rest by explaining the circumstances when doing so is acceptable, and when not.

Voltage Clamp Exercises. The voltage clamp experimental setup provided an innovative platform for determining membrane properties. Understanding the voltage clamp allows one to review carefully the experimental data from which membrane properties are understood, and also allows one to enjoy some concepts that have been abstracted from the voltage clamp experiments.

29. Which one of the following is the objective of the voltage clamp:

a. Confuse students by introducing an unnecessarily large number of electrodes.

b. Hold membrane current constant while measuring changing V_m.

c. Hold V_m constant to stabilize the membrane current and thereby get an accurate measurement.

d. Hold V_m constant and measure membrane current with time.

e. Obtain a dynamic record of changes in V_m with membrane current.

30. In a few sentences, explain how a voltage clamp experiment is conducted: What is controlled, and how, and what is measured?

31. A voltage clamp from rest (Table 13.2) to $v_m = 20$ mV is applied to a resting squid axon. What is the ratio of the potassium conductance that results after a long time divided by the potassium conductance just after the voltage transition. (Provide this ratio based on analysis using the HH equations. A computer simulation is unnecessary.)

32. The voltage clamp experiment allows an evaluation of the dynamics of channels and gates.

   a. In a few sentences, describe what "gates" are all about, and why gating variables such as m are raised to powers.

   b. For current I_K, the differential equation that describes changes in the fraction of open potassium gates (n^4) requires that n satisfy

$$\frac{dn}{dt} = \alpha_n(1 - n) - \beta_n n$$

   In a sentence or so, describe the significance of each variable and term on the right of this equation.

Exercises 33 to 37 refer to Figure 5.7. These exercises provide an opportunity to review the particulars of the experimental arrangement created for the voltage clamp.

33. Current is generated between which pair of electrodes?

34. The membrane *voltage* is measured between which pair of electrodes?

35. The membrane *current* is measured between which pair of electrodes?

36. The measurement between the electrodes used for getting the membrane current actually provides the measured value of the voltage between the electrodes. What other experimental measurement must be available to allow the investigators to learn the membrane current?

37. The value described as being measured in Ex. 35 can also be calculated from the following data, derived from [6]:

a. radius to electrode $C = 2$mm.

b. radius to electrode $D = 12$mm.

c. conductivity of extracellular medium = 0.01 S/cm.

d. width of space-clamped region = 7 mm.

Exercises 38–42 refer to Figure 5.9, which shows ionic current following a voltage clamp. For 37 and 38 focus on the 91-mV trace. For 40–42 focus on the 117-mV trace.

38. The flow of what ion dominates the curve during the period from 1 to 2 msec?

39. In which direction is the net current flow from 1 to 2 msec?

40. Which ion dominates the curve during the period from 3 to 4 msec?

41. Why does this trace fail to fall below the horizontal axis?

42. What is the time period during which Na^+ flow dominates this trace?

Exercises 43–47 picture a voltage clamp that shifts from V_m at –60 mV to $V_m = 15$ mV. At first the clamp is performed with the normal composition of $[\text{Na}]_i = 15$ mM and $[\text{Na}]_o = 180$ mM.

43. The *early* current is carried by which ion, and crossing in which direction?

44. The *steady-state* current is carried by which ion, and crossing in which direction?

45. If the voltage clamp is repeated such that the sodium current is abolished, what extracellular medium changes are necessary? (Be specific and quantitative.)

46. If the voltage clamp at $V_m = 15$ mV is conducted with 10% sodium seawater ($[\text{Na}]_o = 18.0$ mM), how does the sodium current magnitude and time course compare with that under normal conditions? (Be specific and quantitative.)

47. How does the potassium ion flow compare during the voltage clamp under normal, zero sodium flux, 10% sodium seawater conditions?

Analytical and numerical questions arising from the voltage clamp are presented in the next set of exercises.

48. Make plots of all the quantities listed over a v_m range of 0 to 100, with a resolution of 1 mV or less.

   a. α_m versus v_m.

   b. β_m versus v_m.

   c. α_h versus v_m.

d. β_h versus v_m.

Inspect these plots to answer the following questions.

e. The units of α_m are msec^{-1}, e.g., "per msec." Note that sometimes α_m has a value greater than 1. It would seem that, if α_m is greater than 1, m will increase beyond unity within a millisecond. However, because m is a probability, that is not possible. Explain.

f. Explain the significance of the h-related curves having lower magnitudes than the m related curves, in general.

g. For a given value of v_m (choose examples), what is the significance of α having a greater value than β, or vice versa.

49. A membrane that follows the HH rules is subject to a voltage clamp. The clamp begins at the tissues resting potential of $v_m = 0$ mV and then abruptly shifts $v_c = 100$ mV more positive, where it remains constant. Find:

a. The analytic solution of the equation for *dm/dt*.

b. The analytic solution of the equation for *dh/dt*.

c. Write a program that plots m^3h as a function of time. The answer to this exercise is the resulting plot. Note that this plot is proportional to the value of g_{Na} as a function of time. That is, it shows how the sodium conductance changes as a function of time. As such, it should be quite similar to one of the plots shown in the text.

50. This exercise deals with potassium and sodium conductances. In all graphs below, be sure to make a calibrated time scale. The conductance axes need not be calibrated absolutely, but should be consistent in relative magnitude from graph to graph.

a. Draw graphs of voltage versus time that describe the potassium conductance following a voltage step from rest to a clamped transmembrane potential of (a) about 30 mV from rest, and (b) about 90 mV from rest.

b. Draw graphs of voltage versus time that describe the sodium conductance following a voltage step from rest to a clamped transmembrane potential of (a) about 30 mV from rest, and (b) about 90 mV from rest.

In a few sentences, summarize the circumstances when the sodium conductance exceeds the potassium conductance, in terms of voltage level and time.

51. The following whimsical exercise allows one to explore an HH-type mathematical model that has a different outcome: The current flow during the action potential of the very short-lived creature *Giganticus* was studied by George and Gimmy (GG), who developed the GG equations. These started with the familiar form:

$$I = I_{\mathrm{Na}} + I_{\mathrm{K}} + I_{\mathrm{L}} + I_{\mathrm{C}}$$

GG found that in the special habitat of *Giganticus*, the following unusual relationships applied:

$$I, I_{\mathrm{L}}, I_{\mathrm{K}} = 0, \quad I_{\mathrm{Na}} = \overline{g} v_m$$

with $\overline{g} = 2$, and $v_m = -10$ at $t = 0$. Analytically, find the solution for $v_m(t)$.

a. Although GG believed they had solved the puzzle of *Giganticus*, they wished to confirm the result by a numerical method that would begin with the same information used in Ex. 12, get an expression for dv_m/dt, and use it to find $v_m(t)$ numerically.

b. Show the equation for dv_m/dt.

c. Being somewhat naive, GG chose a value of Δt of 2 msec. What were the values of v_m that were computed for 2 and 4 msec?

Electrophysiology Experiments (Exercises 52–60): These exercises explore the electrophysiological characteristics of HH membranes through a series of simulated experiments framed in terms that might have been used in an experimental electrophysiology laboratory. Performing the experiments requires one to have a simulation program for a Hodgkin–Huxley membrane, such as the simulation program developed in the exercises above.

52. *Threshold and strength–duration.* Start with a simulation duration of 10 msec and a stimulus duration of 20 μsec (microseconds). Increase the stimulus amplitude until an action potential is produced.

a. Find the threshold—the amplitude that just produces an action potential, while a decrease to 90% of that amplitude fails. What is this amplitude?

b. What happens to the timing of the action potential for amplitudes just above threshold?

c. Increase the duration to 50 μsec and again find the threshold. Then do the same for a duration of 100, 200, 500, and 1000 μsec. Plot these points as a strength–duration curve.

53. *Membrane conductances.* Determine the membrane conductances as a function of time, from the time of the start of the stimulus to the end of the first action potential.

   a. What are the resting values of the conductances g_{Na} and g_K? (Use the tabular form of output to get numerical values.)

   b. Plot the changes in these conductances during and following the action potential. Be sure to use a large vertical scale. Consider log as well as linear calibration for the scale. What are the largest values the conductances reach?

   c. By what factors does each one change during the action potential? (make a ratio of the largest value over the smallest value).

54. *Refractory period.* Set the stimulus duration to 20 msec. Set the stimulus to start at time zero, with a duration and amplitude adjusted to produce an action potential within 1 msec after the stimulus begins. At 8 msec, initiate a second stimulus pulse with the same duration and amplitude.

   a. What kind of response does the 2nd stimulus produce?

   b. By looking at the time course of the membrane conductances, can you tell why the response is different from the response to the first stimulus?

   c. Find the threshold stimulus amplitude at 8 msec. Also, find the threshold stimulus for delays of 6, 7, 10, and 15 msec.

   d. Does the threshold stimulus for the second action potential have a simple relationship to the delay? What is the relationship?

55. *Temporal summation.* Use a simulation duration of 10 msec. Decrease the amplitude of both stimuli to about 60% of the stimulus threshold (for membrane at rest).

   a. What happens when the delay time of the second stimulus is reduced to 1 msec?

   b. By looking at the time courses of the conductances, can you explain why a second sub-threshold stimulus could cause an action potential after the first one failed?

   c. Increase the time until the onset of the 2nd stimulus. At what interval does neither stimulus produce an action potential?

56. *Anode break excitation.* Shut off both stimulus pulses by setting their amplitudes to zero. Change the initial membrane potential to – – 105 mV.

a. Record what happens during the first 10 msec. This phenomenon is called "anode-break excitation" because excitation occurs after shutting off a hyperpolarizing (anodal) pulse of current if the amount of hyperpolarization is sufficiently large.

b. Study the membrane conductance and describe why this excitation occurs.

57. *Constant stimulus current.* Apply a stimulus that has 50-msec duration and describe what happens during the first 50 msec following the stimulus onset,

a. if the stimulus has an amplitude of 10 μA/cm^2.

b. if the stimulus has an amplitude of 50 μA/cm^2.

c. For part (b), how does the action potential waveform compare with normal ones.

58. *Temperature effects.* observe the effects of increasing the temperature from 6.3°C to 12.6°C, and then to 26°C (approximately room temperature).

a. Does the duration of the action potential change? (Measure the duration as the time interval when $V_m \geq -40mV$.)

b. Does the stimulus threshold change?

c. Does the refractory period change? (Compare the responses to a 2nd stimulus at 8 msec, as done in the refractory period exercise above.)

59. *Ionic concentrations.* Try varying the external medium by changing the parameters. Describe qualitatively the result of each of the following:

a. Double $[\mathrm{Na}]_i$

b. Double $[\mathrm{K}]_e$

c. Increase $[\mathrm{K}]_e$ times 10.

13.6. EXERCISES, CHAPTER 6: IMPULSE PROPAGATION

Exercises 1–3 deal with elements of the core-conductor model, and their units.

1. In the core-conductor model, what units often are used for each of the following:

   a. c_m, the core-conductor's membrane capacitance.

   b. r_m, the core-conductor's membrane resistance.

   c. r_i, the core-conductor's intracellular resistance.

   d. R_i, the intracellular resistivity.

2. Consider a cylindrical HH fiber at rest. The radius of the membrane is 30 micrometers. Extracellular currents flow to twice the membrane radius. The extracellular resistivity is 50 Ωcm. The intracellular resistivity is three times that of the extracellular. The membrane capacitance is 1 μF/cm^2. The HH resting conditions apply. The fiber is passive. Find each of the following, in suitable units, using the linear core-conductor model:

   a. What is the membrane resistivity, R_m?

   b. What is the membrane resistance r_m?

   c. What is the intracellular resistance per unit length r_i?

   d. What is the extracellular resistance per unit length r_e?

3. A cylindrical fiber's membrane has a certain radius. The intracellular volume is within this radius. The extracellular volume is outside the membrane extending to a radius of twice this amount. (The membrane itself is considered to have negligible thickness relative to these dimensions.) The membrane resistance at rest is 2,000 Ωcm^2, and the membrane capacitance is 1.2 μF/cm^2. The intracellular resistivity is 100 Ωcm, and the extracellular resistivity is 40 Ωcm. The radius is 50 μm. Find each of the following, in suitable units. Use the linear core-conductor model.

   a. What is the membrane resistance per unit length?

b. What is the membrane capacitance per unit length?

c. What is the intracellular resistance per unit length?

d. What is the extracellular resistance per unit length?

Exercises 4 and 5 deal with the relations among transmembrane potential, axial current, and transmembrane current.

4. A cylindrical fiber is represented by the core- conductor model. Known quantities are the following: the transmembrane potential, $V_m(x)$, the fiber's radius a, and intracellular and extracellular resistances r_i and r_e. Note that r_i and r_e are in "unit length" form. At the time of interest, there is no stimulus of any kind. Write the mathematical expression by which each of the following can be found, from the known quantities.

   a. Intracellular axial current I_i.

   b. Extracellular axial current I_e.

   c. Transmembrane current (per cm^2) I_m.

5. A cylindrical fiber is represented by the core- conductor model. The upstroke of the transmembrane potential is given by a template function as $V_m(x) = 50\tanh(x)$ (x in mm). The fiber's radius is a, and the intracellular and extracellular resistances are r_i and r_e. (Note that r_i and r_e are in "unit length" form.) V_m is understood to be the spatial distribution of V_m at one moment during propagation of an action potential. At this time, which is after propagation began, there are no stimuli. Give the mathematical expression for the answer, and plot normalized wave shapes, (wave shapes having peaks scaled to ± 1), for each part below.

   a. Plot $V_m(x)$ from x of –4 to 4 mm. Which direction would this action potential be moving?

   b. Find and plot $I_i(x)$. At its peak, which is the direction of the current?

   c. Find and plot $I_e(x)$. At its peak, which is the direction of the current?

d. Find and plot $I_m(x)$. Interpret the sign of the peaks in relation to the direction of AP movement.

Exercises 6–11 focus on finding the transmembrane current.

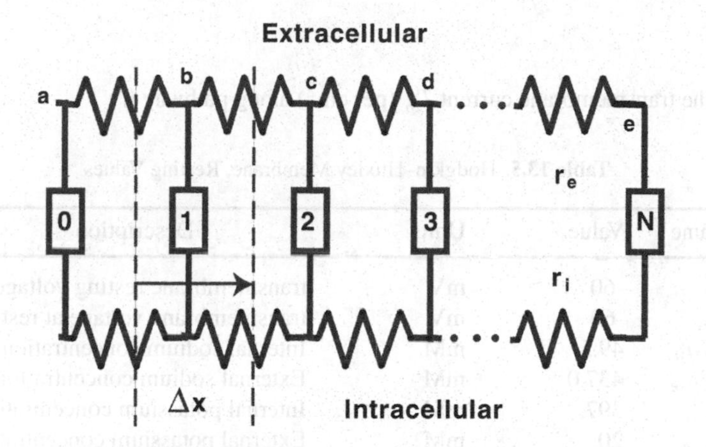

Figure 13.3. Fiber Model for Propagation. The fiber is represented by a network of electrical components. The continuous fiber lies along the x axis. There is a resistive extracellular path (along top line) with resistance per length r_e, a resistive intracellular path (along bottom line) with resistance per length r_i, and a discrete set of membrane crossings. The fiber is divided into N segments, with crossings numbered 0 to N. Often instruction examples limit N to five or fewer, but more realistic analysis frequently uses a much larger number of segments, perhaps 100 to 1000 or more.

6. Figure 13.3 shows a circuit-element representation of a fiber, divided into discrete elements. The fiber is represented with the core-conductor model. In the representation, the transmembrane pathways (rectangular boxes) are numbered starting with 0 at the left end of the fiber. The transmembrane voltages (in units of millivolts) are v_0–v_4. Along the longitudinal intracellular path, each segment's internodal resistance is R ohms. Along the extracellular, it is r ohms. At the time of interest, V_m varies along the fiber. There are no stimuli at this time. Answer each part by giving the mathematical expression for the transmembrane current in terms of R, r, and v_0–v_4. Note the lower-case v, i.e., deviation from baseline.

   a. What is the transmembrane current along transmembrane pathway 1?

   b. What is the transmembrane current along transmembrane pathway 2?

   c. What is the transmembrane current along transmembrane pathway 3?

7. In the fiber of Figure 13.3 and Ex. 6, what is the transmembrane current (per cm^2) along pathway 3?

8. What is the transmembrane current I^0 (mA) along crossing 0?

9. What is the transmembrane current I_m^0 (per cm^2) along pathway 0?

Table 13.5. Hodgkin–Huxley Membrane, Resting Values

Name	Value	Units	Description
V_r	-60	mV	transmembrane resting voltage
V_m	-60	mV	transmembrane voltage at rest
$[Na]_i$	49.5	mM	Internal sodium concentration
$[Na]_e$	437.0	mM	External sodium concentration
$[K]_i$	397.	mM	Internal potassium concentration
$[K]_e$	20.	mM	External potassium concentration
E_K	-72.100	mV	potassium equilibrium potential
E_{Na}	52.4	mV	sodium equilibrium potential
E_L	-49.187	mV	leakage equilibrium potential
C_m	1.0	μF/cm^2	Membrane capacitance
T	6.3	degrees C	Membrane temperature
n	0.31768	–	gating probability n
m	0.05293	–	gating probability m
h	0.59612	–	gating probability h
$\bar{g}_{Na}$	120.0	mS/cm^2	Max Na conductance (a constant)
$\bar{g}_K$	36.0	mS/cm^2	Max K conductance (a constant)
g_L	0.3	mS/cm^2	Leakage conductance (a constant)
g_{Na}	0.011	mS/cm^2	Na conductance
g_K	0.367	mS/cm^2	K conductance

10. Figure 13.3 shows the core-conductor model for a fiber of radius a cm. In this exercise the radius is 10 micrometers. Assume extracellular current flows according to the core- conductor model between radius a and radius 2a. Each node is separated from the next by $\lambda/20$, where $\lambda = \sqrt{r_m/r_i}$. Here the fiber is at rest (Table 13.5). The extracellular specific resistance is 50 Ωcm. The intracellular specific resistance is three times the intracellular. The stimulus current is zero. The transmembrane voltages are:

```
Node          Vm
0,2,4,...    -59 mV
1,3,5,...    -61 mV
```

a. What is the transmembrane current (per cm^2) for crossing 0?

b. For crossing 1?

c. How much is Δx?

11. Suppose one again considers the question asked in Ex. 10 with everything the same except for a new set of transmembrane voltages. Here the transmembrane voltages are

```
Node          Vm
0,2,4,...     -54 mV
1,3,5,...     -66 mV
```

a. What is the transmembrane current (per cm^2) for crossing 0?

b. For crossing 1?

Exercises 12–15 focus on transmembrane potential changes with time.

12. Suppose, in the questions posed in Ex. 4, other quantities are known for the cylindrical fiber, including ionic current, $I_{\text{ion}}(x)$, and the membrane capacitance C_m. For a particular point x_o along the fiber, what is the rate of change (with time) of the transmembrane potential? The answer will be a mathematical expression that is a function of time t.

Exercises 13 and 14: These exercises imagine circumstances built around the grid representation of a fiber, as shown in Figure 13.3. The exercises picture the voltages across all the transmembrane pathways as fixed, except for crossing 3. Those crossings to the left of crossing 3 are clamped to one transmembrane voltage, and those to the right of crossing 3 are clamped to another. At crossing 3 the transmembrane voltage starts at the resting voltage. Then, over time, the voltage changes.

The questions ask, in various forms, what happens at crossing 3. In particular, they ask one first to find the time when the rate of change of the transmembrane potential at crossing 3 is a maximum.[2] Then the values of several other quantities are asked, for the time when $\partial V_m^3/\partial t$ is at its maximum. (Superscripts such as the numeral 3 identify the crossing's number.)

Finding solutions for Exercises 13 and 14 will require the student to have (or create) a computer program that can follow $V_m^3(t)$ as it changes. Such a program will be an extension of a program that can find a patch action potential (Chapter 5). It will not need to be a full-blown program for finding propagating action potentials, as V_m changes, with time, at only one location.

Details of the preparation simulated in exercises 13 and 14 are:

```
distance between xmem crossings   0.10   cm
intra resistance per cm  10,000,000  ohms/cm
extra resistance per cm  10,000,000  ohms/cm
```

and

```
0.32  initial value of n
0.052  initial value of m
0.6  initial value of h
36   mSm/cm^2  Max potassium conductance
120  mSm/cm^2  Max sodium conductance
0.3  mSm/cm^2  Max Leakage conductance
1  $\mu$F/cm^2  membrane capacitance
-72.1  mV  potassium equil potential
52.4  mV  sodium equil potential
-49.2  mV  Leakage equil potential
-60  mV  resting potential
```

As a compromise between solution accuracy and length of calculation, use $\Delta t = 20$ microseconds. A smaller time step, such as $\Delta t = 1$ microsecond will make the calculation, and especially the localization in time of the peak derivative more precise, but also will make the calculation more lengthy. (Often time is kept as an integer value.) Use the HH membrane model and the core-conductor model as a basis for the calculations required to answer the questions.

13. Determine the maximum dV_m/dt under the following conditions.

```
Vm node 2  -18  millivolts  unchanging
Vm node 3  -60  millivolts  initial
Vm node 4  -60  millivolts  unchanging
```

a. Within 0.02 msec, when does the maximum dV_m/dt occur?

b. What is the maximum dV_m/dt value?

[2] The maximum dV_m/dt time point often is measured in experimental studies and used as a fiducial point. Here one is asked to locate the maximum value of dV_m/dt that occurs at any time in the interval from 0.10 to 10 msec, with the time interval 0 to 0.10 excluded so as to avoid misleading numerical transients that may occur during the first time step.

c. What is Vm at the time of the maximum dV_m/dt?

d. What membrane current I_m is present at the time of the maximum dV_m/dt?

e. What membrane ionic current I_{ion} is present at the time of the maximum dV_m/dt?

14. This question is the same as the preceding one, except that the transmembrane voltages are as follows. Note that Vm on the left is much higher in this exercise.

```
Vm  node 2  40   millivolts  unchanging
Vm  node 3  -60  millivolts  initial
Vm  node 4  -60  millivolts  unchanging
```

a. When does the maximum dV_m/dt occur?

b. What is the maximum dV_m/dt value?

c. What is V_m at the time of the maximum dV_m/dt?

d. What membrane current I_m is present at the time of the maximum dV_m/dt?

e. What membrane ionic current I_{ion} is present at the time of the maximum dV_m/dt?

15. *Stability.* In a simulation of a cylindrical HH fiber, investigators have a concern about fluctuations in V_m that are not meaningful. In their problem, the radius of the membrane is 30 micrometers. The fiber is divided into segments having an axial length of 5.761 micrometers. Extracellular currents flow to twice the membrane radius. The extracellular resistivity is 10 Ωcm. The intracellular resistivity is three times that of the extracellular. The membrane capacitance is 1 μF/cm^2. At the start of the simulation, the HH resting conditions apply (Table 13.5).

a. What is the mesh ratio?

b. Does the mesh ratio indicate the likelihood of stability or instability?

c. If the calculation turns out to be unstable, what will be the manifestation of that problem?

d. What step might be taken to improve the likelihood of a stable calculation?

e. Does instability in fact occur?

For Exercises 16–26, assume that the upstroke of a uniformly propagating action potential is described by the V_m template equation (13.3):

$$V_m(x,t) = 50 \tanh\left[t - \frac{(x - x_0)}{\theta}\right] \tag{13.3}$$

In (13.3) V_m is in mV, t is time in msec, and x is distance along the axis in mm. Distance x_0 is a constant, in mm, and θ is the velocity of propagation, in mm/msec (i.e., in m/s). Assume the fiber radius is 50 μm, the intracellular resistivity 100 Ωcm, and the extracellular resistivity 0.

Use of mathematical functions such as hyperbolic tangent to create artificial V_m waveforms as a function of time and space is useful, in that the template function can be used to give one insight across a wide range of responses in space and time, and as calibration waveforms for equipment or display. One can regenerate them much more quickly than experimental measurements, even with HH simulations. Even so, one must keep in mind that the waveforms may be similar to real action potentials in some respects but differ markedly from real action potentials in other respects. For example, the wave shape defined in (13.3) does not show effects at fiber ends, with stimuli, or in response to changes in rate of stimulation. Also, the tanh waveform is unnaturally symmetric. Further, the tanh template function as given models only the action potential upstroke; thus it leaves out the recovery phase. Even so, in part because of the simplification that it embodies, use of the template waveform allows other fundamental relationships in time and space to stand out, and thus to be more evident to the student.

16. Using Eq. (13.3) with the assumptions of the core-conductor model. derive an equation for the longitudinal current $I_i(x,t)$.

17. Using Eq. (13.3) with the assumptions of the core-conductor model, derive an equation for the longitudinal current $I_e(x,t)$.

18. If $x_0 = 2$ mm and $\theta = 2$ m/sec, plot $V_m(t)$ as it will be seen at $x = 10$ mm.

19. If $x_0 = 2$ mm and $\theta = 2$ m/sec, plot $V_m(x)$ as it would exist at $t = 3$ msec.

20. Under the same conditions as in Ex. 19, plot $I_i(x)$.

21. Under the same conditions as in Ex. 19, plot $I_e(x)$.

22. Under the same conditions as in Exs. 19–21, make a diagram of a fiber and draw on it arrows indicating by their position and direction for the following:

    a. membrane current i_m due to *Na*.

    b. I_i.

    c. membrane current through the membrane capacitance.

    d. I_e.

23. If $x_0 = 10$ mm and $\theta = -2$ m/sec, plot $V_m(x)$ as it would exist at $t = 3$ msec.

24. Under the same conditions as in Ex. 23, plot $I_i(x)$.

25. Under the same conditions as in Ex. 23, plot $I_e(x)$.

26. Under the same conditions as in Exs. 23–25, make a diagram of a fiber and draw on it arrows indicating by their position and direction:

    a. membrane current i_m due to Na.

    b. I_i.

    c. membrane current through the membrane capacitance.

d. I_e.

Exercises 27–36, Electrophysiological Experiments: These exercises correspond to possible procedures in an experimental study, and are framed in similar language. Performing the experiments requires a simulation program for propagation using the Hodgkin–Huxley equations.

Replicate the classical Hodgkin–Huxley results for the giant axon of the squid by using the following fiber parameters:

```
Fiber characteristics.
Assume a cylindrical fiber as described in the text.
Length         30 cm
Nodes 0 to 600 spaced .05 cm along axial direction.
Radius a     300 um Extracellular current flows to 2a.
Ri            30 ohm-cm so ri = 0.1061e5 ohm/cm
Re            20 ohm-cm so re = 0.2357e4 ohm/cm
```

Set temperature to 6.3°C initially. For comparative results, recall that 12.6 and 18.9°C correspond to some HH results, and 25°C is roughly room temperature.

Unless given otherwise, stimulate the fiber with a pair of extracellular electrodes, one at the left end (crossing 0) and the other at the right end (crossing 600). Make the stimulus current equal in magnitude, opposite in sign, at every moment. Choose the polarity so that the stimulus at the left end depolarizes the fiber.

One may wish to compare results to the simulation results given in the simulation figures of Chapter 6, which were based on the parameters above.

The following electrophysiological experiments are similar to those for a membrane patch (Chapter 5). It is useful to compare the results, as that demonstrates the powerful changes produced by the fiber's geometry.

27. *Time Step.* Determine a suitable value for Δt, using the fiber parameters above, and the mesh ratio.

28. *Threshold and strength–duration.* Start with a simulation duration of 10 msec and a stimulus duration of 100 microseconds (μsec). Increase the stimulus amplitude until an action potential is produced. Judge the presence of an action potential by whether one later appears at the midpoint of the fiber.

    a. Find the threshold—the amplitude that just produces an action potential, while a decrease to 50% of that amplitude fails. What is this amplitude?

b. What happens to the time interval between the start of the stimulus and the midpoint of the action potential's upstroke, for stimulus amplitudes just above threshold?

c. Increase the duration to 200 μsec and again find the threshold. Then do the same for a duration of 500 and 1000 msec. Plot these points as a strength–duration curve.

d. Compare the strength–duration curve obtained above to that for a patch (Chapter 5). Explain the differences that are present.

29. *Membrane conductances.* Determine the membrane conductances as a function of time, at the left end, center, and right end of a fiber, from the time of the stimulus throughout an action potential.

a. Plot the changes in these conductances during and following the action potential. Use a large vertical scale, and consider a log scale. What are the largest values the conductances reach, at any one of the three sites?

b. Are the conductance waveforms the same at all three sites?

c. By what factors does each one change during the action potential? (make a ratio of the largest value over the smallest value).

30. *Refractory period.* Set the stimulus duration and amplitude to produce an action potential within 1 msec of the start of the first stimulus, and set the stimulus duration to the time for propagation from one end of the fiber to the other, plus 20 msec. Initiate a second stimulus pulse with the same duration and amplitude, but starting at 8 msec.

a. Does the 2nd stimulus produce a propagating action potential, as judged by what happens at the middle of the fiber?

c. If necessary, increase the stimulus amplitude of the stimulus at 8 msec, until it does produce a propagating action potential, as observed at the center of the fiber.

d. With the stimulus determined in part C, is the time interval between the upstroke of the first and of the second action potentials, as observed at the center of the fiber, 8 msec?

31. *Temporal summation.* Use a simulation duration of 10 msec. Decrease the amplitude of both stimuli to about 60% of the stimulus threshold (for membrane at rest).

    a. What happens when the delay time of the second stimulus is reduced to 1 msec?

    b. By looking at the time courses of the conductances, can you explain why a second subthreshold stimulus could cause an action potential after the first one failed?

    c. Increase the time until the onset of the 2nd stimulus. At what interval does neither stimulus produce an action potential?

32. *Anode break excitation.* Set the stimulus duration to 5 msec. Set the stimulus magnitude in such a way that the hyperpolarizing electrode stimulus displaces the transmembrane potential to approximately −105 mV.

    a. Record $V_m(t)$ throughout the hyperpolarizing stimulus and during the first 10 msec after the stimulus ends, at the hyperpolarized end of the fiber. (The phenomenon seen is called "anode-break excitation" because excitation occurs after shutting off a hyperpolarizing (anodal) pulse of current if the amount of hyperpolarization is sufficiently large.)

    b. List values for n, m, and h, and values for conductances g_K and g_{Na}, at the start and at the end of the hyperpolarizing stimulus. Give values for I_K and I_{Na} immediately before and after the stimulus, when the transmembrane is (or returns to) its resting value.

    c. Based on the data in parts A and B, explain why anode-break excitation occurs.

33. *Constant stimulus current.* Apply a stimulus that has an extended duration of 50 msec duration. Set the magnitude high enough to initiate an action potential.

    a. What happens if the stimulus has a minimal amplitude?

    b. What happens if the stimulus has a 5 times this amplitude?

    c. How do the wave shapes compare with those of normal action potentials.

34. *Temperature effects.* Describe the effects of increasing the temperature to temperatures of 6.3, 12.6, 18.9, and 25°C. (All are within the physiological range for squid.)

   a. Make a table showing the propagation velocity at the center of the fiber, as compared to the temperature.

   b. Make a table showing the threshold stimulus amplitude as a function of temperature, for a stimulus duration of 200 μsec.

   c. Describe the results of [A] and [B] in a few sentences. What effects were observed as temperature changed?

35. *Ionic concentrations.* Vary the ionic concentrations in the external medium. Describe qualitatively the result of each of the following. In particular, and describe any changes in initial conditions or in response to a stimulus of 200 μsec duration that would have twice threshold amplitude with standard concentration values.

   a. Double $[\text{Na}]_i$.

   b. Double $[\text{K}]_e$.

   c. Increase $[\text{K}]_e$ times 10.

36. *Spatial extent.* Simulate the propagation of two successive action potentials on a squid axon fiber.

   a. How short an interval can there be and still excite a second action potential, as judged at node 0?

   b. How short an interval can there be and still excite a second action potential, as judged at the center of the fiber?

   c. Use the stimulus interval and amplitude that results in two action potentials at the center of the fiber. Plot the spatial distribution $V_m(x)$ at that time. How many action potentials exist along the fiber at that time?

13.7. EXERCISES, CHAPTER 7: ELECTRICAL STIMULATION OF EXCITABLE TISSUE

Exercises 1–4 deal with stimulation terminology and units. In 1–3 fill in the blank with the one word needed.

1. Examining a stimulus–duration curve, one sees a curve that is high when durations are near zero, but declining to a low value as duration increases to its the longest practical value. The limiting low value is called: ______________.

2. In a preparation with no spatial variation, to a good approximation an action potential will follow if the stimulus brings the voltage above a certain ______________ value.

3. With a long stimulus the transmembrane potential needs to reach only a minimum value, at the end of the stimulus, and only a minimal amplitude of the stimulus current is required to produce this transmembrane potential. Suppose one wishes to reach the same transmembrane voltage, but more quickly, and thus a stimulus current of twice the minimal amount is used. Then ______________ is the time duration required for the stimulus to reach the same transmembrane potential at the end of the stimulus.

4. Give suitable *mks* units for each of the following:

   a. chronaxie

   b. rheobase

   c. threshold

Exercises 5–10 deal with a spherical cell that responds to stimulation according to the model given in the text. An important characteristic of the model is that spherical symmetry causes the intracellular stimulus to produce a uniform transmembrane current everywhere on the membrane. The cell's membrane has membrane resistance of R (Ωcm^2) and membrane capacitance C (μF/cm^2). An experimenter tests the cell and finds that as the stimulus duration becomes very long, the minimal stimulus that produces an action potential has stimulus current magnitude a

μA/cm^2. This applied current produces a transmembrane voltage of W mV at the end of the stimulus. A 2nd stimulus is applied after the membrane has returned to rest. The above conditions apply to all exercises 5 to 10 unless noted otherwise.

5. A stimulus of shorter duration, 0.5 msec, also produces an action potential, if it has a sufficient current magnitude. In the 2nd stimulus, what is the magnitude of the minimal transmembrane current required to produce an action potential? Give the formula, including any conversion factors needed for the answer to be expressed in μA/cm^2.

6. A stimulus of duration 3 msec also produces an action potential, if it has a current magnitude i. In the 2nd stimulus, what is the magnitude of the threshold transmembrane voltage?

7. By changing the stimulus duration over a range of times the investigator finds that the membrane has a time constant of U msec. A stimulus of shorter duration, 1.5 msec, also produces an action potential, if it has a sufficient current magnitude. What is the formula for the membrane capacitance, in terms of U and R? Include any conversion factors needed for the answer to be expressed in μF/cm^2.

8. A stimulus of shorter duration, 4 msec, also produces an action potential, if it has a current magnitude i. What is the magnitude of the transmembrane resistance? Respond with a formula for R. Include any conversion factors necessary for the result to be expressed in Ωcm^2.

9. A stimulus of shorter duration, 1.5 msec, and strength i also produces an action potential. A separate series of measurements by the investigator showed that the time constant of the membrane to be M milliseconds. How much is rheobase? Respond with a formula based on the known quantities. Include any conversion factors needed for an result in millivolts.

10. A stimulus of very short duration d msec requires a stimulus current of magnitude at least i μA/cm^2 to induce an action potential. A stimulus of intermediate duration, D msec also produces an action potential if the current is I μA/cm^2 or more. Following the experiment, the investigator spoke with the flowers on the Duke campus (smile) and learned that the cell's time constant τ could be found from the above data and compared to the expected value. What is the formula that gives the time constant?

Exercises 11–14 are numerical questions about an idealized spherical cell. Investigation of the response to stimuli of the spherical cell used transmembrane stimulation, with a current source inside the cell and current sink outside. Spherical symmetry was preserved. The cell responded according to the model described in the text. Each question provides several items of numerical data and anticipates a numerical answer in specific units.

11. With a long stimulus, the lowest stimulus current that would produce an action potential had magnitude 2 μA/cm^2, and the transmembrane voltage at the end of that stimulus was 20 millivolts. For shorter stimuli, the investigator set a stimulus duration and then carefully tried stimuli of different current magnitudes until the current was found that produced, at the end of the stimulus, the threshold voltage for an action potential. What was rheobase for results of:

duration (msec)	current (μA/cm^2)
1	21.016
3	7.716

12. With a long stimulus, the lowest stimulus current that would produce an action potential had magnitude 10 μA/cm^2, and the transmembrane voltage at the end of that stimulus was 20 mV. For shorter stimuli, the investigator set a stimulus duration and then carefully tried stimuli of different current magnitudes until the current was found that produced, at the end of the stimulus, the threshold voltage for an action potential. What is the membrane resistance (in Ωcm^2), if other results were:

duration (msec)	current (μA/cm^2)
1	25.414
3	12.872

13. With a long stimulus, the lowest stimulus current that would produce an action potential had magnitude 10 μA/cm^2, and the transmembrane voltage at the end of that stimulus was 20 millivolts. Moreover, the membrane was found to have a time constant of 2.4 msec. Using shorter stimuli, the investigator set a stimulus duration and then carefully tried stimuli of different current magnitudes until the current was found that produced, at the end of the stimulus, the threshold voltage for an action potential. What stimulus current is needed for a stimulus duration of 0.2 msec, if other results were as shown in the table. Give a numerical answer, in μA/cm^2.

duration (msec)	current (μA/cm^2)
1	29.346
3	14.015

14. With a long stimulus, the lowest stimulus current that would produce an action potential had magnitude 10 μA/cm^2, and the transmembrane voltage at the end of that stimulus was 20 millivolts. For shorter stimuli, the investigator set a stimulus duration and then carefully tried stimuli of different current magnitudes until the current was found that produced, at the end of the stimulus, the threshold voltage for an action potential. What is the membrane time constant, for the results given in the table. Answer in msec, within 5%.

duration (msec)	current (μA/cm^2)
1	25.414
3	12.872

15. Using a conventional strength–duration curve

$$I_{\text{th}} = \frac{I_R}{1 - e^{-Kt}}$$

determine the expression for minimum charge injection to reach a stimulus threshold V_t.

16. A spherical cell has a radius of 100μm. Its membrane has a specific capacitance of 1.0μ F/cm^2 and a leakage resistance of 2000 Ωcm^2. A current of strength 0.0005 μA is introduced intracellularly and flows outward to a distant grounded electrode in the surrounding uniform unbounded medium. The pulse duration is 2 msec. Calculate and plot the transmembrane potential for the period 0–5 msec.

Cylindrical Fiber Exercises 17–19 deal with parameters of a cylindrical fiber. Solve each one using the core-conductor model. Unless otherwise specified, extracellular currents extends to twice the membrane radius, and the membrane has Hodgkin–Huxley (HH) characteristics.

17. In a cylindrical HH fiber at rest, what is the space constant? The radius of the membrane is 30 micrometers. Extracellular currents flow to twice the membrane radius. The extracellular resistivity is 50 Ωcm. The intracellular resistivity is three times that of the extracellular. The membrane capacitance is 1 μF/cm^2. The standard HH resting conditions apply. The fiber is passive.

18. Find the space constant, for the data given in the table.

150	mem radius	microns
200	width	microns (of segment)
200	intra resistivity	Ωcm
50	extra resistivity	Ωcm
0.9	mem capacitance	μF/cm^2
1,300	mem resistivity	Ωcm^2 (at rest)

19. Find the time constant of the cylindrical fiber that has the following characteristics. Extracellular currents extend to twice the membrane radius.

50	mem radius	microns
500	width	microns (of segment)
100	intra resistivity	Ωcm
50	extra resistivity	Ωcm
1.2	mem capacitance	μF/cm^2
1,500	mem resistivity	Ωcm^2 (at rest)

20. For a fiber under subthreshold conditions, and making reasonable assumptions, how does the time constant depend on fiber radius?

Field Stimulation Exercises deal with stimuli from locations at a point away from the fiber membrane.

21. Within a large 3D volume (Figure 13.4), stimulus electrode E produces extracellular potential $P = S/r$ at position x. Along the fiber:

    a. In terms of S, x, e, h, what is the potential $P(x)$?

    b. What is the activation function $A(x)$?

c. If $e = 0$ and $h = 1$, what is the value of the activating function at $x = 0$? Be sure to verify the sign of the result.

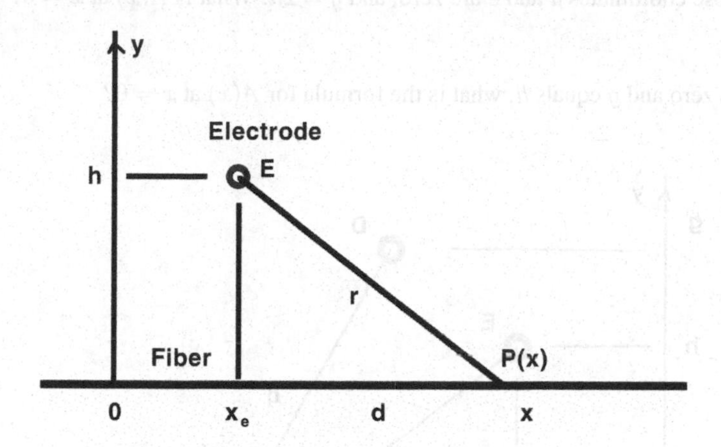

Figure 13.4. One Field-Stimulus Electrode E, located above a fiber. Electrode E is located at coordinates $x = x_e$ and $y = H$. The fiber (thicker line) lies along the horizontal axis at $y = 0$. From the perspective of the stimulus, the fiber is considered so small that it does not affect the potential field created by the stimulus. $P(x)$ is the potential at coordinate position x along the x axis at $y = 0$. Distance r is the straight-line distance between the electrode and the position x, while d is its x component, i.e., the distance between x_e and the site of interest, x.

22. A unipolar electrode is located at $e = 5$ cm, $h = 1$ cm. A small fiber lies along the x axis in a medium of infinite extent. That is, the electrode is a current source. Consider the section of the fiber lying between $x = 0$ and $x = 10$ cm. One approach is to graph $A(x)$ from $x = 0$ to $x = 10$.

    a. The potential from the electrode is $+S/r$. That is, the electrode is a current source. How many cm of the 10 cm length has an activating function that has a positive sign? Give a numerical answer, in centimeters, within 5

    b. The potential from the electrode is $-S/r$. That is, the electrode is a current sink. How many cm of this 10 cm length has an activating function that has a positive sign?

    c. Can one conclude from the results of parts a and b that a source electrode will be a more effective stimulus than a sink?

23. Within a large 3D volume, electrodes D and E are located at (x, y) coordinates (d, g) and (e, h) respectively, as shown in Figure 13.5. The distance from electrode D to position x

along the fiber is q, and from electrode E is r. The extracellular potential P at position x produced by a stimulus current at D is S/r and from E is $-S/r$.

a. Suppose coordinates d and e are zero, and $g = 2h$. What is $A(x)$ at $x = 0$?

b. If e is zero and g equals h, what is the formula for $A(x)$ at $x = 0$?

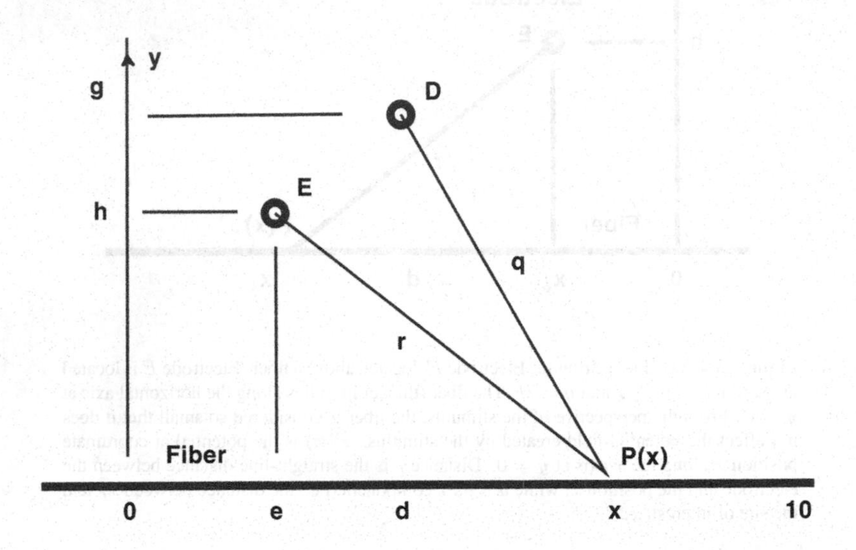

Figure 13.5. Two Electrodes above a Fiber. Electrodes D and E are at (x, y) coordinates (d, g) and (e, h), respectively. Distances q and r separate them from position x, where potential $P(x)$ is to be determined. The fiber to be stimulated is considered to lie along the horizontal axis at $y = 0$, and to be small enough as to have no effect on the externally applied field.

24. Suppose a fiber's transmembrane potentials are all at rest, e.g., at –70 mV. To initiate excitation in the fiber, potentials within a small region of the fiber must be elevated to a higher transmembrane potential, e.g., –25 mV. The activating function is used as an indicator of the effect of a stimulus on the fiber. When looking for a region where excitation is likely to be initiated, one looks for a region where the activating function is (which one): (a) $A(x) > 0$, or (b) $A(x) < 0$.

Exercises 25 and 26 ask about extrema. An extremum is either a local maximum or a local minimum. In these exercises, the ends of the interval are not extrema. An effective way to solve these problems is to graph the function.

25. A bipolar stimulus pair is located perpendicular to the axis of the fiber. The closer pole is the sink, and the more distant pole is the source. The sink is 1 cm from the fiber, and the source sink pair is separated by 1 mm. In other words, electrode E in the figure is the sink and D is the source. Electrode coordinates are: $e = 5$, $d = 5$, $h = 1$, and $g = 1.1$. Between 0 and 10 cm, how many extrema are present?

26. A bipolar stimulus pair is located parallel to the axis of the fiber, centered over $x = 5$. Electrode E, the sink, is closer to the origin. The closer pole is the sink and the more distant pole is the source. The center of the source–sink pair is 1 cm from the fiber, and the source sink pair is separated by 2 mm. That is, in the figure electrode E is the sink and D is the source. Electrode coordinates are: $e = 4.9$, $d = 5.1$, $h = 1$, and $g = 1$, all in cm. Between 0 and 10 cm, how many extrema are present?

27. Consider a unipolar electrode E (Figure 13.4). The electrode is a current sink. Consider the section of the fiber lying between $x = 0$ and $x = 10$ cm.

    a. Electrode E is located first at $e = 5$ cm, $h = 1$ cm. Determine many cm of this 10cm length has an activating function that has a positive sign. Call this amount length one (L_1).

    b. The electrode is now moved away from the fiber, so that $e = 5$ cm and $h = 2$ cm. Again determine how many cm of the 10 cm length has a positive sign. Call this amount L_2.

What is the ratio of L_2/L_1?

28. A bipolar stimulus pair is located perpendicular to the axis of the fiber. The closer pole is the sink and the more distant pole is the source. The sink is 1 cm from the fiber, and the source sink pair is separated by 1 mm. That is, in the figure electrode E is the sink and D is the source.

    a. First locate the electrode such that the electrode coordinates are: $e = 5$, $d = 5$, $h = 0.9$, and $g = 1.1$. Inspect the graph of the activating function along the fiber, between 0 and 10 cm and determine what length of that 10 cm has a positive activating function. Call this amount length 1 (L_1).

    b. Move the electrode so that e and d remain the same but $h = 1.9$ and $g = 2.1$. Again determine the length with a positive activating function. Call this amount length 2 (L_2).

What is the ratio $L_2/L_1 = 2$?

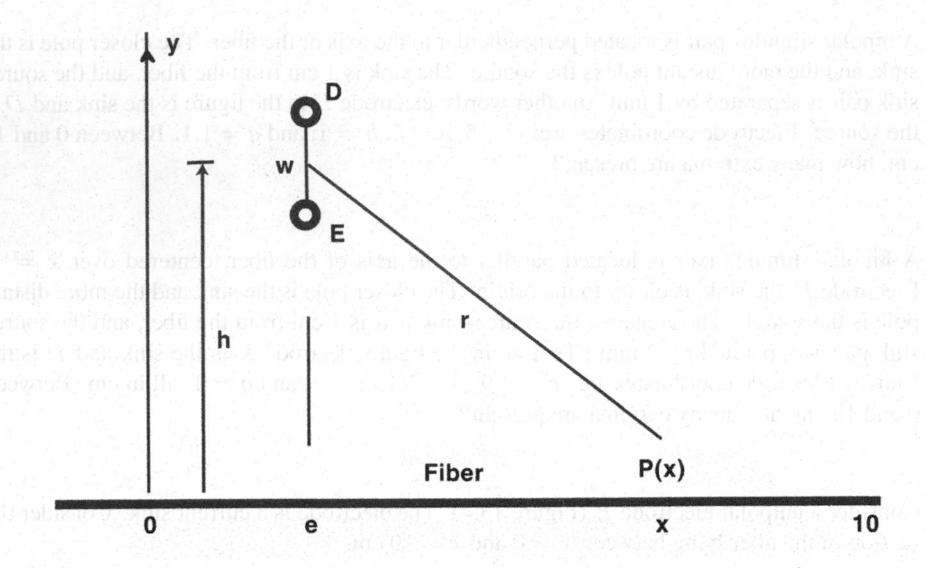

Figure 13.6. Two Electrodes above a Fiber. The inter-electrode axis is normal to the fiber axis. Distance h is the vertical (y-axis) distance to the center of the electrode pair. The electrodes are separated by distance w. Distance r is the distance from the center of the pair to position x along the fiber, which is considered to lie along the x axis at $y = 0$. If the two electrodes are close enough together, relative to the other dimensions of the problem, then their effects approach those of a mathematical dipole.

29. Consider the geometry of a bipolar stimulus oriented normal to a fiber, as given in Figure 13.6. The stimulus electrodes and fiber lie within an unbounded space of resistivity 50 Ωcm. Suppose distance $h = 1$ cm and separation $w = 0.2$ cm. Electrode coordinate e is 0 cm. The electrode closest to the x axis is the sink, and the more distant electrode is the source. The stimulus current is 25 mA. What is $A(x)$ at $x = 0.3$ cm?

The following exercises deal with elements of a stimulator design, which thereafter are brought together into exercises in design format.

30. Batteries often are specified by the voltage and by their Ampere-hour rating. How much energy can this battery supply? For simplicity, ignore some of the complexities of real batteries and assume the battery retains its full voltage until its Ampere-hour capacity is fully used. Give a numeric answer, in Joules.

    a. Suppose a 7.5-volt battery is rated at 0.3 Ampere- hours.

    b. Suppose a 1.5-volt battery is rated at 0.4 Ampere- hours.

31. Stimulators may be implanted in children and used for a lifetime. If a stimulator gives 2 stimuli per second (on average), how many stimuli does it deliver in total over a period of

    a. 50 years.

    b. 90 years.

32. What is the initial change of v_m, with time, at this site, due to the stimulus? The fiber has a 50 μm radius, an intracellular resistivity of $R_i = 100\ \Omega$cm, and a capacitance of 1 μF/cm^2.

    a. A stimulus produces an activating function value of 7 mV/cm^2 at a certain site on a fiber of 50 μm radius. The fiber is immersed in a large volume having resistivity of 50 Ωcm.

    b. A stimulus produces an activating function value of 6 mV/cm^2 at a certain site. The fiber is immersed in a large volume having resistivity of 140 Ωcm.

33. A bipolar stimulus electrode (as in Figure 13.6) has two electrodes, a source and a sink. What voltage between the electrodes is required to produce this stimulus current? Current is injected at the source electrode and removed at the sink electrode. The medium is effectively unbounded. (For simplicity, ignore electrode–electrolyte voltages, resistances, or other effects, assume the each electrode is not large enough to modify the field produced by the other, and ignore any complications from whatever structures hold the electrodes in place.)

    a. Each electrode is a spherical conductor that has a radius of 0.15 centimeters. The two electrodes are separated by 4 cm, center to center. The stimulus produces a rectangular current pulse of duration 4 milliseconds and current amplitude 25 milliamperes. The medium has a resistivity of 50 Ωcm.

    b. Each electrode is a spherical conductor that has a radius of 0.1 centimeters. The two electrodes are separated by 3 cm, center to center. The stimulus produces a rectangular current pulse of duration 1 milliseconds and current amplitude 25 milliamperes. The medium has a resistivity of 25 Ωcm.

34. What is the magnitude of the stimulus current that the stimulator gives on each stimulus, if the battery is able to energize 1,000,000,000 stimuli? Suppose the battery can supply 1,000 Joules before the energy in the battery is consumed. Suppose further that the battery is used to provide energy to a bipolar stimulus electrode. (Such a battery might exist in a pacemaker.) The voltage between the electrodes produces the stimulus current into the conductive medium.

Ignore electrode–electrolyte aspects. Batteries tend to change their voltage over time, as they discharge. Ignore this aspect and assume the battery retains the same voltage throughout its life until all its energy is used. The stimulus produces a rectangular current pulse of duration 4 milliseconds and amplitude I_o milliamperes, into an effectively unbounded medium. Every current pulse has the same current as every other. Thus, since each stimulus interval consumes a certain amount of energy, there are only so many stimuli that can be given before the energy in the battery is entirely consumed.

a. The bipolar stimulus electrode uses a voltage of 210 millivolts between the source and sink electrodes, which have a radius of 0.1 centimeters. The voltage between the electrodes produces the stimulus current into the conductive medium.

b. Suppose the battery can supply 1,500 Joules. The bipolar stimulus electrode uses a voltage of 150 millivolts between the source and sink electrodes, which have a radius of 0.1 centimeters.

Stimulator Design Exercises: Consider each of the stimulator designs given in Exercises 32–34. In each case, the electrode geometry is that of a bipolar stimulus oriented normal to a fiber, as given in Figure 13.6. Each stimulus electrode is spherical.
In each design, state whether the design meets requirement 1, requirement 2, both, or neither. A design meets a requirement if the design fulfills the condition.

35. The electrode radius is 0.01 cm. The stimulus electrodes and fiber lie within an unbounded space of resistivity 50 Ωcm. Distance h is 0.6 cm and separation w is 0.3 cm. Electrode coordinate e is 0 cm. The electrode closest to the x axis is the sink, and the more distant electrode is the source. The stimulus current is 0.01 A and the stimulus duration is 0.002 sec, and the number of stimuli per second is 2. Function $P(x)$ is the electric potential produced by the stimulus, as a function of coordinate x.

    The fiber to be stimulated lies along the x axis. The fiber has a radius of 0.02 cm, an intracellular resistivity of 100 Ωcm, and a membrane capacitance of 1 μF/cm^2.

    Requirement 1: The stimulus voltage (voltage required between the stimulus electrodes to produce the specified stimulus current) must be less than 10 volts.

    Requirement 2: Initial dv_m/dt directly under the stimulus electrode must be 25 mV/msec or more.

36. Consider the following stimulator design. Its geometry is that of a bipolar stimulus oriented normal to a fiber, as given in Figure 13.6. Each stimulus electrode is spherical, with a radius of 0.01 cm. The stimulus electrodes and fiber lie within an unbounded space of resistivity 50 Ωcm. Suppose distance h is 0.6 cm and separation w is 0.3 cm. Electrode coordinate e is 0 cm. The electrode closest to the x axis is the sink, and the more distant electrode is the source. The stimulus current is 0.01 A and the stimulus duration is 0.002 sec, and the number of stimuli per second is 2. Function $P(x)$ is the electric potential produced by the stimulus, as a function of coordinate x.

    The fiber to be stimulated lies along the x axis. The fiber has a radius of 0.02 cm, an intracellular resistivity of 100 Ωcm, and a membrane capacitance of 1 μF/cm^2.

    Requirement 1: The stimulus voltage (voltage required between the stimulus electrodes to produce the specified stimulus current) must be less than 5 volts.

    Requirement 2: Initial dv_m/dt directly under the stimulus electrode must be 25 mV/msec or more.

37. Consider the following stimulator design. Its geometry is that of a bipolar stimulus oriented normal to a fiber, as given in Figure 13.6. Each stimulus electrode is spherical, with a radius of 0.01 cm. The stimulus electrodes and fiber lie within an unbounded space of resistivity 50 Ωcm. Suppose distance h is 1 cm and separation w is 0.3 cm. Electrode coordinate e is 0 cm. The electrode closest to the x axis is the sink, and the more distant electrode is the source. The stimulus current is 0.01 A and the stimulus duration is 0.002 sec, and the number of stimuli per second is 2. Function $P(x)$ is the electric potential produced by the stimulus, as a function of coordinate x.

    The fiber to be stimulated lies along the x axis. The fiber has a radius of 0.0025 cm, an intracellular resistivity of 100 Ωcm, and a membrane capacitance of 1 μF/cm^2.

    The power source is a 1 Ampere-hour battery.

    Requirement 1: An initially positive dV_m/dt occurs from directly under the stimulus electrode at least half a space constant along the fiber in both $+x$ and $-x$ directions.

    Requirement 2: The stimulator's will have lifetime of 1 year or more.

38. Consider the following stimulator design. Its geometry is that of a bipolar stimulus oriented normal to a fiber, as given in Figure 13.6. Each stimulus electrode is spherical, with a radius of 0.01 cm. The stimulus electrodes and fiber lie within an unbounded space of resistivity 50 Ωcm. Suppose distance h is 1 cm and separation w is 0.3 cm. Electrode coordinate e is 0 cm. The electrode closest to the x axis is the sink, and the more distant electrode is the

source. The stimulus current is 0.01 A and the stimulus duration is 0.002 sec, and the number of stimuli per second is 2. Function $P(x)$ is the electric potential produced by the stimulus, as a function of coordinate x.

The fiber to be stimulated lies along the x axis. The fiber has a radius of 0.0025 cm, an intracellular resistivity of 100 Ωcm, and a membrane capacitance of 1 μF/cm^2.

The power source is a 3 Ampere-hour battery.

Requirement 1: An initially positive dV_m/dt occurs from directly under the stimulus electrode at least half a space constant along the fiber in both $+x$ and $-x$ directions.

Requirement 2: The stimulator's will have lifetime of 1 year or more.

39. Waveform dilemma. An early derivation ("derivation one") in this chapter finds the pattern of transmembrane voltage versus distance that results from the stimulus, at steady state. The pattern shows that a single extracellular stimulus just outside the membrane of a 1D cylindrical fiber results in monophasic wave shape $v_m(x)$. A derivation in a later section of the chapter ("derivation two") gets quite a different result after staring from a beginning point that seems only slightly different. Specifically, derivation two shows that an single point-source extracellular stimulus a distance from the fiber produces a biphasic waveform (triphasic if the stimulus is away from fiber ends) for $v_m(x)$.

    The two waveforms resulting from the two derivations are not compatible, i.e., they are qualitatively different to the degree that there is no possibility that one can become the other through scaling. At first, it would seem that the difference might be explained as the consequence, in the second derivation, of moving the stimulus away from the fiber. However, note that in derivation two the distance from the fiber is a parameter h, and h may be chosen to be a low value, so that the stimulus is placed just outside the fiber membrane, a location that seems very similar to that of derivation one. When the stimulus of derivation two is located just outside the membrane, the multiphasic nature of the resulting $v_m(x)$ pattern not only does not disappear but becomes **more pronounced,** thus emphasizing its incompatibility with the result of derivation one.

    Are both of these derivations correct? If so, explain what it is about the fundamental assumptions that are the starting points of each derivation that causes such qualitatively different results to come about.

Pacemaker Design: Stimulators are used for a large number of experimental and clinical purposes. The following design problem asks you to design a stimulator for clinical use. Some of the considerations of real stimulator designs are present. Conversely, there are many ways in which the clinical problem has been simplified to make it easier to address using material presented here. The first exercise asks that any solution be found that meets the requirements, while the following exercise asks for the lowest cost solution, where "cost" is defined within the exercise.

40. You are to design a nerve pacemaker. The nerve pacemaker is to be implanted, and must function correctly for a minimum of one year without requiring battery replacement. To perform satisfactorily, the pacemaker must deliver a stimulus to the nerve that will cause a nerve action potential to result 10 times each second throughout the year of use. The pacemaker must not miss an interval when it should deliver a stimulus.

Your design should consist of values for the following:

1. The initial battery energy, a single number in units of Joules.

2. The initial battery voltage, a single number in units of Volts.

3. The nominal stimulus duration, in seconds.

Parameters. Since the pacemaker must be designed and built before the patient whom it will serve has been identified, the precise characteristics of the nerve and its environment are not known precisely. Parameters have the nominal values given in Table 13.6.

Table 13.6. Parameters for Pacemaker Design

R_i	100	Ωcm
R_e	10	Ωcm
R_m	10^4	Ωcm^2
C_m	1	μF/cm^2
a	50	μm (radius)

Because the nerve fiber of the person in whom the pacemaker will be implanted may vary appreciably from the nominal characteristics above, the pacemaker must allow for them. In a particular patient, it may be that no abnormalities (parameter variations) occur, that abnormalities occur one at a time, or that abnormalities occur in some combination. The pacemaker characteristics must allow for any of these possibilities. For simplicity, assume that if a variation occurs, then any one parameter has one of the following values: (1) its nominal value; (2) its nominal value plus 20%; (3) its nominal value minus 20%.

Conditions. The threshold voltage V_T is assumed to be 25 mV. To pace the fiber, v_m must rise to V_T at a point one λ from the pacing site within 0.05 sec.

The fiber to be paced is infinitely long, and pacing is in the center.

Assume that all battery energy is delivered as pacing pulses, i.e., none is used for other purposes.

Any abnormality or combination of abnormalities remains the same for the whole year.

Assume that $r_e << r_i$.

The battery voltage for any pacing pulse should be computed as the initial battery voltage times the fraction of the initial battery energy left after all previous pulses have been completed. Assume the battery voltage to be constant throughout any one pacing pulse.

The stimulus current for any pacing pulse should be computed as the current that would leave a spherical electrode of 25μm radius if the electrode was placed in an infinite medium having resistivity R_e and energized by the battery voltage. Note that the battery voltage will be a function of time, so the stimulus current will be also.

Duration. Stimulus duration will be assumed to be the nominal value assigned by you. However, the design of the pacemaker to be used is "smart" and can determine when the stimulus will not be strong enough to cause an action potential (i.e., will not reach the threshold voltage before chronaxie). When the pacemaker so determines, it will change the duration of the stimulus so that the new stimulus duration is 2 times what it was before. If the stimulus is still insufficient, the stimulus duration will be doubled again, etc., until the duration is sufficient to make an action potential.

Find any stimulator design that meets the requirements above. A stimulus design consists of the specification of its battery capacity and its stimulus voltage,

41. Continue with the stimulator design started in the preceding exercise.

Cost. Two major features affect the cost of the pacemaker device you will design. These are battery capacity (in Joules), and stimulus current (specified in amperes). The cost is proportional to C_1 times battery capacity plus C_2 times stimulus current. (For simplicity in this exercise, both battery capacity and stimulus current are to be constants specified as part of the design, not functions of time or functions of other membrane parameters.)

The cost of the stimulus is related to the stimulus voltage because higher voltages require physically larger components. Not only are these more expensive in themselves, but more expensive surgical procedures are required for pacemaker implantation. To take all these costs into account, the pacemaker's cost is, for the purpose of this problem, made proportional to the initial stimulus voltage.

Assignment. Consider how to design the best stimulator (i.e., the one that has the lowest cost but still meets all the requirements). Use the specific cost equation given by the instructor, which will include values for constants C_1 and C_2. Determine its battery capacity and its stimulus voltage, and enter these in the space provided for them on the design sheet for your team.

Evaluation. Evaluation of your design will be by means of the following procedure:

At the deadline for receiving designs, the parameters of your design and its cost will be listed. After the deadline, the design team will no longer be allowed to make any further revisions in the design that was submitted.

The teacher (or assistants) may or may not independently check the design submitted. The teacher will accept suggestions from students in the class or other members of the teaching staff as to ways in which particular designs may have design errors.

Design Failures

Battery failure. This error will be considered to occur if, under any allowed sequence of events, the battery capacity is insufficient to power the stimulator for a full year.

Insufficient stimulus voltage. This failure will be considered to occur if, under any allowed sequence of events, the design is found to provide insufficient stimulus voltage to pace the nerve.

Rankings. Designs that have no failures will be ranked in order of cost. Designs with failures will be ranked below those with no failures in an order determined by the instructor's judgment of overall quality.

13.8. EXERCISES, CHAPTER 8: EXTRACELLULAR FIELDS

Prologue: As instructed in the exercises that follow, use the following template function to define $V_m(x, t)$:

$$V_m(x, t) = b + a[\tanh(u_1) - \tanh(u_2)] \tag{13.4}$$

where u_1 and u_2 are

$$u_1 = s_1[(t - t1) - |x - x_0|/\theta] \tag{13.5}$$

$$u_2 = s_2[(t - t2) - |x - x_0|/\theta] \tag{13.6}$$

Parameter values are given in Table 13.7.

Table 13.7. Action Potential Template

Name	Value	Units	Description
t_1	2	milliseconds	upstroke center time delay
t_2	5	milliseconds	downstroke center time delay
x_0	0	millimeters	site of excitation origin
θ	4	mm/ms	speed of propagation
s_1	2	ms^{-1}	rate of upstroke
s_2	0.5	ms^{-1}	rate of downstroke
a	50	millivolts	AP amplitude
b	-60	millivolts	AP baseline

The radius of the fiber is 50 μm. The fiber's axis lies along the x axis and extends a long distance on both directions. The fiber membrane resting resistance is $R_m = 1500\ \Omega\text{cm}^2$. R_i is 1000 Ωmm, and R_e is 400 Ωmm. The extracellular space is unbounded.

In the exercises below, for simplicity include only the portion of the fiber for which $x > 0$, unless otherwise instructed. In all plots, include a calibrated vertical axis, and be sure the plot size is large enough to allow reading the magnitude of every extrema in the curve. If tabulating V_m, use about 20 points per millimeter (spatial) or 20 points per millisecond (temporal) to have enough resolution to compute derivatives accurately. Answering many of the questions without undue tedium requires a computing environment that allows both calculation and plotting.

Comment for Students: Past experience suggests that no single step in answering these exercises is especially difficulty, yet often students observe that finding the answers takes longer and is much more confusing than it seems at first that it should. Sometimes one has the feeling of never quite being in control of all the different parts of the question and how the parts come together. In part the intrinsic difficulty occurs because one has to evaluate changes in space and changes in time, which are linked but different. Additionally, usually there are a series of steps that have to join together in just the right way. Finally, the actual amount of computation often is greater than at first is apparent, especially when many input values combine to produce a single point on an output curve. The best strategy is to think about and write out the underlying mathematics (not just the computational procedure), to follow the math systematically, and to be patient.

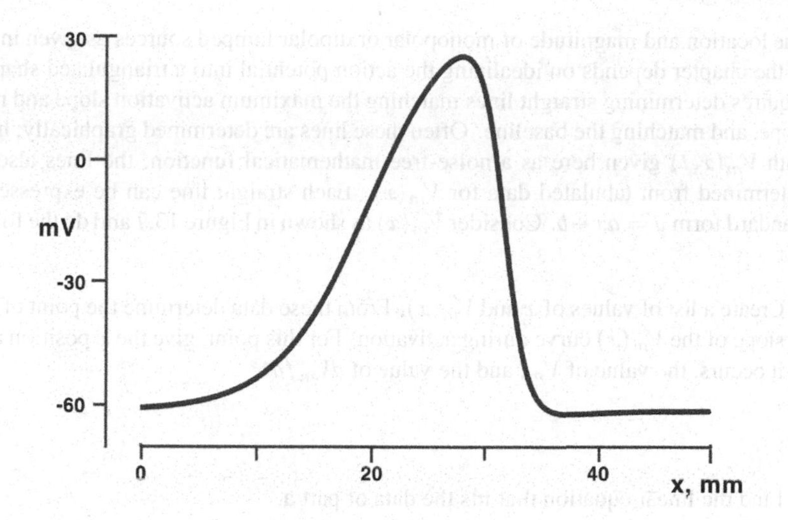

Figure 13.7. Action potential as a function of distance x along the fiber. The triangle was made by overlaying the action potential with three lines, one for the baseline and the other two at peak activation and repolarization slope.

Exercises 1–6 involve $V_m(t)$ and $V_m(x)$. These exercises compare the one to the other in several ways, and also allow practice in manipulating the tabulated data of each. Completing these exercises requires a computer system to tabulate the function and make the plots.

1. Inspect $V_m(x)$ in Figure 13.7. This figure plots the function defined in the prologue. For what time t does this plot result? Estimate by looking at the equations in the prologue and verify by making a confirming plot.

2. Plot $V_m(t)$ at $x = 24$ millimeters. Make the horizontal axis extend from zero to 20 milliseconds.

3. In a few sentences, explain the differences seen in the wave shape of $V_m(x)$ as compared to that of $V_m(t)$, as plotted.

4. Plot $\dot{V}_m(t) = dV_m(t)/dt$ at $x = 24$ millimeters, from zero to 20 milliseconds, and give the number value for $\dot{V}_m^{\max}$, the peak value of the time derivative. (The use of the dot notation for time derivatives is customary in electrophysiology.)

5. The location and magnitude of monopolar or dipolar lumped sources as given in the text of the chapter depends on idealizing the action potential into a triangulated shape. That requires determining straight lines matching the maximum activation slope and recovery slope, and matching the baseline. Often these lines are determined graphically; however, with $V_m(x,t)$ given here as a noise-free mathematical function, the lines also can be determined from tabulated data for $V_m(x)$. Each straight line can be expressed in the standard form $y = ax+b$. Consider $V_m(x)$ as shown in Figure 13.7 and do the following:

   a. Create a list of values of x and $V_m(x)$. From these data determine the point of greatest slope of the $V_m(x)$ curve during activation. For this point, give the x position at which it occurs, the value of V_m, and the value of dV_m/dt.

   b. Find the linear equation that fits the data of part a.

   c. Find the same data as in a, for the point of greatest slope during recovery.

   d. Find the linear equation that fits the data of part c.

   e. Find the point of intersection of the two lines. Specifically, find the x and V_m coordinates of the point of crossing.

   f. Find the x coordinates of the points where the activation line and the recovery line cross the baseline (–60 mV).

   You may be able to answer these questions with purely analytical methods instead of numerical ones.

6. Make the value of θ twice that given in the Table 13.7 for the template function for $V_m(x,t)$ of the prologue. (Although $V_m(x,t)$ here is given as a mathematical function, measured action potentials show similar kinds of changes.) Then answer each of the following. With θ doubled:

   a. Find and plot $V_m(x)$ for 10 msec.

b. Compare the waveform from part A to that of exercise 1. In a sentence or two, describe the differences observed.

c. Find and plot $V_m(t)$ for the same position used in exercise 2.

d. Compare the waveform from part C to that of exercise 2. In a sentence or two, describe the differences observed.

In Exercises 7–13, begin with the action potential template as defined in the prologue.

7. Plot $I_i(x)$ at $t = 10$ msec on a horizontal axis from 0 to 50 millimeters. One way to do this exercise is to differentiate $V_m(x)$ analytically and find $i_m(x)$ numerically; another is to differentiate $V_m(x)$ numerically.

8. Plot $i_m(x)$ at $t = 10$ msec on a horizontal axis from 0 to 50 millimeters.

9. Interpret the results of Exercises 7 and 8 in terms of the locations and relative magnitudes of monopolar sources and sinks. Draw two horizontal lines and label them 0 to 50 mm. (a) Over one, place arrows indicating distributed dipole sources. (b) Over the other, place plus and minus signs to show the source and sink distribution. In each case, a nice touch is to make the source larger where its magnitude is greater.

In Exercises 10–13, find extracellular potentials by using the distributed monopole sources (i.e., not lumped sources).

10. For $t = 10$ msec, find the extracellular potential Φ_e. Include units.

a. At the point $x = 34$, $y = 0.100$ millimeters.

b. At the point $x = 34$, $y = 1.00$ millimeters.

11. Plot $\Phi_e(x)$ [note: function of distance] along a line parallel to the fiber axis but distance h away. Let the horizontal axis extend from –50 to 50 millimeters.

    a. For $h = 100$ micrometers.

    b. For $h = 1000$ micrometers.

12. Plot $\Phi_e(t)$ [note: function of time] for a point at $x = 24$, $y = h$ mm. Let the horizontal axis extend from 0 to 25 msec.

    a. For $h = 100$ micrometers.

    b. For $h = 1000$ micrometers.

13. In a few sentences, explain the relationship between the solutions of Exercises 10 and 11.

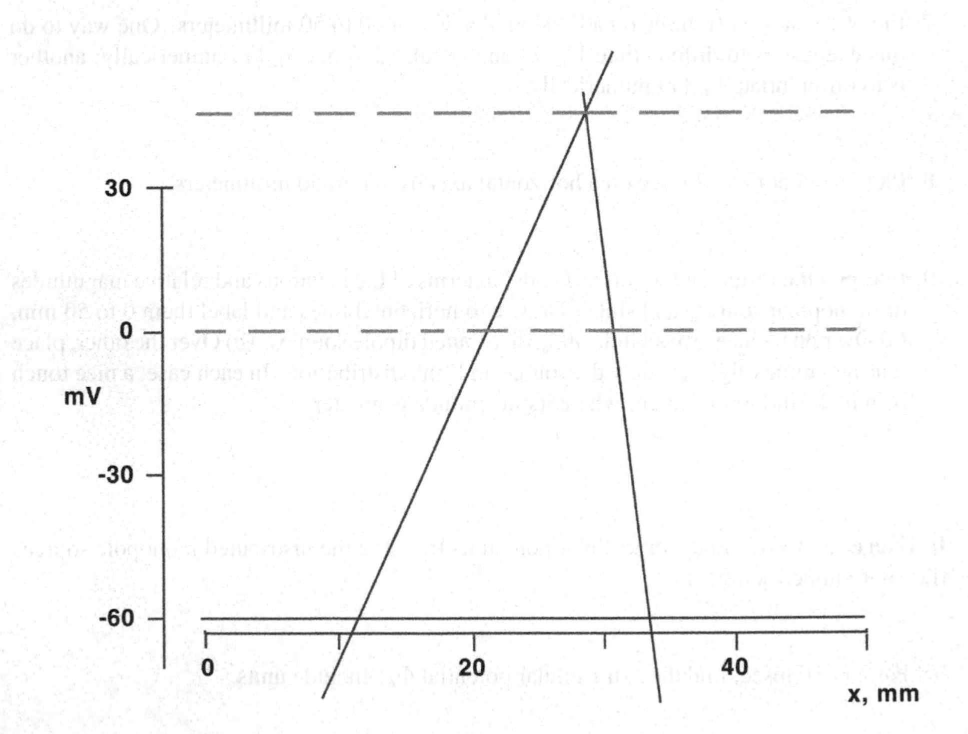

Figure 13.8. Action potential as a function of distance x along the fiber. The action potential was constructed using the template function. This figure is a triangulation of Figure 13.7.

For Exercises 14–21 use Figure 13.8, which is a triangulated version of Figure 13.7.

In Exercises 14–17, use lumped monopolar sources.

14. Carefully examine the triangulated action potential as shown in Figure 13.8 and give a value for each of the following quantities, with units. As these values are used in subsequent exercises, you may wish to enlarge the figure so as to measure it more precisely. [Because there is a template function for V_m, another possibility for answering each part is to examine the data for a digitized version of the action potential, to locate the points of maximum slope during the activation and recovery phases, and to find the points of intersection of the lines thereby defined. (See Ex. 5 above.) The figure then can be used to check the answers obtained analytically.]

    a. Activation slope $A_{\max}$.

    b. Recovery slope $B_{\max}$.

    c. Peak-to-peak voltage of triangularized AP, V_{pp}.

    d. Width w_a.

    e. Width w_r.

    f. Points of intersection x_1, x_2, x_3.

15. Determine the strength and location of each of the three monopole lumped sources. Be sure to include the location (x coordinate) magnitude, sign, and units of each one.

16. Computations of extracellular potentials from membrane currents can be placed into a matrix format:

$$[P_e] = [H][I_m] \tag{13.7}$$

Here matrix $[I_m]$ is a column vector containing the values of lumped sources 1 to 3 for time t. Matrix $[H]$ is set of coefficients by which each source must be multiplied to find one of the extracellular potentials. $[P_e]$ is a column vector giving the set of extracellular potentials. Matrices $[H]$ and $[P_e]$ have one row for each extracellular potential to be found. Suppose there are three extracellular potentials to be found, at three field points. The three field points have x positions 30, 31, and 32 mm. They are located at a distance h away from the fiber axis.

a. Define each of the 9 elements of $[H]$ defined in terms of r_{ij}, where r_{ij} is the distance from field point i to monopolar source j.

b. Find matrix $[H]$ (numerical values) for $h = 100$ micrometers.

c. If the monopole sources have values of 3, –4, and 1 mA, respectively, what are the values of the extracellular potentials at each of the 3 field points?

17. Plot $\Phi_e(x)$ along a line parallel to the x axis at distance $y = h$ with $z = 0$ (i.e., plot $\Phi_e(x)$, as it is generated at distance h away from the fiber axis, from the three lumped monopole sources). Let the horizontal axis extend from $x = 0$ to $x = 50$ millimeters. Calibrate the vertical axis, and make each plot large enough that the magnitudes of the peaks can be read from the graph. Compare to the plots for $\Phi_e(x)$ as determined from the distributed sources, as in Ex. 11. Make plots for:

a. $h = 100$ micrometers.

b. $h = 1000$ micrometers.

c. $h = 10000$ micrometers.

In Exercises 18–20, use lumped dipolar sources.

18. What are the magnitude, location, orientation of:

a. a single dipole that lumps activation?

b. a single dipole that lumps recovery?

19. Plot $\Phi_e(x)$, as it is generated from the two lumped dipoles that were found in Ex. 17. Determine values along a line parallel to the x axis at $y = h$ millimeters and $z = 0$ (i.e., plot $\Phi_e(x)$ along a line parallel to the fiber axis, but distance h away from the axis). Make the plots from $x = 0$ to $x = 50$ millimeters. Calibrate the vertical axis, and make the vertical size large enough that the magnitudes of each peak can be read from the graph.

 a. For $h = 100$ micrometers.

 b. For $h = 1000$ micrometers.

20. Plot $\Phi_e(t)$, as it is generated from the dipole sources, for a point at $x = 12$, $y = h$ mm. Let the horizontal axis extend from 0 to 25 msec [note: function of time].

 a. For $h = 100$ micrometers.

 b. For $h = 1000$ micrometers.

Extensions: Exercises 21–24 involve questions that can be answered through straightforward extensions of the material examined so far.

21. Collision: For this exercise, modify the description in the prologue so that action potentials are correctly generated as functions of time and space if the fiber described in the prologue is simultaneously stimulated at $x = \pm 100$ mm (leading to a collision of excitation waves near $x = 0$). Note that θ is 4 mm/msec for the fiber.

 a. Plot $V_m(t)$ at $x = \pm 8$ mm and $x = 0$ mm.

 b. Plot $V_m(x)$ at $t = 20$ ms and $t = 24$ ms.

 c. Plot $I_i(t)$ at $x = \pm 8$ mm and at $x = 0$ mm.

 d. Plot $\phi_e(t)$ at $x = \pm 8$ mm and at $x = 0$ mm.

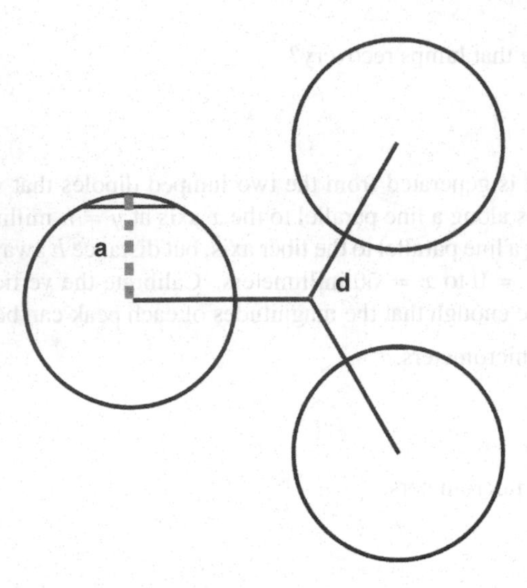

Figure 13.9. A monophasic action potential is approximated as triangular; the latter is described above. Along the horizontal axis, x is given in millimeters. The fiber extends well beyond the region shown and is cylindrical with a radius of 50 micrometers. The intracellular and extracellular resistivities are 90 and 30 Ωcm, respectively. The action potential is propagating along the fiber with a velocity of 4 m/sec.

22. Many times measurements of nerves and muscles show the effects of multiple simultaneously active fibers. Here suppose three long cylindrical fibers of the kind described in the prologue have parallel axes and are arranged at equal angles around a central axis, within a large volume conductor, as illustrated in Figure 13.9. The axis of each fiber is $d = 75$ micrometers from a mathematical axis at the center of the group. Fiber 1 is the AP of the prologue, while fibers 2 and 3 are delayed by 3 and 6 milliseconds, respectively. All three fibers have the same propagation velocity. What is $\Phi_e(t)$ at a point on the central axis? To make the question easier, suppose there is no interaction among the fibers, so that the observed waveform is the summation of the potentials from each fiber found as if that fiber were in the volume alone.

23. Examine Figure 13.10. What is the corresponding plot of $\Phi_e(x)$ at a distance of 1 mm from the axis of this fiber? Assume the fiber parameters (though not the action potential) are those given in the prologue.

24. Examine Figure 13.11. Assume the peak of this fiber is at $x = -100$ mm at $t = 0$ ms.

    a. Plot $\Phi_e(t)$ at $x = 0$ mm at a distance of 1 mm from the axis of the fiber.

    b. Make a table that shows the magnitude and timing of every extrema (positive or negative peak) of the waveform plotted in part A.

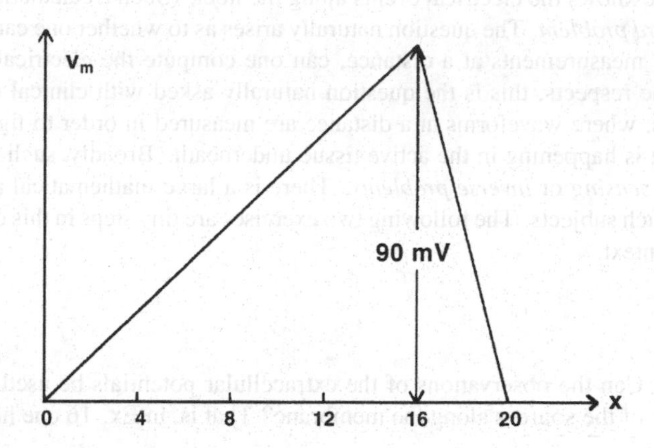

Figure 13.10. A monophasic action potential is approximated as triangular; the latter is described above. Along the horizontal axis, x is given in millimeters. The fiber extends well beyond the region shown and is cylindrical with a radius of 50 micrometers. The intracellular and extracellular resistivities are 90 and 30 Ωcm, respectively. The action potential is propagating along the fiber with a velocity of 4 m/sec.

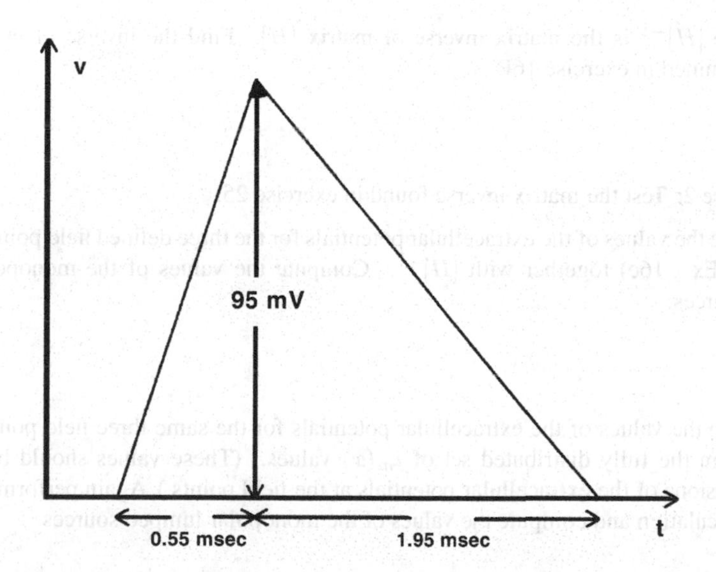

Figure 13.11. This propagating action potential is approximated with a triangular waveform, as shown. The axon on which it is propagating has a diameter of 45μm. Propagation is in the $+x$ direction with a velocity of 5 m/sec. The extracellular conductivity $\sigma_e = 0.05$ S/cm, while the intracellular conductivity is $\sigma_i = 0.01$ S/cm. (The extracellular medium is unbounded.) Assume that the same $V_m(t)$ occurs at every position, offset in time.

The text and exercises above have shown that potentials at points away from the fiber can be found if one knows the electrical events along the fiber. (Such a calculation is sometimes called a *forward problem*. The question naturally arises as to whether one can do the reverse. That is, given measurements at a distance, can one compute the electrical sources of the fiber? In some respects, this is the question naturally asked with clinical or experimental measurements, where waveforms at a distance are measured in order to figure out, as best possible, what is happening in the active tissue underneath. Broadly, such calculations are called *remote sensing* or *inverse problems*. There is a large mathematical and engineering literature on such subjects. The following two exercises are tiny steps in this direction, within the present context.

25. Inverse 1: Can the observations of the extracellular potentials be used to compute the magnitude of the sources along the membrane? That is, in Ex. 16 one had

$$[P_e] = [H][I_m] \tag{13.8}$$

so by inverting matrix $[H]$ one gets

$$[I_m] = [H]^{-1}[P_e] \tag{13.9}$$

where $[H]^{-1}$ is the matrix inverse of matrix $[H]$. Find the inverse of matrix $[H]$ as determined in exercise 16B.

26. Inverse 2: Test the matrix inverse found in exercise 25.

a. Use the values of the extracellular potentials for the three defined field points (as found in Ex. 16c) together with $[H]^{-1}$. Compute the values of the monopolar lumped sources.

b. Use the values of the extracellular potentials for the same three field points as found from the fully distributed set of $i_m(x)$ values. (These values should be improved versions of the extracellular potentials at the field points.) Again perform the inverse calculation and compute the values of the monopolar lumped sources.

c. In a few sentences, comment on the differences (if any) in the solutions to parts a and b.

Exercises 27–30 extend the theory developed in the text.

27. Evaluate the discontinuity in the electric field and its normal derivative along the axis of a single-layer disc of radius a. [Follow the method developed for a double-layer disc.] Describe the membrane source of an active cell as the sum of a single- plus double-layer component. (Note that these are true and not equivalent sources, as they give both intracellular and extracellular fields.)

28. Change a and b in the derivation in the text for the potential from a cell. Rather than (8.62) use

$$a = 1/r \qquad b = \Phi \tag{13.10}$$

What is the new equation for ϕ_e that results?

29. Lone Cell's Potentials: A lone cell is immersed within an medium of infinite extent. The conductivity inside the cell is 0.005 S/cm, and the conductivity outside the cell is 0.02 S/cm. The cell has a cross-section that is square (on the x, y axes) to an excellent approximation, with an edge length of 15 micrometers on the sides of the square. The cell length (along z) is 100 micrometers. The cell is centered at the coordinate origin.

Suppose the cell depolarizes sequentially in the $+z$ direction. At the moment of interest the cell is in the process of becoming depolarized. At the more depolarized end (the portion with z negative), the intracellular potential is 0.02 V. At the less depolarized end (the portion with z positive) the intracellular potential is –0.06 V. The intracellular potential changes linearly, to a good approximation, from one end to the other. At all points around the cell, approximate the potential just outside the cell as 0 V.

There are two extracellular electrodes. Electrode A is located at coordinates $(0, 0, 0.01)$ cm, while electrode B is at $(0, 0, -0.01)$ cm. What is the voltage between the two extracellular electrodes?

Voltage polarity: the voltage is positive when the potential at electrode A is greater than the potential at electrode B.

30. Use the results of the preceding exercise to approximate ϕ_e just outside the cell. Then use these values to correct the calculation of V_{AB}. What is the peak magnitude of the correction?

31. Design Problem: Extracellular Detection Design

Overview. A cylindrical fiber of radius a is excited at both ends simultaneously. The fiber is more than 100λ long. Excitation propagates toward the center. At any point along the fiber, the time course of the action potential's upstroke has the shape of

$$V_m(t) = 50\tanh(t) \tag{13.11}$$

where V_m is in mV and t is in msec. Of course, the absolute timing of the upstroke depends on the x position along the strand.

The cylindrical fiber is located in a conducting medium of infinite extent. At the center of the fiber ($x = 0$), the positive electrode of a voltmeter is placed on the outer surface of the fiber, and at the same x position the negative electrode is placed at a distance of $10a$ from the axis of the fiber. Voltage Φ_e is measured between these two electrodes.

The electrodes will be used to monitor Φ_e and thereby detect when excitation reaches the center. Since little noise is present, a simple signal detection plan will be used. Specifically, the monitor will decide that excitation has reached the fiber's center when the detected voltage reaches 3 mV.

Other Information. The fiber has $R_m = 10^3\ \Omega\text{cm}^2$, intracellular conductivity 0.02 S/cm, and extracellular conductivity 0.04 S/cm.

An experimental trial under the conditions described in this design found that measured velocity θ was 0.4 m/sec when radius a was 20 μm.

Cost. The cost of the fiber is \$1 per micron times the radius of the fiber.

Objective. Design a fiber (by selecting its radius) that minimizes the cost while meeting the signal detection criterion.

Supporting Information. To provide confidence in your design, supply on the answer sheet a calibrated graph that gives Φ_e as a function of time.

13.9. EXERCISES, CHAPTER 9: CARDIAC ELECTROPHYSIOLOGY

1. Cardiac electrophysiology (as presented in this chapter) has some aspects in common with the electrophysiology of nerve (as presented in earlier chapters), but also has substantial differences. Which of the following statements are correct?

   a. Nerve analysis often deals with membrane excitation in one spatial dimension, whereas cardiac often deals with excitation in two or three spatial dimensions.

   b. Nerve analysis includes electrically active membrane (generates action potentials), whereas cardiac membrane is passive (generates no action potentials).

   c. Nerve cells often are much longer than are cardiac cells.

   d. Potentials around nerve can be measured outside nerve cells, whereas potentials outside the heart are too small to be measured.

   e. Nerve cells generate action potentials through movement across the membrane of sodium and potassium ions, as well as other ions, and action potentials in cardiac cells also involve the movement of sodium and potassium, as well as others.

2. Cardiac structure: Which statements are correct?

   a. Human hearts have four chambers.

   b. Action potentials, an electrical event, trigger contraction, a mechanical event.

   c. Excitation in the ventricles is more nearly apex to base than base to apex.

   d. The base of the ventricles is the part that adjoins the atria.

e. In a normal heart, blood flow is principally from the ventricles to the atria.

f. Each of the cardiac chambers initiates its own excitation.

g. Connexons serve to join the ventricles to each other.

h. Gap junctions serve mainly to replace missing cells.

i. The SA node is a specialized region of the left ventricle.

3. Figure 9.1 shows the atrioventricular node as one part of the overall cardiac structure. Compared to the rate in the atria, excitation moves through the AV node:

a. just slightly slower.

b. much more slowly.

c. about the same speed.

d. slightly faster.

e. much faster.

4. Cardiac action potentials: Please refer to Figure 9.2 and the associated text. Check the responses that are correct. The terminology is standard for action potential description.

a. Phase 0 corresponds to activation.

b. Phase 1 corresponds to rapid recovery.

c. Phase 2 corresponds to the plateau.

d. Phase 3 corresponds to recovery.

e. Phase 4 corresponds to rest or slow depolarization.

f. All action potentials have all 5 phases.

g. The two action potentials in parts a and b extend across the same range, peak to peak.

h. Phase 1 corresponds to activation.

i. It is significant that thc sinus node action potential has a rising baseline.

5. Cardiac cell connections. Please refer to Figure 9.3 and the associated text. To how many cells does cell A connect?

6. Intercellular junctions: Please consider Figure 9.4 and the associated text. Suppose the channel connecting one cell to the next is cylindrical. Suppose it has the diameter given in the text and the length given in the figure. Suppose the fluid in the channel has a resistivity 100 Ωcm. What is the resistance from one cell to the next?

7. Isochrones: Isochrones of ventricular excitation are shown in Figure 9.12. Each isochrone is marked by a number. By what factor must the isochrone number be multiplied to convert it to a time, in seconds?

Figure Precision. A number of the questions that follow ask for values to be read from Figure 9.10 in the text. At the printed size here, the figure may be too small to read the calibration accurately, so large variances are given in the answer key. However, the experimental uncertainty is much less. Here one can achieve better resolution by enlarging the figure (perhaps with a photocopier) so as to better see the size of calibration marks, or (even better) by consulting the figures in the original sources, as cited.

In the next two exercises, refer to Figure 9.10 and respond based on those data.

8. Time of first: Examine the sequence of excitation. The time of excitation for the first lead to be excited is what? (Assume scale starts at time zero.)

9. Time of last: The time of excitation for the last lead to be excited is what? (Assume scale starts at time zero.)

10. Examine Figure 9.5 showing stimulation of a Purkinje strand. Check each box that is correct.

    a. Part b of the figure shows that membrane capacitance plays little role in the response.

    b. Part a of the figure involves the length of the fiber affected by a point stimulus.

    c. Part b of the figure involves time of response to the stimulus.

    d. Part b of the figure shows the amount of inductance in the fiber loop.

    e. Part a of the figure shows the temperature where effects are maximal.

11. Membrane resistance: A cell is 100 μm long. It has a square cross-section with a side of 8 μm. Excluding the ends, what is the resistance between the inside and outside of the cell if the membrane has R_m of 10,000 Ωcm^2?

12. Magnitude of v_m: By inspecting a suitable figure, estimate the peak-to-peak magnitude of a cardiac action potential.

In Exercises 13–17, consult Figure 9.10 showing waveforms from a plunge electrode across the cardiac wall.

13. Velocity in the ventricular wall: Assume the electrode is perpendicular to the direction of propagation. Evaluate the recordings at a site that is a good approximation to outward, uniform propagation. Estimate the outward velocity of propagation,

14. Wave thickness: Assume the electrode is perpendicular to the direction of propagation. Evaluate the recordings at a site that is a good approximation to outward, uniform propagation. Estimate the thickness of the excitation wave.

15. Extracellular potential: Assume the electrode is perpendicular to the direction of propagation. Evaluate the recordings at a site that is a good approximation to outward, uniform propagation. Estimate the magnitude of the change in extracellular potential from the leading edge to the trailing edge of the excitation wave.

16. Unipolar magnitude: What is the largest voltage magnitude, relative to baseline, observed on any unipolar recording?

17. Bipolar magnitude: What is the largest voltage observed on any bipolar recording?

18. Differential equation: Two ventricular cells are connected by junctional resistance q. The cells are surrounded by a large, common extracellular volume of fluid that has a near-

zero potential. Both cells have membrane resistance p and capacitance c for the cell as a whole. The first cell has transmembrane potential u, and a constant voltage well above baseline. The second has transmembrane potential v, which is zero at time $t = 0$. Because of the transmembrane potential difference, potential v will rise as time passes. Write the differential equation for v in the form

$$dv/dt + av = b$$

where a and b are constants. For the questions below, determine the constants in the expression. Define the ratio $R = p/(p+q)$ and use R in the expressions requested if the ratio is present.

a. What is constant a?

b. What is constant b?

19. Solution to DifEQ: Again consider two ventricular cells connected by junctional resistance q, as in the preceding exercise. Note that the second has transmembrane potential v, which is zero at time $t = 0$. Because of the transmembrane potential difference, potential v will rise as time passes. Write the differential equation for v, and solve the equation. Write the solution in the form

$$v = a[1 - \exp(-t/b)]$$

where a and b are constants. Enter the expression required by each of the parts that follow. Define the ratio $R = p/(p+q)$ and use R in the expression if the ratio is present.

a. What is constant a?

b. What is constant b?

20. Purkinje fibers are (select those that apply):

a. located on the epicardium.

b. nerve cells.

c. conduct action potentials.

d. contract when stimulated.

e. modified cardiac cells.

21. Connexon selectivity: When current flows from one cell to another through gap junctions, which charged particles carry the charge? Select all eligible.

a. K+ ions.

b. Na+ ions.

c. Cl- ions.

d. Ca++ ions.

e. Charged DNA.

22. Consider a cardiac cell that has the shape of a brick. The cell lies within an extensive extracellular medium that remains near 0 Volts. The cell has capacitance and resistance with respect to the extracellular medium, and an intracellular connection between the two, within an extensive extracellular medium that remains near 0 Volts. The cell has capacitance and resistance with respect to the extracellular medium. Suppose the cell has the following characteristics.

a. The cell length is 100 μm.

b. Both cell width and height are 10 μm..

c. C_m is 1E-6 F/cm^2..

d. R_m is 10,000 Ωcm^2.

e. The cell's threshold voltage is 0.01 Volts.

Which of the following statements apply?

a. At threshold, the required charge is 4.2E-13 Coulombs.

b. The cell's volume is 1E-8 cm^3.

c. The cell's surface area is 4.2E-5 cm^2.

d. The resistance (of the whole cell) across the membrane is 2.381E8 ohms.

e. The capacitance of the cell is 4.2E-11 Farads.

The next several exercises assume the following context. One or more cardiac cells are within an extensive extracellular medium that remains near 0 Volts. Each cell is shaped like a brick, with square cross-section. The cells have capacitance and resistance with respect to the extracellular medium. When the number of cells is greater than one, the cells have the same dimensions and are side by side, and there are intracellular connection between them. When the membrane resistance is not given, the resistance is assumed large enough that its presence may be ignored. Conversely, when a membrane resistance is given, it should be taken into account. The membrane capacitance is 1E-6 Farads per cm^2. In some exercises an intervention is applied to the first cell, and the question asks about the second or subsequent cell. In such cases, assume that the transmembrane voltage of the second (or subsequent) cell is initially zero (relative to rest).

23. Suppose the cell length is 100 μm, and an edge of the cross-section is 8 μm long. Suppose a heart has a volume equal to that of a cube of length 8 cm on a side. How many cells are in the heart? Estimate the number of cells as the number required to make the aggregate cell volume equal to 80% the heart volume.

24. Picture a number of cells each having cell length 100 μm and edge length 42 μm. Suppose cells are placed side by side. Suppose excitation moves in the transverse direction, and the transverse velocity is 10 cm/sec. On the average, how much time is required for excitation to move from one of these cardiac cells to the next?

25. Rise of v_m^2: The capacitance of each cell is 7.1E-11 Farads. Suppose there is a constant current of 2.032E-8 Amperes along the intracellular path between the first and the second cell. The current lasts for 1.4E-4 seconds. By how much does v_m of cell 2 rise?

26. Junction R: The capacitance of each cell is 7.65E-11 Farads. The cell length is 120 μm. Both cells have a square cross-section, with 15 μm edges. There is a constant voltage difference between the first and second cells of 0.05 Volts. That is, the transmembrane voltage of the first cell rises along with that of the 2nd cell, so as to maintain this voltage difference. The current lasts for a time equal to the average transverse cell-to-cell propagation time, for a transverse velocity of 10 cm/sec. What is the necessary junctional resistance for the voltage in a second cell to rise from zero to a triggering level of 0.04 Volts in the time available?

27. Cell 2 v_m with lower R_m: Two brick- shaped cells have length 100 μm and a square cross-section of edge length 10 μm. Both cells have C_m of 1E-6 F/cm^2 and R_m of 500 Ωcm^2. Cell 1 has transmembrane voltage 0.1 V. At time $t = 0$, cell 2 has voltage v of 0 Volts. Cell 2 is connected to cell 1 through a junctional resistance of 10,000,000 Ω. Cell 2 remains passive. At the end of time 1E-4 sec, what is its transmembrane voltage?

28. Cell 2 v_m with higher R_m: Two brick- shaped cells have length 100 μm and a square cross-section of edge length 10 μm. Both cells have C_m of 1E-6 F/cm^2 and R_m of 10,000 Ωcm^2. Cell 1 has transmembrane voltage 0.1 V. At time $t = 0$ cell 2 has voltage v of 0 Volts. Cell 2 is connected to cell 1 through a junctional resistance of 10,000,000 Ω. Cell 2 remains passive. At the end of time 1E-4 sec, what is its transmembrane voltage?

29. Cell 2 time to threshold: Two brick-shaped cells have length 100 μm and a square cross-section of edge length 10 μm. Both cells have C_m of 1E-6 F/cm^2 and R_m of 1,500 Ωcm^2. Cell 1 has transmembrane voltage 0.1 V. At time $t = 0$ cell 2 has voltage v_m of 0 Volts. Cell 2 is connected to cell 1 through a junctional resistance of 10,000,000 Ω. Cell 2 remains passive until it reaches its voltage threshold, which is 0.04 Volts. How long is required for cell 2 to reach its threshold?

30. Nominal velocity: Two brick-shaped cells have length 80 μm and a square cross-section of edge length 10 μm. Both cells have C_m of 1E-6 F/cm^2 and R_m of 20,000 Ωcm^2. Cell 1

has transmembrane voltage 0.1 V. At time $t = 0$ cell 2 has voltage v_m of 0 Volts. Cell 2 is connected to cell 1 through a junctional resistance of 2,000,000 Ω. Cell 2 remains passive until it reaches its voltage threshold, which is 0.04 Volts. The "propagation time" for cell 2 is the time required to reach this threshold. What is the average transverse propagation velocity from cell 1 to cell 2? Estimate the average transverse velocity as the cell's edge length divided by cell 2's propagation time.

31. Velocity variations: In this exercise examine changes in velocity that result from each of a series of variations in cell parameters. In each part, estimate the velocity if other parameters the estimation procedure remain the same as in the preceding question, except for the value specified.

    a. Increase cell size. Cell length is 160 μm and edge length 20 μm.

    b. Decrease R_m to 1,000 Ωcm^2.

    c. Increase junctional resistance to 4,000,000 Ω.

    d. Decrease voltage threshold to 0.02 Volts.

32. Which (one or more) of the following increases the velocity? Assume that other factors remain constant, and that each item is changed alone.

    a. Increasing membrane resistance R_m.

    b. Increasing junctional resistance R_j.

    c. Increasing the voltage threshold V_t.

    d. Increasing the temperature.

e. Increasing cell length L.

33. Math terms: A series of statements about math operations "del" (∇) and "del dot" ($\nabla\cdot$) follow. Check the statements that are true.

a. In the heart, current has a direction.

b. The equation $V = r$, where r is the 3D distance from the origin, is a scalar function.

c. The math operation "del dot" can be used on V, if $V = r$.

d. Geometrically, the "del" and "del dot" functions refer to the same thing.

e. In electric fields, the potential function is usually a vector.

f. In typesetting, the upside-down triangle is called "nabla."

g. If the current is described as a vector function, then the "del dot" operation is larger at points where the current originates.

h. The math operation "del" can be used on V, if $V = r$.

i. Given a vector function, one can use either "del" or "del dot" on it.

34. Bidomain equations: Refer to the equations below. Then select those of the following statements that are true:

$$\overline{J}_i = g_{ix}\frac{\partial \Phi_i}{\partial x}\overline{a}_x + g_{iy}\frac{\partial \Phi_i}{\partial y}\overline{a}_y + g_{iz}\frac{\partial \Phi_i}{\partial z}\overline{a}_z \tag{13.12}$$

$$\overline{J}_e = g_{ex}\frac{\partial \Phi_e}{\partial x}\overline{a}_x + g_{ey}\frac{\partial \Phi_e}{\partial y}\overline{a}_y + g_{ez}\frac{\partial \Phi_i}{\partial z}\overline{a}_z \tag{13.13}$$

$$-\nabla \cdot \overline{J}_i = \nabla \cdot \overline{J}_e = I_v \tag{13.14}$$

$$g_{iz} = \sigma_i F \quad \text{and} \quad g_{ez} = \sigma_e(1-F) \tag{13.15}$$

a. The "divergence" operation on (13.12) will locate where current enters or leaves the intracellular space.

b. The bar over the J on the left of (13.12) and (13.13) indicate that these are vector functions.

c. Use of the "del" math operation on (13.12) and (13.13) is sensible.

d. Equation (13.12) is about the intracellular current.

e. Equation (13.13) is about the extracellular current.

f. The equality in (13.14) implies that currents in the intracellular and extracellular regions are equal and opposite.

g. Equation (13.15) says that voltages in the bidomain are less than those in the actual tissue.

Context for Exercises 35–42. Consider a three-dimensional grid of spherical cells, each surrounded by an interstitial region. The cells all have radius a and membrane resistance R_m.

Assume that the interstitial space has, on average, the same volume as that left between a spherical cell and a surrounding cube of edge length $2a$. For the physical (as distinct from bidomain) regions, the extracellular resistivity is R_e, and the intracellular resistivity is R_i. A unipolar electrode injects current at the coordinate origin. (Unipolar injection means that in concept current is injected at one point and flows in all directions to a current sink a long distance away.)

35. g_i: For $a = 12\ \mu$m, $R_m = 8,000\ \Omega$cm^2, $R_e = 100\ \Omega$cm, and $R_i = 200\ \Omega$cm. What is the intracellular conductivity g_i?

36. F: For $a = 20\ \mu$m, $R_m = 8,000\ \Omega$cm^2, $R_e = 100\ \Omega$cm, and $R_i = 200\ \Omega$cm. What is fraction F?

37. β: For $a = 17\ \mu$m, $R_m = 8,000\ \Omega$cm^2, $R_e = 100\ \Omega$cm, and $R_i = 200\ \Omega$cm, what is β, the surface-to-volume ratio? Note that the volume used to find β is the total membrane surface area of all cells divided by the total volume, i.e., including both intracellular plus interstitial (extracellular) volumes.

38. ρ_e: For $a = 12\ \mu$m, $R_m = 8,000\ \Omega$cm^2, $R_e = 100\ \Omega$cm, and $R_i = 200\ \Omega$cm, what is the bidomain interstitial resistivity ρ_e?

39. λ: For $a = 8\ \mu$m, $R_m = 8,000\ \Omega$cm^2, $R_e = 100\ \Omega$cm, and $R_i = 200\ \Omega$cm, what is the space constant, λ?

In Exercises 40–47, use the following values: cells have radius $a = 10\ \mu$m, and membrane resistance $R_m = 20,000\ \Omega$cm^2. The resistivities are $R_i = 200\ \Omega$cm and $R_e = 50\ \Omega$cm. Use the bidomain model for analysis.

40. λ: In this 3D cellular structure, what is λ?

41. ϕ_e: If the amount of current introduced at the electrode is 0.002 Amperes, what is the extracellular potential at a point 0.5 cm from the site of current injection?

42. ϕ_e^2: If the amount of current introduced at the electrode is 0.002 Amperes, what is the extracellular potential at a point 0.2 cm from the site of current injection?

43. v_m: If the amount of current introduced at the electrode is 0.002 Amperes, what is the transmembrane potential at a point 0.1 cm from the site of current injection?

44. v_m^2: If the amount of current introduced at the electrode is 0.002 Amperes, what is the transmembrane potential at a point 0.2 cm from the site of current injection?

45. ϕ_i: If the amount of current introduced at the electrode is 0.002 Amperes, what is the intracellular potential at a point 0.1 cm from the site of current injection?

46. ϕ_i^2: If the amount of current introduced at the electrode is 0.002 Amperes, what is the intracellular potential at a point 0.2 cm from the site of current injection?

47. Two Electrodes: Two electrodes, separated by a distance of 0.5 cm, have been introduced into the interstitial space of the bidomain. The electrodes function as a source–sink pair. The amount of current introduced at the source electrode is 0.002 Amperes, and the amount of current introduced at the sink electrode is –0.002 Amperes. Consider the transmembrane voltages along the line connecting the two electrodes. What is the transmembrane potential at a point 0.1 cm from the source electrode? Use the bidomain model, and assume the solution is a linear combination.

48. Compare values of v_m, ϕ_i, and ϕ_e made in the vicinity of the point of unipolar injection of current I_o into a 3D bidomain of large extent. I_o is positive, and measurement position 1 is closer to the electrode tip than is measurement position 2. Consider the sign and the magnitude. Select the statements that are true.

a. v_m is more positive at the closer measurement position.

b. ϕ_i is more positive at the closer measurement position.

c. At the closer measurement position, ϕ_i is greater in magnitude than ϕ_e.

d. The magnitude of ϕ_e is larger at the closer measurement position.

e. The magnitude of v_m is larger at the closer measurement position.

f. At all positions, $v_m = \phi_i - \phi_e$.

g. The magnitude of ϕ_i is larger at the closer measurement position.

49. Design 1 OK? Imagine that a three-dimensional mesh of cardiac cells is composed of brick-shaped cells, each surrounded by an interstitial region. Cells have a square cross-section, with an edge E of 12 micrometers and a length L of 100 μm. Cells are (on the average) within a geometrical block that also is brick shaped, with dimensions 1 μm greater in L and E. The membrane resistance for a unit area is 20,000 Ωcm^2. For the physical (as distinct from bidomain) space, the extracellular resistivity is 100 Ωcm, and the intracellular resistivity is 200 Ωcm. A stimulus design calls for a stimulus current (unipolar electrode) of –0.002 Amperes. (An important simplification is this exercise is the use of a stimulus of long duration and resulting v_m values for steady state.) The design is judged successful if all of the following requirements are satisfied:

a. Requirement 1: In the region of the tissue with radii r of 0.1 and 0.2 cm, the magnitude of v_m is nowhere greater than 300 mV, so as to avoid tissue damage from electroporation.

b. Requirement 2: In the region of the tissue between r of 0.1 and 0.2 cm from the stimulus site, the magnitude of v_m is everywhere greater than 40 mV, so as to produce an action potential.

c. Requirement 3: The change in transmembrane voltage due to the stimulus, v_m, has a positive sign.

Select which of the following statements are true, and state whether the design is successful:

1. Requirement 1 is satisfied.

2. Requirement 2 is satisfied.

3. Requirement 3 is satisfied.

50. Design 2 OK? Reconsider the previous design when the membrane resistance has decreased to 2,000 Ωcm^2. Again select which requirements are satisfied, and whether the design is successful.

 1. Requirement 1 is satisfied.

 2. Requirement 2 is satisfied.

 3. Requirement 3 is satisfied.

Exercises 51–56: Coordinate systems used to represent cardiac activity within the human torso often are defined as follows. The origin is located at the center of the heart. The x axis is the direction between the left and right arms (positive toward the left arm). The y axis is perpendicular, in the direction from the feet toward the head. The z axis is perpendicular to the x and the y axes, and positive z is in the direction frontwards from the chest. Suppose, at a certain moment, a sheet of cardiac excitation in the right ventricle has the shape of a triangle. One corner is located near the base at $(0, 0, 1)$ cm. Another corner is near the cardiac apex, at $(4, 0, 0)$ cm. A third corner is located at $(0, 3, 0)$ cm. The direction of excitation is perpendicular to the surface, generally outward, roughly in the positive direction.

Potentials from the excitation wave are to be observed at a "precordial lead" located at point is located on the chest at $(0, 0, 2)$ cm.

Assume potentials can be found as those arising from such an excitation wave when it is within an infinite medium with extracellular resistivity 60 Ωcm. The transmembrane voltage is –90 mV prior to excitation and +30 mV after excitation. That is, it will be –90 mV on the leading edge of the excitation wave and +30 mV on the trailing edge. The intracellular resistivity is 200 Ωcm. About 80% of the cardiac volume (along the excitation wave) is intracellular.

For reference, standard results give:

$$\phi_e(P) = \frac{1}{4\pi F} \frac{\rho_e}{\rho_i + \rho_e} \Delta v_m \int_S \frac{\vec{a}_r \cdot \vec{a}_n}{r^2} \, dS \tag{13.16}$$

51. F: What is the value of parameter F?

52. Δv_m: What is the value of Δv_m as defined in the reference equation?

53. Ω: What is the value of the solid angle of the excitation wave, with respect to observer position?

54. ρ_i: What is the value of ρ_i?

55. ρ_e: What is the value of ρ_e?

56. ϕ_e: What is the value of ϕ_e?

ECG Values: Coordinate systems used to represent cardiac activity within the human torso often are defined as follows. The origin is located at the mid-chest level, in the center of the torso. The x axis extends in the direction from the right toward the left arm. The y axis extends perpendicularly from foot to head. The z axis extends outward from the chest.

Suppose that within such a coordinate system an electrode on the right arm has (x, y, z) coordinates $(-20, 4, 0)$, the left arm $(20, 4, -0)$, and an electrode on the left leg has (x, y, z) coordinates $(20, -50, 0)$.

Suppose that at a certain moment an excitation wave has the form of a sheet of area 25 cm^2, located in the LV base. The sheet is centered at (x, y, z) coordinates $(3, 0, -3)$ and the vector perpendicular to the sheet has components $(0, -1, 0)$.

Assume the excitation wave is located within an infinite medium having intracellular resistivity of 200 Ωcm and extracellular resistivity of 50 Ωcm. (Ignore conductivity boundaries.) The intracellular volume uses fraction 0.8 of the total space.

The polarized region of the LV has a transmembrane voltage of –0.08 Volts, and the depolarized region behind the excitation wave has a transmembrane voltage of 0.04 V.

Estimates of voltages at points on the body surface based on information about excitation within the heart may be based on the following equation, which includes the most fundamental aspects of the cardiac excitation wave, and the pertinent geometry:

$$\phi_e(P) = \frac{1}{4\pi F} \frac{\rho_e}{\rho_i + \rho_e} \Delta v_m \int_S \frac{\vec{a}_r \cdot \vec{a}_n}{r^2} \, dS \tag{13.17}$$

It is important to keep in mind some substantial simplifications that use of the above equation, in the context of electrocardiography, implicitly adopts. Among these limits are the relatively simple form used to represent the cardiac excitation wave, the use of an infinite conductive medium for the volume conductor (thus avoiding taking into account the limited conductive region of the torso), and the omission of homogeneity boundaries within the torso (such as the lungs). These approximations have a substantial effect on the computed voltages, so they are not good estimates of values that would be measured. Nonetheless, they do provide insight into some of the underlying determinants of ECG voltages, and the relative potentials seen in one circumstance as compared to another.

57. LL: Estimate the potential at the left leg electrode, for this moment in time, relative to the potential at infinity.

58. Lead II, LV: Estimate the voltage of lead II of the ECG, for this moment in time.

59. Lead II, RV: Suppose that at another moment an excitation wave, located in the anterior RV, has the form of a sheet of area 10 cm^2. The sheet is centered at (x, y, z) coordinates $(1, 2, -3)$ and the vector perpendicular to the sheet has components $(-0.58, 0.58, 0.58)$.

60. Provide a computer code that evaluates equation (13.17) and computes the numerical value of V_{II} for the preceding exercise.

Electrocardiography nomenclature and practice.

61. Make a table. On different rows, show each of the lettered waveforms of an ECG. In different columns, give the cardiac source (e.g., "atrial repolarization"), the approximate magnitude when observed on the body surface (in microvolts), and approximate duration.

62. What is the approximate peak-to-peak noise magnitude of a measured ECG trace under good (but not extraordinary) laboratory conditions?

63. What is the RMS noise magnitude that corresponds to the peak-to-peak value given in Ex. 62?

64. In a few sentences, explain the advantages and the potential problems of using amplifier filters to reduce the noise.

65. Why is safety a special concern in ECG recording?

66. Explain in a few sentences what Wilson's central terminal is.

Exercises 67–70 require an ECG model based on cross- sectional anatomy. Determine the relevant data from Figure 9.23.

67. Suppose you are at point AN on the surface. Suppose that the RV free wall, as seen from point AN, appears to be an elliptical surface with one diameter as shown in the figure, and another in the head-to-toe (superior–inferior) direction. Compute the (approximate) solid angle formed by the plate as seen from AN.

68. Again for the RV free wall with the same assumptions as Ex. 67, compute the solid angle at point LM.

69. Suppose an electrode pair is connected with the negative lead to Wilson's central terminal and the positive lead to the chest at point AN. A mostly positive deflection is observed. It is known that the deflection comes either from the RV free wall or the LV free wall. Which does it come from, and why?

70. In the same context as Ex. 69, suppose a mostly positive deflection is observed at point LM. It is known that the deflection comes either from the RV free wall or the LV free wall. Which does it come from, and why?

71. A cardiac strand consists of 8 parallel "fibers," each containing 15 cells, and lies in an unbounded homogeneous uniformly conducting medium. Each cell is 100 μm long and 10 μm in diameter. Since each fiber is close to the unbounded extracellular medium, it can be described by a linear core-conductor model with $r_e = 0$. The myoplasmic resistivity of each cell is 300 Ωcm and the specific resistance of the end-to-end gap junction is 3 Ωcm^2. If an electrotonic measurement of the Colatsky–Tsein type is made, what effective single fiber intracellular resistivity would be obtained? Assuming $R_m = 1000\ \Omega\text{cm}^2$, what is the effective r_m Ωcm for the strand? What value of λ should be measured?

72. A dipole whose strength and direction is $1.5\overline{a}_x$ (mA-cm) is located at $x = 1.0$ cm, $y = z = 0$. A similar dipole (same strength and orientation) is located at $x = -1.0$, $y = z = 0$. Consider that each represents the net electrical activity in different regions of the heart (possibly the right and left ventricles). The heart dipole model assumes that the same field could be generated by a single dipole equal to the vector sum of the components. In this case its magnitude would be $3.0\overline{a}_x$ and its position would be the origin (i.e., a mean position).

    a. Calculate the field as a function of r for $\theta = 30°$ (measured from the x axis) from the two aforementioned component dipoles and from the approximating single dipole. Plot these fields for comparison.

    b. At what range of r is the error less than 5%? Assume the extracellular medium is uniform, infinite in extent, and that the conductivity is 0.01 S/cm.

73. Consider the data of van Oosterom (Figure 9.10), where the electrode separation is 1 mm.

a. What is the velocity of propagation?

b. The magnitude of the bipolar signal is seen to be around 15 mV while the potential profile shows an intrinsic deflection of around 30 mV. What is a possible explanation?

74. The heart vector has the values given in Table 13.8 for patient A.

a. Plot the frontal (xy) plane vector loop for the given data.

b. Plot the sagittal (yz) plane vector loop for the given data.

c. What would be the value of V_I at each sequential instant of time? (Make an estimate of the lead vector first.)

Table 13.8. The Heart Vector, with Time

Time(sec)	H_x	H_y	H_z
0.01	−0.08	−0.06	−0.22
0.02	−0.10	−0.02	−0.62
0.03	1.12	0.70	−0.4
0.04	2.1	1.72	0.68
0.05	1.3	1.48	1.54
0.06	0.62	1.13	0.61
0.07	−0.05	0.43	0.25
0.08	0	0	0

75. *Distributed cardiac potentials.* The following exercise is devoted to the introduction of equations that describe the relationship between fields on the surface of the heart and those on the surface of the body. These equations form the basis of the forward problem (i.e., the determination of body surface potentials from heart potentials) and the inverse problem (finding heart potentials from body surface potentials). The execution of solutions based on realistic geometry, regularization (for the inverse), using finite- or boundary-element methods lies outside the scope of this introductory text. Several additional references to the literature may be consulted (e.g., [29]–[31]).

Green's theorem applied to body volume. Green's theorem is

$$\int_{\forall}(A\nabla^2 B - B\nabla^2 A)dV = \int_{s}(A\nabla B - B\nabla A)\,d\overline{s} \qquad (13.18)$$

where A and B are any twice-differentiable scalar functions. Surface S encloses volume V, and vector $d\overline{S}$ points outward from the volume. (Note that if $A = 1$, Green's theorem reduces to Gauss's law, from whence it came.) Let us set

$$A = 1/r \quad B = \sigma\phi \tag{13.19}$$

where r is the magnitude of a vector directed from an arbitrary field point b to an element of integration, ϕ is an electric potential function, and σ is the conductivity. Note that ϕ, r, and σ vary with position throughout the volume. Substituting the choices for A and B in (13.19) into Green's theorem (13.18) and choosing V to be the volume bounded by the body surface (S_B), lungs (S_L), and heart (S_H), as illustrated in Figure 13.12, what is the equation that results?

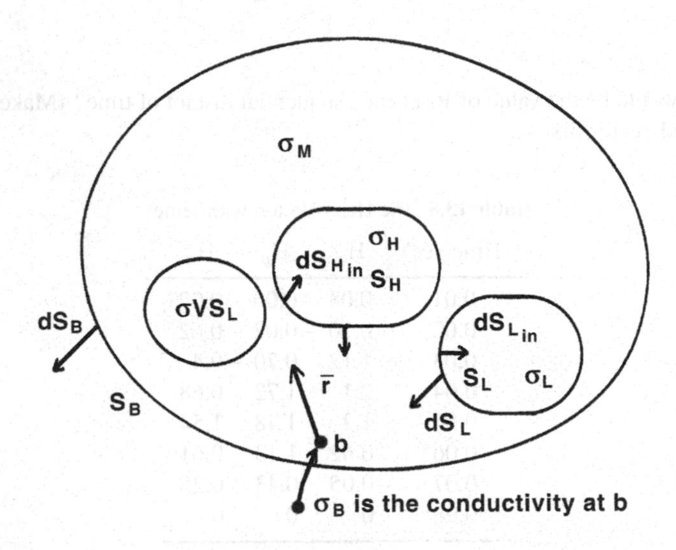

Figure 13.12. Human cross-sectional anatomy in schematic form, with math symbols added.

Simplification of volume integral. Simplify the volume integral of the answer to part a to place it in the form of the potential at a particular point, and a delta function. What equation results?

Simply surface integrals. The two integrals over the lung surface in the result of part a can be combined, since they are taken over the same surface. In doing so we select the surface normal to be outward. What equation results?

Place in solid-angle form. Further simplify the result of part C so that

a. In the integrals over H choose the surface normal to be outward,

b. Expand the $\nabla(1/r)$ operations as

$$\nabla\left(\frac{1}{r}\right) = -\frac{\overline{a}_r}{r^2} \tag{13.20}$$

where the negative sign on the right arises because the vector $\overline{a}_r$ extends from the field point b to the (integration) source point.

c. Present the result more compactly by making use of the definition of the solid angle:

$$d\Omega = \frac{\overline{r} \cdot \overline{n}}{r^2} dS \tag{13.21}$$

What equation for the body surface potential is the result? This equation should give the potential at point b on the body surface as a function of integrals of potential over the body, lung, and heart surface, and an additional integral of the gradient of potential over the heart surface.

Lung conductivity. How does the result of part d simplify if the conductivity inside and outside the lung are the same?

13.10. EXERCISES, CHAPTER 10: THE NEUROMUSCULAR JUNCTION

1. It is stated in the text that calcium influx is terminated when V_m is raised beyond 130 mV. What is the mechanism? Assuming the external calcium concentration to be 10.5 mM, what is the internal concentration?

2. In an experiment on frog muscle, the average EPP magnitude was determined to be 0.40 mV, while the average MEPP was 0.25 mV. What was the mean number of quanta released per impulse?

3. Assuming a Poisson distribution for the experiment described in Ex. 2, make a table showing the number of instances of 0, 1, 2, and 3 quanta released after 250 trials.

4. Boyd and Martin [7] performed 198 trials of nerve stimulation and measured the quantal release for each. These are given in the table below.

Table 13.9. Table for Exercise 4

Quanta released per stimulus	Number of cases observed
0	18
1	44
2	55
3	36
4	25
5	12
6	5
7	2
8	1
9	0

a. Determine the mean number of quanta released per stimulus.

b. With the above value, determine the number of cases expected from a Poisson process for each quanta release case. Do the Boyd and Martin data fit the Poisson distribution?

5. A vesicle in the neuromuscular junction can be considered spherical with a radius of 250 Å. Its ACh content has a concentration of 150 mM/liter. How many ACh molecules does this correspond to?

6. For a frog neuromuscular junction, the parameters in Figure 10.7 could be $E_r = -90$ mV, $g_r = 5 \times 10^{-6}$ S, $E^s = 0.20$ mV, and $g_s = 5 \times 10^{-5}$ S.

   a. Calculate the amplitude of the end-plate potentials under these conditions.

   b. What is the reversal potential?

   c. If $E_{\mathrm{K}} = -90$ mV, what is E_{Na}, assuming $g_{\mathrm{K}} = g_{\mathrm{Na}}$ in the activated synaptic membrane?

7. For $g_{\mathrm{K}} = g_{\mathrm{Na}}$ in the end-plate region activated by ACh and with $E_{\mathrm{K}} = -95$ mV and $E_{\mathrm{Na}} = 50$ mV,

   a. what is the reversal potential?

   b. If $g_{\mathrm{Na}}/g_{\mathrm{K}} = 1.29$ (rather than unity), what is the reversal potential?

8. Using Eq. (10.22), plot the EPP amplitude as a function of $[\mathrm{Ca}^{++}]$ (for concentration 0 to 1.0 mm) and for $[\mathrm{Mg}^{++}]$ = 0.5, 2.0, and 4.0 mM. In (10.22) take $K_1 = 1.1$ mM, $K_2 = 3.0$ mM, $k_2 = 1.0$, and $W = 1.14$ ($V^{1/4}$ mM). Compare these results with those measured by Dodge and Rahamimoff [4].

13.11. EXERCISES, CHAPTER 11: SKELETAL MUSCLE

1. Describe the differences between a whole muscle, a muscle bundle, muscle fibers, and muscle fibrils.

2. Describe the difference between a muscle contraction carried out under isometric as compared to *isotonic* conditions.

3. Enumerate the specific experimental findings that support the sliding filament theory. Describe briefly each finding and then summarize why it supports the theory.

4. From the viewpoint of the sliding filament theory, explain whether (and if so, why) isometric tension should vary as a function of sarcomere length. (That is does can be seen in Figure 11.12.)

5. This chapter mentions several roles played by ATP. Describe each briefly.

6. Describe the transverse tubular system, and make a sketch of its structure.

7. Describe the sarcoplasmic reticulum, and make a sketch of its structure. What is the apparent role of excitation–contraction coupling?

13.12. EXERCISES, CHAPTER 12: FUNCTIONAL ELECTRICAL STIMULATION

1. Consider an external point-source stimulus to a nerve trunk, which includes nerve fibers of various diameters.

   a. Which size fiber is the first to be excited, as the stimulus current is increased in magnitude?

   b. Will the stimulus first initiate contraction in an FO or an SO muscle fiber?

   c. How does the order found in the preceding part compare to the natural order?

2. Consider the cleft separating pre- and post- junctional membranes in the neuromuscular junction.

   a. The cleft is around ______________ in width?

   b. The main effect of complexing of transmitter at the post-junctional site on the sodium, potassium, and chloride conductance is (roughly) ______________?

3. An action potential is elicited on a space-clamped squid axon. Based on Hodgkin–Huxley theory,

   a. Plot the changes that take place in $V_m(t)$, $g_{Na}(t)$, and $g_K(t)$.

   b. In a few sentences, describe the time relationship of the rise and fall of $g_{Na}(t)$ as compared to timing of the other two.

4. In the post-junctional membrane, if the Nernst potentials of sodium and potassium are $E_{Na} = 60$ m V and $E_K = -80$ m V, what will be the value of V_m as a result of transmitter action?

For Exercises 5–8, consider a bipolar nerve cuff electrode with an electrode spacing of 6 mm. We wish to estimate the induced stimulating voltage in a selected fiber of a nerve bundle (trunk). We assume that the external path can be neglected and that only the transmembrane nodal elements directly under the electrodes need be considered.

The extracellular resistance per internode (within the cuff) = 6000 Ω/internode, while the interstitial resistance per internode = 8000 Ω/internode. The intracellular resistivity is 120 Ωcm, the transmembrane nodal resistance for the single axon of interest, whose intracellular diameter is 18 μm, is 18.5 MΩ. The internodal distance is 2 mm and a node lies under each electrode. The bipolar electrode is driven to 50 mV.

5. Consider the sequence of depolarization for the described electrode and stimulus.
   a. Under which electrode, anode (+), or cathode (–), will depolarization first occur?

   b. Will depolarization ever occur under the other electrode?

6. Draw an equivalent circuit and label elements with the proper resistances.

7. Calculate the transmembrane potential depolarization produced.

8. If an accurate study of strength–duration were desired, what additional details should be added to the model?

In Exercises 9–14, use the following data unless other values are given. A coiled stainless steel electrode has an area of 0.4 mm^2 and is to be used at a stimulus rate of 50 Hz. The current stimulus pulse has a magnitude of 15 mA.

9. If the maximum safe charge injection is 0.35 μC/mm^2, what maximum pulse duration is permitted for balanced-charge biphasic conditions?

10. From the strength–duration data given below, will the design conditions in Ex. 9 be satisfactory? If not, what changes could be made?

11. The width of the reversible region of stainless steel is about 0.6 μC/mm^2. If loss of reversibility is associated with a voltage that breaks down the electrode capacitive dielectric and the latter is 4.0 volts, what is the effective capacitance/cm^2?

12. Schaldach[3] designed a cardiac pacemaker electrode formed of tantalum–tantalum pentoxide (2 mm diameter and 3 mm long), and placed it at the end of a transvenous catheter. The roughness factor is 12.5 and the electrode–tissue capacitance is 0.75 μF/mm^2 (geometrical area). Breakdown voltage is 2 volts and reversibility is assumed in the absence of breakdown for balance charge biphasic rectangular pulses.

    a. What is the maximum current (pulse duration is 250 μsec)?

    b. Correspondingly, what is the charge density?

    c. What is the total charge per phase?

    d. How does the capacitance compare with the double-layer capacitance of an untreated tantalum? Assume that the same data given above are applicable.

13. A coiled stainless steel electrode has an area of 2 mm^2 (and is to be used at a stimulus rate of 40 Hz).

    a. What average (primary) safe current can be used, assuming the primary pulse duration is to be 225 μsec?

    b. What maximum duration of secondary pulse will lie within the buffering capacity of the system? (Assume values from Mortimer in Chapter 12.)

14. To avoid dissolution of iron, stainless steel electrodes should be operated so that the charge density per phase is less than around 0.4 μC/mm^2. But, according to McHardy et al. [McHardy J, Geller D, Brummer SB. 1977. An approach to corrosion control during electrical stimulation. *Ann Biomed Eng* **5**:144–149.], a stimulating current of 1.6 mA/mm^2 (average primary current, assumed rectangular) can be used (cathodic then anodic) at 50 Hz with no corrosion if an excess of cathodic charge of (only) 0.002 μC/mn^2 per phase is used! They remind us that reversibility could be maintained even in the region where electrochemical reactions take place, were it not for loss of reaction products by diffusion. The excess cathodic reaction arising in their system offsets this diffusion (provides additional products) to

[3] Schaldach M. 1971. New pacemaker electrodes. *Trans Am Soc Artif Internal Organs* **17**:29.

obtain reversibility in the anodic region—the frank lack of reversibility in the cathodic region is offset by buffering.

a. What is the net cathodic current density and what percentage is it of the average cathodic current?

b. By what factor is the charge density per phase improved over balanced biphasic stimulation?

For Exercises 15–18, use Table 13.10 along with the following data. The maximum anodic charge injection is 0.35 μC/mm^2, and the maximum excess cathodic average current is 10 μA/mm^2. A coiled stainless steel electrode has an area of 0.4 mm^2 and is to be used at a stimulus rate of 40 Hz. Maximum safe charge injection is 0.375 μC/mm^2.

Table 13.10. Threshold Strength–Duration

I_{th} (mA)	Duration(μ sec)
15.000	10
9.470	16
6.150	25
3.940	40
2.720	60
1.740	100
1.190	160
0.875	250
0.674	400
0.575	600
0.520	1000

15. For a design with balanced-charge biphasic stimulation:

a. What pulse amplitude and duration will give good operation?

b. Explain your choice in part a.

16. Reconsider the design of the preceding exercise.

a. Can operation be improved with imbalanced-charge biphasic conditions?

b. If so, what are the design conditions?

17. For a bipolar cuff electrode within which a nerve trunk lies, as the stimulus current is increased:

a. Which size fiber is the earliest to be excited?

b. Will this initiate contraction in an FO or an SO muscle fiber?

c. How does the order of the previous part compare to the natural order?

18. If a single motor neuron is excited by a single stimulus:

a. Then a mechanical response from the innervated motor unit is known as ______________.

b. Electrical propagation along the membrane of the fiber reaches its target by spreading through ______________; ______________ ions are released to catalyze the contractile machinery.

Exercises 19–22 consider a bipolar nerve cuff electrode with electrode spacing of 7.5 mm. We wish to examine the induced voltage in a selected fiber in the nerve trunk. The extracellular resistance per internode within the cuff is 7500 Ω per internode. The total interstitial resistance per internode is 10.000 Ω/internode. The intracellular resistivity is 140 Ωcm. The transmembrane resistance at a single node of the fiber of interest, with an intracellular diameter of 25 μm, is 20 MΩ. The bipolar (cuff) electrode delivers 40 mV. The internodal distance is 2.5 mm. Assume that the external current pathway can be neglected, and that the transmembrane current is mainly through the internodes directly beneath the electrodes.

19. Under which electrode [anode (+) or cathode (–)] will hyperpolarization take place?

20. Draw an equivalent circuit and label elements with proper resistance values.

21. Calculate the magnitude of the hyperpolarization that is produced.

22. If the model was improved so that the external current pathways were included, what interesting transmembrane potential(s) could be determined?

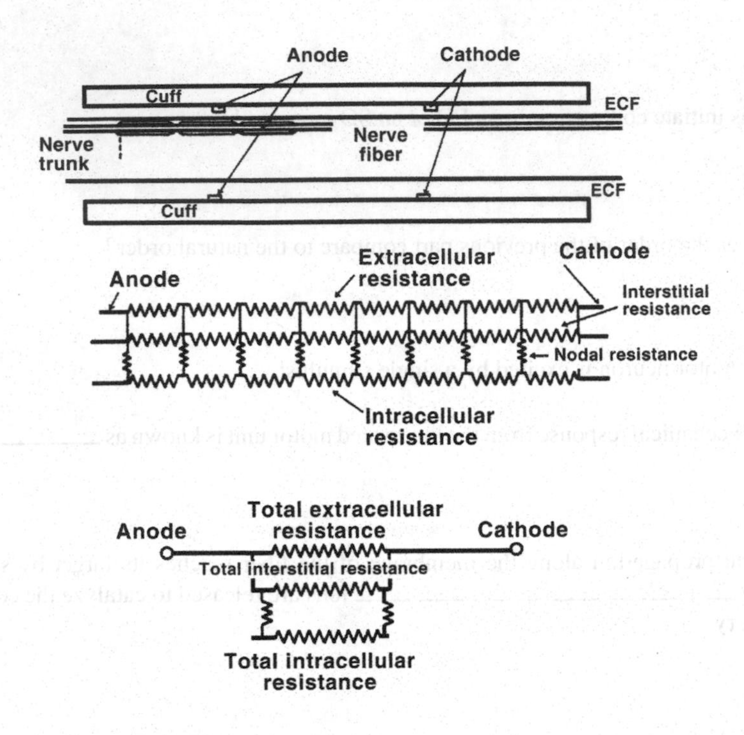

Figure 13.13. The upper portion is the physical arrangement of the cuff electrode surrounding a whole nerve (consisting of myelinated fibers). The lower portion is the suggested equivalent electrical circuit.

23. A cuff electrode is depicted in Figure 13.13. A resistive model can give some order-of-magnitude insight into its operation, as follows. The extracellular and interstitial resistance may be found from the cylindrical resistance formula (both are assumed to have the same resistivity), while current can enter and leave the intracellular space only through the nodal resistance (see Figure 12.27). The results are the approximate circuit shown. The data are given in Table 13.11.

    a. For cuff diameters of 1.05, 1.1, 1.3, 1.5, and 1.9 mm, determine the electrode voltage needed to produce an excitatory voltage (assume 10 mV).

    b. Determine the electrode current and the total charge (200 μsec pulse) in each case.

    c. If the electrode area is 2.5 mm^2. determine the charge density.

Table 13.11. Data for Cuff Electrode

Value	Units	Description
−44	%	of nerve trunk that is intracellular area
70	%	of axon is intracellular area (remainder is myelin sheath)
18	mm	nerve trunk diameter
1.0	mm	electrode separation
2.0	mm	internodal distance
300	Ω-cm	ρ_e
110	Ω-cm	ρ_i
30	mS/cm^2	nodal membrane conductance per unit area
2.5	μm	nodal gap width
20	μm	single fiber (axon) diameter

d. Would this be satisfactory for reversible operation? [*Hint*: Determine the number of fibers (closest integer)].

e. Evaluate the extracellular resistance for an 18-mm path.

f. Evaluate total interstitial resistance for an 18-mm path,

g. Evaluate total intracellular resistance for an 18-mm path.

h. Assume the transmembrane current occurs only at nodes beneath electrodes, and evaluate total transmembrane resistance.

24. Write computer code, and use a computer language suited for your environment, for the membrane model portions of a fiber simulation, assuming the membrane follows the Frankenhaeuser–Huxley equations. Specifically, write code to:

a. Initialize the necessary FH variables.

b. Compute the membrane currents for the current membrane state.

c. Advance the values of the gating variables.

INDEX